Essentials of Meteorology

Essentials of Meteorology

Edited by Ben Perry

SYRAWOOD
PUBLISHING HOUSE
New York

Published by Syrawood Publishing House,
750 Third Avenue, 9th Floor,
New York, NY 10017, USA
www.syrawoodpublishinghouse.com

Essentials of Meteorology
Edited by Ben Perry

International Standard Book Number: 978-1-64740-108-5 (Hardback)

Cataloging-in-Publication Data

Essentials of meteorology / edited by Ben Perry.
p. cm.
Includes bibliographical references and index.
ISBN 978-1-64740-108-5
1. Meteorology. 2. Atmospheric science. I. Perry, Ben.

QC861.3 .E87 2022
551.5--dc23

TABLE OF CONTENTS

PREFACE

This book aims to highlight the current researches and provides a platform to further the scope of innovations in this area. This book is a product of the combined efforts of many researchers and scientists, after going through thorough studies and analysis from different parts of the world. The objective of this book is to provide the readers with the latest information of the field.

Meteorology is the science which studies the atmosphere and atmospheric phenomena, and how these influence the weather. It specializes in the study of atmospheric events occurring in the troposphere and lower stratosphere. A major area of meteorology is atmospheric research or operational weather forecasting. Meteorology also explores the relationship between the atmosphere and the Earth's oceans, climates and biological life. Such research guides the assessment of the current state of the atmosphere and allows a future prediction of the climate to be developed. A number of technologies aid in examining, describing, modeling and predicting weather systems with a high degree of accuracy and efficiency. Radar and satellites are two such technologies that are of utmost importance to meteorological studies. This book discusses the fundamentals as well as modern approaches of meteorology. It traces the progress of this field and highlights some of its key concepts and applications. It is appropriate for students seeking detailed information in this area as well as for experts.

I would like to express my sincere thanks to the authors for their dedicated efforts in the completion of this book. I acknowledge the efforts of the publisher for providing constant support. Lastly, I would like to thank my family for their support in all academic endeavors.

Editor

1

CFD Analysis of Urban Canopy Flows Employing the V2F Model: Impact of Different Aspect Ratios and Relative Heights

Fabio Nardecchia,[1] Annalisa Di Bernardino,[2] Francesca Pagliaro,[1] Paolo Monti,[2] Giovanni Leuzzi,[2] and Luca Gugliermetti[1]

[1] *Sapienza University of Rome, DAEEE, Via Eudossiana 18, Rome 00184, Italy*
[2] *Sapienza University of Rome, DICEA, Via Eudossiana 18, Rome 00184, Italy*

Correspondence should be addressed to Paolo Monti; paolo.monti@uniroma1.it

Academic Editor: Jorge E. Gonzalez

Computational fluid dynamics (CFD) is currently used in the environmental field to simulate flow and dispersion of pollutants around buildings. However, the closure assumptions of the turbulence usually employed in CFD codes are not always physically based and adequate for all the flow regimes relating to practical applications. The starting point of this work is the performance assessment of the V2F (i.e., $\overline{v^2} - f$) model implemented in Ansys Fluent for simulating the flow field in an idealized array of two-dimensional canyons. The V2F model has been used in the past to predict low-speed and wall-bounded flows, but it has never been used to simulate airflows in urban street canyons. The numerical results are validated against experimental data collected in the water channel and compared with other turbulence models incorporated in Ansys Fluent (i.e., variations of both k-ε and k-ω models and the Reynolds stress model). The results show that the V2F model provides the best prediction of the flow field for two flow regimes commonly found in urban canopies. The V2F model is also employed to quantify the air-exchange rate (ACH) for a series of two-dimensional building arrangements, such as step-up and step-down configurations, having different aspect ratios and relative heights of the buildings. The results show a clear dependence of the ACH on the latter two parameters and highlight the role played by the turbulence in the exchange of air mass, particularly important for the step-down configurations, when the ventilation associated with the mean flow is generally poor.

1. Introduction

The continuous growth of large cities occurred in the last decades has prompted the scientific community towards the understanding of the urban environment [1, 2]. Great attention has been paid especially in predicting the flow field within and outside the urban street canyon, which is the space delimited by the street and the facades of the surrounding buildings. Knowledge on wind and temperature distributions within the street canyon is crucial, for example, in the design of the urban geometry with the aim of achieving an energy-optimized architecture of the city [3–5] as well as determining the concentration of pollutants emitted at the street level by vehicular traffic [6–8].

One of the parameters that mostly influence the gross features of the flow over urban canopies is the aspect ratio, AR, which is defined as the ratio between the average height of the buildings, H, and the spacing, W, between two consecutive buildings. Oke [3] introduced three kinds of flow regimes as a function of AR: isolated obstacle, wake interference, and skimming flow. In the isolated-obstacle regime (AR < 0.4), the flow around each building is not affected by disturbances coming from other obstacles. In the wake-interference flow (0.4 < AR < 0.67), two counter-rotating vortices form within the canyon, and the wake of each building interacts with the subsequent building. The skimming flow (AR > 0.67) corresponds to narrow urban canyons, where the wind circulation is characterized by a vortex that occupies a large part of the canyon. Besides the three flow regimes defined by Oke, there is a fourth flow pattern, the multivortex flow regime (AR > 1.54), which is a variant of the skimming flow [9]. Another important parameter influencing the street canyon is the relative height of the buildings, H_2/H_1, where H_1 and H_2 are the heights of the leeward and the windward buildings, respectively.

Thanks to the increasing computational power of computers, computational fluid dynamics (CFD) has recently supported laboratory and field experiments, improving the knowledge of street-canyon flows. Much effort has been done in recent years to analyze urban canopy flows by means of CFD, often using Reynolds-averaged Navier–Stokes (RANS) simulations of two-dimensional (2D) arrays of buildings. The interest of the scientific community for such a simplified building arrangement is justified by the fact that the 2D array can be considered as an archetype for more complex geometries [10–13]. Huang et al. [14] carried out 2D simulations to investigate the effect of wedge-shaped roofs on the flow in an urban street canyon and found that they have significant influence on the vortex structure and pollutant distribution pattern. Memon et al. [15] analyzed heating in 2D isolated street canyons applying the RNG k-ε model (here, k is the turbulent kinetic energy, while ε is its rate of dissipation). Those authors compared their results with wind-tunnel data and showed that the nighttime and daytime air temperature difference between urban and rural areas closely resembles each other. Murena and Mele [16] analyzed an ideal deep street canyon with 2D unsteady RANS simulations using the shear stress transport (SST) k-ω model. They observed that short-time variations of wind velocity can greatly influence the mass transfer rate between the canyon and the overlying boundary layer. Allegrini et al. [17] carried out 2D steady RANS simulations with different near-wall treatments in order to validate numerical results for buoyant flows in urban street canyons by comparison with wind-tunnel measurements. They compared the results of different turbulence models (STD k-ε, realizable k-ε, k-ω, Spalart–Allmaras, and Reynolds stress model (RSM)), showing a better agreement of the STD k-ε model with the NEWFs (nonequilibrium wall functions) than the LRNM (low-Reynolds number modeling). Ho et al. [18] studied idealized 2D urban street canyons of different ARs and urban boundary-layer depths using the RNG k-ε model. They found that the atmospheric turbulence contributes most to street-level ventilation because the turbulent component of the air-exchange rate (ACH) dominates the transport process. Xie et al. [19] investigated the impact of the urban street layout on the local atmospheric environment through numerical simulation and wind-tunnel experiments. The authors found that the vortex structure in the canyon and, consequently, the street layout strongly influence the wind field and the pollutant dispersion in the canopy.

A well-known CFD approach alternative to RANS simulation is the large eddy simulation (LES), which explicitly resolves the larger structures of the turbulence, while it models the finer ones by adopting suitable closure assumptions [20–22]. It is believed that the RANS approach provides reasonable accurate predictions of mean flow quantities and that it is still an appropriate methodology considering the low CPU cost. However, in some applications such as the analysis of transient features of the flow like vortex shedding in the wake, LES performs generally better than RANS simulation [6]. In any case, LES resolves the large-scale turbulent eddies, which are 3D by nature. Therefore, since in this work a 2D simulation has been used, the most suitable CFD approach is the RANS one.

Based on the previous literature, the k-ε turbulence model appears to be the most widely employed one in CFD simulations of urban canopy flows. However, uncertainties still exist regarding the capability of CFD codes in simulating velocity and turbulence fields in different flow regimes. For this reason, a comparison between numerical results obtained through Ansys Fluent v.14.5 [23] and experimental data taken in the water channel has been carried out in this work. In addition to the most known turbulence models, the comparison has also taken into account the V2F model, based on the $k-\varepsilon-\overline{v^2}$ closure developed by Durbin [24]. The V2F model is similar to the STD k-ε model but includes an additional transport equation that models the velocity scale, $\overline{v^2}$, and its source term, f [25]. Since the V2F model incorporates both near-wall turbulence anisotropies and nonlocal pressure-strain effects, it is usually employed for low-speed and wall-bounded flows. This implies that wall functions are not required, and consequently, lower computational costs are needed. The V2F model has been developed for attached or mildly separated boundary layers and used mainly for studying three-dimensional (3D) boundary layers [26, 27] and heat transfer problems in jet impingement [28–30] and in ribbed-channel flows [31, 32], subsonic and transonic flows for aerospace applications [33, 34], and flow physical phenomena in enclosed environments [35–37]. To the best of our knowledge, this paper is the first one to deal with numerical simulations of 2D street canyons by means of the V2F model. Here, the effectiveness of the V2F model in predicting the flow field for two typical building arrangements (AR = 0.5 and 1) has been investigated. The V2F model is also employed to analyze the air-exchange rate (ACH) for a series of two-dimensional building arrangements, such as step-up and step-down configurations, having different aspect ratios and relative heights of the buildings, a design quite underexplored in the literature.

This paper is organized as follows: firstly, the experimental setup used in the water channel and the numerical approach followed in the simulation are described. Secondly, tests of the V2F model through comparisons with the experimental data and results obtained employing other turbulence models are presented and discussed together with the analyses of several flow regimes referred to several ARs and H_2/H_1. Particular attention is also paid to the analysis of canyon ventilation as well as to its dependence on the canyon geometry. This paper concludes with a summary of the main results.

2. The Water-Channel Experiments

The numerical simulations have been validated with a series of experiments conducted in the close-loop water channel located at the Laboratory of Hydraulics of the

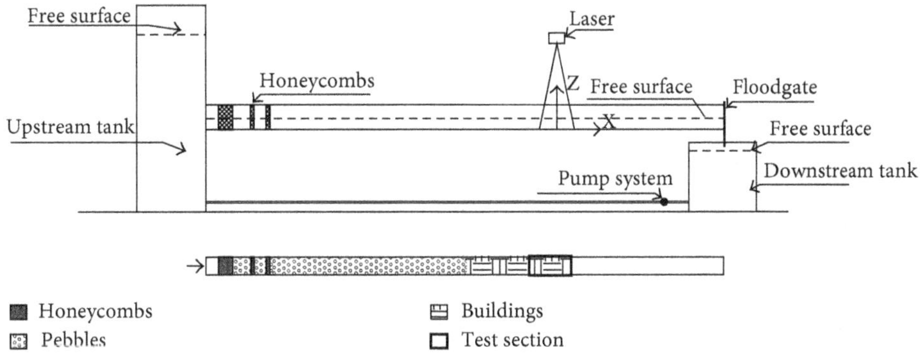

FIGURE 1: Side and top views of the experimental apparatus. The x-axis refers to the longitudinal axis of the channel (streamwise), while the z-axis is parallel to the vertical.

University of Rome "La Sapienza." The water channel allows the reproduction of the atmospheric boundary layer with several advantages [38–41]. One of them is that image analysis techniques, such as particle tracking velocimetry, can be easily employed. These permit accurate spatial measurements, which generally allow a clearer understanding of complex flows such as the one under investigation.

The water channel has a rectangular cross section of 0.35 m height and 0.25 m width and 7.4 m length (Figure 1). The flow rate is set by a floodgate placed at the closing section of the channel, and the water depth, $h = 0.16$ m, is maintained constant throughout the experiments (more information about the facility can be found in [42]). The reference frame has been defined with the x-axis aligned with the streamwise velocity and the z-axis vertical. The water is seeded with nonbuoyant particles (2 μm in diameter), which were assumed to be passively transported by the flow. Upwind of the buildings, the channel bottom is covered by unevenly spaced, roughness elements (pebbles with an average diameter of 5 mm) in order to reproduce the logarithmic vertical profile of the undisturbed streamwise velocity as well as the (nearly) constant Reynolds stress profile typically observed in the atmospheric boundary layer. The roughness Reynolds number, $\mathrm{Re}_\tau = u_* H/\nu$ (here, $u_* = \sqrt{-\overline{u'w'}}$ is the friction velocity, $\nu = 10^{-6}\cdot\mathrm{m}^2\cdot\mathrm{s}^{-1}$ is the kinematic viscosity of water, and H is the obstacle height), ranges from 340 up to 470; that is, it is well above the critical value of 70 given by Snyder [43], which guarantees the independence of the investigated large-scale structures and the mean flow of Reynolds number effects [44]. Therefore, in our experiments, Re_τ is large enough to ensure both the conditions of full turbulence of the simulated boundary layer and the dynamic similarity between experiments and real conditions.

The urban canopy is composed of a 2D regular array of urban-like obstacles with square sections of $B = H = 20$ mm and length 25 cm glued onto the channel bottom (Figure 2). Two geometrical configurations have been investigated, one referred to the skimming flow (AR = 1) and one to the wake-interference regime (AR = 0.5). To this end, the distance between the obstacles, W, has been varied from 20 mm to 40 mm.

The framed area is 99 mm long (x-axis) and 72 mm high (z-axis) and is located in correspondence with the 20th canyon, where the flow can be considered fully developed. The velocity components along the x- and z-axes, respectively, u and w, have been measured using the feature tracking (FT), a technique based on the image analysis [45]. A high-speed camera (CMOS with a resolution of 1280 × 1024 pixels) acquires images at 250 frames per second for 40 s during each experiment, while a green laser light sheet (wavelength of 532 nm) illuminates the acquisition area. Velocities have been determined by the FT algorithm from the displacements of the seeding particles between successive frames. A Gaussian interpolation algorithm [46] was applied to the scattered velocity samples to obtain a two-dimensional, Eulerian description of the motion on the x-z plane. After this procedure, 10000 instantaneous flow fields (each 1/250 s) with a spatial resolution of 1 mm have been obtained. Details of the undisturbed approaching flow are given in [47].

3. The Numerical Approach

3.1. Mathematical Formulation. In the case of incompressible, turbulent flows, Ansys Fluent solves the balance equations of mass and momentum and additional transport equations related to closure assumptions. In particular, the first two equations can be expressed as follows (Einstein summation rule applies):

$$\frac{\partial \overline{u}_i}{\partial x_i}, \tag{1}$$

$$\overline{u}_j \frac{\partial \overline{u}_i}{\partial x_j} = -\frac{1}{\rho}\frac{\partial \overline{p}}{\partial x_i} + \frac{\mu}{\rho}\frac{\partial^2 \overline{u}_i}{\partial x_j^2} - \frac{\partial}{\partial x_j}\left(\overline{u_i' u_j'}\right) + g_i, \tag{2}$$

where g_i is the acceleration due to gravity, p the pressure, μ the viscosity, and ρ the density. The Reynolds stress tensor $\overline{u_i' u_j'}$ (here, prime indicates fluctuation around the mean) is usually modeled using a linear proportionality to the rate of strain (Boussinesq eddy-viscosity model):

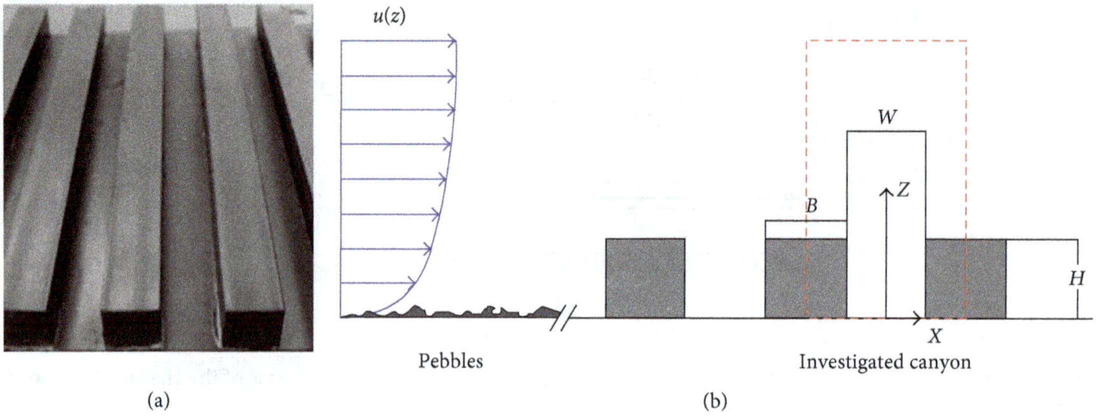

FIGURE 2: Obstacles set for AR = 1 (a). Sketch of the vertical section (b).

$$-\overline{u_i'u_j'} = \frac{\mu_t}{\rho}\left(\frac{\partial \overline{u}_i}{\partial x_j} + \frac{\partial \overline{u}_j}{\partial x_i}\right) - \frac{2}{3}\delta_{ij}k, \tag{3}$$

where δ_{ij} is the Kronecker delta. The turbulent kinetic energy k and the eddy viscosity μ_t are determined by adding to (1) and (2) two or more additional equations [48]. Since the three k-ε models, the two k-ω models and the RSM, implemented in Ansys Fluent are well known and widely employed in the literature, only the V2F model is briefly described below.

3.2. The V2F Model. Standard eddy-viscosity models use specific damping functions to simulate the region close to the solid boundary. This is because the k-ε closure for μ_t is isotropic, while near-wall turbulence is strongly anisotropic [49]. The V2F model was developed to avoid the use of damping functions and correctly reproduce the attenuation of the turbulence near solid boundaries [25, 31]. This model solves an elliptic relaxation function, f, and three transport equations, respectively, for k, ε, and $\overline{v^2}$, where the latter is the velocity scale. The V2F model is based on a turbulent viscosity hypothesis proposed by Durbin for the region close to the solid boundary [49]:

$$\mu_t = \rho C_\mu \overline{v^2} T, \tag{4}$$

where $T = k/\varepsilon$ and C_μ is a constant. Information on the anisotropy of the flow in the near-wall region is taken through the transport equation of $\overline{v^2}$, which, in turn, is derived from the transport equation of the Reynolds stress normal to the wall. Summarizing, the transport equations for k, ε, and $\overline{v^2}$ are, respectively, read as follows:

$$\frac{\partial}{\partial t}(\rho k) + \frac{\partial}{\partial x_i}(\rho k \overline{u}_i) = P - \rho\varepsilon + \frac{\partial}{\partial x_j}\left[\left(\mu + \frac{\mu_t}{\sigma_k}\right)\frac{\partial k}{\partial x_j}\right] + S_k, \tag{5}$$

$$\frac{\partial}{\partial t}(\rho\varepsilon) + \frac{\partial}{\partial x_i}(\rho\varepsilon\overline{u}_i) = \frac{C_{\varepsilon 1}'P - C_{\varepsilon 2}'\rho\varepsilon}{T} + \frac{\partial}{\partial x_j}\left[\left(\mu + \frac{\mu_t}{\sigma_\varepsilon}\right)\frac{\partial \varepsilon}{\partial x_j}\right] + S_\varepsilon, \tag{6}$$

$$\frac{\partial}{\partial t}\left(\rho\overline{v^2}\right) + \frac{\partial}{\partial x_i}\left(\rho\overline{v^2}\overline{u}_i\right) = \rho k f - 6\rho\overline{v^2}\frac{\varepsilon}{k} + \frac{\partial}{\partial x_j}\left[\left(\mu + \frac{\mu_t}{\sigma_{\overline{v^2}}}\right)\frac{\partial \overline{v^2}}{\partial x_j}\right] + S_{\overline{v^2}}, \tag{7}$$

where S_k, S_ε, and $S_{\overline{v^2}}$ are source terms, while

$$P = 2\mu_t S^2, \quad S^2 = S_{ij}S_{ij}, \quad S_{ij} = \frac{1}{2}\left(\frac{\partial \overline{u}_i}{\partial x_j} + \frac{\partial \overline{u}_j}{\partial x_i}\right). \tag{8}$$

The function f, included to take into account the anisotropic wall effects, is modeled by solving an elliptic Helmholtz-type equation:

$$f - L^2\frac{\partial^2 f}{\partial x_j^2} = (C_1 - 1)\frac{(2/3)\left(\overline{v^2}/k\right)}{T} + C_2\frac{P}{\rho k}\frac{5\left(\overline{v^2}/k\right)}{T} + S_f, \tag{9}$$

where L is a length scale, C_1 and C_2 are constants, and S_f is a source term. More information on the V2F model can be found in [48].

3.3. Geometry and Numerical Domain. The building array described in Section 2 has been numerically modeled by considering a 3D domain (Figure 3). It is composed of twelve buildings and, consequently, eleven street canyons. According to [50], the computational domain has been extended in the streamwise direction to $10H$ between the inflow boundary and the first building and to $20H$ between the last building and the outflow boundary. Its height has been set equal to $10H$. The numerical runs have also been conducted considering a 2D computational domain corresponding to the vertical section passing through the channel axis (Figure 4 and the grey plane in Figure 3). This procedure has been followed since the investigated domain is symmetric along the y-axis. The choice of a 2D model allows lower computational costs without compromising the accuracy of the results [14–18].

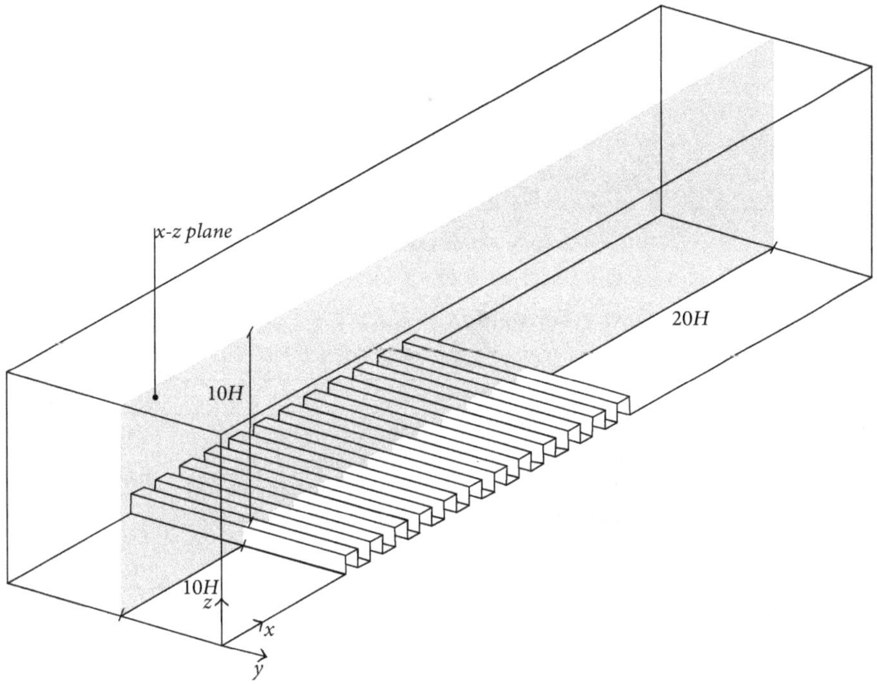

FIGURE 3: The computational domain for the 3D simulation. H indicates the building height.

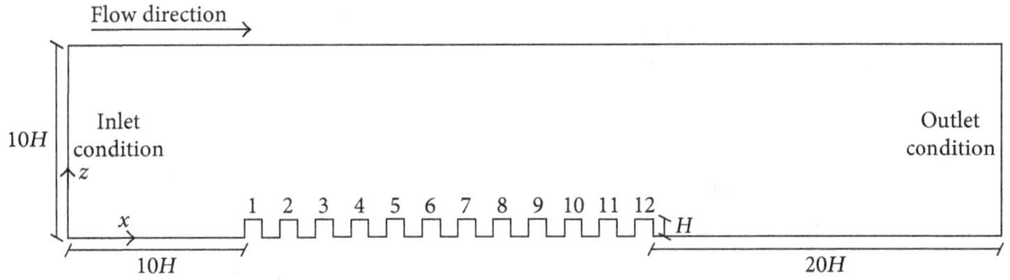

FIGURE 4: The computational domain for the 2D case.

In order to test the effectiveness of this choice, both the 3D and 2D models have been run and compared with the experimental data. The results of this test will be shown below. In addition to the cases AR = 0.5 and 1 with buildings of equal height used to test the model capability, other geometries have been taken into account. In particular, five arrays of buildings with different ARs (0.5, 1, 1.33, 2, and 4) have also been investigated varying H_2/H_1 (0.4, 0.5, 0.7, 1.5, 2, and 2.5) for each of the five ARs.

3.4. Boundary Conditions and Simulation Settings. The vertical profile of the magnitude of the undisturbed mean velocity, V, measured in the water channel upwind of the building array (interpolated by using a fitting function implemented in Ansys Fluent) has been set as the velocity inlet boundary condition at the inflow. Here, V assumes the meaning of time average of the 10000 instantaneous velocity magnitudes collected during the experiment. The flow has been considered not affected by any obstacle at its boundary since the inlet velocity profile is fully developed. For this reason, a free-slip condition has been applied to the bottom surface before and after the building array. A zero (relative) pressure has been imposed as the outlet boundary condition. It is used to force the flow in the direction normal to the outlet without any backflow. A free-slip condition has been imposed at the water-air boundary, while a no-slip condition has been applied at the surfaces of the buildings and at the bottom of the canyon as well. This choice is justified on the basis of several tests we conducted, which showed a better agreement between measured and simulated flows inside the canyon.

The numerical solver has employed a structured, non-orthogonal, fully collocated, cell-centered, finite-volume approach for the discretization of the computational domain, and the velocity vector has been decomposed into its Cartesian components. Physical diffusive cell fluxes have been approximated using a conventional second-order central differencing scheme, and the SIMPLE algorithm [51] has been used for pressure correction. The convergence target based on the root mean square has been set to 10^{-7}. The quantities of interest, such as velocity and turbulent kinetic energy, have been monitored at several grid

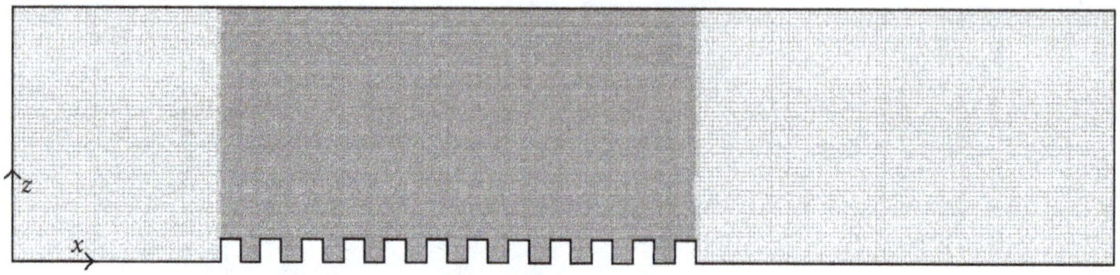

FIGURE 5: Mesh used for AR = 1.

points during the solving process to check whether stable levels before convergence were met or not.

3.5. Mesh and Grid Independence Test. Both the 3D and 2D domains have been discretized by using orthogonal grids (Figure 5). Since the geometry is relatively simple, a block-based hexahedral mesh has been used to enhance the quality of the mesh. The grid lines have been refined near the solid surfaces (bottom, rooftops, and building walls).

The choice of the most suitable mesh spacing is not a trivial task since the use of a too coarse mesh can give rise to considerable errors, while an excessively fine mesh costs in terms of computing time. This is the reason why any CFD simulation should be preceded by a series of grid independence tests. The velocity magnitude V computed at $z/H = 7.5$ in correspondence with the vertical profiles passing through the center of the ninth canyon (the reason for this choice is clarified in the next section) has been analyzed as a function of four grid meshes of different densities (Table 1). Assuming Mesh A (interval size equal to 0.001 m) as the pivot case, the percentage difference Δ (%) between Mesh A and the finest of the four (Mesh D) is only 0.08%. Therefore, in the remainder of this work, Mesh A is used for all the CFD analyses. The same interval size of the mesh has been employed for the 3D model, for a total amount of nearly 1.2 million cells.

The vertical profiles of the nondimensional velocity, $V/\overline{u}_{ref}$, passing through the center of the ninth canyon are given in Figure 6 for the case AR = 1. The results show an overall good agreement between the experiment and simulations. The percentage difference between the 2D and 3D cases is nearly 5.30% for the entire profile. Therefore, it is possible to assume that the 2D model describes the flow field with the same accuracy as the 3D model.

3.6. Stabilization Analysis. Since the interrogation area adopted during the experiments is located far enough from the inlet to assure the flow independence along the streamwise direction, before starting the comparisons, it is essential to verify whether the same condition holds for the CFD simulations or not. This test has been conducted for AR = 1 by analyzing the $V/\overline{u}_{ref}$ vertical profiles passing through the center of each of the 11 canyons. The percentage differences between the velocity magnitudes calculated for two contiguous canyons have been evaluated within the canyon ($0 < z/H \leq 1$) and in the boundary layer above ($z/H > 1$) (Table 2). Such differences become small from the fourth building onward, after which a well-defined trend towards the equilibrium occurs. This trend stabilizes after the ninth canyon, where Δ is almost constant (≈0.35%). For this reason, from now on, it is implicit that the vertical profile considered for comparisons is the one passing through the center of the ninth canyon.

TABLE 1: Characteristics of the four meshes used for the grid independence test and corresponding percentage differences Δ.

	Mesh A	Mesh B	Mesh C	Mesh D
Number of cells	104896	32350	14275	394576
Mesh interval size within the canyon (m)	0.001	0.002	0.004	0.0005
Velocity magnitude (ms^{-1})	0.225996	0.225531	0.225273	0.225800
Δ (%)	0.08	0.20	0.32	—

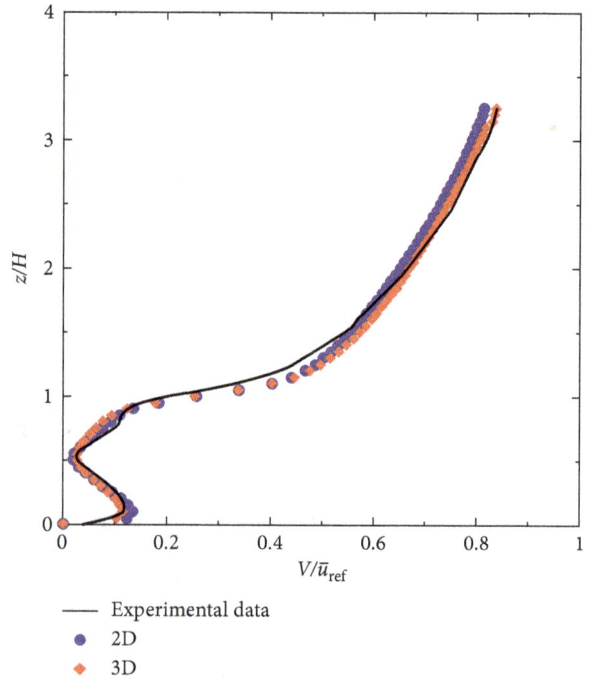

FIGURE 6: Vertical profiles, passing through the center of the ninth canyon, of the nondimensional velocity magnitude calculated using the 2D (blue circles) and the 3D (red diamonds) models for AR = 1. The continuous line refers to the experimental data. The height is normalized by the building height, H.

3.7. Turbulence Model Evaluation. In this section, the average velocities calculated by means of numerical simulations conducted using seven turbulence models implemented in Ansys

TABLE 2: Percentage differences between the velocity magnitudes calculated along the vertical profiles passing through the center of two contiguous canyons.

	Canyon	1	2	3	4	5	6	7	8	9	10
$0 < z/H \le 1$	Δ (%)	34.20	11.63	3.47	1.61	0.90	0.67	0.57	0.35	0.34	0.33
$z/H > 1$	Δ (%)	4.54	2.07	1.13	0.73	0.51	0.37	0.27	0.25	0.23	0.21

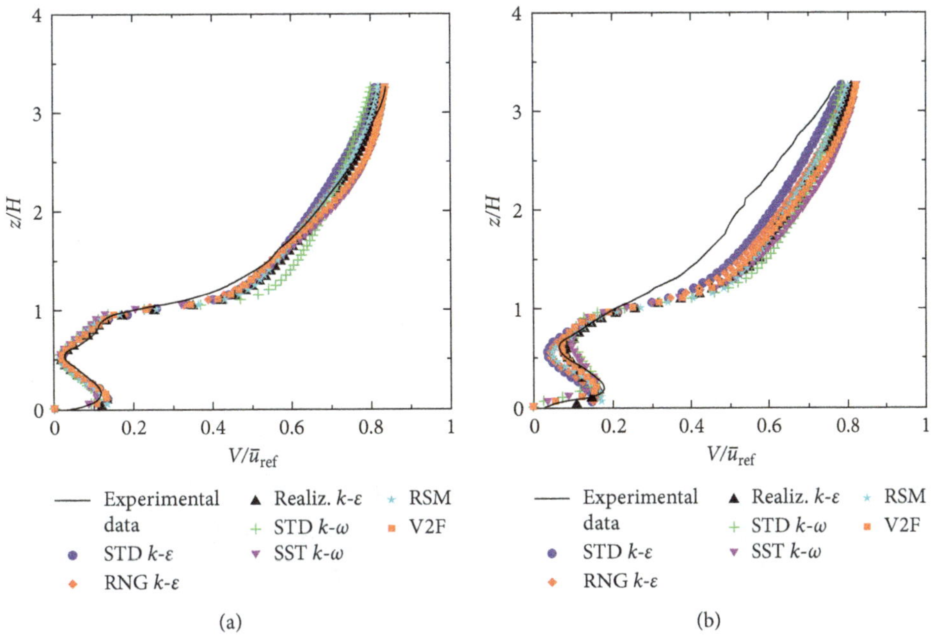

FIGURE 7: Comparison between simulated (symbols) and measured (line) vertical profiles of $V/\overline{u}_{\mathrm{ref}}$ passing through the center of the canyon for AR = 1 (a) and AR = 0.5 (b).

Fluent are compared with those measured in the water channel. The turbulence models employed in the analysis are the standard (STD) k-ε model [52], renormalization group (RNG) k-ε model [53], realizable k-ε model [54], standard (STD) k-ω model [55], shear stress transport (SST) k-ω model [56], Reynolds stress model (RSM) [57], and $\overline{v^2} - f$ (V2F) model [24]. The aim is to assess the accuracy of the V2F model in estimating the velocity profiles and the flow field within the canyon for both the skimming flow (AR = 1) and the wake-interference regime (AR = 0.5).

Figure 7 shows the comparisons between the observed (line) and simulated (symbols) vertical profiles of $V/\overline{u}_{\mathrm{ref}}$ for AR = 1 (Figure 7(a)) and 0.5 (Figure 7(b)). Overall, the velocity above the canyon is lower for AR = 0.5, that is, for the wake-interference regime, in agreement with the field campaign measurements [54]. The simulated profiles do not differ considerably, and the differences with the measured profiles are reasonably small, even though a general underestimation occurs within the canyon. In contrast, the model generally overestimates the velocity above the canyons. However, the V2F model gives velocity profiles closer to those observed in both the analyzed flow regimes (see the percentage differences listed in Tables 3 and 4 among the seven turbulence models obtained for the two ARs).

The V2F model results have been compared with the experimental data also to assess its capability to capture global flow characteristics such as the number and location of the vortex structures formed within the canyon. The correct simulation of the vortex topology is of great importance [58], for example, for the determination of the concentration of

TABLE 3: Percentage differences between measured and modeled velocities within and above the canyon (AR = 1).

	STD k-ε	RNG k-ε	Real k-ε	STD k-ω	SST k-ω	RSM	V2F
$z/H > 0$, Δ (%)	8.88	7.05	8.55	13.42	11.25	10.60	8.01
$0 < z/H \le 1$, Δ (%)	18.82	17.47	17.78	23.16	24.30	24.40	15.72
$z/H > 1$, Δ (%)	4.46	2.42	4.44	9.09	5.45	4.46	4.59

TABLE 4: Percentage differences between measured and modeled velocities within and above the canyon (AR = 0.5).

	STD k-ε	RNG k-ε	Real k-ε	STD k-ω	SST k-ω	RSM	V2F
$z/H > 0$, Δ (%)	19.53	20.70	19.30	25.72	24.38	22.66	19.30
$0 < z/H \le 1$, Δ (%)	31.89	26.21	12.14	26.15	20.71	22.72	10.56
$z/H > 1$, Δ (%)	14.04	18.26	22.48	25.53	26.01	22.63	23.19

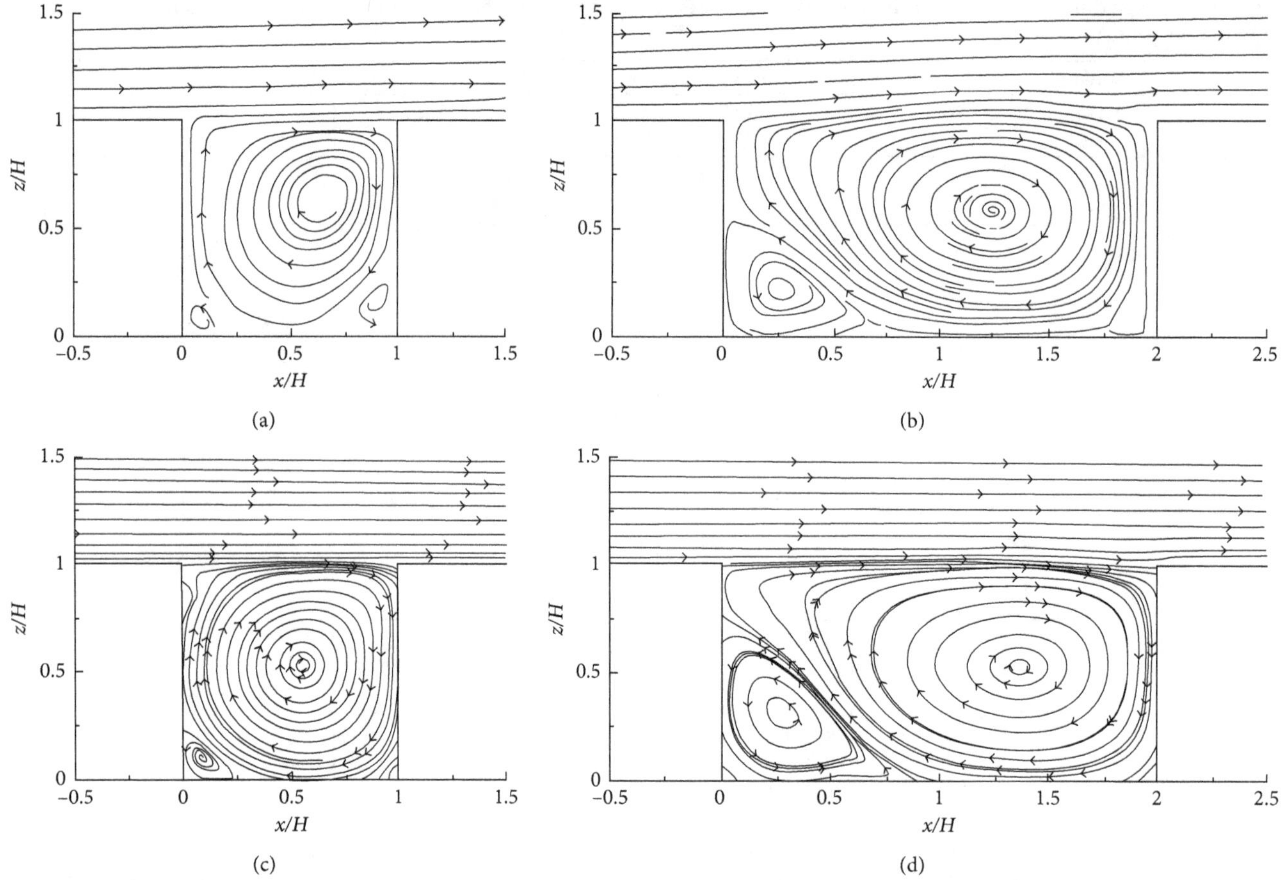

FIGURE 8: Streamlines of the mean velocity magnitude for AR = 1 measured in the laboratory (a) and simulated (c) and for AR = 0.5 measured in the laboratory (b) and simulated (d).

pollutants emitted within the canyon [59, 60], particularly when the source is located at the street level.

The three k-ε models results also show a good agreement with the experiments, especially for AR = 1, while the two k-ω models and the RSM, in particular, always show the larger differences. This is understandable in that the Reynolds number within the canyon is not large and the k-ε models are more accurate in these conditions [55]. In contrast, the V2F models are recognized as giving better performance for both low-Reynolds number and wall-bounded flows. Overall, it is possible to conclude that the V2F model reproduces the velocity profiles inside the canyon better than the other turbulence models.

Figure 8 shows the maps for the two ARs of the streamlines associated with the measured (Figures 8(a) and 8(b)) and simulated (Figures 8(c) and 8(d)) velocities. All of them conform to the canonical configuration of the canopy flow, that is, a current nearly parallel to the streamwise direction above the canopy and a main vortex within the canyon, characterized by lower velocity. For AR = 1, the CFD simulates also a counterclockwise recirculating region located in the upper part of the facet of the leeward building and other two smaller vortices, located at the bottom corners. These data also match other results reported in the literature [60–64]. Both measurements and simulations show that the size of the secondary vortex located at the bottom of the leeward building grows with AR. At AR = 0.5, indeed, it shows two adjacent vortices: the downstream one is by far the larger and rotates clockwise, while the upstream one is smaller, occupying nearly 1/4 of the canyon and rotates counterclockwise. This pattern is in agreement with experimental data and numerical simulations performed in [9, 60, 61, 63–65].

In conclusion, among the seven turbulence models considered here, the model V2F shows the best agreement with the experimental data, particularly within the canyon. Furthermore, it requires the shortest calculation time.

3.8. Effects of Aspect Ratio and Building Height Variations on the Canyon Ventilation. Once the V2F model performance has been verified against experimental data, the same model has been used to investigate urban street canyons characterized by variants of the skimming flow for narrow canyons (AR > 1) and for variations of the relative height of the buildings $H_2/H_1 \neq 1$. The goal is to quantify the air ventilation properties of the canyon making use of the air-exchange rate (ACH). The latter is a measure of the rate of air removal from the street canyon [66]:

$$\mathrm{ACH} = \int_W w_+ dx, \tag{10}$$

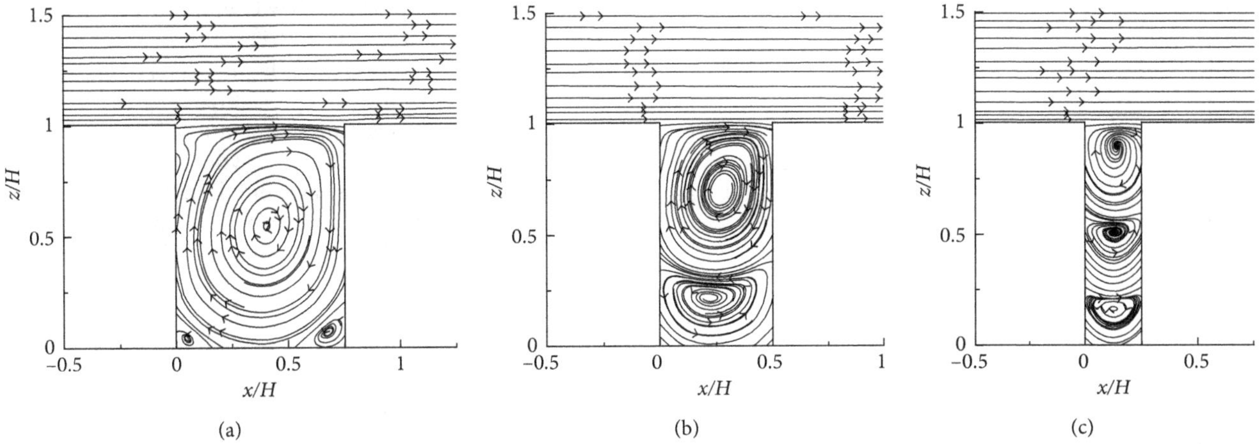

FIGURE 9: Streamlines of the simulated velocity magnitude for AR = 1.33 (a), AR = 2 (b), and AR = 4 (c).

where the subscript "+" indicates positive (upward) vertical velocity, while W is the canyon length. Since RANS models do not calculate the instantaneous velocity components, according to [67], ACH has been estimated as the sum of its average and fluctuating parts:

$$\mathrm{ACH} = \int_W \overline{w}_+ dx + \int_W w'_+ dx = \overline{\mathrm{ACH}} + \mathrm{ACH}', \qquad (11)$$

where the contribution w' is obtained from μ_t and k by using (3) and assuming isotropic conditions. The computational parameters and settings adopted in the previous section have also been employed for these additional analyses.

3.9. Effect of Aspect Ratio Variations. Street canyons with AR > 1.54 are characterized by the multivortex flow regime. Compared to AR = 1, this variant of the skimming flow involves a higher reduction of the wind speed within the canyon and lower vertical diffusion of pollutants emitted within the cavity [9]. Three narrower street canyons with AR > 1 (AR = 1.33, AR = 2, and AR = 4) have therefore been investigated to analyze their ventilation properties.

The multivortex configurations in the skimming flow have firstly been analyzed in terms of streamlines (Figure 9). The results show the transition from the one-vortex regime to the multivortex regime as AR increases. The case AR = 1.33 (Figure 9(a)) still shows the main clockwise vortex as seen for AR = 1 (Figure 8(c)), but the two recirculation zones at the canyon bottom are more noticeable. For AR = 2, these two vortices merge to form a larger counterclockwise structure (Figure 9(b)). The canyon is therefore divided into two regions, one lying above the other, where the upper recirculation is stronger and drives the lower vortex. The upper recirculation is still shifted downstream ($x = 0.26H$, $z = 0.71H$), according to AR = 1, while the lower, flatter vortex is centered in the cavity ($x = 0.26H$, $z = 0.22H$). These results agree reasonably well with the water-channel experiments performed by Baik et al. [68] for AR = 2, which showed ($x = 0.32H$, $z = 0.75H$) and ($x = 0.29H$, $z = 0.17H$) as the locations of the centers of the upper and lower recirculation regions, respectively.

Further increases in AR lead to the formation of additional vortices within the cavity. For example, for AR = 4, three vertically aligned vortices are formed (Figure 9(c)), with increasing dimensions upwards. The configuration of narrow buildings is particularly interesting for the investigation of dispersion phenomena. In fact, pollutants typically emitted by vehicular traffic at the canyon bottom through linear sources are trapped in the lower part of the canyon, where strong values of mean and standard deviation of concentration occur near the sidewalks [7, 69], directly affecting the final receptor. Furthermore, the external wind flows above the canopy almost parallel to the roofs, resulting in a poor canyon ventilation process, are strongly hampered by the structure of the vortices. This corroborates the idea that, for the skimming flow, the fluid has difficulty in penetrating the interelement spaces, and therefore it skims, remaining nearly parallel to the roofs [68, 70]. For this reason, it is fundamental to consider a correct urban planning to minimize unwanted effects of pollutant accumulation.

By comparing the vertical profile of $V/\overline{u}_{\mathrm{ref}}$ calculated for all the ARs (Figure 10), it can be seen that, above the canyon, it depends appreciably on AR. On the contrary, $V/\overline{u}_{\mathrm{ref}}$ changes considerably with AR within the canyon, especially going from the standard skimming flow to the multivortex regime. While the velocity magnitude for AR = 1.33 is similar to that seen for AR = 1, it drastically drops for AR = 2 and 4, indicating that the multivortex flow is characterized by very poor ventilation, particularly at the street level. The results presented above conform to those presented in [9, 58, 61, 65–67, 71], which simulated canopy flows through CFD, employing different turbulence models.

The nondimensional air ventilation components for the five ARs are shown in Figure 13 (see data points referring to $H_2/H_1 = 1$). While the mean contribution $\overline{\mathrm{ACH}}/(\overline{u}_{\mathrm{ref}}W)$ does not change appreciably with AR, the lower the aspect ratio, the higher the $\mathrm{ACH}'/(\overline{u}_{\mathrm{ref}}W)$. This suggests that turbulence plays a major role in air exchanges between the canyon and the overlying layer. This is particularly true for the wake-interference regime, where

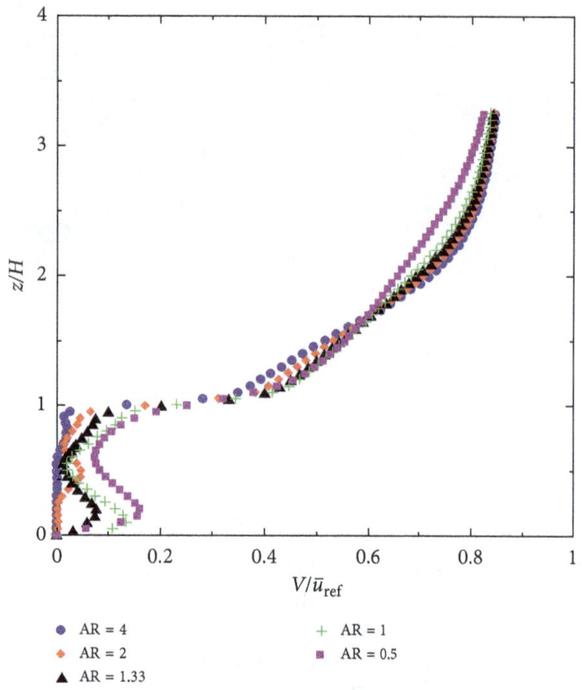

FIGURE 10: Comparison among the simulated vertical profiles of $V/\overline{u}_{ref}$ passing through the center of the canyon for AR = 4, AR = 2, AR = 1.33, AR = 1, and AR = 0.5.

$\overline{ACH}$ is nearly an order of magnitude greater than ACH′. However, a question here arises regarding the significance of the air-exchange rate for the multivortex flow. In fact, since ACH depends only on the flow kinematics around the upper part of the canyon, it cannot provide any information on the actual air exchange occurring at the cavity bottom, where the second (or third) vortex is located. Therefore, the use of ACH for evaluating the ventilation performance of a street canyon and its relation with pollutant removal mechanisms must be considered with circumspection for multivortex regimes.

3.10. Effect of Building Height Variations. Another geometrical factor that considerably influences the ventilation in the urban street canyon is the relative height of the buildings, H_2/H_1. This parameter has already been investigated in other works [62, 72–77] for 2D flows, while useful insights into the effects of building height variations for arrays of 3D buildings have recently been reported in [78–80]. The additional analysis we provide here focuses on the combined effect of the variability of both building heights and AR. Six H_2/H_1 have been considered for each of the five ARs, in particular the step-up configurations, where the leeward building (H_1) is shorter than the windward building ($H_2/H_1 = 1.5$, 2, and 2.5), and the step-down configurations, where the leeward building is taller than the windward building ($H_2/H_1 = 0.4$, 0.5, and 0.7).

Figure 11 shows the streamlines obtained for the step-up geometry ($H_2/H_1 > 1$). For AR = 0.5 and $H_2/H_1 = 1.5$ (Figure 11(a)), the flow field does not depart significantly from that observed for $H_2/H_1 = 1$ (Figure 9(a)), in agreement with the numerical results in [64]. The size of the main vortex increases as H_2/H_1 increases (Figures 11(f) and 11(k)), and its center does not move appreciably along the x-axis, while it moves upward, reaching about the height of the leeward building. At the bottom of the canyon, the recirculation zone at the corner of the leeward building becomes smaller as H_2/H_1 increases, while the anticlockwise vortex at the corner with the windward building progressively increases in size. A similar behavior occurs for AR = 1 (Figures 11(b), 11(g), and 11(l)).

Regarding the other skimming flows, their dependence on H_2/H_1 is somehow greater. The progressive ejection of the upper vortex from the canyon into the overlying layer observed for AR = 1.33 (Figures 11(c), 11(h), and 11 (m)), in fact, is increasingly evident going from AR = 2 (Figures 11(d), 11(i), and 11(n)) to 4 (Figures 11(e), 11(j), and 11(o)). While for AR = 2 and $H_2/H_1 = 1.5$, the center of the upper vortex is located at $z \approx H_1$ (Figure 11(d)) and it progressively moves upward as H_2/H_1 grows, and for $H_2/H_1 = 2.5$, the vortex is practically outside of the canyon. Similar considerations can be drawn for the case AR = 1.33, which shows the transition from the standard skimming flow when $H_2/H_1 = 1$ (Figure 9(a)) to the multivortex regime for $H_2/H_1 > 1$ (Figures 11(c), 11(h), and 11(m)). The recirculation zones at the bottom of the canyon are combined together, and two counterrotating vortices occupy the canyon.

In terms of air ventilation, $ACH'/(\overline{u}_{ref}W)$ always exceeds its average counterpart, $\overline{ACH}/(\overline{u}_{ref}W)$, even though not to a large extent as for $H_2/H_1 = 1$. Furthermore, ACH does not depend significantly on H_2/H_1 when AR = 1 and 2, while a clear decrease in ACH for increasing H_2/H_1 takes place for the other aspect ratios. In particular, taller windward buildings allow lower vertical mass transfer between the canyon and the overlying region.

Finally, Figure 12 shows the flow patterns for the step-down configurations ($H_2/H_1 < 1$). They are characterized by a wide clockwise vortex placed over the canyon and the top of the windward building. Overall, the lower the H_2/H_1, the smaller the ACH (Figure 13(c)), with the exception of the case (AR = 2, $H_2/H_1 = 0.67$), when there is only a large vertical structure occupying both the canyon and the overlying region up to $z \approx H_1$ (Figure 12(a)). The latter configuration corresponds with the largest ACH calculated in the present analysis and is mainly associated with large $\overline{ACH}$. In contrast, for all the other step-down configurations investigated here, the main vortex (or the two or more vortices, when AR ≥ 1.33) remains confined within the canyon. The latter represents the main difference between step-up and step-down configurations, and it might have great influence on the concentration of pollutants emitted within the canyon, particularly at the street level.

Lastly, we note that, from the point of view of air quality analysis, the development of secondary vortices in the lower corners of the canyon for AR = 0.5 and 1 should determine an accumulation of pollutants near the sidewalks in the case of vehicular traffic emissions, whatever be the value of the height ratio. For AR = 1.33 and 2, the presence of the two

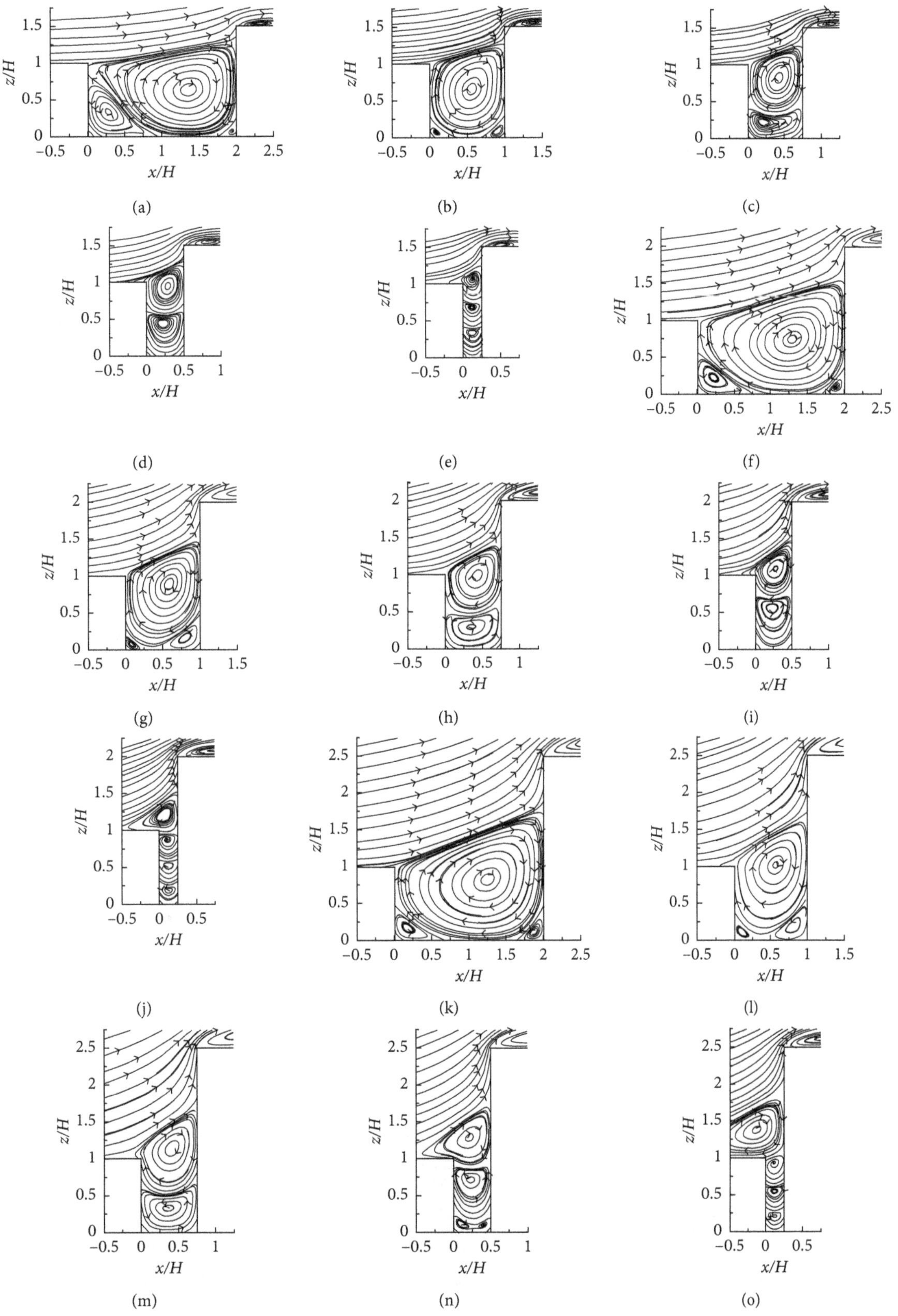

FIGURE 11: Streamlines of the simulated velocity magnitude V for AR = 0.5 with $H_2/H_1 = 1.5$ (a), $H_2/H_1 = 2$ (f), and $H_2/H_1 = 2.5$ (k); for AR = 1 with $H_2/H_1 = 1.5$ (b), $H_2/H_1 = 2$ (g), and $H_2/H_1 = 2.5$ (l); for AR = 1.33 with $H_2/H_1 = 1.5$ (c), $H_2/H_1 = 2$ (h), and $H_2/H_1 = 2.5$ (m); for AR = 2 with $H_2/H_1 = 1.5$ (d), $H_2/H_1 = 2$ (i), and $H_2/H_1 = 2.5$ (n); for AR = 4 with $H_2/H_1 = 1.5$ (e), $H_2/H_1 = 2$ (j), and $H_2/H_1 = 2.5$ (o).

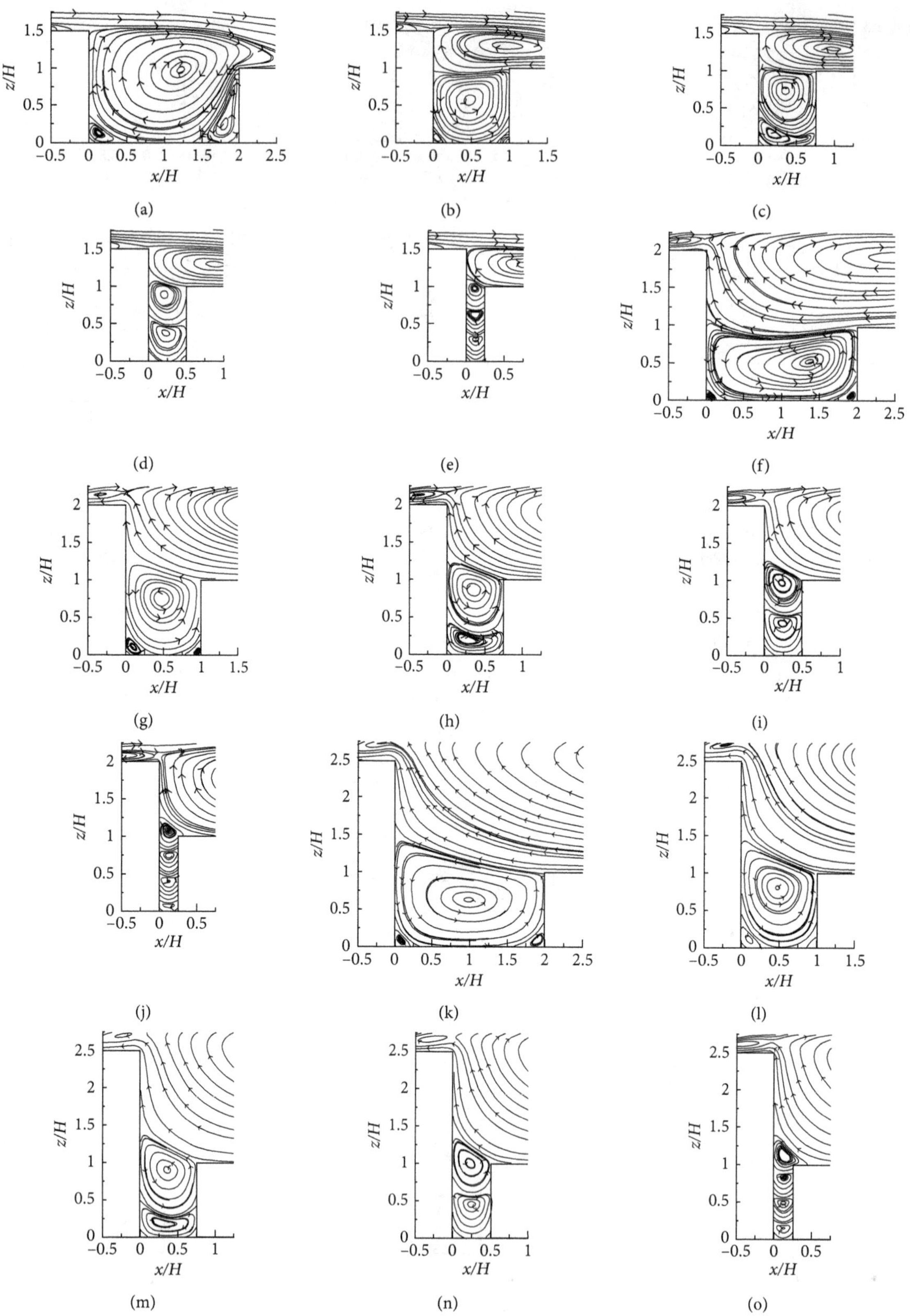

FIGURE 12: Streamlines of the simulated velocity magnitude V for AR = 0.5 with $H_2/H_1 = 0.67$ (a), $H_2/H_1 = 0.5$ (f), and $H_2/H_1 = 0.4$ (k); for AR = 1 with $H_2/H_1 = 0.67$ (b), $H_2/H_1 = 0.5$ (g), and $H_2/H_1 = 0.4$ (l); for AR = 1.33 with $H_2/H_1 = 0.67$ (c), $H_2/H_1 = 0.5$ (h), and $H_2/H_1 = 0.4$ (m); for AR = 2 with $H_2/H_1 = 0.67$ (d), $H_2/H_1 = 0.5$ (i), and $H_2/H_1 = 0.4$ (n); for AR = 4 with $H_2/H_1 = 0.67$ (e), $H_2/H_1 = 0.5$ (j), and $H_2/H_1 = 0.4$ (o).

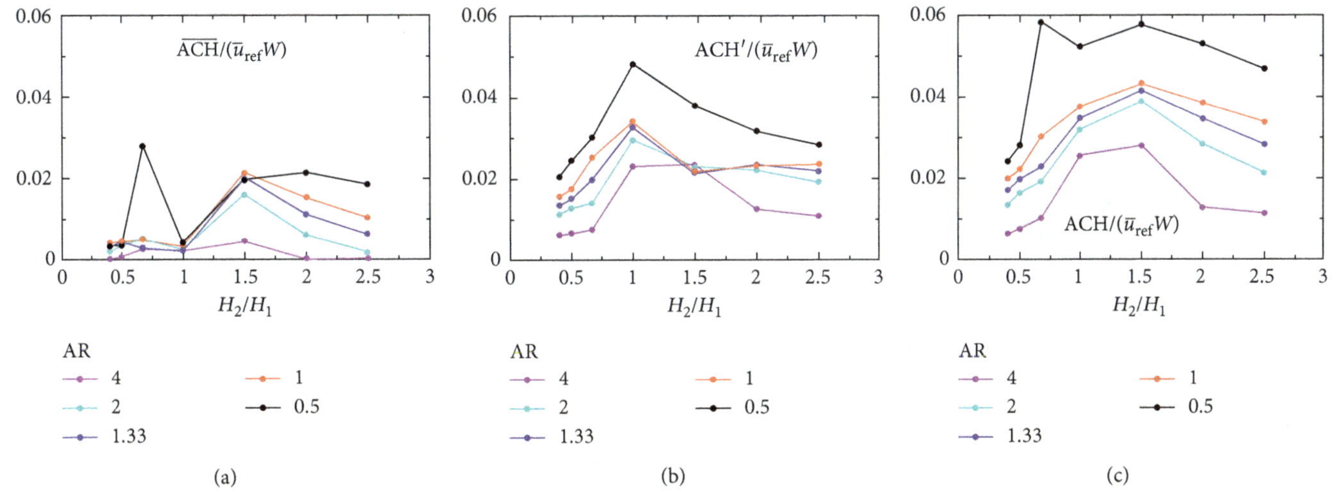

FIGURE 13: Nondimensional ACH components as a function of the relative building heights for five aspect ratios: (a) mean component, (b) turbulent component, and (c) total ACH.

counterrotating vortices further limits the ventilation in the canyon, especially in the portion closest to the ground. For AR = 4, the vertically aligned multiple vortices configuration strongly inhibits the exchange of air with the higher levels and paves the way to the stagnation of pollutants at the pedestrian level.

4. Summary and Conclusions

Water-channel data have been used to diagnose the capability of the $\overline{v^2} - f$ (V2F) turbulence model, implemented in Ansys Fluent, to reproduce the flow field in a regular array of 2D buildings. The experiments refer to two very common geometrical configurations, that is, the skimming flow (AR = 1) and the wake-interference regime (AR = 0.5). One of the strengths of the V2F model is the equation of the turbulent viscosity, which takes into account the anisotropy of the flow in the near-wall region through the modeling of a velocity scale. The performances of the V2F model have been compared with those of other six turbulence models implemented in Ansys Fluent. The results have shown that the V2F model gives the best results with shorter computational time.

Further simulations conducted using the V2F model have made it possible to analyze canyon ventilation for a variety of aspect ratios and step-up and step-down configurations by calculating the air-exchange rate (ACH). For the step-up configurations ($H_2/H_1 > 1$), the increase of the relative height of the buildings does not appreciably change the total ACH for both the wake-interference and the skimming flows, while a certain decrease of ACH occurs for AR > 1.33. On the contrary, step-down configurations ($H_2/H_1 < 1$) appear to be in general less ventilated and therefore more prone to pollutant recirculation. For all the geometries investigated, the air ventilation is mainly determined by turbulent motions with the exception of the wake-interference regime for $H_2/H_1 = 0.67$, the latter canyon geometry being characterized by the largest contribution of the mean ACH.

In conclusion, the V2F turbulence model has proved to be a useful tool for wind engineers as well as for investigations concerning air quality control and urban planning.

Conflicts of Interest

The authors declare that there are no conflicts of interest regarding the publication of this paper.

Acknowledgments

This research was partially supported by the BRiC (ID 22) fund from INAIL (Project VIEPI: "Integrated evaluation of particulate pollutant for indoor air quality") and the RG11715C7D43B2B6 fund from the University of Rome "La Sapienza".

References

[1] C. W. Kent, S. Grimmond, J. Barlow et al., "Evaluation of urban local-scale aerodynamic parameters: implications for the vertical profile of wind speed and for source areas," *Boundary-Layer Meteorology*, vol. 164, no. 2, pp. 183–213, 2017.

[2] A. Pelliccioni, P. Monti, and G. Leuzzi, "Wind-speed profile and roughness sublayer depth modelling in urban boundary layers," *Boundary-Layer Meteorology*, vol. 160, no. 2, pp. 225–248, 2016.

[3] T. R. Oke, "Street design and urban canopy layer climate," *Energy and Buildings*, vol. 11, no. 1-3, pp. 103–113, 1988.

[4] F. Nardecchia, F. Gugliermetti, and F. Bisegna, "A novel approach to CFD analysis of the urban environment," *Journal of Physics: Conference Series*, vol. 655, article 012013, 2015.

[5] A. Salvati, H. Coch Roura, and C. Cecere, "Assessing the urban heat island and its energy impact on residential buildings in Mediterranean climate: Barcelona case study," *Energy and Buildings*, vol. 146, pp. 38–54, 2017.

[6] B. Blocken, T. Stathopoulos, J. Carmeliet, and L. M. Hensen, "Application of computational fluid dynamics in building performance simulation for the outdoor environment: an

overview," *Journal of Building Performance Simulation*, vol. 4, no. 2, pp. 157–184, 2011.
[7] A. Di Bernardino, P. Monti, G. Leuzzi, and G. Querzoli, "Pollutant fluxes in two-dimensional street canyons," *Urban Climate*, vol. 24, pp. 80–93, 2018.
[8] S. Ferrari, M. G. Badas, M. Garau, A. Seoni, and G. Querzoli, "The air quality in narrow two-dimensional urban canyons with pitched and flat roof buildings," *International Journal of Environment and Pollution*, vol. 62, no. 2–4, pp. 347–368, 2017.
[9] J. F. Sini, S. Anquetin, and P. G. Mestayer, "Pollutant dispersion and thermal effects in urban street canyon," *Atmospheric Environment*, vol. 30, no. 15, pp. 2659–2677, 1996.
[10] Y.-K. Ho and C.-H. Liu, "Street-level ventilation in hypothetical urban areas," *Atmosphere*, vol. 8, no. 12, p. 124, 2017.
[11] M. Immer, J. Allegrini, and J. Carmeliet, "Time-resolved and time-averaged stereo-PIV measurements of a unit-ratio cavity," *Experiments in Fluids*, vol. 57, no. 6, pp. 57–101, 2016.
[12] J. J. O'Neill, X.-M. Cai, and R. Kinnersley, "Stochastic backscatter modelling for the prediction of pollutant removal from an urban street canyon: a large-eddy simulation," *Atmospheric Environment*, vol. 142, pp. 9–18, 2016.
[13] M. Garau, M. G. Badas, S. Ferrari, A. Seoni, and G. Querzoli, "Turbulence and air exchange in a two-dimensional urban street canyon between gable roof buildings," *Boundary-Layer Meteorology*, vol. 167, no. 1, pp. 123–143, 2018.
[14] Y. Huang, X. Hu, and N. Zeng, "Impact of wedge-shaped roofs on airflow and pollutant dispersion inside urban street canyons," *Building and Environment*, vol. 44, no. 12, pp. 2335–2347, 2009.
[15] R. A. Memon, D. Y. C. Leung, and C. H. Liu, "Effects of building aspect ratio and wind speed on air temperatures in urban-like street canyons," *Building and Environment*, vol. 45, no. 1, pp. 176–188, 2010.
[16] F. Murena and B. Mele, "Effect of short–time variations of wind velocity on mass transfer rate between street canyons and the atmospheric boundary layer," *Atmospheric Pollution Research*, vol. 5, no. 3, pp. 484–490, 2014.
[17] J. Allegrini, V. Dorer, and J. Carmeliet, "Buoyant flows in street canyons: validation of CFD simulations with wind tunnel measurements," *Building and Environment*, vol. 72, no. 63, pp. 63–74, 2014.
[18] Y.-K. Ho, C.-H. Liu, and M. S. Wong, "Preliminary study of the parameterisation of street-level ventilation in idealised two-dimensional simulations," *Building and Environment*, vol. 89, pp. 345–355, 2015.
[19] G. Xie, S. Zheng, W. Zhang, and B. Sundén, "A numerical study of flow structure and heat transfer in a square channel with ribs combined downstream half-size or same-size ribs," *Applied Thermal Engineering*, vol. 61, no. 2, pp. 289–300, 2013.
[20] X. M. Cai, J. F. Barlow, and S. E. Belche, "Dispersion and transfer of passive scalars in and above street canyons-large-eddy simulations," *Atmospheric Environment*, vol. 42, no. 23, pp. 5885–5895, 2008.
[21] D. Hamlyn and R. A. Britter, "Numerical study of the flow field and exchange processes within a canopy of urban-type roughness," *Atmospheric Environment*, vol. 39, no. 18, pp. 3243–3254, 2005.
[22] R. Ramponi and B. Blocken, *A Computational Study on the Influence of Urban Morphology on Wind-Induced Outdoor Ventilation*, International Environmental Modelling and Software Society (iEMSs), Manno, Switzerland, 2012.
[23] ANSYS Inc., *ANSYS Fluent Software Package, v. 14.5, User's Guide*, ANSYS Inc., Canonsburg, PA, USA, 2009.
[24] P. A. Durbin, "Separated flow computations with the k-epsilon-v-squared model," *AIAA Journal*, vol. 33, no. 4, pp. 659–664, 1995.
[25] D. R. Laurence, J. C. Uribe, and S. V. Utyuzhnikov, "A robust formulation of the $v^2 - f$ model," *Flow, Turbulence and Combustion*, vol. 73, no. 3-4, pp. 169–185, 2004.
[26] S. Parneix and P. A. Durbin, *Numerical Simulation of 3D Turbulent Boundary Layers Using the $v^2 - f$ Model*, Annual Research Briefs Center For Turbulence Research, NASA/Stanford University, Stanford, CA, USA, 1997.
[27] R. Rossi and G. Iaccarino, "Numerical analysis and modeling of plume meandering in passive scalar dispersion downstream of a wall-mounted cube," *International Journal of Heat and Fluid Flow*, vol. 43, pp. 137–148, 2013.
[28] M. Behnia, S. Parneix, and P. A. Durbin, "Prediction of heat transfer in a jet impinging on a flat plate," *International Journal of Heat and Mass Transfer*, vol. 41, no. 12, pp. 1845–1855, 1998.
[29] A. B. Bhagwat and A. Sridharan, "Convective heat transfer from a heated plate to the orthogonally impinging air jet," *Journal of Thermal Science and Engineering Applications*, vol. 8, no. 4, article 041009, 2016.
[30] A. K. Pujari, B. V. S. S. S. Prasad, and N. Sitaram, "Effect of blowing ratio on the internal heat transfer of a cooled nozzle guide vane in a linear cascade," *Journal of Thermal Science and Engineering Applications*, vol. 8, no. 4, article 041004, 2016.
[31] R. Manceau, S. Parneix, and D. Laurence, "Turbulent heat transfer predictions using the $v^2 - f$ model on unstructured meshes," *International Journal of Heat and Fluid Flow*, vol. 21, no. 3, pp. 320–328, 2000.
[32] L. Marrocco and A. Franco, "Direct numerical simulation and RANS comparison of turbulent convective heat transfer in staggered ribbed channel with high blockage," *Journal of Heat Transfer*, vol. 139, no. 2, article 021701, 2017.
[33] G. Kalitzin, *Application of the $v^2 - f$ Model to Aerospace Configurations*, Center for Turbulence Research Ann Res Briefs, Stanford, CA, USA, 1999.
[34] F.-S. Lien and G. Kalitzin, "Computations of transonic flow with the $v^2 - f$ turbulence model," *International Journal of Heat and Fluid Flow*, vol. 22, no. 1, pp. 53–61, 2001.
[35] Z. Zhang, W. Zhang, Z. Zhai, and Q. Chen, "Evaluation of various turbulence models in predicting airflow and turbulence in enclosed environments by CFD: part-2: comparison with experimental data from literature," *HVAC&R Research*, vol. 13, no. 6, pp. 871–886, 2007.
[36] C. Heschl, Y. Tao, K. Inthavong, and J. Tu, "Improving predictions of heat transfer in indoor environments with eddy viscosity turbulence models," *Building Simulation*, vol. 9, no. 2, pp. 213–220, 2016.
[37] Z. Zhang and Q. Chen, "Prediction of particle deposition onto indoor surfaces by CFD with a modified Lagrangian method," *Atmospheric Environment*, vol. 43, no. 2, pp. 319–328, 2009.
[38] J.-J. Baik, R. E. Park, H. Y. Chun, and J.-J. Kim, "A laboratory model of urban street-canyon flows," *Journal of Applied Meteorology*, vol. 39, no. 9, pp. 1592–1600, 2000.
[39] H. Liu, B. Liang, F. Zhu, B. Zhang, and J. Sang, "A laboratory model for the flow in urban street canyons induced by bottom heating," *Advances in Atmospheric Sciences*, vol. 20, no. 4, pp. 554–564, 2003.
[40] P. Huq and P. Franzese, "Measurements of turbulence and dispersion in three idealized urban canopies with different aspect ratios and comparison with a Gaussian plume model," *Boundary-Layer Meteorology*, vol. 147, no. 1, pp. 103–121, 2013.

[41] X.-X. Li, D. Y. C. Leung, and C.-H Liu, "Physical modeling of flow field inside urban street canyons," *Journal of Applied Meteorology and Climatology*, vol. 47, no. 7, pp. 2058–2067, 2007.

[42] A. Di Bernardino, P. Monti, G. Leuzzi, and G. Querzoli, "Water-channel study of flow and turbulence past a two-dimensional array of obstacles," *Boundary-Layer Meteorology*, vol. 155, no. 1, pp. 73–85, 2015.

[43] W. H. Snyder, "Guideline for fluid modeling of atmospheric diffusion," EPA Tech. Rep. EPA-600/8-81-009, Environmental Protection Agency, Washington, DC, USA, 1981.

[44] K. Uehara, S. Wakamatsu, and R. Ooka, "Studies on critical Reynolds number indices for wind-tunnel experiments on flow within urban areas," *Boundary-Layer Meteorology*, vol. 107, no. 2, pp. 353–370, 2003.

[45] A. Cenedese, Z. Del Prete, M. Miozzi, and G. Querzoli, "A laboratory investigation of the flow in the left ventricle of a human heart with prosthetic, tilting-disk valves," *Experiments in Fluids*, vol. 39, no. 2, pp. 322–335, 2005.

[46] P. Monti, G. Querzoli, A. Cenedese, and S. Piccinini, "Mixing properties of a stably stratified parallel shear layer," *Physics of Fluids*, vol. 19, no. 8, article 085104, 2007.

[47] A. Di Bernardino, P. Monti, G. Leuzzi, and G. Querzoli, "Water-channel estimation of Eulerian and Lagrangian time scales of the turbulence in idealized two-dimensional urban canopies," *Boundary-Layer Meteorology*, vol. 165, no. 2, pp. 251–276, 2017.

[48] ANSYS Inc., *ANSYS Fluent Software Package, v. 14.5, Theory Guide*, ANSYS Inc., Canonsburg, PA, USA, 2013.

[49] P. A. Durbin, "Near-wall turbulence closure modeling without "damping functions"," *Theoretical and Computational Fluid Dynamics*, vol. 3, pp. 1–13, 1991.

[50] J. Franke, A. Hellsten, H. Schlünzen, and B. Carissimo, *Best Practice Guideline for the CFD Simulation of Flows in the Urban Environment, COST Action 732*, Meteorological Inst., Norway, 2007.

[51] S. V. Patankar, *Numerical Heat Transfer and Fluid Flow*, Hemisphere Publishing Corporation, New York, NY, USA, 1980.

[52] B. E. Launder and D. B. Spalding, *Lectures in Mathematical Models of Turbulence*, Academic Press, London, UK, 1972.

[53] D. Choudhury, *Introduction to the Renormalization Group Method and Turbulence Modeling*, Fluent Inc., New York, NY, USA, Technical Memorandum TM-107, 1983.

[54] T.-H. Shih, W. W. Liou, A. Shabbir, Z. Yang, and J. Zhu, "A new k-ε eddy-viscosity model for high Reynolds number turbulent flows: model development and validation," *Computer and Fluids*, vol. 24, no. 3, pp. 227–238, 1995.

[55] D. C. Wilcox, *Turbulence Modeling for CFD*, DCW Industries, Inc., La Cañada Flintridge, CA, USA, 1998.

[56] F. R. Menter, "Two-equation eddy-viscosity turbulence models for engineering applications," *AIAA Journal*, vol. 32, no. 8, pp. 1598–1605, 1994.

[57] M. M. Gibson and B. E. Launder, "Ground effects on pressure fluctuations in the atmospheric boundary layer," *Journal of Fluid Mechanics*, vol. 86, no. 3, pp. 491–511, 1978.

[58] F. Nardecchia, F. Gugliermetti, and F. Bisegna, "How temperature affects the airflow around a single-block isolated building," *Energy and Buildings*, vol. 118, pp. 142–151, 2016.

[59] A. Amicarelli, P. Salizzoni, G. Leuzzi et al., "Sensitivity analysis of a concentration fluctuation model to dissipation rate estimates," *International Journal of Environment and Pollution*, vol. 48, no. 1–4, pp. 164–173, 2012.

[60] J.-J. Baik and J.-J. Kim, "A numerical study of flow and pollutant dispersion characteristics in urban street canyons," *Journal of Applied Meteorology*, vol. 38, no. 11, pp. 1576–1589, 1999.

[61] A. Kovar-Panskus, P. Louka, J. F. Sini et al., "Influence of geometry on the mean flow within urban street canyons–a comparison of wind tunnel experiments and numerical simulations," *Water, Air, and Soil Pollution*, vol. 2, no. 5-6, pp. 365–380, 2002.

[62] X. Xie, Z. Huang, and J.-S. Wang, "The impact of urban street layout on local atmospheric environment," *Building and Environment*, vol. 41, no. 10, pp. 1352–1363, 2006.

[63] X.-X. Li, R. E. Britter, L. K. Norford, T.-Y. Koh, and D. Entekhabi, "Flow and pollutant transport in urban street canyons of different aspect ratios with ground heating: large-eddy simulation," *Boundary-Layer Meteorology*, vol. 142, no. 2, pp. 289–304, 2012.

[64] A. W. M. Yazid, N. A. C. Sidik, S. M. Salim, and N. H. M. Yusoff, "Numerical prediction of air flow within street canyons based on different two-equation k-ε models," *IOP Conference Series: Materials Science and Engineering*, vol. 50, article 012012, 2013.

[65] L. J. Hunter, G. T. Johnson, and I. D. Watson, "An investigation of three-dimensional characteristics of flow regimes within the urban canyon," *Atmospheric Environment Part B Urban Atmosphere*, vol. 26, no. 4, pp. 425–432, 1992.

[66] C.-H. Liu, D. Y. C. Leung, and M. C. Barth, "On the prediction of air and pollutant exchange rates in street canyons of different aspect ratios using large-eddy simulation," *Atmospheric Environment*, vol. 39, no. 38, pp. 1567–1574, 2005.

[67] M. G. Badas, S. Ferrari, M. Garau, and G. Querzoli, "On the effect of gable roof on natural ventilation in two-dimensional urban canyons," *Journal of Wind Engineering and Industrial Aerodynamic*, vol. 162, pp. 24–34, 2017.

[68] J. J. Baik, R. S. Park, H. Y. Chun, and J. J. Kim, "A laboratory model of urban street-canyon flows," *Journal of Applied Meteorology*, vol. 39, no. 9, pp. 1592–1600, 2000.

[69] W. C. Cheng and C.-H. Liu, "Large-eddy simulation of flow and pollutant transports in and above two-dimensional idealised street canyons," *Boundary-Layer Meteorology*, vol. 139, no. 3, pp. 411–437, 2011.

[70] C. S. B. Grimmond and T. R. Oke, "Aerodynamic properties of urban areas derived from analysis of urban surface form," *Journal of Applied Meteorology*, vol. 38, no. 9, pp. 1261–1292, 1999.

[71] S. J. Jeong and M. J. Andrews, "Application of the k-ε turbulence model to the high Reynolds number skimming flow field of an urban street canyon," *Atmospheric Environment*, vol. 36, no. 7, pp. 1137–1145, 2002.

[72] X.-X. Li, C.-H. Liu, and D. Y. C. Leung, "Large-eddy simulation of flow and pollutant dispersion in high-aspect-ratio urban street canyons with wall model," *Boundary-Layer Meteorology*, vol. 129, no. 2, pp. 249–268, 2008.

[73] M. Scungio, F. Arpino, L. Stabile, and G. Buonanno, "Numerical simulation of ultrafine particle dispersion in urban street canyons with the Spalart-Allmaras turbulence model," *Aerosol and Air Quality Research*, vol. 13, pp. 1423–1437, 2013.

[74] V. D. Assimakopoulos, H. M. ApSimon, and N. Moussiopoulos, "A numerical study of atmospheric pollutant dispersion in different two-dimensional street canyon configurations," *Atmospheric Environment*, vol. 37, no. 29, pp. 4037–4049, 2003.

[75] X. Xie, Z. Huang, and J.-S. Wang, "Impact of building configuration on air quality in street canyon," *Atmospheric Environment*, vol. 39, no. 25, pp. 4519–4530, 2005.
[76] H. Zhang, T. Xu, Y. Wang, Y. Zong, S. Li, and H. Tang, "Study on the influence of meteorological conditions and the street side buildings on the pollutant dispersion in the street canyon," *Building Simulation*, vol. 9, no. 6, pp. 717–727, 2016.
[77] Y. Miao, S. Liu, T. Zheng, S. Wang, and Y. Li, "Numerical study of traffic pollutant dispersion within different street canyon configurations," *Advances in Meteorology*, vol. 2014, Article ID 458671, 14 pages, 2014.
[78] M. Princevac, J.-J. Baik, X. Li, H. Pan, and S. B. Park, "Lateral channeling within rectangular arrays of cubical obstacles," *Journal of Wind Engineering and Industrial Aerodynamics*, vol. 98, no. 8-9, pp. 377–385, 2010.
[79] J. Hang, Q. Wang, X. Chen et al., "City breathability in medium density urban-like geometries evaluated through the pollutant transport rate and the net escape velocity," *Building and Environment*, vol. 94, pp. 166–182, 2015.
[80] S. Kenjereš, S. de Wildt, and T. Busking, "Capturing transient effects in turbulent flows over complex urban areas with passive pollutants," *International Journal of Heat and Fluid Flow*, vol. 51, pp. 120–137, 2015.

2

Multitemporal Soil Moisture Retrieval over Bare Agricultural Areas by Means of Alpha Model with Multisensor SAR Data

Xiang Zhang,[1] Xinming Tang,[1] Xiaoming Gao,[1] and Hui Zhao[2]

[1]*Satellite Surveying and Mapping Application Center, National Administration of Surveying, Mapping and Geo-information, Beijing 100048, China*
[2]*National Geomatics Center of China, Beijing 100080, China*

Correspondence should be addressed to Xiang Zhang; zhangxiangcumt@126.com

Academic Editor: Qingyan Meng

The objective of this research is to optimize the Alpha approximation model for soil moisture retrieval using multitemporal SAR data. The Alpha model requires prior knowledge of soil moisture range to constrain soil moisture estimation. The solution of the Alpha model is an undetermined problem due to the fact that the number of observation equations is less than the number of unknown parameters. This research primarily focused on the optimization of Alpha model by employing multisensor and multitemporal SAR data. The disadvantage of the Alpha model can be eliminated by the combination of multisensor SAR data. The optimized Alpha model was evaluated on the basis of a comprehensive campaign for soil moisture retrieval, which acquired multisensor time series SAR data and coincident field measurements. The agreement between the estimated and measured soil moisture was within a root mean square error of 0.08 cm^3/cm^3 for both methods. The optimized Alpha model shows an obvious improvement for soil moisture retrieval. The results demonstrated that multisensor and multitemporal SAR data are favorable for time series soil moisture retrieval over bare agricultural areas.

1. Introduction

Soil moisture is an essential parameter controlling many biophysical processes that impact water, energy, and carbon exchanges at the land-atmosphere interface. Synthetic aperture radar (SAR) is one of the most promising techniques for measuring surface soil moisture at moderate-to-high spatial resolution required by hydrological, meteorological, ecological, and agricultural applications [1–3]. However, accurate soil moisture retrieval from SAR data is still a challenging task due to the fact that the radar backscatter is influenced by multiple parameters such as soil dielectric constant (related to soil moisture), surface roughness, and vegetation conditions [4–10]. Therefore, soil moisture retrieval from SAR data is an ill-posed problem, and thus, it requires either prior knowledge of vegetation and soil surface parameters or multiple configuration SAR data. The multitemporal [11, 12], multi-incidence angle [13–15], multipolarization [16–19], and multifrequency [20] SAR data are increasingly applied for soil moisture retrieval to avoid using less observations than the number of unknown parameters [21, 22].

The availability of SAR data characterized by short repeating cycles such as Radarsat-2, Sentinel-1, ALOS-2, Cosmo-SkyMed, and TerraSAR-X/TanDEM-X provides possible alternatives for monitoring soil moisture change at fine spatial scales through change detection methods [23]. The rationale of such approach is that temporal changes of surface roughness and vegetation take place at longer temporal scales than soil moisture changes [24–27]. Therefore, time series SAR data acquired with short repeat cycles are expected to obtain the soil moisture change. A change detection method referred to as the Alpha approximation model was initially developed under a simplified theoretical assumption [28], being that the ratio of two consecutive backscatter measurements could be approximately represented as the squared ratio of corresponding Alpha coefficients. The Alpha approximation model has been tested using different SAR datasets with different radar frequencies [29–32].

The Alpha approximation model is appealing for soil moisture retrieval due to its simplicity, and this method requires the initial estimates of soil moisture boundary. Such bounds can be obtained from climate models, calibration on a specific dataset [31–34], or juxtaposition method [35, 36]. Furthermore, the system of equations constructed using the Alpha model has more unknowns than equations; thus, there exist an infinite number of solutions. Therefore, these issues hampered the accurate estimation of soil moisture content using the Alpha model.

This paper aims at developing an optimized Alpha model by transforming the original underdetermined system of the Alpha approximation model into an overdetermined system. The contribution is an extension of the Alpha model to multisensor configurations. With the application of time series Radarsat-2 and Sentinel-1A SAR data, the number of the observation equations is more than the number of unknown parameters. Thus, the comprehensive cost function can be structured to estimate the optimized soil moisture content within a valid bound. The developed approach was quantitatively evaluated according to the field soil moisture measurements over Hebei agricultural areas.

This paper is organized as follows. Section 2 introduces the comprehensive soil moisture retrieval campaign, including the experimental area, acquisitions of multisensor time series SAR data, and continuous field measurements. Section 3 provides an overview of the Alpha approximation model and its extension to multisensor configurations. Section 4 firstly evaluates the Alpha approximation model based on the forward scattering model and time series SAR data. Then, the comparison between the Alpha approximation model and the developed method is implemented, and the results are further analyzed and discussed. Section 5 presents the conclusions and discusses potential future applications.

2. Study Area and Datasets

From Oct. 2015 to Mar. 2016, a comprehensive scientific campaign for soil moisture retrieval was implemented over Hebei agricultural areas. The campaign encompassed multisensor time series spaceborne SAR acquisitions and continuous field measurements for vegetation and soil surface parameters during 13 Oct. 2015 to 5 Mar. 2016.

2.1. Study Area. The agricultural area chosen for this study is part of the North China Plain located in the south of Hebei, China (114°5'–114°35'E, 36°25'–36°55'N). Field measurements were implemented over the JiuLong (114°10′–114°20′E, 36°25′–36°35′N) and WanNian (114°5′–114°15′E, 36°35′–36°45′N) experimental areas, as shown in Figure 1. The topography of the experimental area is relatively flat. The soil texture is dominantly characterized as loam soil, and the sand and clay proportions are approximate 50% and 15%, respectively. The main crops are wheat and corn over the study area.

Since Oct 2015, the dominated crop corn has been totally harvested over the study area. Partial of the agricultural areas were seeded with winter wheat, and partial of the agricultural areas were languished. From Oct. 2015 to Mar. 2016, the study area was mainly characterized by bare soil or sparse winter wheat. During this period, the temperature is relatively low, and thus, the winter wheat grows slowly and the biomass retains a relatively low level (less than $0.5\,\text{kg/m}^2$). Figure 2 shows the different winter wheat growth stages from 13 Oct. 2015 to 5 Mar. 2016. Therefore, the study area can be considered as bare soil surfaces, and the influence of winter wheat to backscattering coefficients can be ignored [1]. In addition, there is no agricultural activity over the study area during this period, the soil surface roughness can be considered as constant, and thus, it is suitable for the application of change detection methods.

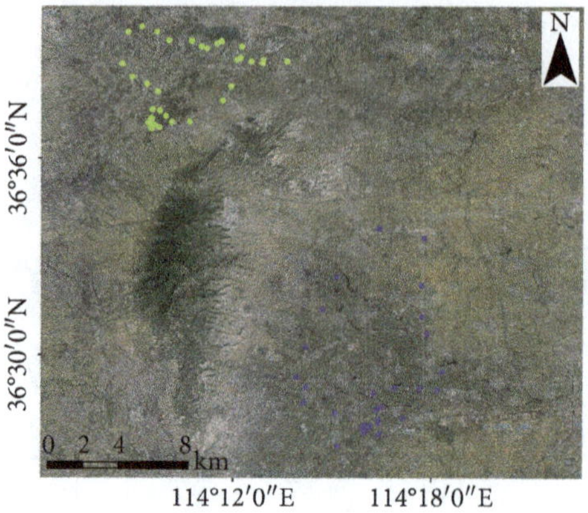

FIGURE 1: Study area and sampling sites. The upper left and lower right represent the WanNian and JiuLong experimental areas, respectively. The round dots indicate the sampling sites.

2.2. SAR Data. Time series Radarsat-2, TerraSAR-X, and Sentinel-1A images were acquired, and continuous field measurements were implemented over the study area. The time interval among TerraSAR-X acquisitions and other SAR data acquisitions is long; thus, only the time series Radarsat-2 and Sentinel-1A data were selected as experimental dataset. Table 1 lists the configuration parameters and acquisition time of multisensor SAR data.

During the scientific campaign, a total of five Radarsat-2 and three Sentinel-1A images were acquired from Oct. 2015 to Mar. 2016. To transfer the intensity to backscattering coefficients, a standard preprocessing phase is performed. The entire process, including the speckle filter, radiometric correction, and range-Doppler terrain correction, is conducted using the NEST and SNAP software provided by ESA. First, the images are filtered using the Lee filter with a 5×5 window size [37]. Radiometric calibrations are then conducted to derive the backscattering coefficients. Finally, the images are georeferenced using SRTM as an external DEM. Considering the different spatial resolutions of Radarsat-2 and Sentinel-1A images, the resampling process was conducted to obtain the same spatial resolution. Table 2 lists the time of SAR data acquisitions and field measurements.

During two days interval of Radarsat-2 and Sentinel-1A acquisitions, there was no precipitation, large temperature

(a) (b) (c) (d) (e)

FIGURE 2: Wheat growth stage in Hebei study area from Oct. 2015 to Mar. 2016. (a) 13 Oct. 2015, (b) 6 Nov. 2015, (c) 24 Dec. 2015, (d) 17 Jan. 2016, and (e) 5 Mar. 2016.

TABLE 1: Configuration parameters and acquisition time of multisensor SAR data.

SAR data	Acquisition date	Band frequency	Polarization	Incidence angle	Imaging mode	Resolution	Revisit period
Radarsat-2	Oct. 2015 to Mar. 2016	C 5.4 GHz	HH	36°	Multilook fine	5 m	24 days
Sentinel-1A	Oct. 2015 to Mar. 2016	C 5.4 GHz	VV/VH	38°	Interferometric wide swath	20 m	12 days

TABLE 2: Time of SAR data acquisitions and field measurements.

Acquisition time	Radarsat-2	Sentinel-1A	Field measurements
T_1	13 Oct. 2015	15 Oct. 2015	13 Oct. 2015
T_2	6 Nov. 2015	—	6 Nov. 2015
T_3	24 Dec. 2015	26 Dec. 2015	24 Dec. 2015
T_4	17 Jan. 2016	—	17 Jan. 2016
T_5	5 Mar. 2016	7 Mar. 2016	5 Mar. 2016

variation, and agricultural activity over the experimental areas. Therefore, the soil moisture and surface roughness are considered to be constant between each of the multisensor SAR data acquisitions, which provide the potential of multisensor SAR for soil moisture retrieval.

2.3. Field Measurements. Simultaneously with Radarsat-2 acquisitions, continuous field measurements were implemented over the study area. Different sampling sites were selected over JiuLong and WanNian experimental areas, respectively. Soil surface parameters were measured, including volumetric soil moisture content (0–5 cm) and surface roughness parameters. For each sampling site, three sample points were selected as representatives within an area of 30 m × 30 m. The distance between the sample points is approximately ten meters. Soil moisture content was collected by a calibrated TDR (time domain reflectometry) probe, and the measured soil moisture was calibrated on the basis of gravimetric method.

Soil moisture content of each sampling site was obtained by the average of three sample measurements. Over the JiuLong experimental area, the soil moisture content varied from 0.03 to 0.46 cm^3/cm^3, and the soil moisture ranged from 0.06 to 0.51 cm^3/cm^3 over the WanNian experimental area. A portable global positioning system (GPS) was used to identify and register the sampling positions with one meter accuracy. The surface roughness was measured using a needle profilometer with a length of one meter (2 cm sampling interval). At each sample point, four surface profiles (two parallel and two perpendicular to the row direction) were recorded. These profiles were photographed and then digitized. The root mean square height (h) and correlation length (l) were calculated using a MATLAB program [38]. The value of h varied from 0.51 to 1.79 cm over the JiuLong experimental area, while in WanNian experimental area, h changed from 0.50 to 1.90 cm. The value of l varied from 5.2 to 21.6 cm over the JiuLong experimental area, while in WanNian experimental area, l changed from 6.1 to 23.1 cm. Furthermore, considering the presence of positioning errors, geographical registration between SAR data and field measurement sites was implemented based on ten corner reflectors located in the experimental areas. The field measurements collected from the experimental areas

Table 3: The statistics results of the field measurements over the study area.

Field measurement	Experimental area	Sampling number	Soil moisture (cm^3/cm^3)			h (cm)			l (cm)		
			Min	Max	Average	Min	Max	Average	Min	Max	Average
13 Oct. 2015	JiuLong	27	0.03	0.345	0.116	0.51	1.79	1.13	7.3	20.2	15.1
	WanNian	26	0.06	0.307	0.15	0.50	1.88	1.12	7.6	21.8	16.3
6 Nov. 2015	JiuLong	21	0.249	0.427	0.34	0.61	1.67	1.07	5.6	17.7	13.6
	WanNian	23	0.224	0.508	0.297	0.57	1.73	1.11	6.7	19.7	15.4
24 Dec. 2015	JiuLong	30	0.15	0.464	0.243	0.54	1.77	1.09	7.7	18.5	14.3
	WanNian	27	0.13	0.421	0.194	0.60	1.90	1.17	8.3	21.9	16.7
17 Jan. 2016	JiuLong	30	0.118	0.322	0.193	0.66	1.64	1.03	5.2	18.9	15.3
	WanNian	30	0.11	0.452	0.202	0.59	1.78	1.13	7.9	23.1	17.6
5 Mar. 2016	JiuLong	30	0.067	0.46	0.176	0.63	1.73	1.17	6.4	21.6	16.1
	WanNian	30	0.06	0.29	0.147	0.60	1.71	1.11	6.1	22.3	16.8

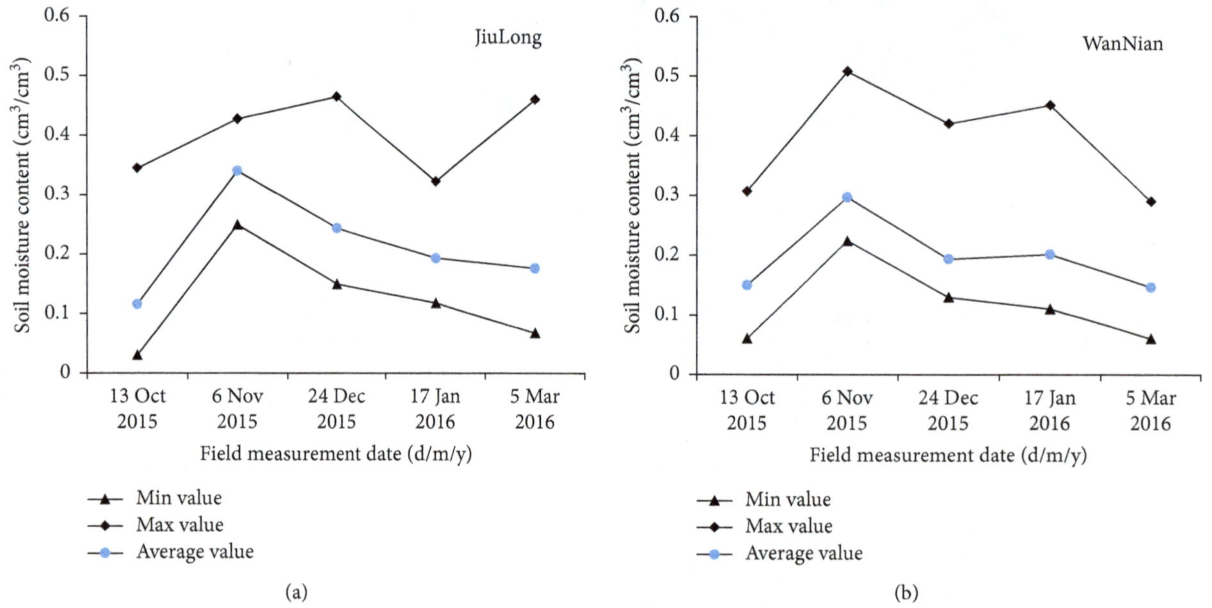

Figure 3: Statistics graphs of the time series soil moisture measurements. (a) JiuLong experimental area. (b) WanNian experimental area.

allow for the validation of feasibility and effectiveness of the developed soil moisture retrieval method. Table 3 lists the statistic results of the time series field measurements over JiuLong and WanNian experimental areas.

For time series soil moisture measurements, the minimum, maximum, and average values of soil moisture content were graphed as Figure 3.

There is obvious precipitation on 5 Nov. 2015 over the study area. Thus, the average soil moisture content measured on 6 Nov. 2015 is higher than the other field measurements. There is no agricultural activity during Oct. 2015 to Mar. 2016 over the study area, and the soil surface roughness shows relatively stable for the continuous field measurements. Therefore, the assumptions of the change detection methods are fulfilled. Figure 4 shows part of the field photos for in situ measurements over the study area.

3. Methodologies

3.1. Overview of the Alpha Model. Soil surface roughness has significant influence on the backscattering coefficient. For soil moisture retrieval, surface roughness can be considered as noise; thus, it is essential to eliminate the noise to obtain reliable soil moisture retrievals. Generally, soil surface roughness is more stable than soil moisture. The variation of soil surface roughness is originated from the seasonal agricultural activities over the agricultural areas. Therefore, soil surface roughness can be considered as constant during a certain period, and soil moisture change is the crucial factor for the variation of backscattering coefficients for bare soil surfaces. Based on the abovementioned theory and assumption, the application of multitemporal SAR data can effectively remove the influence of surface roughness, thus to obtain accurate soil moisture content. Therefore, the change detection methods for soil moisture retrieval have been widely developed.

Balenzano et al. proposed the Alpha approximation approach for soil moisture retrieval from multitemporal SAR data [28]. When time series SAR measurements are available for bare soil surface, assuming no variation of surface roughness during SAR acquisitions, the backscatter change is related to soil moisture change only. Furthermore,

(a)

(b)

FIGURE 4: Field photos of the in situ measurements over the study area. (a) Corner reflectors. (b) Surface roughness measurements.

the ratio of two consecutive backscatter measurements can be approximately represented as the squared ratio of corresponding Alpha coefficients [28].

$$\frac{\sigma_{pp}^{T_2}}{\sigma_{pp}^{T_1}} \approx \left|\frac{\alpha_{pp}^{T_2}(\varepsilon, \theta)}{\alpha_{pp}^{T_1}(\varepsilon, \theta)}\right|^2, \tag{1}$$

where σ is the backscattering coefficient represented as intensity, θ is the incidence angle, ε is the soil dielectric constant, pp denotes the polarization mode (i.e., HH or VV), and T_1 and T_2 represent the acquisition time of multitemporal SAR data. The Alpha coefficient α_{pp} is a function of dielectric constant ε and incidence angle θ, and it is given by [39]

$$\begin{aligned} \left|\alpha_{HH}(\varepsilon, \theta)\right| &= \left|\frac{(\varepsilon - 1)}{\left(\cos\theta + \sqrt{\varepsilon - \sin^2\theta}\right)^2}\right|, \\ \left|\alpha_{VV}(\varepsilon, \theta)\right| &= \left|\frac{(\varepsilon - 1)\left(\sin^2\theta - \varepsilon\left(1 + \sin^2\theta\right)\right)}{\left(\varepsilon\cos\theta + \sqrt{\varepsilon - \sin^2\theta}\right)^2}\right|. \end{aligned} \tag{2}$$

According to T_1 and T_2 SAR acquisitions, the observation equation can be established based on (1).

$$\left|\alpha_{pp}^{T_1}(\varepsilon, \theta)\right| - \sqrt{\frac{\sigma_{pp}^{T_1}}{\sigma_{pp}^{T_2}}} \cdot \left|\alpha_{pp}^{T_2}(\varepsilon, \theta)\right| = 0. \tag{3}$$

For the consecutive N SAR acquisitions, the number of the observation equations can reach to $N \times (N-1)/2$. Due to the ratio relationship between different temporal SAR data, there is redundancy among these observation equations. Therefore, based on the N SAR acquisitions, the number of the effective observation equations is $N-1$. In order to maintain the soil surface roughness to be constant between different temporal SAR acquisitions, the adjacent SAR acquisitions (T_{N-1} and T_N) were utilized to structure the observation equations. Therefore, the observation equations of the Alpha model can be expressed as follows:

$$\begin{bmatrix} 1 & -\sqrt{\frac{\sigma_{pp}^{T_1}}{\sigma_{pp}^{T_2}}} & 0 & 0 & \cdots & 0 & 0 \\ 0 & 1 & -\sqrt{\frac{\sigma_{pp}^{T_2}}{\sigma_{pp}^{T_3}}} & 0 & \cdots & 0 & 0 \\ \cdots & \cdots & \cdots & \cdots & \cdots & \cdots & \cdots \\ \cdots & \cdots & \cdots & \cdots & \cdots & 1 & -\sqrt{\frac{\sigma_{pp}^{T_{N-1}}}{\sigma_{pp}^{T_N}}} \end{bmatrix} \cdot \begin{bmatrix} \left|\alpha_{pp}^{T_1}(\varepsilon, \theta)\right| \\ \left|\alpha_{pp}^{T_2}(\varepsilon, \theta)\right| \\ \cdots \\ \left|\alpha_{pp}^{T_N}(\varepsilon, \theta)\right| \end{bmatrix} = \begin{bmatrix} 0 \\ 0 \\ \cdots \\ 0 \end{bmatrix}. \tag{4}$$

If the ratios between consecutive backscatter values are considered according to (1), N SAR acquisitions result in $(N-1)$ observation equations and N unknown dielectric constants (for single polarization case), leading to a system of equations having more unknowns than equations. To solve this underdetermined system of equations, the bounded least-squares optimization is applied [28] to estimate the dielectric constant values. In the case where multitemporal SAR data are used, the Alpha coefficients can be derived in a least-squares sense. Thus, the soil moisture content can be derived on the basis of the dielectric mixing model [40].

3.2. Developed Alpha Model. The system of equations constructed using (1) has more unknowns than the number of equations, and thus there exist infinite number of solutions. In this paper, the multisensor SAR data are employed to solve the underdetermined system of equations. The original underdetermined solution was transformed into the solution of overdetermined equation. A technique by minimizing the comprehensive cost function was employed to obtain the optimized soil dielectric constant. During this processing, soil dielectric constant bounds should be given according to the field measurements. The consecutive field measurement of soil moisture content is within $0.03\,\text{cm}^3/\text{cm}^3$ and $0.51\,\text{cm}^3/\text{cm}^3$. Thus, the solution of dielectric constant can be restricted in a valid bound [40]. The retrieval schemes of the developed Alpha model are detailed in the following sections.

Firstly, time series Radarsat-2 SAR data with HH polarization can be used to establish the independent observation equations.

$$\begin{bmatrix} 1 & -\sqrt{\frac{\sigma_{\text{hhR2}}^{T_1}}{\sigma_{\text{hhR2}}^{T_2}}} & 0 & 0 & 0 \\ 0 & 1 & -\sqrt{\frac{\sigma_{\text{hhR2}}^{T_2}}{\sigma_{\text{hhR2}}^{T_3}}} & 0 & 0 \\ 0 & 0 & 1 & -\sqrt{\frac{\sigma_{\text{hhR2}}^{T_3}}{\sigma_{\text{hhR2}}^{T_4}}} & 0 \\ 0 & 0 & 0 & 1 & -\sqrt{\frac{\sigma_{\text{hhR2}}^{T_4}}{\sigma_{\text{hhR2}}^{T_5}}} \end{bmatrix} \cdot \begin{bmatrix} \left|\alpha_{\text{hh}}^{T_1}(\varepsilon_1, \theta_{\text{R2}})\right| \\ \left|\alpha_{\text{hh}}^{T_2}(\varepsilon_2, \theta_{\text{R2}})\right| \\ \left|\alpha_{\text{hh}}^{T_3}(\varepsilon_3, \theta_{\text{R2}})\right| \\ \left|\alpha_{\text{hh}}^{T_4}(\varepsilon_4, \theta_{\text{R2}})\right| \\ \left|\alpha_{\text{hh}}^{T_5}(\varepsilon_5, \theta_{\text{R2}})\right| \end{bmatrix} = \begin{bmatrix} 0 \\ 0 \\ 0 \\ 0 \end{bmatrix}. \tag{5}$$

According to the observation equations established by the multitemporal Radarsat-2 data, the estimation of soil moisture content is an underdetermined solution in combination with the effective range of soil moisture content. Accounting for the ill-posed problem, the optimized Alpha model combining the multisensor SAR data was developed to overcome the uncertainty of the estimated results.

The multisensor time series SAR data are utilized to estimate the soil moisture content by means of the Alpha model. The time series Radarsat-2 and Sentinel-1A SAR data are respectively employed to construct the observation equations, which transform the underdetermined solution into overdetermined solution. Thus, the multiple solutions of soil moisture content can be constrained to the unique solution. Therefore, the uncertainty of the estimated soil moisture was significantly reduced.

The time series Sentinel-1A SAR data with VV polarization can be utilized to construct the following observation equations.

$$\begin{bmatrix} 1 & -\sqrt{\frac{\sigma_{\text{vvS1A}}^{T_1}}{\sigma_{\text{vvS1A}}^{T_3}}} & 0 \\ 0 & 1 & -\sqrt{\frac{\sigma_{\text{vvS1A}}^{T_3}}{\sigma_{\text{vvS1A}}^{T_5}}} \end{bmatrix} \begin{bmatrix} \left|\alpha_{\text{vv}}^{T_1}(\varepsilon_1, \theta_{\text{S1A}})\right| \\ \left|\alpha_{\text{vv}}^{T_3}(\varepsilon_3, \theta_{\text{S1A}})\right| \\ \left|\alpha_{\text{vv}}^{T_5}(\varepsilon_5, \theta_{\text{S1A}})\right| \end{bmatrix} = \begin{bmatrix} 0 \\ 0 \end{bmatrix}. \tag{6}$$

Therefore, the integrated observation equations can be structured in combination with multitemporal Radarsat-2 and Sentinel-1A data.

$$\begin{bmatrix} M_{\text{hhR2}} & M_{\text{o}} \\ M_{\text{o}} & M_{\text{vvS1A}} \end{bmatrix} \begin{bmatrix} \alpha_{\text{hhR2}} \\ \alpha_{\text{vvS1A}} \end{bmatrix} = \begin{bmatrix} 0 \\ 0 \end{bmatrix}, \tag{7}$$

where M_{hhR2} and M_{vvS1A} are the coefficient matrixes of (5) and (6), derived from time series Radarsat-2 and Sentinel-1A data respectively. α_{hhR2} and α_{vvS1A} are the Alpha coefficient corresponding to HH and VV polarization, respectively.

For the above observation equations, the unknown parameters $|\alpha_{\text{pp}}^{T_N}(\varepsilon, \theta)|$ just depend on the soil dielectric constant and incidence angle. For the approximately simultaneous acquisitions of multisensor SAR data, the soil dielectric constant is equivalent, and the incidence angle is a known parameter. Therefore, there are five unknown dielectric constants in observation (7). The original underdetermined system was transformed into an overdetermined system. Thus, the unique solution of soil moisture can be obtained with multisensor time series SAR data as inputs.

The observation (7) constructed by multitemporal Radarsat-2 and Sentinel-1A data can be expressed as $M_{\text{hhR2}}\alpha_{\text{hhR2}} = 0$ and $M_{\text{vvS1A}}\alpha_{\text{vvS1A}} = 0$. Thus, the target function of optimal solution can be obtained by combining the time series Radarsat-2 and Sentinel-1A data. According to the above analysis, the comprehensive cost function can be expressed as

$$\Delta = \min\sqrt{\left(M_{\text{hhR2}}\alpha_{\text{hhR2}}\right)^2 + \left(M_{\text{vvS1A}}\alpha_{\text{vvS1A}}\right)^2}. \tag{8}$$

This expression not only contains the constraint of time series Radarsat-2 data but also is constrained by time series Sentinel-1A data. Based on the given range of soil moisture content, the numerical solution of soil moisture can be obtained by minimizing the comprehensive cost function.

4. Results and Discussions

Firstly, the rationality of Alpha model was evaluated on the basis of IEM (integral equation model) and Oh model within a wide range of soil surface parameters. Then, the applicability of Alpha model was further assessed in combination with time series SAR data and field measurements. The multitemporal Radarsat-2 data and measured soil moisture of the same sampling sites were employed for the theoretical analysis.

After the evaluation and validation for the forward model, the Alpha approximation method was applied for soil moisture retrieval over the experimental areas. Firstly, the observation equations based on the Alpha model were constructed using multitemporal Radarsat-2 data. Then, soil moisture was estimated in combination with the valid range of soil moisture content. Against the shortcoming of the underdetermined system, the application of multisensor SAR data was developed to transform the underdetermined system into an overdetermined system by providing independent observation equations.

4.1. Alpha Model Evaluation

4.1.1. Alpha Model Evaluation Based on the Oh and IEM Simulation Data. Oh [16] and IEM [41] simulation data were utilized to evaluate the rationality of Alpha model. The input parameters of radar backscattering model include the soil dielectric constant (related to soil moisture content), root mean square height, correlation length, incidence angle, and radar wavelength, which can be determined according to the field measurements and SAR configuration parameters. The configuration parameters (incidence angle, radar wavelength, and polarization) of Radarsat-2 and Sentinel-1A images were used as inputs for Oh and IEM simulation. Soil surface root mean square height changes from 0.1 cm to 2.0 cm with an interval of 0.1 cm, and correlation length varies from 1 cm to 20 cm with an interval of 1 cm. Backscattering coefficients were simulated based on the Oh and IEM model with different soil moisture contents as inputs. Thus, the time series backscattering coefficients can be obtained corresponding to the same soil roughness parameters and different soil moisture contents. The relationship between the ratio of backscattering coefficients $\sigma_{pp}^{T_2}/\sigma_{pp}^{T_1}$ and squared ratio of Alpha coefficients $|\alpha_{pp}^{T_2}/\alpha_{pp}^{T_1}|^2$ were quantitatively assessed. Figure 5 shows the relationship between $|\alpha_{pp}^{T_2}/\alpha_{pp}^{T_1}|^2$ and $\sigma_{pp}^{T_2}/\sigma_{pp}^{T_1}$ simulated by the IEM and Oh model, respectively.

The simulation results show good agreement between the squared ratio of Alpha coefficients and the ratio of backscattering coefficients simulated by IEM for HH and VV polarization. And, good correlation between $|\alpha_{pp}^{T_2}/\alpha_{pp}^{T_1}|^2$ and $\sigma_{pp}^{T_2}/\sigma_{pp}^{T_1}$ was obtained for the Oh model. The aforementioned results theoretically demonstrated the rationality of Alpha model. The influence of surface roughness can be effectively eliminated by the ratio model. Therefore, the relationship between the ratio of backscattering coefficients and the squared ratio of Alpha coefficients can be structured to estimate the soil moisture content using time series SAR data.

4.1.2. Alpha Model Evaluation Based on Multitemporal Radarsat-2 Data and Field Measurements. Time series Radarsat-2 data and measured soil moisture were employed to further evaluate the availability of Alpha model. The following acquisitions of SAR data and measured soil moisture were utilized over JiuLong and WanNian experimental areas, including 24 Dec. 2015, 17 Jan. 2016, and 5 Mar. 2016. Figure 6 shows the relationship between the ratio of time series backscattering coefficients and the squared ratio of Alpha coefficients derived from measured soil moisture.

The availability of Alpha model was evaluated using time series Radarsat-2 data and field soil moisture measurements. The results show good correlation between the $|\alpha_{pp}^{T_2}/\alpha_{pp}^{T_1}|^2$ and $\sigma_{pp}^{T_2}/\sigma_{pp}^{T_1}$, which further demonstrate the effectiveness of Alpha model for time series soil moisture retrieval.

Based on the aforementioned evaluation of rationality and availability for the Alpha model, time series Radarsat-2 data and multisensor SAR data were employed to estimate the soil moisture content over JiuLong and WanNian experimental areas, respectively.

4.2. Soil Moisture Retrieval Using Alpha Model. Time series Radarsat-2 data were applied for soil moisture retrieval over JiuLong and WanNian experimental areas. Firstly, the observation equations were constructed to eliminate the influence of surface roughness. Then, in combination with the valid range of soil moisture content, the soil moisture can be derived by the minimization of cost function. The estimated soil moisture was evaluated based on the field measurements. Figure 7 shows the comparison between the measured and estimated soil moisture using the Alpha model with multitemporal Radarsat-2 data as inputs over JiuLong and WanNian experimental areas.

The quantitative evaluation for the Alpha model was implemented on the basis of time series field measurements. The root mean square error (RMSE) and correlation coefficient (R) were selected as statistical indicators. Table 4 presents the statistical results between the estimated and measured soil moisture over JiuLong and WanNian experimental areas.

The statistical results indicated that relatively accurate soil moisture was obtained over JiuLong and WanNian experimental areas, with RMSE ranging from 0.052 cm^3/cm^3 to 0.082 cm^3/cm^3 and R changing from 0.51 to 0.92. The results demonstrated the practicability of Alpha model for agricultural area soil moisture retrieval with time series SAR data as inputs. Since October 2015, the dominated crop corn has been harvested over the study area, and the average soil moisture content is relatively low due to little precipitation during this period. There is obvious precipitation over the study area on 5 Nov. 2015; thus, the measured average soil moisture is high on 6 Nov. 2015. During the whole winter, there is relatively less precipitation over the study area. The average soil moisture content is medium over the study area on 24 Dec. 2015, 17 Jan. 2016, and 5 March 2016 due to low temperature and less evaporation. In addition, partial of the farmland was irrigated before the field measurements on 24 Dec. 2015, and 5 Mar. 2016; thus, the soil moisture content of the irrigated sampling sites is high on 24 Dec. 2015, and 5 Mar. 2016. In conclusion, the estimated soil moisture preliminarily captured the change trend of the measured time series soil moisture. However, the estimated soil moisture

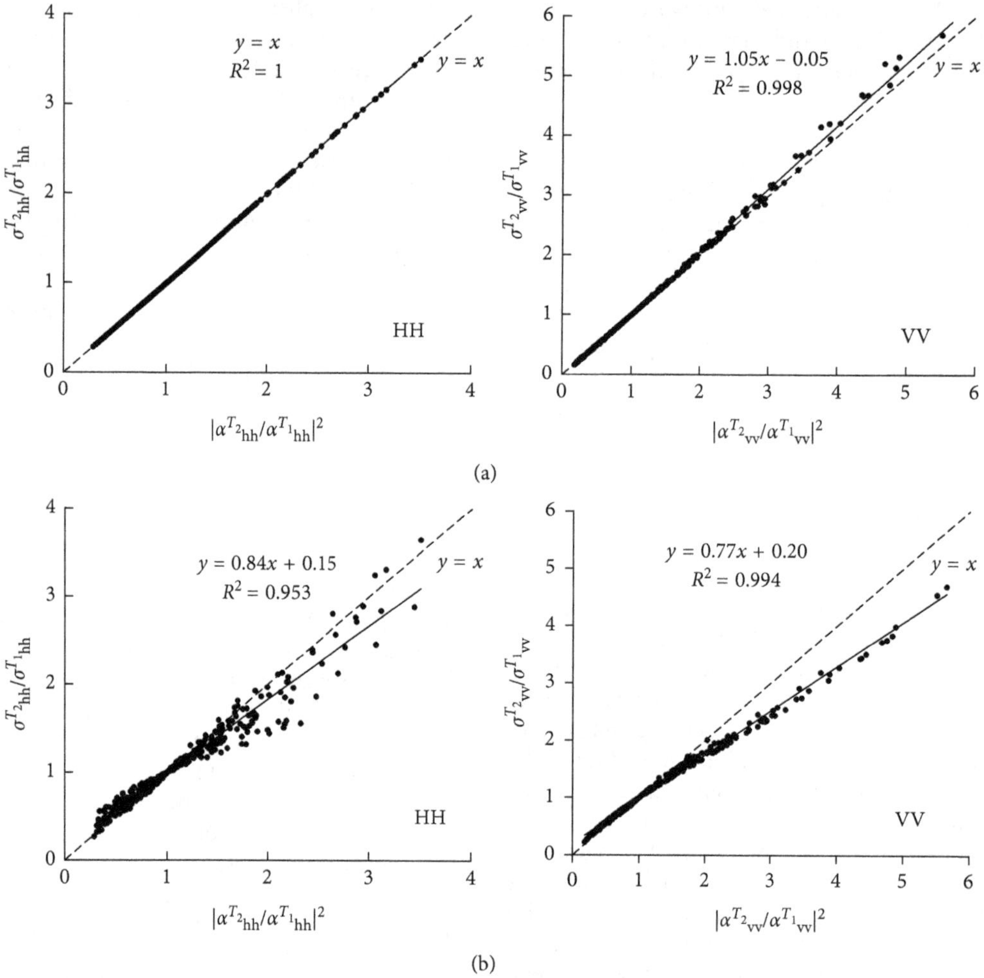

FIGURE 5: Relationships between $|\alpha_{pp}^{T_2}/\alpha_{pp}^{T_1}|^2$ and $\sigma_{pp}^{T_2}/\sigma_{pp}^{T_1}$ simulated by IEM and Oh models. (a) Comparison between $\sigma_{pp}^{T_2}/\sigma_{pp}^{T_1}$ simulated by IEM and the squared ratio of Alpha coefficient $|\alpha_{pp}^{T_2}/\alpha_{pp}^{T_1}|^2$. (b) Comparison between $\sigma_{pp}^{T_2}/\sigma_{pp}^{T_1}$ simulated by the Oh model and the squared ratio of Alpha coefficient $|\alpha_{pp}^{T_2}/\alpha_{pp}^{T_1}|^2$.

showed relatively large error over JiuLong and WanNian experimental areas on 6 Nov. 2015. The reason may be that the sensitivity of backscattering coefficients to soil moisture decreased when the soil moisture content is high.

4.3. Soil Moisture Retrieval Using the Developed Alpha Model. The influence of surface roughness can be effectively removed by means of the Alpha model. Observation equations can be constructed using time series SAR data; thus, soil moisture can be estimated in combination with the boundary constraint of soil moisture. Against the underdetermined system, the Alpha model fusing multi-sensor SAR data was developed. The independent observation equations can be constructed based on time series Radarsat-2 and Sentinel-1A data, respectively. The number of unknown parameters does not alter, while the number of effective observation equations is increased. Therefore, the underdetermined system can be transformed into an overdetermined system for soil moisture retrieval using the developed Alpha model.

During the two days interval of Radarsat-2 and Sentinel-1A acquisitions, there was no precipitation, significant temperature change, and agricultural activity, thus soil moisture content and surface roughness can be considered as constant. Therefore, time series Radarsat-2 and Sentinel-1A data can be combined to estimate the soil moisture content by means of Alpha model. The validity of the developed Alpha model was quantitatively assessed based on continuous field measurements.

Time series Radarsat-2 data with HH polarization and Sentinel-1A data with VV polarization were employed to construct the observation (6) and (7), respectively. In combination with the valid soil moisture range, soil moisture content can be derived based on the comprehensive cost function (8). Figure 8 shows the comparison between the measured and estimated soil moisture using the developed Alpha model with multitemporal Radarsat-2 and Sentinel-1A data as inputs.

The developed Alpha model was quantitatively evaluated based on the measured soil moisture. Table 5 shows the quantitative statistics of the estimated soil moisture using the developed Alpha model.

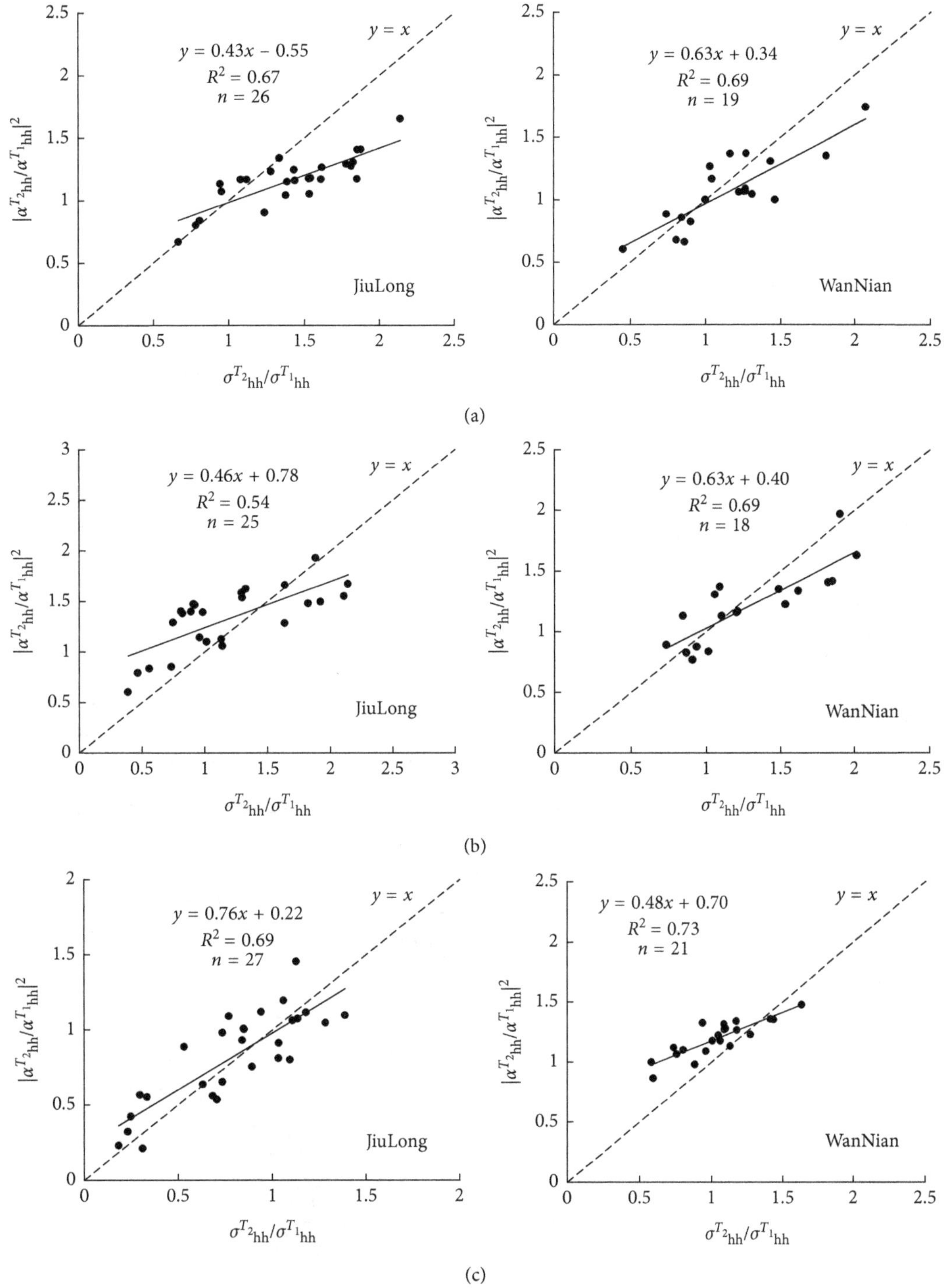

FIGURE 6: Evaluation for Alpha model using time series Radarsat-2 data and measured soil moisture. (a) Dataset with 24 Dec. 2015 and 17 Jan. 2016. (b) Dataset with 24 Dec. 2015 and 5 Mar. 2016. (c) Dataset with 17 Jan. 2016 and 5 Mar. 2016.

The results indicated that accurate time series soil moisture was obtained over JiuLong and WanNian experimental areas with RMSE ranging from 0.048 cm^3/cm^3 to 0.068 cm^3/cm^3 and R changing from 0.54 to 0.91. The estimated soil moisture shows good agreement with the field measurements. The quantitative results demonstrated the practicability of the developed Alpha model for soil moisture retrieval over bare agricultural areas.

In addition, the comparison between the Alpha approximation model and the developed Alpha model was implemented based on the measured soil moisture. Figure 9 shows the comparison of soil moisture retrieval between the Alpha model and the developed method.

The quantitative comparison between the Alpha model and the developed method indicated that time series soil moisture retrieval performance can be improved when multisensor SAR data was combined to estimate soil moisture. In addition, the comparison of the average accuracy between the Alpha approximation model and the

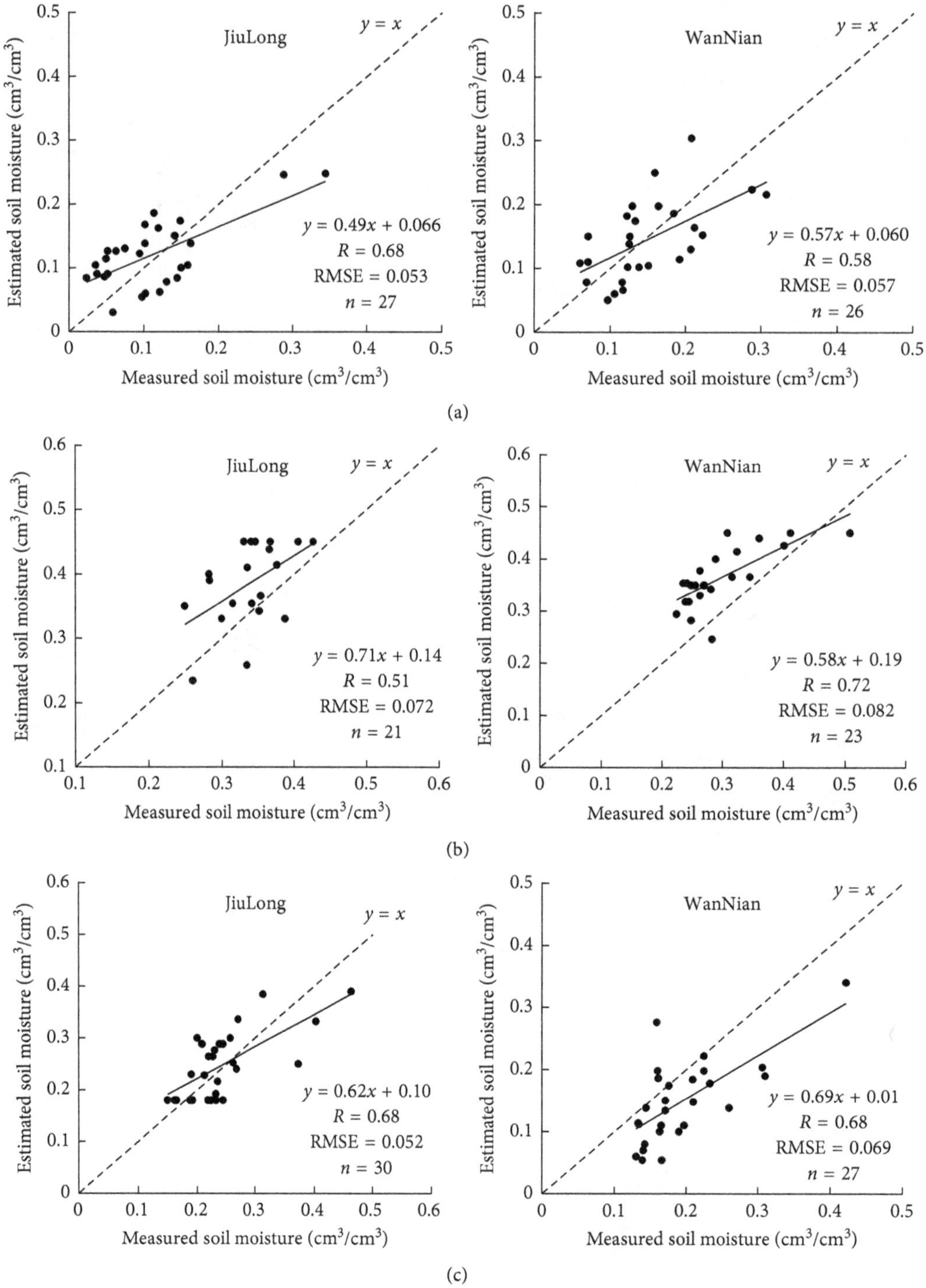

FIGURE 7: Continued.

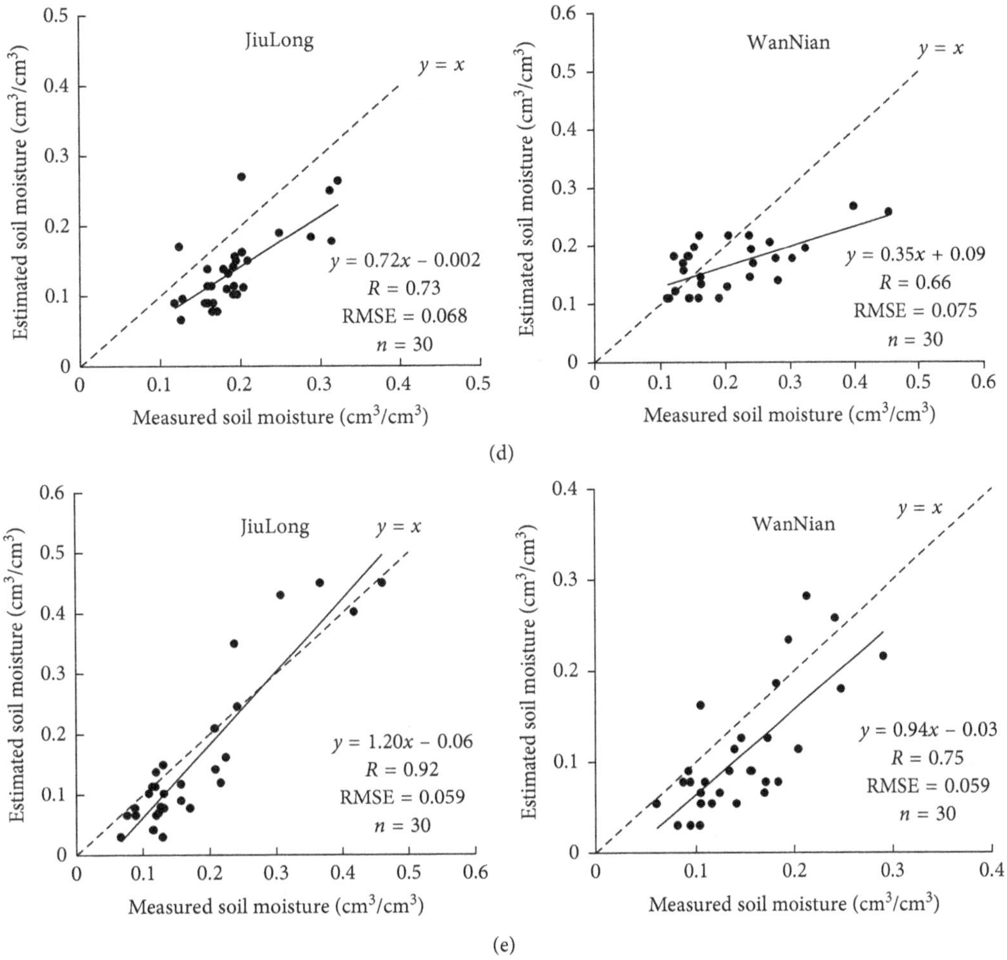

FIGURE 7: Comparison between the measured and estimated soil moisture using the Alpha model. (a) Comparison between the estimated and measured soil moisture on 13 Oct. 2015. (b) Comparison between the estimated and measured soil moisture on 6 Nov. 2015. (c) Comparison between the estimated and measured soil moisture on 24 Dec. 2015. (d) Comparison between the estimated and measured soil moisture on 17 Jan. 2016. (e) Comparison between the estimated and measured soil moisture on 5 Mar. 2016.

developed method was implemented. The average RMSE of the developed method with 0.051 cm^3/cm^3 is less than the original Alpha approximation model with 0.065 cm^3/cm^3. The effectiveness of the Alpha model fusing multisensor SAR data for soil moisture retrieval was further verified.

4.4. Discussion. In order to evaluate the spatial characteristics of the estimated soil moisture using the developed Alpha model, the soil moisture content at regional scale was estimated over the study area. Soil moisture maps from 13 Oct. 2015 to 5 Mar. 2016 were obtained, respectively, shown as Figures 10(a)–10(e). The qualitative results can also capture the soil moisture changes from 13 Oct. 2015 to 5 Mar. 2016. Therefore, the developed Alpha model can be used for soil moisture retrieval over agricultural areas.

For the developed Alpha model, time series multisensor SAR data are required for soil moisture retrieval. However, with the development of SAR technique, the concurrent acquisition of SAR data with different frequency, polarization, and incidence angle can be achieved and hence provide the basis of soil moisture retrieval by means of multisensor SAR

TABLE 4: Quantitative evaluation of the estimated soil moisture using Alpha model.

Date	Statistical indicators	JiuLong	WanNian
	RMSE (cm^3/cm^3)	0.053	0.057
13 Oct. 2015	R	0.68	0.58
	Sampling number	27	26
	RMSE (cm^3/cm^3)	0.072	0.082
6 Nov. 2015	R	0.51	0.72
	Sampling number	21	23
	RMSE (cm^3/cm^3)	0.052	0.069
24 Dec. 2015	R	0.68	0.68
	Sampling number	30	27
	RMSE (cm^3/cm^3)	0.068	0.075
17 Jan. 2016	R	0.73	0.66
	Sampling number	30	30
	RMSE (cm^3/cm^3)	0.059	0.059
5 Mar. 2016	R	0.92	0.75
	Sampling number	30	30

data with different system configuration parameters. The rationale is that the SAR signal is sensitive to soil surface properties in a different way and to a different extent

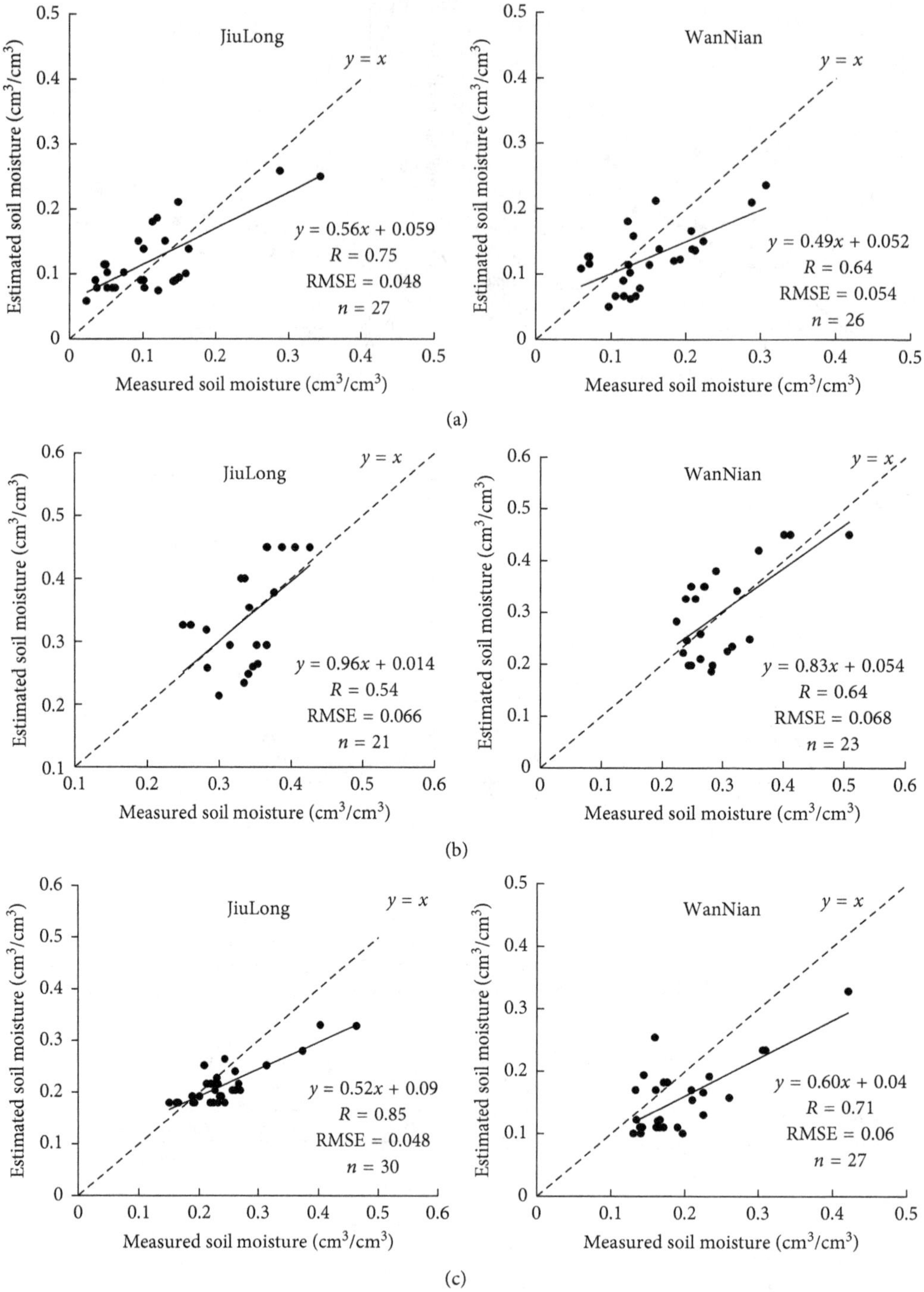

Figure 8: Continued.

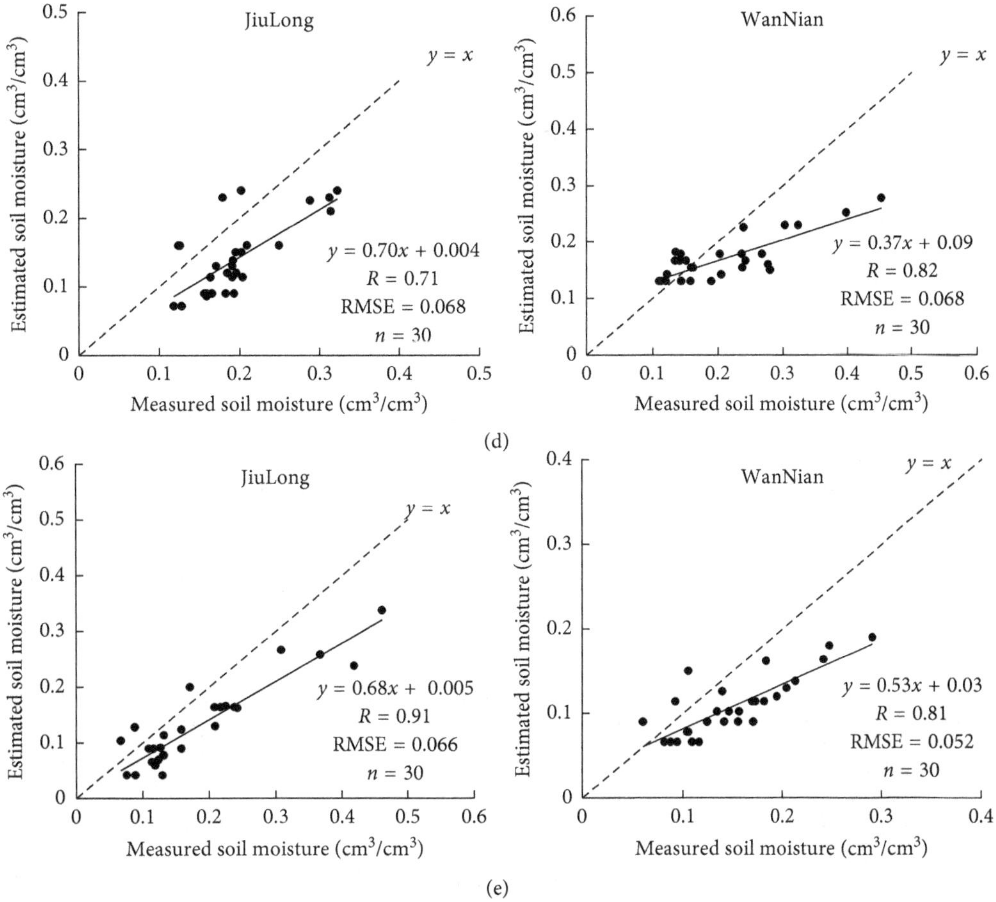

FIGURE 8: Comparison between the measured and estimated soil moisture using the developed Alpha model. (a) Comparison between the estimated and measured soil moisture on 13 Oct. 2015. (b) Comparison between the estimated and measured soil moisture on 6 Nov. 2015. (c) Comparison between the estimated and measured soil moisture on 24 Dec. 2015. (d) Comparison between the estimated and measured soil moisture on 17 Jan. 2016. (e) Comparison between the estimated and measured soil moisture on 5 Mar. 2016.

TABLE 5: Quantitative evaluation of estimated soil moisture using the developed Alpha model.

Date	Statistical indicators	JiuLong	WanNian
	RMSE (cm^3/cm^3)	0.048	0.054
13 Oct. 2015	R	0.75	0.64
	Sampling number	27	26
	RMSE (cm^3/cm^3)	0.066	0.068
6 Nov. 2015	R	0.54	0.64
	Sampling number	21	23
	RMSE (cm^3/cm^3)	0.048	0.06
24 Dec. 2015	R	0.85	0.71
	Sampling number	30	27
	RMSE (cm^3/cm^3)	0.068	0.068
17 Jan. 2016	R	0.71	0.82
	Sampling number	30	30
	RMSE (cm^3/cm^3)	0.066	0.052
5 Mar. 2016	R	0.91	0.81
	Sampling number	30	30

depending on the SAR frequency, polarization, and incidence angle. Accordingly, the confounding effect of soil moisture and surface roughness may be decoupled by means of the multiangle, multipolarization, and multifrequency SAR data. The combination of SAR information acquired with different system configurations leads to a better characterization of the parameters affecting the SAR signal and thus to an accurate estimation of soil moisture content [42]. Therefore, the developed Alpha model is promising for time series soil moisture retrieval over agricultural areas. In addition, the HH and VV polarization backscatter information for single SAR sensor may have certain correlation; thus, the combination of HH and VV backscatter may not improve soil moisture estimation obviously [32]. Consequently, the combination of time series Radarsat-2 and Sentinel-1A data with different incidence angle and polarization improved the soil moisture estimation.

It should be noticed that the uncertainties of soil moisture retrieval originated from different aspects. Firstly, the uncertainties aroused from SAR data were unpredicted, especially for different SAR satellites. Speckle noise of the SAR image, radiometric calibration and registration errors from SAR data, and inevitable field measurements errors would induce the deviation of soil moisture retrieval. Secondly, the uncertainties resulted from the inversion model was inevitable. The Alpha approximation model may not be robust enough for specific experimental areas. Thirdly,

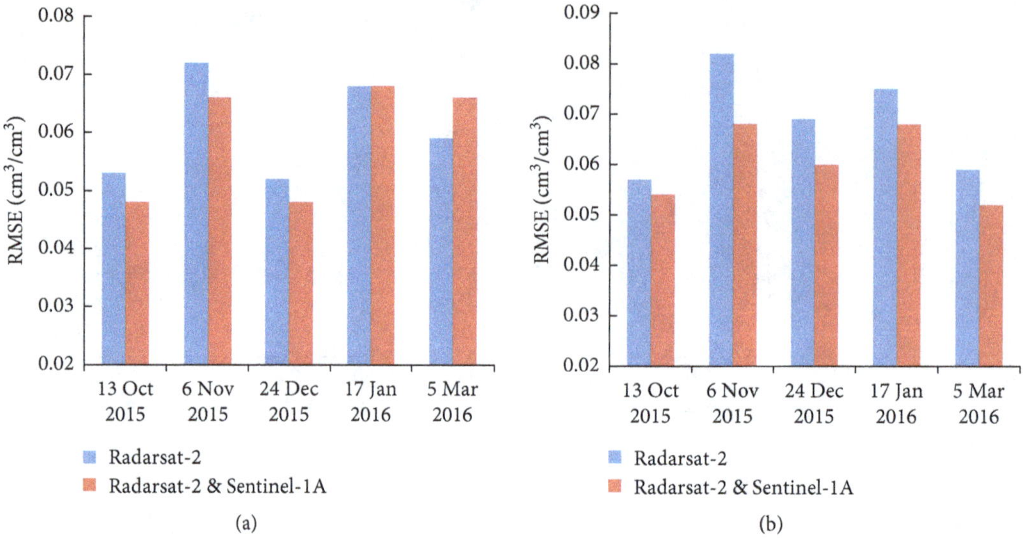

FIGURE 9: Comparison of soil moisture retrieval between the Alpha model and the developed method. (a) JiuLong experimental dataset. (b) WanNian experimental dataset.

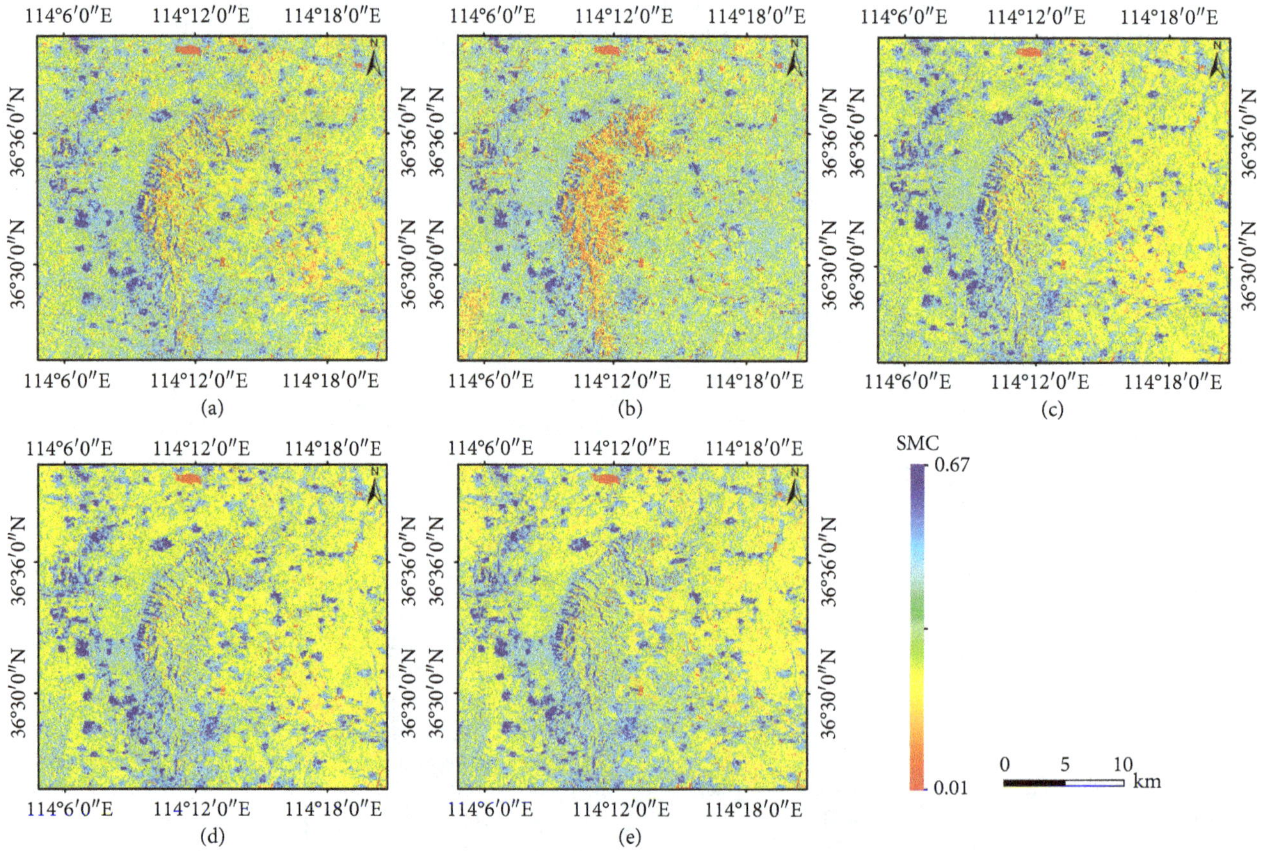

FIGURE 10: Soil moisture maps of the study area using the developed Alpha model. (a)–(e) represent the soil moisture maps obtained on 13 Oct. 2015, 6 Nov. 2015, 24 Dec. 2015, 17 Jan. 2016, and 5 Mar. 2016, respectively.

the assumptions were that soil surface roughness is constant during the period of SAR data acquisitions, and soil moisture does not change during two days interval of Radarsat-2 and Sentinel-1A acquisitions. All of the above aspects may induce the uncertainties for soil moisture retrieval. An accurate uncertainty assessment for soil moisture estimation contributed to evaluate the applicability and validity of the developed method.

5. Conclusions

Based on time series SAR data, the Alpha approximation model was quantitatively assessed for soil moisture retrieval. This approach has certain theoretical foundation structuring the relationship between the ratio of backscattering coefficients and squared ratio of Alpha coefficients and does not require the prior knowledge of surface roughness. Therefore, this research focused on the Alpha model for soil moisture retrieval. Considering the deficiency of this method, the Alpha model fusing multisensor SAR data was developed to solve the underdetermined system. The evaluation for the Alpha model and the developed method was implemented based on the multitemporal Radarsat-2 and Sentinel-1A data, as well as continuous field measurements. The following conclusions can be derived:

(1) The Alpha model can capture the change trend of time series soil moisture content. However, the underdetermined system may deteriorate the accuracy and reliability of estimated soil moisture.

(2) Against the deficiency of Alpha model, the Alpha model fusing time series Radarsat-2 and Sentinel-1A data was developed to transform the underdetermined system into overdetermined system. The effectiveness of the developed Alpha model was demonstrated based on time series Radarsat-2, Sentinel-1A data, and field measurements over bare agricultural areas.

It should be remarked that the experimental area is relatively small, and the available dataset is not very sufficient for drawing global considerations. However, the developed approach is promising for obtaining an improved accuracy for soil moisture retrieval with respect to single-sensor SAR data. The main limitation for generalizing this approach is the simultaneous acquisitions of multisensor SAR data. However, with the development of SAR techniques and increasing research on SAR satellites, novel SAR systems may provide abundant data sources for soil moisture retrieval. Therefore, the research on multisensor and multitemporal SAR data for soil moisture retrieval is promising.

Data Availability

The data used in our research include time series Radarsat-2, Sentinel-1A data, and field-measured soil surface parameters. The Sentinel-1A data are available on the ESA website (https://scihub.copernicus.eu/dhus/#/home). The Radarsat-2 data were acquired from the commercial channels on http://www.ev-image.com/. The soil surface parameters were measured during the SAR data acquisitions. Extensive expenses and efforts were expended on the data acquisitions. Furthermore, the data used in our research are not only used for academic research but also for project application (Special Fund for Public Projects of National Administration of Surveying, Mapping, and Geoinformation of China). Therefore, at present, sections of the data are not available. Statistical aspects of the data used to support the findings of this study are included within the article.

Conflicts of Interest

The authors declare that they have no conflicts of interest.

Acknowledgments

The authors wish to thank the ESA for providing the NEST, SNAP software, and Sentinel-1A data. The authors would like to thank Jilei Huang, Junkai Yang, Hua Chen, Lei Wang, Da Li, Meinan Zheng, and Sen Du for their help with extensive field measurements. This research was supported by the National Key R&D Programme of China (no. 2017YFB0502700), Civilian Space Programme of China (no. D010102), National Basic Surveying and Mapping Science and Technology Plan (no. 2018KJ0204/2018KJ0304), and the Special Fund for Public Projects of National Administration of Surveying, Mapping, and Geoinformation of China (no. 201412016).

References

[1] F. T. Ulaby, P. C. Dubois, and J. van Zyl, "Radar mapping of surface soil moisture," *Journal of Hydrology*, vol. 184, no. 1-2, pp. 57–84, 1996.

[2] P. C. Dubois, J. van Zyl, and T. Engman, "Measuring soil moisture with imaging radars," *IEEE Transactions on Geoscience and Remote Sensing*, vol. 33, no. 4, pp. 915–926, 1995.

[3] D. Entekhabi, E. G. Njoku, P. E. O'Neill et al., "The Soil Moisture Active Passive (SMAP) mission," in *Proceedings of the IEEE International Geoscience and Remote Sensing Symposium*, pp. 704–716, Honolulu, HA, USA, July 2010.

[4] M. C. Dobson and F. T. Ulaby, "Active microwave soil moisture research," *IEEE Transactions on Geoscience and Remote Sensing*, vol. 24, no. 1, pp. 23–36, 1986.

[5] S. Paloscia, P. Pampaloni, S. Pettinato, and E. Santi, "A comparison of algorithms for retrieving soil moisture from ENVISAT/ASAR images," *IEEE Transactions on Geoscience and Remote Sensing*, vol. 46, no. 10, pp. 3274–3284, 2008.

[6] S. Paloscia, S. Pettinato, and E. Santi, "Soil moisture mapping using Sentinel-1 images: algorithm and preliminary validation," *Remote Sensing of Environment*, vol. 134, pp. 234–248, 2013.

[7] J. Y. Zeng, Z. Li, Q. Chen, H. Y. Bi, J. X. Qiu, and P. F. Zou, "Evaluation of remotely sensed and reanalysis soil moisture products over the Tibetan Plateau using in-situ observations," *Remote Sensing of Environment*, vol. 163, pp. 91–110, 2015.

[8] J. Y. Zeng, Z. Li, Q. Chen, and H. Y. Bi, "Method for soil moisture and surface temperature estimation in the Tibetan Plateau using spaceborne radiometer observations," *IEEE Geoscience and Remote Sensing Letters*, vol. 12, no. 1, pp. 97–101, 2015.

[9] Q. Y. Meng, Q. X. Xie, C. M. Wang, J. X. Ma, Y. X. Sun, and L. L. Zhang, "A fusion approach of the improved Dubois model and best canopy water retrieval models to retrieve soil moisture through all maize growth stages from Radarsat-2 and Landsat-8 data," *Environmental Earth Sciences*, vol. 75, no. 20, p. 1377, 2016.

[10] S. G. Lei, H. Q. Chen, Z. F. Bian, and Z. G. Liu, "Evaluation of integrating topographic wetness index with backscattering coefficient of TerraSAR-X image for soil moisture estimation in a mountainous region," *Ecological Indicators*, vol. 61, pp. 624–633, 2016.

[11] S. B. Kim, L. Tsang, and J. T. Johnson, "Soil moisture retrieval using time-series radar observations over bare surfaces," *IEEE Transactions on Geoscience and Remote Sensing*, vol. 50, no. 5, pp. 1853–1863, 2012.

[12] N. Pierdicca, L. Pulvirenti, and G. Pace, "A prototype software package to retrieve soil moisture from Sentinel-1 data by using a bayesian multitemporal algorithm," *IEEE Journal of Selected Topics in Applied Earth Observations and Remote Sensing*, vol. 7, no. 1, pp. 153–166, 2014.

[13] H. S. Srivastava, P. Patel, M. L. Manchanda, and S. Adiga, "Use of multi-incidence angle RADARSAT-1 SAR data to incorporate the effect of surface roughness in soil moisture estimation," *IEEE Transactions on Geoscience and Remote Sensing*, vol. 41, no. 7, pp. 1638–1640, 2003.

[14] H. S. Srivastava, P. Patel, Y. Sharma, and R. R. Navalgund, "Large-area soil moisture estimation using multi-incidence angle RADARSAT-1 SAR data," *IEEE Transactions on Geoscience and Remote Sensing*, vol. 47, no. 8, pp. 2528–2535, 2009.

[15] M. Zribi, N. Baghdadi, N. Holah, and O. Fafin, "New methodology for soil surface moisture estimation and its application to ENVISAT-ASAR multi-incidence data inversion," *Remote Sensing of Environment*, vol. 96, no. 3-4, pp. 485–496, 2005.

[16] Y. Oh, K. Sarabandi, and F. T. Ulaby, "An empirical model and inversion technique for radar scattering from bare soil surfaces," *IEEE Transactions on Geoscience and Remote Sensing*, vol. 30, no. 2, pp. 370–381, 1992.

[17] J. C. Shi, J. Wang, A. Y. Hsu, P. E. O'Neill, and E. T. Engman, "Estimation of bare surface soil moisture and surface roughness parameter using L-band SAR image data," *IEEE Transactions on Geoscience and Remote Sensing*, vol. 35, no. 5, pp. 1254–1266, 1997.

[18] A. Merzouki and H. McNairn, "A hybrid (multi-angle and multipolarization) approach to soil moisture retrieval using the integral equation model: preparing for the RADARSAT constellation mission," *Canadian Journal of Remote Sensing*, vol. 41, no. 5, pp. 349–362, 2015.

[19] Q. X. Xie, Q. Y. Meng, L. L. Zhang, C. M. Wang, Y. X. Sun, and Z. H. Sun, "A soil moisture retrieval method based on typical polarization decomposition techniques for a maize field from full-polarization Radarsat-2 data," *Remote Sensing*, vol. 9, no. 2, p. 168, 2017.

[20] R. Bindlish and A. P. Barros, "Multifrequency soil moisture inversion from SAR measurements with the use of IEM," *Remote Sensing of Environment*, vol. 71, no. 1, pp. 67–88, 2000.

[21] N. Baghdadi, S. Gaultier, and C. King, "Retrieving surface roughness and soil moisture from synthetic aperture radar (SAR) data using neural networks," *Canadian Journal of Remote Sensing*, vol. 28, no. 5, pp. 701–711, 2002.

[22] N. Pierdicca, L. Pulvirenti, and C. Bignami, "Soil moisture estimation over vegetated terrains using multitemporal remote sensing data," *Remote Sensing of Environment*, vol. 114, no. 2, pp. 440–448, 2010.

[23] E. J. M. Rignot and J. van Zyl, "Change detection techniques for ERS-1 SAR data," *IEEE Transactions on Geoscience and Remote Sensing*, vol. 31, no. 4, pp. 896–906, 1993.

[24] W. Wagner, G. Lemoine, and H. Rott, "A method for estimating soil moisture from ERS scatterometer and soil data," *Remote Sensing of Environment*, vol. 70, no. 2, pp. 191–207, 1999.

[25] H. Yang, J. C. Shi, Z. Li, and H. D. Guo, "Temporal and spatial soil moisture change pattern detection in an agricultural area using multi-temporal Radarsat ScanSAR data," *International Journal of Remote Sensing*, vol. 27, no. 19, pp. 4199–4212, 2006.

[26] Z. Bartalis, W. Wagner, V. Naeimi et al., "Initial soil moisture retrievals from the METOP-A Advanced Scatterometer (ASCAT)," *Geophysical Research Letters*, vol. 34, no. 20, pp. 5–9, 2007.

[27] C. Pathe, W. Wagner, D. Sabel, M. Doubkova, and J. B. Basara, "Using ENVISAT ASAR global mode data for surface soil moisture retrieval over Oklahoma, USA," *IEEE Transactions on Geoscience and Remote Sensing*, vol. 47, no. 2, pp. 468–480, 2009.

[28] A. Balenzano, F. Mattia, G. Satalino, and M. W. J. Davidson, "Dense temporal series of C- and L-band SAR data for soil moisture retrieval over agricultural crops," *IEEE Journal of Selected Topics in Applied Earth Observations and Remote Sensing*, vol. 4, no. 2, pp. 439–450, 2011.

[29] A. Balenzano, F. Mattia, G. Satalino, V. Pauwels, and P. Snoeij, "SMOSAR algorithm for soil moisture retrieval using Sentinel-1 data," in *Proceedings of IEEE International Geoscience and Remote Sensing Symposium*, pp. 1200–1203, Munich, Germany, July 2012.

[30] G. Satalino, F. Mattia, A. Balenzano, R. Panciera, and J. Walker, "Soil moisture maps from time series of PALSAR-1 scansar data over Australia," in *Proceedings of IEEE International Geoscience and Remote Sensing Symposium*, pp. 719–722, Melbourne, VIC, Australia, 2013.

[31] J. D. Ouellette, J. T. Johnson, A. Balenzano, and F. Mattia, "A study of soil moisture estimation from multitemporal L-band radar observations of vegetated surfaces," in *Proceedings of EUSAR*, Berlin, Germany, 2014.

[32] J. D. Ouellette, J. T. Johnson, A. Balenzano et al., "A time-series approach to estimating soil moisture from vegetated surfaces using L-band radar backscatter," *IEEE Transactions on Geoscience and Remote Sensing*, vol. 55, no. 6, pp. 3186–3193, 2017.

[33] A. Balenzano, G. Satalino, F. Lovergine, M. Rinaldi, and V. Iacobellis, "On the use of temporal series of L- and X-band SAR data for soil moisture retrieval. Capitanata plain case study," *European Journal of Remote Sensing*, vol. 46, no. 1, pp. 721–737, 2013.

[34] F. Mattia, A. Balenzano, G. Satalino, and V. Pauwels, "A SAR soil moisture retrieval algorithm inverting the temporal changes of radar backscatter," in *Proceedings of EUSAR*, Berlin, Germany, 2014.

[35] S. Kweon and Y. Oh, "Estimation of soil moisture and surface roughness from single-polarized radar data for bare soil surface and comparison with dual- and quad-polarization cases," *IEEE Transactions on Geoscience and Remote Sensing*, vol. 52, no. 7, pp. 4056–4064, 2014.

[36] L. He, Q. M. Qin, R. Panciera, M. Tanase, J. P. Walker, and Y. Hong, "An extension of the alpha approximation method for soil moisture estimation using time-series SAR data over bare soil surfaces," *IEEE Geoscience and Remote Sensing Letters*, vol. 14, no. 8, pp. 1328–1332, 2017.

[37] J. S. Lee, "Digital image enhancement and noise filtering by use of local statistics," *IEEE Transactions on Pattern Analysis and Machine Intelligence*, vol. 2, no. 2, pp. 165–168, 1980.

[38] M. Trudel, F. Charbonneau, F. Avendano, and R. Leconte, "Quick Profiler (QuiP): a friendly tool to extract roughness statistical parameters using a needle profiler," *Canadian Journal of Remote Sensing*, vol. 36, no. 4, pp. 391–396, 2010.

[39] A. G. Voronovich, *Wave Scattering from Rough Surfaces*, Springer, Heidelberg, Germany, 1994.

[40] M. C. Dobson, F. T. Ulaby, and M. T. Hallikainen, "Microwave dielectric behavior of wet soil-part II: dielectric mixing models," *IEEE Transactions on Geoscience and Remote Sensing*, vol. 23, no. 4, pp. 35–46, 1985.

[41] A. K. Fung, Z. Li, and K. S. Chen, "Backscattering from a randomly rough dielectric surface," *IEEE Geoscience and Remote Sensing Letters*, vol. 30, no. 2, pp. 356–369, 1992.

[42] L. Pasolli, C. Notarnicola, L. Bruzzone et al., "Polarimetric RADARSAT-2 imagery for soil moisture retrieval in alpine areas," *Canadian Journal of Remote Sensing*, vol. 37, no. 5, pp. 535–547, 2011.

3

An Analysis of Anomalous Winter and Spring Tornado Frequency by Phase of the El Niño/Southern Oscillation, the Global Wind Oscillation, and the Madden-Julian Oscillation

Todd W. Moore, Jennifer M. St. Clair, and Tiffany A. DeBoer

Department of Geography and Environmental Planning, Towson University, Towson, MD 21252, USA

Correspondence should be addressed to Todd W. Moore; tmoore@towson.edu

Academic Editor: Anthony R. Lupo

Winter and spring tornado activity tends to be heightened during the La Niña phase of the El Niño/Southern Oscillation and suppressed during the El Niño phase. Despite these tendencies, some La Niña seasons have fewer tornadoes than expected and some El Niño seasons have more than expected. To gain insight into such anomalous seasons, the two La Niña winters and springs with the fewest tornadoes and the two El Niño winters and springs with the most tornadoes between 1979 and 2016 are identified and analyzed in this study. The relationships between daily tornado count and the Global Wind Oscillation and Madden-Julian Oscillation in these anomalous seasons are also explored. Lastly, seasonal and daily composites of upper-level flow, low-level flow and humidity, and atmospheric instability are generated to describe the environmental conditions in the anomalous seasons. The results of this study highlight the potential for large numbers of tornadoes to occur in a season if favorable conditions emerge in association with individual synoptic-scale events, even during phases of the El Niño/Southern Oscillation, Global Wind Oscillation, and Madden-Julian Oscillation that seem to be unfavorable for tornadoes. They also highlight the potential for anomalously few tornadoes in a season even when the oscillations are in favorable phases.

1. Introduction

More tornadoes occur in the United States (US) per year than in any other country [1], and these tornadoes are capable of producing incredible economic and human loss [2–5]. Despite their relatively common occurrence, there is notable intra- and interannual variability in the number of US tornadoes [6–10]. Due to the danger and variability of tornado activity, there is a need to improve seasonal outlook capabilities. One approach is to identify relationships between seasonal tornado count and climate oscillations, such as the El Niño/Southern Oscillation (ENSO) (e.g., [11–19]).

Numerous studies over the past couple of decades have analyzed the relationship between tornadoes and ENSO. An early study by Monfredo [15] reported that strong and violent tornadoes were more common during the La Niña (LN) phase of ENSO and less common during the El Niño (EN) phase. Cook and Schaefer [16] later reported that winter tornado outbreaks were stronger and more frequent during the neutral phase (N) of ENSO, followed by the LN then EN phases. The most recent studies report that tornado activity is heightened during the LN phase [12–14, 17–19]. Recent studies also illustrate that the seasons with the most tornadoes tend to be classified as LN. Lee et al. [17], for example, analyzed the number of tornadoes rated 3 or higher on the Fujita or Enhanced Fujita damage scales (hereafter referred to as E(F)) and reported that the five most extreme tornado outbreak years were characterized as persistent LN events or LN events that were transitioning to a different phase. Allen et al. [12] similarly reported that most of the seasons with a tornado count >100% of climatology occurred during the LN phase, whereas most that were <75% of climatology occurred during the EN phase.

The relatively consistent reporting of heightened tornado activity with the LN phase of ENSO elicits the possibility of predicting the likelihood of below- or above-normal tornado activity based on ENSO conditions. Elsner et al. [18] developed a statistical model to evaluate variability in tornado activity in which ENSO was the most important predictor. Their model also indicated that tornado activity was heightened in the Midwest and Southeast regions of the US during LN and in the Great Plains region during EN. Allen et al. [12] and Lepore et al. [13] developed extended logistic regression models using ENSO to predict the likelihood of below-normal, normal, or above-normal tornado activity in spring. Their models similarly showed that tornado activity is increased during the LN phase, particularly in the south central US.

Anomalous seasons occur. Some LN seasons have fewer than expected tornadoes and some EN seasons have more than expected. Lee et al. [19], for example, noted that an anomalously large number of tornadoes occurred in the 2015-2016 EN winter. Elsner et al. [18] also noted the occurrence of anomalous seasons and attributed them to additional unknown factors. While previous studies have noted these anomalous seasons, none have provided detailed descriptions and analyses of them. More research, including work focusing on the anomalous seasons, is needed to improve our understanding of the tornado-ENSO link and to refine statistical models.

A case study approach was taken here to describe and analyze the two EN winters and springs with the most tornadoes, and the two LN winters and springs with the fewest tornadoes. The states of the Madden-Julian Oscillation (MJO) and Global Wind Oscillation (GWO) are of particular interest here. Both of these oscillations vary over subseasonal timescales (i.e., shorter than the timescale over which ENSO varies) and have been linked to variations in tornado activity [20–23]. Seasonal and daily composites are also generated to document the synoptic patterns that accompanied the anomalous seasons.

The specific objectives of this study are to

(1) identify and describe the two EN winters and springs with the most tornadoes and the two LN winters and springs with the fewest tornadoes;
(2) analyze GWO and MJO activity during these anomalous seasons;
(3) document the synoptic patterns that accompany these anomalous seasons.

2. Data and Methods

Tornado data were taken from the Storm Prediction Center's Severe Weather Database (specifically, the *1950–2016_actual_tornadoes.csv* file) [24]. This dataset was a subset to the 18,251 E(F)1+ tornadoes that occurred over the period December 1978–November 2016 in the contiguous US. The time series of E(F)1+ tornadoes is less influenced by detection and reporting changes over time than is the E(F)0+ series, which includes a notable upward trend [25, 26]. The beginning of the period of record is consistent with recent studies [12, 13] and corresponds to the first year of the North American Regional Reanalysis (NARR) product [27], which is used to generate seasonal and daily atmospheric composites. Seasonal tornado counts were generated for winter (December of previous year, January, and February (DJF)) and spring (March, April, and May (MAM)). DJF and MAM are the focus of this study because previous studies illustrate that tornado counts and ENSO are related in these seasons [12–14, 17–19].

Multiple indices represent ENSO conditions. Some represent mostly the oceanic component of ENSO (e.g., the Niño 1.2, 3, 3.4, and 4 regions and the Oceanic Niño Index (ONI)), some represent mostly the atmospheric component (e.g., the Southern Oscillation Index (SOI)), and some combine both (e.g., the Multivariate ENSO Index (MEI) and Bivariate ENSO Index (BEST)). The classification of seasons as LN or EN will likely change in some cases, depending on the index. In this study, ONI data were obtained from the Climate Prediction Center [28] to represent ENSO conditions, which is consistent with recent efforts to predict the likelihood of an active or inactive season based on the state of ENSO [12, 13]. ONI is also one of the commonly used indices by the Climate Prediction Center to diagnose ENSO conditions and to place current events into historical perspective. It is given as anomalies of 3-month running means of sea surface temperature in the Niño 3.4 region (5°N–5°S, 120°W–170°W). Series of ONI for DJF and MAM were extracted and merged with the seasonal tornado counts. Once merged, seasons were classified as LN if ONI ≤ -0.5 and EN if ONI ≥ 0.5. Seasons with $-0.5 < \text{ONI} < 0.5$ were classified as neutral (N). The two most active EN DJFs and MAMs (i.e., with the most tornadoes) and two least active LN DJFs and MAMs (i.e., with the fewest tornadoes) were extracted, yielding eight seasons as the focus of this study.

Climate oscillations other than ENSO have been linked to tornado activity. Those that vary over subseasonal timescales may provide insight into the intraseasonal distribution of tornadoes in the anomalous seasons. Two such oscillations are the MJO and GWO. The MJO is a tropical wave of enhanced convection that propagates the globe from west to east at approximately $5\,\text{m}\cdot\text{s}^{-1}$ [29]. It is linked to variability in tornado activity in the midlatitudes through changes in global relative atmospheric angular momentum (AAM), which arise from tropical convective forcing and teleconnected Rossby wave propagation in the midlatitudes [20, 23, 30–32]. Other factors, such as friction and mountain torque, affect AAM in addition to tropical convection [30–32]. The GWO combines these factors and, therefore, provides a comprehensive representation of AAM and midlatitude circulation patterns [31, 32]. Daily GWO data were obtained from the Earth System Research Laboratory's GWO dataset [33]. The attributes of the GWO used in this study are AAM anomalies, AAM time tendency, and GWO phase. The relationship between AAM anomalies and AAM time tendency determines the phase of GWO, which run from 1 to 8. Phases 2 and 3 represent anomalously low AAM, whereas phases 6 and 7 represent anomalously high AAM. Phases 1, 4, 5, and 8 represent transition states. Daily MJO data were obtained from the Australian Bureau of Meteorology's Real-Time Multivariate (RMM) MJO dataset [34, 35]. The attributes used here are the first two principal components of empirical orthogonal functions that consist of a meridional average of the 200 mb zonal wind, the 850 mb zonal wind, and outgoing longwave radiation between 15°N and 15°S (referred to as RMM1 and RMM2) and the MJO phase. Similar to GWO, the MJO index has eight

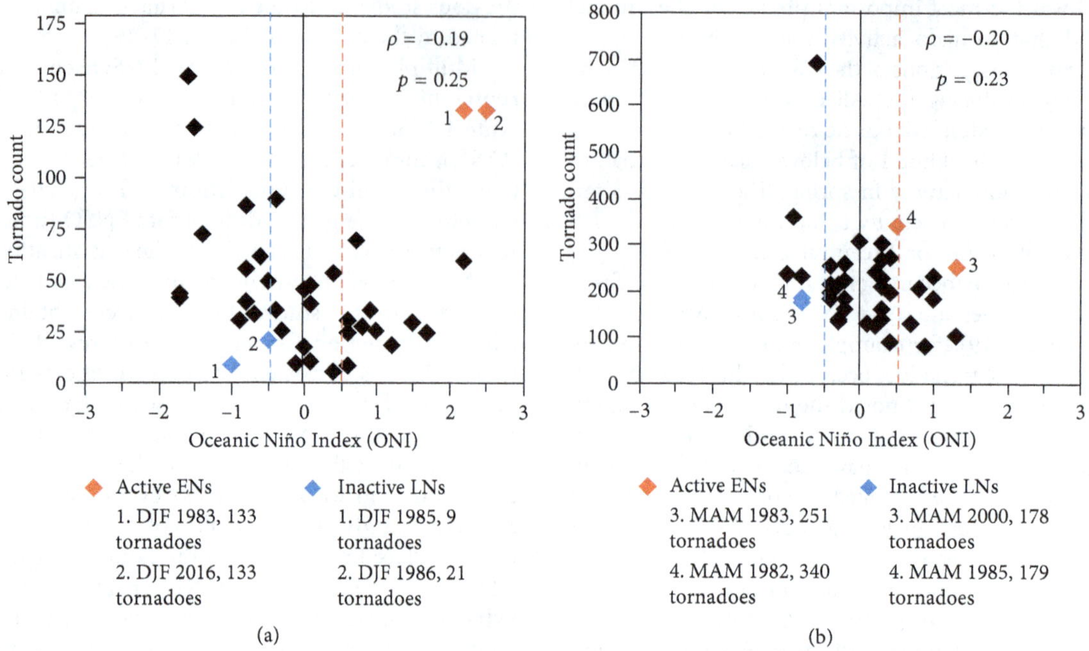

Figure 1: Number of E(F)1+ tornadoes in (a) DJF and (b) MAM and the concurrent Oceanic Niño Index (ONI). Vertical dashed blue and red lines are placed at ONI = −0.5 and 0.5 and demarcate LN and EN, respectively. The blue and red diamonds represent the two most inactive and active LNs and ENs, respectively. Spearman's ρ rank correlation coefficients and p values are provided in the top-right corners of the graphs.

phases, but they represent the location of enhanced tropical convection. Phase 1 of the MJO indicates that the enhanced convection is located near eastern Africa and the subsequent phases represent its eastward progression.

AAM anomalies, AAM time tendency, RMM1, and RMM2 were used to construct GWO and MJO phase space diagrams for each of the anomalous seasons. Daily tornado counts were generated (where a day is defined as 0000–2359 UTC) and plotted in the phase space diagrams along with the progression of GWO and MJO. Daily tornado counts were also assessed across the phases of MJO and GWO, and Kruskal–Wallis tests were used to determine if the mean rank of the tornado counts significantly varied across the phases of the oscillations. This nonparametric test was chosen because of the skewed nature of daily tornado counts.

The synoptic patterns associated with the anomalous seasons were characterized using gridded composites from NARR [27]. The following variables were chosen to be comparable to composites in previous research (e.g., [12, 20, 21]): upper-level flow was characterized using 300 mb geopotential heights; low-level flow was characterized using 850 mb geopotential heights; low-level moisture was characterized using 850 mb specific humidity; and atmospheric stability was characterized using surface-based convective available potential energy (CAPE). Surface-based CAPE was used because Gensini et al. [36] illustrate that surface-based parcels are more accurate than 100 mb mixed layer parcels, largely due to errors in low-level moisture fields. The seasonal mean composites are averages of the variables for the dates under consideration (e.g., mean 300 mb geopotential height in DJF 2016 is generated by averaging the 300 mb geopotential height for the days in DJF 2016). The seasonal anomaly composites represent the difference between the mean composites and the 1981–2010 climatology. Daily composites are averages of the 3 hr NARR data. All composites were generated with NOAA's Earth System Research Laboratory's online plotting tool [37, 38]. Once generated, the Network Common Data Form (NetCDF) of the composites was imported into ArcMAP [39] for plotting and display.

Table 1: Minimum, mean, and maximum values of global atmospheric angular momentum (AAM; kg·m^2·s^{-2}) anomalies during the active EN and inactive LN seasons.

Seasons	Minimum	Mean	Maximum
Active EN seasons			
DJF 1983	1.3	2.6	4.4
DJF 2016	0.4	1.9	4.0
MAM 1982	−1.3	0.1	1.4
MAM 1983	0.1	1.8	3.5
Inactive LN seasons			
DJF 1985	−2.0	−0.9	0.1
DJF 1986	−1.1	0.1	1.3
MAM 1985	−1.0	0.1	1.1
MAM 2000	−2.7	−1.5	−0.4

3. Results and Discussion

3.1. Identification and Description of the Anomalous Seasons. The relationships between the number of E(F)1+ tornadoes and ONI in DJF and MAM are depicted in Figure 1. The negative relationship reported by others, whereby tornado frequency tends to be greater during the LN phase and lesser

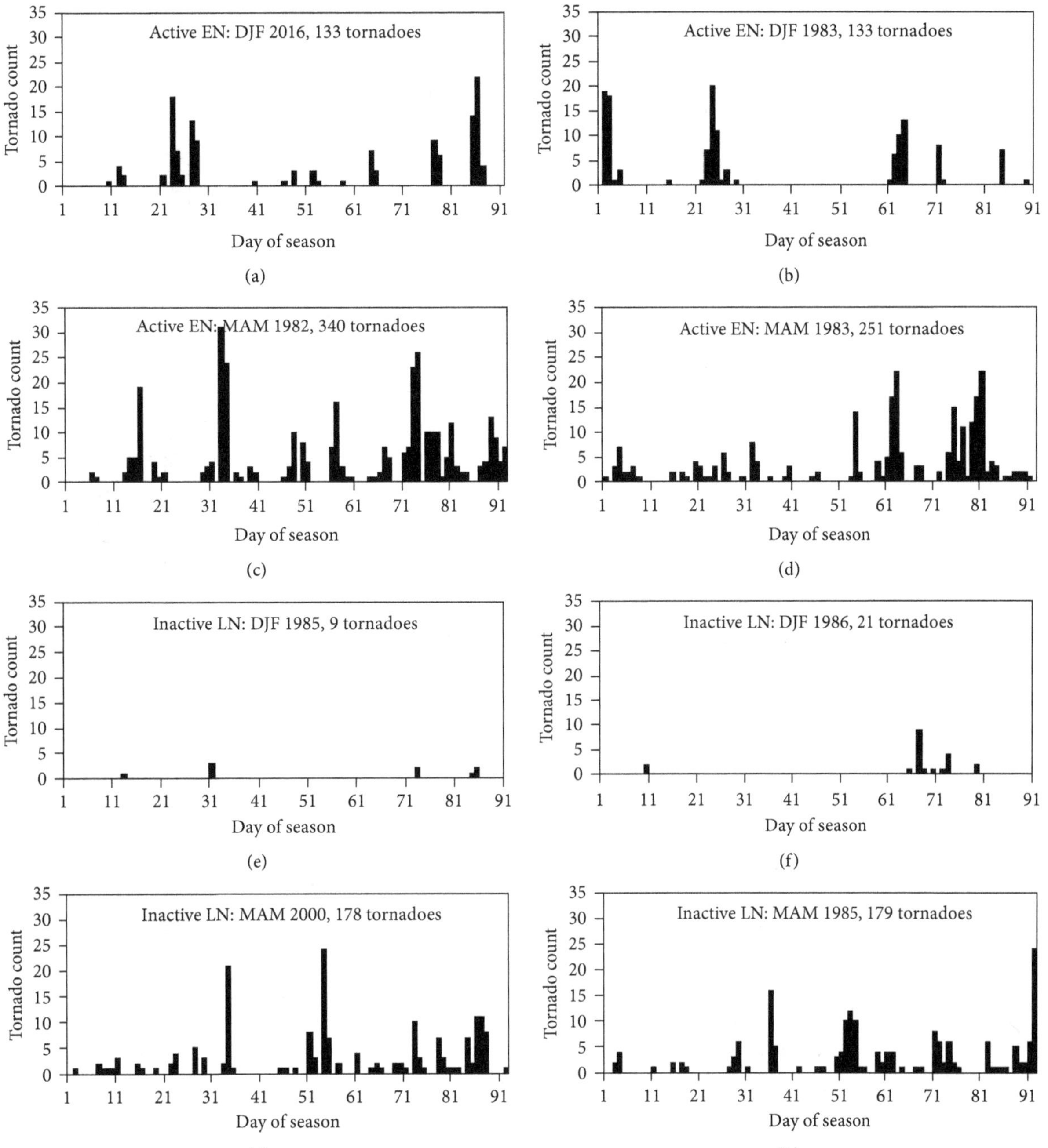

FIGURE 2: Daily tornado count of the two EN DJFs (a, b) and MAMs (c, d) with the most tornadoes and of the two LN DJFs (e, f) and MAMs (g, h) with the fewest tornadoes.

during the EN phase [12–14, 17–19], is visually apparent, but the correlation is not statistically significant in this subset. The central tendencies in Table 1 also illustrate the negative association by showing that the mean and median tornado counts of the LN seasons exceed those of the EN seasons.

The two most inactive LN DJF and MAM seasons are marked with blue diamonds in Figure 1, and the two most active ENs are marked with red diamonds. The two most active EN DJFs, which occurred in 2016 and 1983, are especially noticeable. There were 133 E(F)1+ tornadoes in both seasons. These counts are 183% greater than the mean number of tornadoes in EN DJFs and 359% greater than the median. The two most active EN MAM seasons were 1982 and 1983, when 340 and 251 E(F)1+ tornadoes occurred, respectively. MAM 1982 had approximately 79% more E(F)1+ tornadoes than the mean EN MAMs (76% greater than the median). MAM 1983 exceeded the mean and median counts of EN MAM seasons by approximately 30%.

The two DJF seasons with the fewest E(F)1+ tornadoes were 1985 and 1986, with 9 and 21 tornadoes, respectively. DJF 1985 had approximately 80% fewer tornadoes than the mean and median LN DJF, and DJF 1986 had 64% and 54%

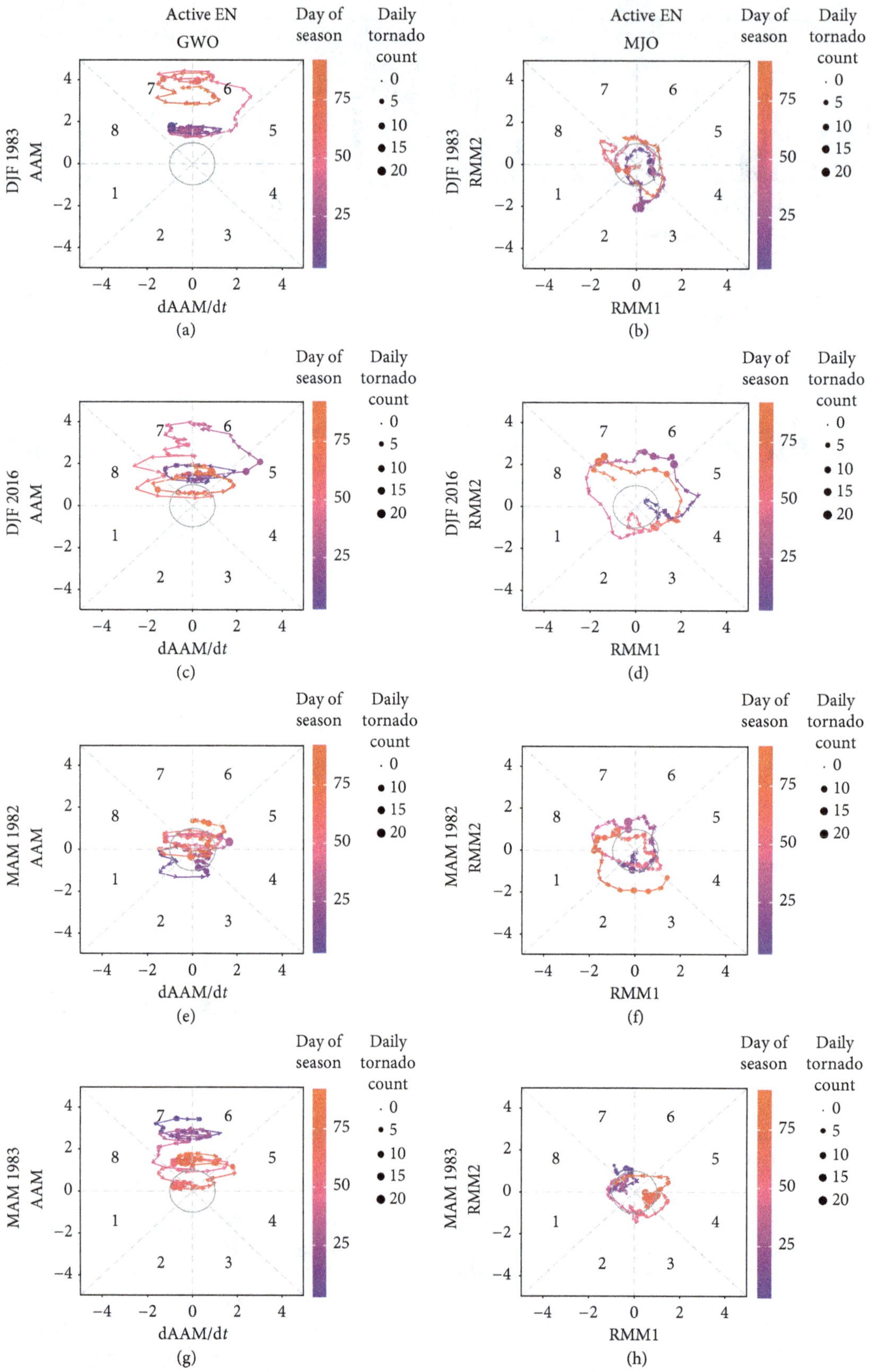

Figure 3: Phase space diagrams for the GWO (a, c, e, and g) and MJO (b, d, f, and h) during the two EN DJFs and MAMs with the most tornadoes.

Table 2: Mean (mean rank) of daily tornado count by GWO and MJO phase.

	GWO phase							
	1	2	3	4	5	6	7	8
Active EN DJF	0.0 (70.0)	—	—	—	—	1.5 (83.3)	1.3 (92.8)	2.3 (96.3)
Active EN MAM	3.8 (100.5)	2.9 (86.8)	7.5 (106.9)	3.2 (81.1)	3.6 (83.9)	3.3 (89.5)	2.7 (98.3)	2.8 (88.2)
Inactive LN DJF	0.1 (89.4)	0.1 (90.5)	0.1 (86.4)	0.1 (87.2)	0.0 (84.0)	0.5 (100.5)	0.8 (106.5)	0.1 (90.2)
Inactive LN MAM	2.3 (115.4)	3.6 (100.0)	1.4 (88.8)	1.0 (79.8)	2.1 (93.6)	1.8 (111.0)	2.6 (79.8)	1.8 (84.8)
	MJO phase							
Active EN DJF[a]	3.5 (113.2)	0.4 (90.0)	1.9 (101.4)	0.7 (81.2)	1.1 (80.6)	4.0 (119.9)	1.3 (80.5)	0.3 (80.6)
Active EN MAM[b]	1.1 (58.0)	2.5 (85.4)	2.6 (87.3)	5.9 (116.1)	1.2 (72.5)	1.5 (83.3)	4.5 (101.9)	3.1 (97.8)
Inactive LN DJF	0.4 (100.3)	0.0 (84.0)	0.1 (88.0)	<0.1 (87.3)	0.1 (91.4)	0.2 (91.7)	0.5 (92.7)	0.1 (87.9)
Inactive LN MAM	2.0 (96.6)	2.2 (87.3)	2.7 (92.3)	0.5 (70.1)	2.3 (103.0)	2.6 (96.7)	2.0 (95.6)	1.7 (94.5)

[a]A Kruskal–Wallis test indicates that the mean rank of tornado count varies across the phases of MJO ($X^2 = 23.9$; df = 7; $p = 0.001$). Post hoc comparisons show that the mean rank of tornado count is greater in phase 6 than in phases 4, 7, and 8. [b]A Kruskal–Wallis test indicates that the mean rank of tornado count varies across the phases of MJO ($X^2 = 17.9$; df = 7; $p = 0.012$). Post hoc comparisons show that the mean rank of tornado count is greater in phase 4 than in phase 1.

fewer than the mean and median counts, respectively. The least active LN MAM seasons were 1985 when 179 E(F)1+ tornadoes occurred and 2000 when 178 E(F)1+ occurred. These seasons had approximately 43% fewer tornadoes than the mean LN MAM tornado count and approximately 24% fewer than the median count.

As with most seasons, the tornadoes in the anomalous seasons were not uniformly distributed (Figure 2). In DJF 2016, for example, 50% of the tornadoes occurred on 4 days, each with 10+ tornadoes. There were 6 days in DJF 1983 with 10+ tornadoes that account for 68% of the tornadoes in that season. The tornadoes in the anomalous MAM seasons were spread over a larger number of days, but there were still clusters of activity. Twelve days in MAM 1982 and eight in MAM 1983 had 10+ tornadoes. The tornadoes on these days account for 60% and 52% of the seasonal count, respectively.

3.2. Role of GWO and MJO in the Anomalous Seasons. Previous studies illustrate that tornado activity is heightened during certain phases of the GWO and MJO. Gensini and Marinaro [21] reported that daily tornado anomalies in spring (March–June) are greatest on GWO phase 1 and 2 days when AAM is negative. Moore [22] similarly reported a tendency for tornado frequency to be greater in MAM seasons when GWO phase 2, 3, and 4 days are more common. Moore [22] also showed this to be true in DJF. Barrett and Gensini [23] reported that tornado days in April are most common on phase 6 and 8 days of the MJO and less common on phase 3, 4, and 7 days. They also reported that tornado days are most common with phases 5 and 8 and less common with phases 2 and 3 in May. Thompson and Roundy [20] reported that violent tornado outbreaks in MAM are most common on MJO phase 2 days and least common on phase 8 days. GWO and MJO vary on subseasonal timescales. They are, therefore, capable of modulating tornado activity within a given season and may provide insight into some of the subseasonal periods of suppressed and heightened tornado activity during these anomalous seasons (as seen in Figure 2).

AAM anomalies were positive throughout both of the anomalously active EN DJFs (Figures 3(a) and 3(c)). The mean AAM anomaly was 2.6 $kg{\cdot}m^2{\cdot}s^{-2}$ in DJF 1983 and 1.9 $kg{\cdot}m^2{\cdot}s^{-2}$ in DJF 2016 (Table 1). AAM anomalies were also positive throughout the active EN MAM 1983 season, when the mean was 1.8 $kg{\cdot}m^2{\cdot}s^{-2}$ (Figure 3(g), Table 1). Tornadoes in these seasons, therefore, occurred on GWO phase 5–8 days when AAM was anomalously high and the amplitude was most often >1. Positive anomalies were expected in these seasons because AAM tends to be heightened during the EN phase of ENSO [40], but it is unexpected, based on previous studies linking enhanced tornado activity to anomalously low AAM [21, 22], that all of the days with tornadoes in DJF 1983, DJF 2016, and MAM 1983 had anomalously high AAM. Despite the positive AAM anomalies throughout these seasons, the tendency of AAM was volatile, and tornadoes occurred on days when AAM tendency was increasing and decreasing.

In MAM 1982, there were periods when AAM was anomalously low, which resulted in a lower seasonal mean of 0.1 $kg{\cdot}m^2{\cdot}s^{-2}$ (Figure 3(e), Table 1). In this season, numerous tornadoes occurred on GWO phase 1–4 in addition to 5–8 days. However, there is not a significant difference in daily tornado counts across the phases of GWO in this season or any of the others (Table 2).

AAM is often relatively low during the LN phase of ENSO [40]. It is, therefore, not surprising that AAM was anomalously low throughout most of the inactive LN seasons (Figure 4). The mean AAM anomaly was −1.5·$kg{\cdot}m^2{\cdot}s^{-2}$ in MAM 2000 and −0.9 $kg{\cdot}m^2{\cdot}s^{-2}$ in DJF 1985 (Table 1). AAM fluctuated between negative and positive anomalies in MAM 1985 and DJF 1986, which led to higher mean values of 0.1 $kg{\cdot}m^2{\cdot}s^{-2}$ in each season (Figures 4(c) and 4(e); Table 2). Similar to the active EN seasons, there is not a significant difference in daily tornado counts across the GWO phases in any of the inactive LN seasons. Also, similar to the tornadoes in the active EN seasons, those in the inactive LN seasons occurred during periods of increasing and decreasing AAM (Figure 4).

MJO varied more than GWO throughout the anomalous seasons (Figures 3 and 4). The progression from phase 1 through 8 is apparent, with multiple oscillations in most

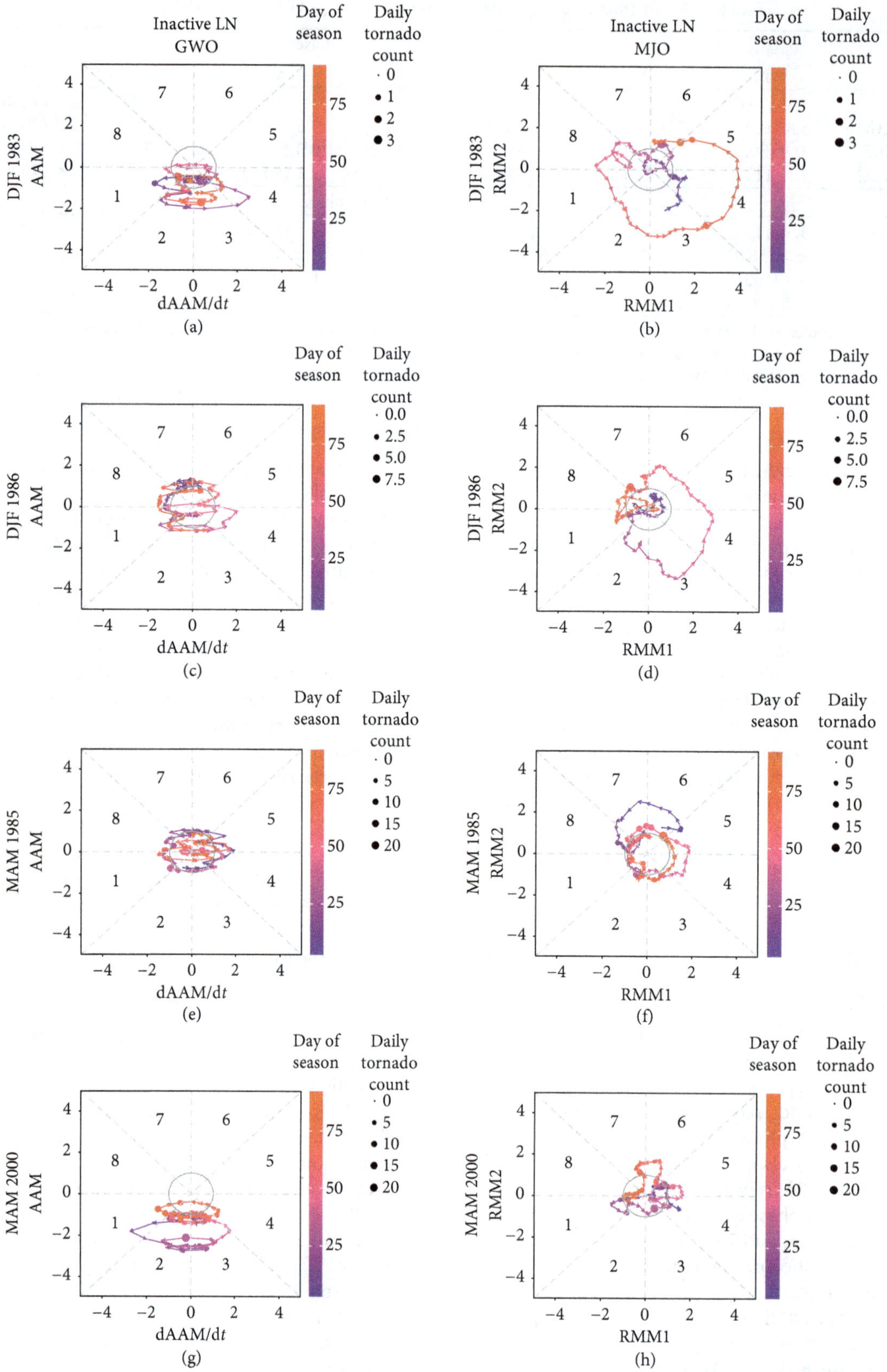

FIGURE 4: Phase space diagrams for the GWO (a, c, e, and g) and MJO (b, d, f, and h) during the two LN DJFs and MAMs with the fewest tornadoes.

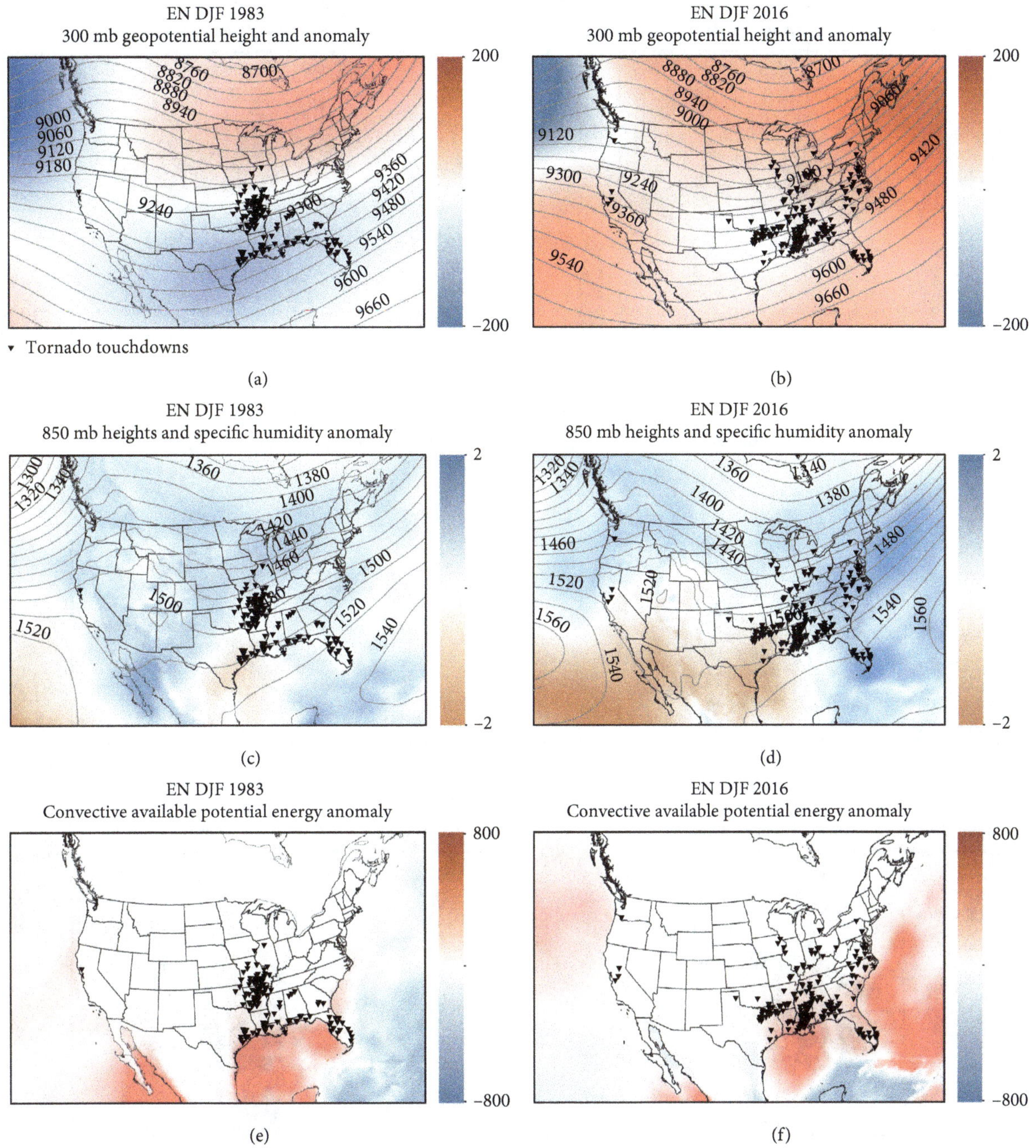

FIGURE 5: (a, b) 300 mb geopotential heights (contours (m)) and anomalies (color (m)), (c, d) 850 mb geopotential heights (contours (m)) and specific humidity anomalies (color ($g \cdot kg^{-1}$)), and (e, f) CAPE anomalies ($J \cdot kg^{-1}$) for the two anomalously active EN DJFs.

seasons. Tornadoes concentrate on certain MJO phase days more so than with GWO phases, which led to significant differences in the mean number of tornadoes per day across the phases. In the two active EN DJFs, for example, the mean and mean ranks of the tornado counts were greatest with phases 1 and 6 of the MJO (Table 2). A Kruskal–Wallis test and subsequent post hoc comparisons indicate that the mean rank of phase 6 is significantly greater than the mean ranks of phases 4, 7, and 8; remaining comparisons yielded insignificant differences (see the subscript below Table 2). The percentage of days with tornadoes was also greatest with phase 6—tornadoes occurred on 10 of the 19 (53%) phase 6 days. In the two EN MAMs, the mean and mean ranks of the tornado counts are greatest with phase 4 (Table 2). The statistical tests indicated that the mean rank of phase 4 is significantly greater than that of phase 1 (see the subscript below Table 2). The percentage of days with tornadoes was also greatest with phase 4 (23 of 31 (74%) phase 4 days had tornadoes). The percentage of days with tornadoes was also high with phases 7 and 8 (69% and 68%, resp.). There were not any significant differences in daily tornado counts across the phases of MJO in the inactive LN seasons (Table 2).

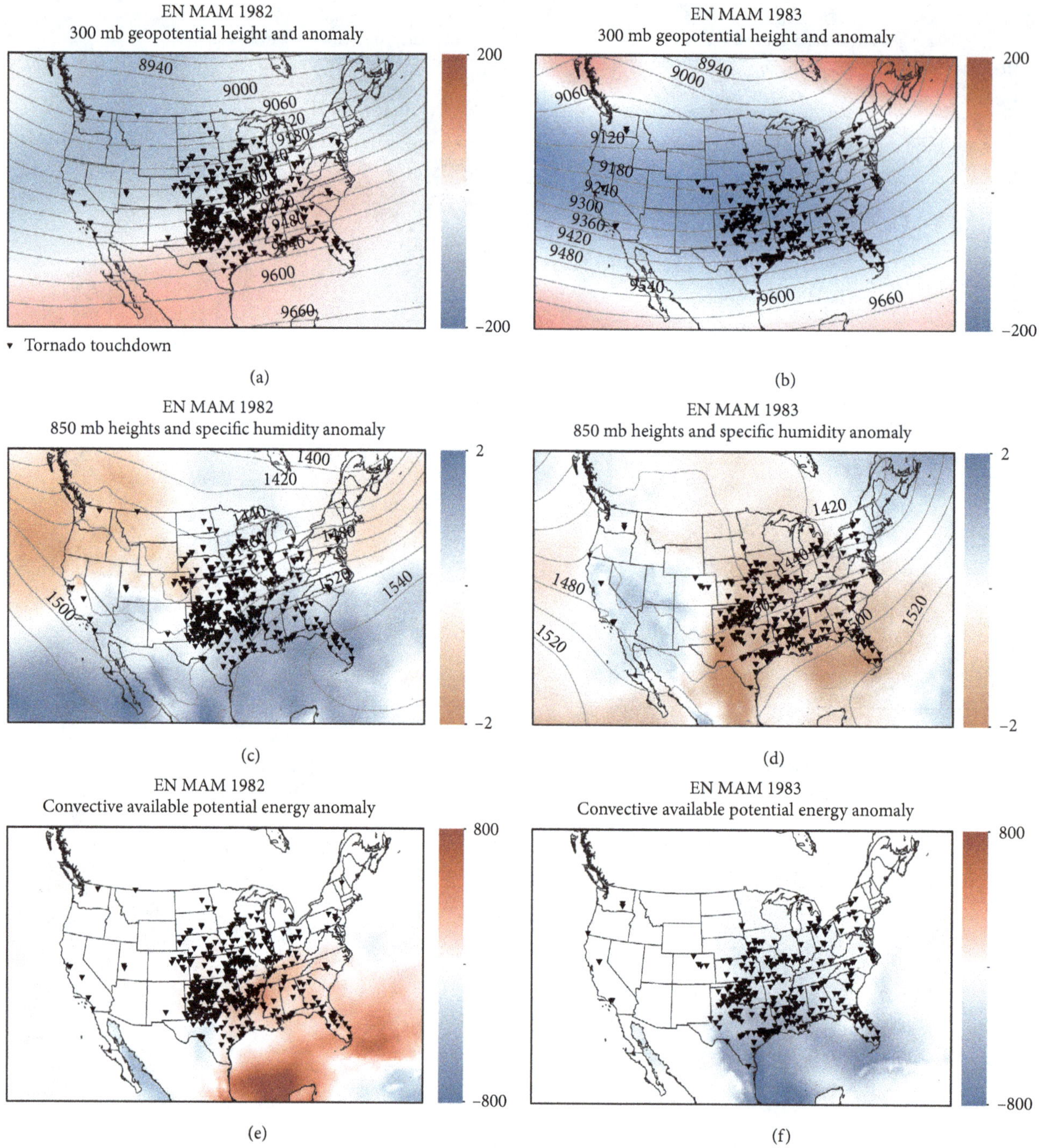

FIGURE 6: (a, b) 300 mb geopotential heights (contours (m)) and anomalies (color (m)), (c, d) 850 mb geopotential heights (contours (m)) and specific humidity anomalies (color (g·kg^{-1})), and (e, f) CAPE anomalies (J·kg^{-1}) for the two anomalously active EN MAMs.

3.3. Atmospheric Composites of the Anomalous Seasons. The seasonal composites of 300 mb geopotential height, 850 mb geopotential height and specific humidity, and CAPE are shown in Figures 5–8. Anomalously high geopotential heights at 300 mb were present across most of the contiguous US in DJF 2016 and across the north central and northeast US in DJF 1983 (Figures 5(a) and 5(b)). The presence of higher than normal heights is more similar to the LN composite reported by Allen et al. [12] rather than their EN composite. Upper-level ridges and troughs were present over the western and central US, respectively, in both of the active DJFs. Anomalously high heights were also present during the MAM 1982 seasons, but only over the southern and eastern US (Figure 6(a)). Low height anomalies were present over the contiguous US during MAM 1983 (Figure 6(b)), which is consistent with the EN composite reported by Allen et al. [12]. The patterns seen in the composites of low-level moisture are inconsistent across the active seasons. For example, anomalously high specific humidity was present in the southeastern US during MAM

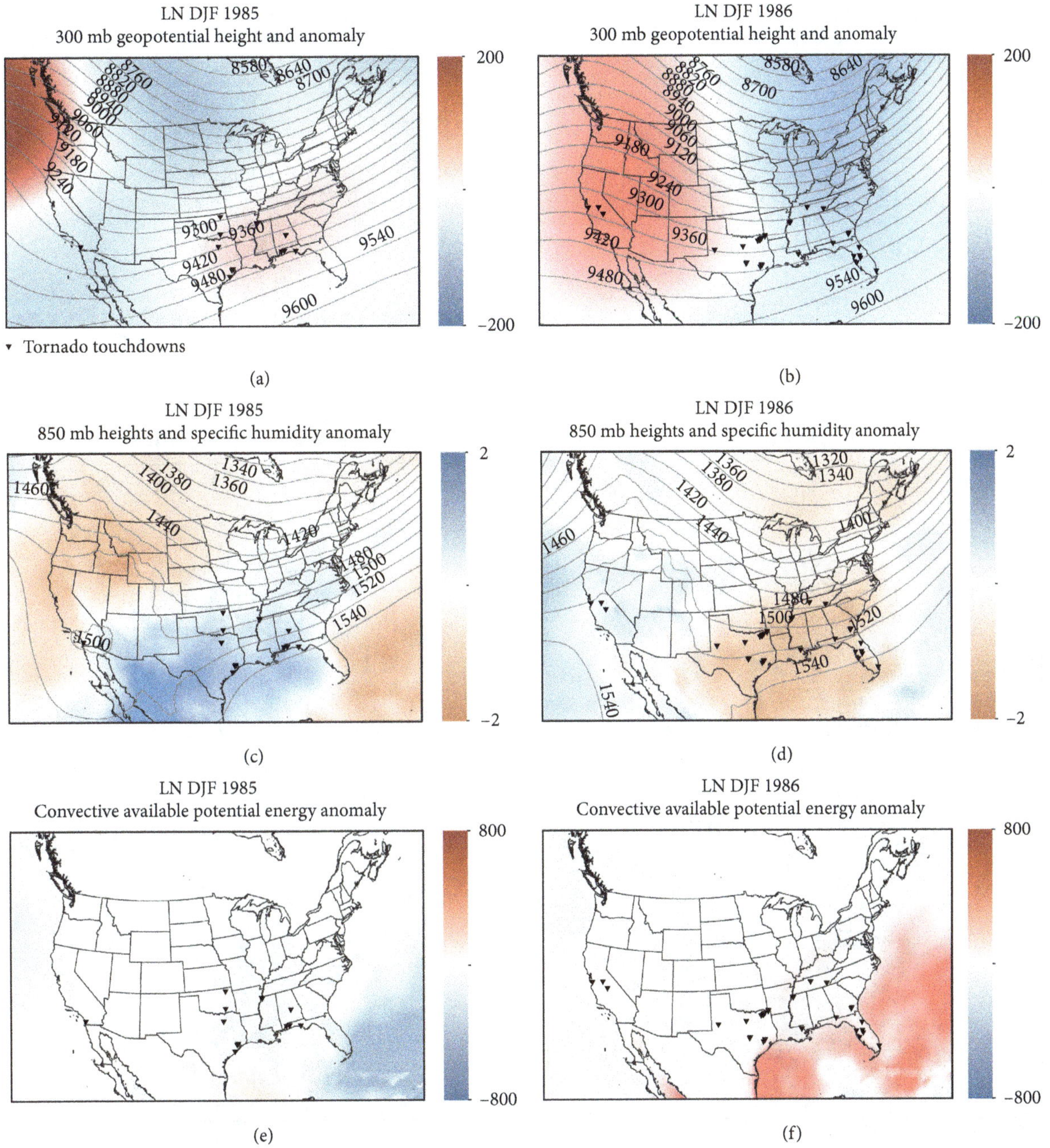

FIGURE 7: (a, b) 300 mb geopotential heights (contours (m)) and anomalies (color (m)), (c, d) 850 mb geopotential heights (contours (m)) and specific humidity anomalies (color (g·kg^{-1})), and (e, f) CAPE anomalies (J·kg^{-1}) for the two anomalously inactive LN DJFs.

1982, but the humidity in this region was anomalously low during MAM 1983 (Figures 6(c) and 6(d)). The humidity was also anomalously low in southeast Texas and Louisiana during both DJF seasons, but was higher to the north and east where most of the tornadoes occurred (Figures 5(c) and 5(d)). CAPE was near normal or anomalously high over most of the eastern US during the DJF 1983, DJF 2016, and MAM 1982 active EN seasons (Figures 5(e), 5(f), and 6(e)). Similar to some of the patterns of 300 mb geopotential height, these patterns of elevated CAPE are more similar to the CAPE composites shown by Allen et al. [12] in association with the LN phase rather than the EN phase.

The patterns of 300 mb geopotential height varied between the two LN DJFs (Figures 7(a) and 7(b)). A dipole pattern was present in DJF 1985, with anomalously low heights spanning the northern Great Plains southwestward to the Southwest and high heights across the Pacific Northwest and Southeast US. The US was split in DJF 1986, with anomalously high heights across the western region and low heights across the eastern. The low-level humidity anomaly patterns also varied between the DJFs (Figures 7(c) and 7(d)). The upper-level height and low-level humidity patterns were more similar in the two inactive LN MAMs—near normal or anomalously high heights most of

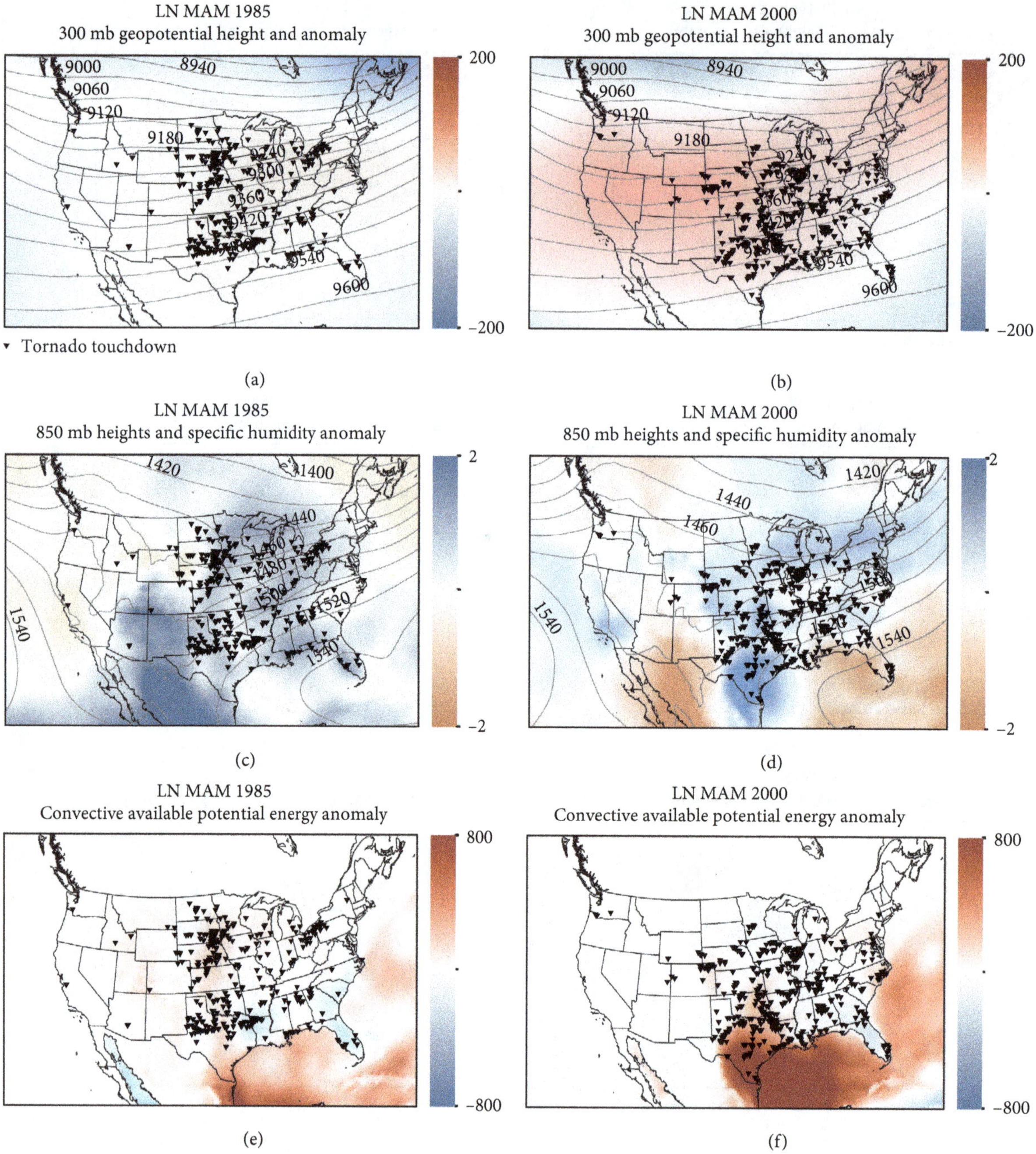

FIGURE 8: (a, b) 300 mb geopotential heights (contours (m)) and anomalies (color (m)), (c, d) 850 mb geopotential heights (contours (m)) and specific humidity anomalies (color ($g \cdot kg^{-1}$)), and (e, f) CAPE anomalies ($J \cdot kg^{-1}$) for the two anomalously active LN MAMs.

the US and anomalously humid conditions across most of its eastern half (Figures 8(a)–8(d)). Anomalously low CAPE was present across the Southeast US in DJF 1985 when the fewest tornadoes occurred (Figure 7(e)). Anomalously high CAPE was present across the portions of the US with the most tornadoes in the remaining seasons (Figures 7(f), 8(e), and 8(f)).

Some of the seasonal composites illustrate reasonable patterns. Anomalously high CAPE is shown in both of the active EN DJFs, for example, in Figures 5(e) and 5(f). Most of the eastern US was also anomalously humid in these seasons (Figures 5(c) and 5(d)). Other seasonal composites are unexpected and do not capture the environment on the days when many tornadoes occurred. For example, the negative CAPE anomalies across the eastern US in MAM 1983 are unexpected, given that anomalously many tornadoes occurred in this season (Figure 6(f)). Examination of CAPE on 1-2 May 1983 and 18–20 May 1983, when 39 and 51 tornadoes occurred, respectively, illustrates that elevated CAPE spread from the Gulf of Mexico northward into the eastern US (Figure 9). The negative anomalies of low-level humidity across the south central US in DJF 2016 also did not represent well the environments that were present when many of the

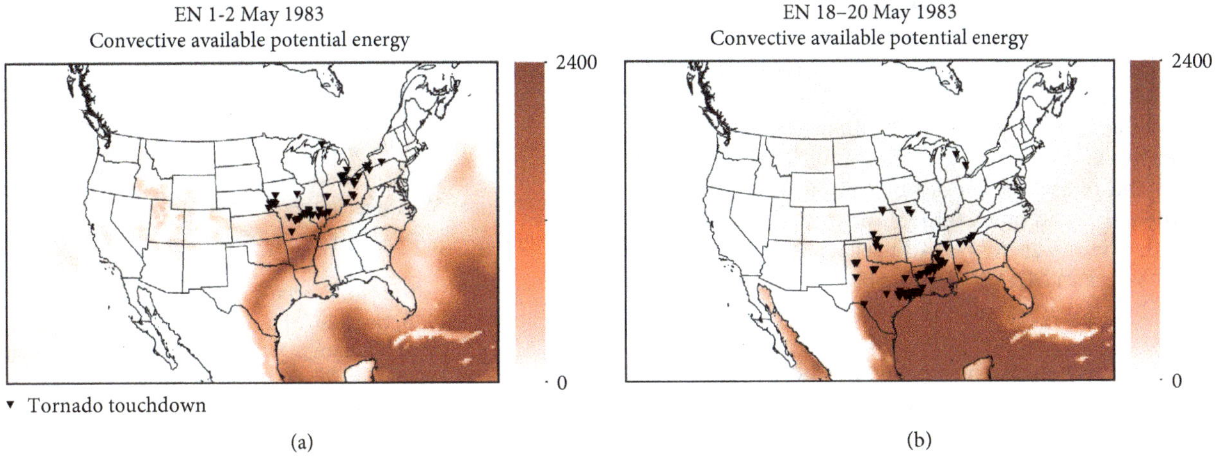

FIGURE 9: CAPE ($J{\cdot}kg^{-1}$) on (a) 1-2 May 1983 and (b) 18–20 May 1983, when 39 and 51 tornadoes occurred, respectively.

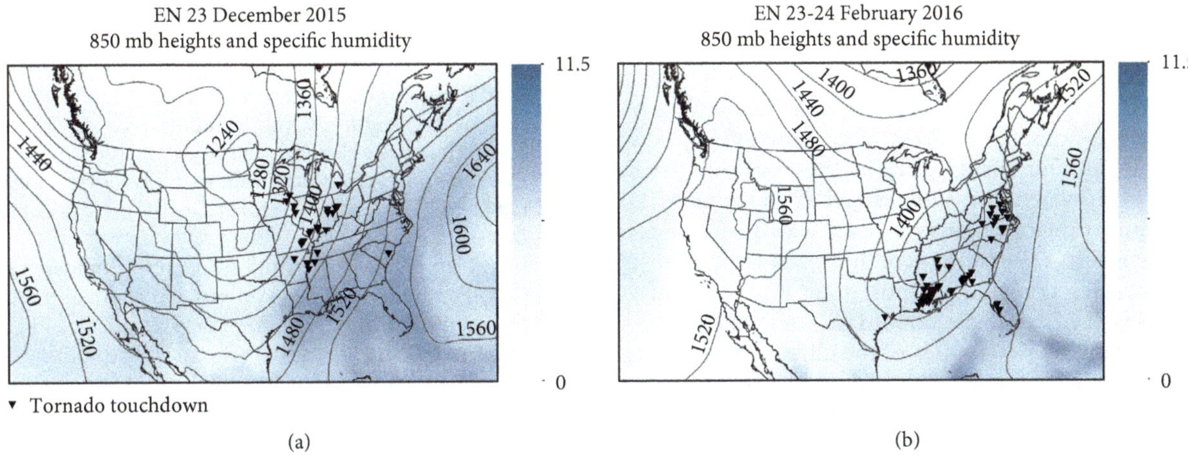

FIGURE 10: 850 mb geopotential heights (contours (m)) and specific humidity (color ($g{\cdot}kg^{-1}$)) on (a) 23 December 2015 and (b) 23-24 February 2016, when 18 and 36 tornadoes occurred, respectively.

tornadoes occurred (Figure 5(d)). Closed shortwave troughs with upstream southerly moisture advection into the US were present on 23 December 2015 and 23-24 February 2016, when 18 and 36 tornadoes occurred, respectively (Figure 10). These cases illustrate that anomalously active seasons can have seemingly unfavorable seasonal composites.

4. Conclusions

Previous studies have established a relationship between tornado and ENSO in DJF and MAM, generally with more tornadoes during the LN phase and fewer during the EN phase [12–14, 17–19]. This study was focused on the seasons that do not fit this relationship—EN seasons with many tornadoes and LN seasons with few tornadoes. Specifically, the two EN DJFs and MAMs with the most tornadoes and the two LN DJFs and MAMs with the fewest tornadoes were described and analyzed. The most anomalous seasons were DJF 1983 and 2016, both of which were active EN seasons with 133 E(F)1+ tornadoes. They were, therefore, 183% (359%) above the mean (median) EN DJF. These seasons illustrate that large numbers of tornadoes are possible even during EN seasons when such large numbers might be unexpected. The other seasons were far less anomalous.

GWO does not explain the anomalous nature of the seasons. Climatological studies show that tornado activity in DJF and MAM tends to be heightened during GWO phases 1–4 when AAM is anomalously low [21, 22], but nearly all of the tornadoes in the active EN seasons occurred on GWO phase 5–8 days when AAM was anomalously high. Furthermore, daily tornado count did not significantly vary across the phases of GWO in any of the seasons. The concentration of tornadoes on high AAM days (GWO phase 5–8 days) during EN seasons, as suggested here, would undoubtedly weaken the statistical relationship between tornadoes and GWO that was reported by others, whereby tornadoes are most common on GWO phase 1–4 days when AAM is anomalously low [21, 22]. Analyzing the tornado-GWO relationship by ENSO phase might amend this relationship and provide additional insight into the interactions between tornado activity, ENSO, and GWO. Another consideration is that the GWO data used in this study are based on globally integrated AAM. As noted by Gensini and Allen [41], this may confound the results. Therefore, it would also be worthwhile to reassess the relationship between tornado activity and GWO with AAM calculated on a hemispheric or latitudinal basis.

Unlike with GWO, daily tornado count did significantly vary across the phases of MJO, but only in the active EN seasons. The mean and mean rank of the tornado counts was greatest with phase 4 in MAM. Clustering of tornadoes on MJO phase 4 days in MAM is unique because climatological studies [20, 23] that aggregated data over a larger number of seasons reported heightened tornado activity with phase 2, 5, and 7 of the MJO in boreal spring months. While only representative of MAM 1982 and 1983, the clustering of tornadoes on MJO phase 4 days in these EN seasons highlights the possibility for a large number of tornadoes when subseasonal climate oscillations are in unfavorable phases. The mean and mean rank of the tornado counts was also greatest with phase 6 of the MJO in the anomalously active EN DJFs. Additional study is needed to determine if the tendency for DJF tornadoes to cluster on phase 6 days is common across a larger number of seasons.

This study highlights EN seasons with anomalously many tornadoes and LN seasons with anomalously few tornadoes. It also highlights days in these seasons with many tornadoes when the subseasonal GWO and MJO were in phases previously unassociated with tornado activity. These results suggest that conditions at the synoptic scale and smaller contributed most to the anomalously active seasons (i.e., favorable conditions emerged despite the seemingly unfavorable states of the climate patterns in the active EN DJFs). Active tornado days in seasons when climate patterns are more favorable are also often driven by synoptic- and subsynoptic-scale conditions, but the tornado-favorable conditions are likely to emerge more often in these seasons. This study only considered the two most anomalously active EN MAMs and DJFs and the two most anomalously inactive LN MAMs and DJFs. Additional study of a larger number of seasons is needed. Study of additional EN seasons is particularly warranted to see if the weakened association between tornado activity and GWO that was seen here is common in other EN seasons. A better understanding of the ways in which ENSO, GWO, and MJO interact across seasons to influence tornado activity will likely improve seasonal and subseasonal outlooks.

Conflicts of Interest

The authors declare that there are no conflicts of interest regarding the publication of this paper.

References

[1] National Centers for Environmental Information, "U.S. Tornado climatology," January 2018, https://www.ncdc.noaa.gov/climate-information/extreme-events/us-tornado-climatology.

[2] S. Changnon, "Tornado losses in the United States," *Natural Hazards Review*, vol. 10, no. 4, pp. 145–150, 2009.

[3] K. M. Simmons and D. Sutter, *Economic and Societal Impacts of Tornadoes*, American Meteorological Society, Boston, MA, USA, 2011.

[4] K. R. Knupp, T. A. Murphy, T. A. Coleman et al., "Meteorological overview of the devastating 27 April 2011 tornado outbreak," *Bulletin of the American Meteorological Society*, vol. 95, no. 7, pp. 1041–1062, 2014.

[5] National Weather Service, "Natural hazards statistics," January 2018, http://www.nws.noaa.gov/om/hazstats.shtml.

[6] H. E. Brooks, G. W. Carbin, and P. T. Marsh, "Increased variability of tornado occurrence in the United States," *Science*, vol. 346, no. 6207, pp. 349–352, 2014.

[7] M. K. Tippett, "Changing volatility of U.S. annual tornado reports," *Geophysical Research Letters*, vol. 41, no. 19, pp. 6956–6961, 2014.

[8] L. Guo, K. Wang, and H. B. Bluestein, "Variability of tornado occurrence over the continental United States since 1950," *Journal of Geophysical Research: Atmospheres*, vol. 121, no. 12, pp. 6943–6953, 2016.

[9] T. W. Moore, "On the temporal and spatial characteristics of tornado days in the United States," *Atmospheric Research*, vol. 184, pp. 56–65, 2017.

[10] T. W. Moore, "Annual and season tornado trends in the contiguous United States and its regions," *International Journal of Climatology*, vol. 38, no. 3, pp. 1582–1594, 2017.

[11] M. K. Tippett, A. H. Sobel, and S. J. Camargo, "Association of U.S. tornado occurrence with monthly environmental parameters," *Geophysical Research Letters*, vol. 39, no. 2, article L02801, 2012.

[12] J. T. Allen, M. K. Tippett, and A. H. Sobel, "Influence of the El Niño/Southern Oscillation on tornado and hail frequency in the United States," *Nature Geoscience*, vol. 8, no. 4, pp. 278–283, 2015.

[13] C. Lepore, M. K. Tippett, and J. T. Allen, "ENSO-based probabilistic forecasts of March–May U.S. tornado and hail activity," *Geophysical Research Letters*, vol. 44, no. 17, pp. 9093–9101, 2017.

[14] A. R. Cook, L. M. Leslie, D. B. Parsons et al., "The impact of El Niño-Southern Oscillation (ENSO) on winter and early spring U.S. tornado outbreaks," *Journal of Applied Meteorology and Climatology*, vol. 56, no. 9, pp. 2455–2478, 2017.

[15] W. Monfredo, "Relationships between phases of the El Niño-Southern Oscillation and character of the tornado season in the south-central United States," *Physical Geography*, vol. 20, pp. 413–421, 1999.

[16] A. R. Cook and J. T. Schaefer, "The relation of El Niño-Southern Oscillation (ENSO) to winter tornado activity," *Monthly Weather Review*, vol. 136, no. 8, pp. 3121–3137, 2008.

[17] S.-K. Lee, R. Atlas, D. B. Enfield, C. Wang, and L. Hailong, "Is there an optimal ENSO pattern that enhances large-scale atmospheric processes conducive to major tornado outbreaks in the US?," *Journal of Climate*, vol. 26, no. 5, pp. 1626–1642, 2013.

[18] J. B. Elsner, T. H. Jagger, and T. Fricker, "Statistical models for tornado climatology: long and short-term views," *PLoS One*, vol. 11, no. 11, Article ID e0166895, 2016.

[19] S.-K. Lee, A. T. Wittenberg, D. B. Enfield, S. J. Weaver, C. Wang, and R. Atlas, "US regional tornado outbreaks and their links to spring ENSO phases and North Atlantic SST variability," *Environmental Research Letters*, vol. 11, no. 4, article 044008, 2016.

[20] D. B. Thompson and P. E. Roundy, "The relationship between the Madden-Julian Oscillation and US violent tornado

outbreaks in the spring," *Monthly Weather Review*, vol. 141, no. 6, pp. 2087–2095, 2013.

[21] V. A. Gensini and A. Marinaro, "Tornado frequency in the United States related to global relative angular momentum," *Monthly Weather Review*, vol. 144, no. 2, pp. 801–810, 2016.

[22] T. W. Moore, "Annual and seasonal tornado activity in the United States and the global wind oscillation," *Climate Dynamics*, vol. 50, no. 11-12, pp. 4323–4334, 2017.

[23] B. S. Barrett and V. A. Gensini, "Variability of central United States April–May tornado day likelihood by phase of the Madden-Julian Oscillation," *Geophysical Research Letters*, vol. 40, no. 11, pp. 2790–2795, 2013.

[24] Storm Prediction Center, "Severe weather database files (1950–2016)," January 2018, http://www.spc.noaa.gov/wcm/.

[25] S. M. Verbout, H. E. Brooks, L. M. Leslie, and S. M. Schultz, "Evolution of the U.S. tornado database: 1954–2003," *Weather and Forecasting*, vol. 21, no. 1, pp. 86–93, 2006.

[26] K. E. Kunkel, T. R. Karl, H. Brooks et al., "Monitoring and understanding trends in extreme storms," *Bulletin of the American Meteorological Society*, vol. 94, no. 4, pp. 499–514, 2013.

[27] F. Mesinger, G. DiMego, E. Kalnay et al., "North American Regional Reanalysis," *Bulletin of the American Meteorological Society*, vol. 87, no. 3, pp. 343–360, 2006.

[28] Climate Prediction Center, "Cold & warm episodes by season," January 2018, http://www.cpc.ncep.noaa.gov/products/analysis_monitoring/ensostuff/ONI_v5.php.

[29] R. A. Madden and P. R. Julian, "Description of global-scale circulation cells in the tropics with a 40–50 day period," *Journal of Atmospheric Sciences*, vol. 29, no. 6, pp. 1109–1123, 1972.

[30] K. M. Weickmann, S. Khalsa, and J. Eischeid, "The atmospheric angular momentum cycle during the tropical Madden-Julian Oscillation," *Monthly Weather Review*, vol. 120, no. 10, pp. 2252–2263, 1992.

[31] K. M. Weickmann and E. Berry, "A synoptic-dynamic model of sub-seasonal atmospheric variability," *Monthly Weather Review*, vol. 135, no. 2, pp. 449–474, 2007.

[32] K. M. Weickmann and E. Berry, "The tropical Madden-Julian oscillation and the global wind oscillation," *Monthly Weather Review*, vol. 137, no. 5, pp. 1601–1614, 2009.

[33] Earth System Research Laboratory, "GWO dataset," February 2018, https://www.esrl.noaa.gov/psd/map/clim/gwo.data.txt.

[34] M. Wheeler and H. H. Hendon, "An all-season real-time multivariate MJO index: development of an index for monitoring and prediction," *Monthly Weather Review*, vol. 132, no. 8, pp. 1917–1932, 2004.

[35] Australian Bureau of Meteorology, "Madden-julian oscillation (MJO)," February 2018, http://www.bom.gov.au/climate/mjo/.

[36] V. A. Gensini, T. L. Mote, and H. E. Brooks, "Severe-thunderstorm reanalysis environments and collocated radiosonde observations," *Journal of Applied Meteorology and Climatology*, vol. 53, no. 3, pp. 742–751, 2014.

[37] National Oceanic and Atmospheric Administration, "Daily average NCEP NARR composites," February 2018, https://www.esrl.noaa.gov/psd/cgi-bin/data/narr/plotday.pl/.

[38] National Oceanic and Atmospheric Administration, "NARR monthly/seasonal climate composites," February 2018, https://www.esrl.noaa.gov/psd/cgi-bin/data/narr/plotmonth.pl.

[39] ESRI, *ArcMAP Desktop, Release 10.5.1*, Environmental Systems Research Institute, Redlands, CA, USA, 2017.

[40] J. O. Dickey, S. L. Marcus, and T. M. Chin, "Thermal wind forcing and atmospheric angular momentum: origin of the Earth's delayed response to ENSO," *Geophysical Research Letters*, vol. 34, no. 17, article L17803, 2007.

[41] V. A. Gensini and J. T. Allen, "U.S. hail frequency and the global wind oscillation," *Geophysical Research Letters*, vol. 45, no. 3, pp. 1611–1620, 2018.

4

Historical Spatiotemporal Trends in Snowfall Extremes over the Canadian Domain of the Great Lakes Basin

Janine A. Baijnath-Rodino and Claude R. Duguay

Department of Geography and Environmental Management and Interdisciplinary Centre on Climate Change, University of Waterloo, 200 University Avenue West, Waterloo, Ontario, Canada N2L 3G1,

Correspondence should be addressed to Janine A. Baijnath-Rodino; janineannb@gmail.com

Academic Editor: Stefano Dietrich

The Laurentian Great Lakes Basin (GLB) is prone to snowfall events developed from extratropical cyclones or lake-effect processes. Monitoring extreme snowfall trends in response to climate change is essential for sustainability and adaptation studies because climate change could significantly influence variability in precipitation during the 21st century. Many studies investigating snowfall within the GLB have focused on specific case study events with apparent under examinations of regional extreme snowfall trends. The current research explores the historical extremes in snowfall by assessing the intensity, frequency, and duration of snowfall within Ontario's GLB. Spatiotemporal snowfall and precipitation trends are computed for the 1980 to 2015 period using Daymet (Version 3) monthly gridded interpolated datasets from the Oak Ridge National Laboratory. Results show that extreme snowfall intensity, frequency, and duration have significantly decreased, at the 90% confidence level, more so for the Canadian leeward shores of Lake Superior than that of Lake Huron, for the months of December and January. To help discern the spatiotemporal trends is snowfall extremes, several trend analyses for lake-induced predictor variables were analysed for two cities, Wawa and Wiarton, along the snowbelts of Lakes Superior and Huron, respectively. These variables include monthly maximum and minimum air temperature, maximum wind gust velocity, lake surface temperature, and maximum annual ice cover concentration. Resultant significant increase in December's maximum and minimum air temperature for the city of Wawa may be a potential reason for the decreased extreme snowfall trends.

1. Introduction

During the winter season, heavy snowfall is a prominent meteorological phenomenon in the Great Lakes Basin (GLB) and is derived from either extratropical cyclones or lake-induced snowfall processes. Extratropical cyclones are generated by quasigeostrophic forcing from positive temperature or vorticity advection [1] and are associated with a low-pressure centre tracking zonally following the jet stream. Thus, frequent wintertime extratropical storms, such as the Alberta Clipper, Colorado Low, and Nor'easter, track from west to east, affecting surface-atmosphere conditions within the GLB region. Contrary to these large-scale synoptic systems are shallow meso-beta scale lake-effect snowfall (LES). LES ranges from approximately 5 km to a few hundred km in length and is triggered by turbulent energy and moisture fluxes off lakes [1].

The shear spatial extent and geographic location of the Laurentian Great Lakes make LES a prominent meteorological phenomenon during the late autumn and winter months [1]. LES forms when boundary layer convection is initiated as a result of a cold and dry continental air mass advecting over the relatively warm lake, generating turbulent moisture and heat fluxes that destabilize the lower part of the planetary boundary layer (PBL). The increase of moisture into the lower atmosphere enhances cloud formation and precipitation along the leeward shores of the Great Lakes [1–12]. The lake-effect vertical atmospheric profile often features a moist-neutral or unstable convective boundary layer that extends 1 to 4 km above lake surfaces, where a capping stable layer or inversion limits the vertical extent of convection [12–16]. In addition to LES, lake-enhanced snowfall can occur when synoptic scale systems, driven by quasigeostrophic forcing, move over the GLB. The synoptic system can increase the altitude of the capping inversion, producing deeper cloud convection, and as a result, increase snowfall. Thus, the term lake-induced snowfall is used to describe both lake-effect and lake-enhanced snowfall processes, as defined by [17]. Lake-induced snowfall can organize into a spectrum of LES morphologies ranging from discrete and disorganized cells to organized mesoscale bands [16] and include widespread coverage, shoreline bands, midlake bands, and mesoscale vortices [1, 11, 18–20]. Convective bands are typically long and narrow, between 50 and 300 km in length and 5 to 20 km in width [1, 10], and can produce highly localized snowfall in cities that are along the snowbelts of the GLB.

The highly populated GLB is home to over 33 million people including 90% of Ontario's population. Over 1.5 million people reside along the Canadian shores of Lake Huron and 200,000 along Lake Superior's [21]. The Laurentian Great Lakes support 40% of Canada's economic activities, 25% of its agricultural capacity, and 45% of its industrial capacity. These lakes also support approximately 400 million dollars in cumulative recreational and commercial fishing industries and 180 billion dollars in Canada and US trade [22]. Thus, heavy lake-induced snowfall in the GLB can impact local residents and numerous industries [23–27]. The GLB's provision on the surrounding communities and its climatologically favourable snowfall location gives credence to assessing the historical spatiotemporal trends in snowfall extremes within the context of climate change.

According to the Intergovernmental Panel on Climate Change, Fifth Assessment Report, (IPCC AR5), climate change, driven by either natural or human forcings can lead to changes in the occurrence or strength of extreme weather and climate events such as extreme precipitation [28]. Therefore, assessing and monitoring spatiotemporal trends in snowfall extremes to a historically changing climate can provide knowledge as to future behavioural spatiotemporal trends in snowfall extremes within the densely populated Canadian GLB domain and can be useful for sustainability and adaptation studies.

However, only a few lake-induced snowfall studies focused on the Canadian GLB while many studies, including [25, 27, 29–32], concentrated on the United States snowbelts. Furthermore, underexaminations of historically observed LES extremes compared to LES case studies are apparent. The majority of LES investigations have analysed specific LES events and associated features as opposed to LES climatology, for example, [9, 10, 33–42].

Several studies have examined trends in snowfall patterns across the GLB over the 20th century and found a snowfall trend reversal, in which there was an increase in snowfall prior to the 1980s, followed by a decrease in snowfall over the past few decades. A study by [43] investigated snowfall trends over North America from 1948 to 2001 and observed a snowfall increase over a narrow band from Colorado to the lee of Lake Erie and Lake Ontario. Furthermore, [44] found that the area-averaged snowfall across Northern Canada (55°N to 88°N) significantly increased at a rate of 8.8 cm/decade in the late 20th century. However, in Southern Canada (below 55°N), a negative trend of 0.65 cm/decade during this period was discovered, with the most decrease occurring in the 1980s [45]. Study [17] suggests that while some studies have shown a general increase in snowfall over the GLB [8, 23, 25, 29, 46, 47], recent studies have seen a decline in LES through the later half of the 20th century and early 21st century. For example, [48] showed a decrease in snowfall along the leeward shores of Lake Michigan between 1980 and 2005 and [32] showed a decrease in Central New York between 1971 and 2012. While these studies outline trends in North American snowfall, there is still a lack in the examination of climatological snowfall extremes over the GLB. Little is still known about the physical processes influencing the past changes in daily snowfall extremes on the global and hemispheric scales. On a regional scale, observational studies show large interdecadal variations in snowfall extreme measures; however, long-term trends remain unclear [49–51].

The objective of this study was to assess historical (1980–2015) spatiotemporal trends in snowfall extremes for the Canadian snowbelt zones of Lake Superior and Lake Huron-Georgian Bay (Figure 1). This will be conducted by examining snowfall intensity, frequency, and duration and provide potential explanations of the results in the context of lake-induced predictor variables, including monthly extreme maximum and minimum air temperature, maximum wind gust velocity, lake surface temperature (LST), and annual maximum ice cover concentration. The study is divided into two folds: the first will assess the statistical extremes of snowfall intensity, frequency, and duration over the 36-year period, while the second will examine the trends in these extremes over the given duration. The term "snowfall extremes" will be used when referring to all extremes in intensity, frequency, and duration, unless otherwise specified. In this paper, Section 1 presents the introduction that provides a background on the development of snowfall within the GLB region; Section 2 outlines the data and methodology; Section 3 presents the results; Section 4 provides discussions and analyses on the results; and Section 5 summarizes the study and suggests future potential research avenues.

2. Data and Methods

2.1. Datasets. The datasets used in this study include gridded interpolated precipitation and temperature data from Daymet. Predictor lake-enhanced meteorological variables, such as temperature and winds, are from weather observation stations for the cities of Wawa and Wiarton Ontario and are provided by Environment and Climate Change Canada (ECCC). In addition, the National Oceanic and Atmospheric Administration (NOAA) provides lake ice concentration and lake surface temperature (LST) datasets. Table 1 lists the datasets and sources used in this study.

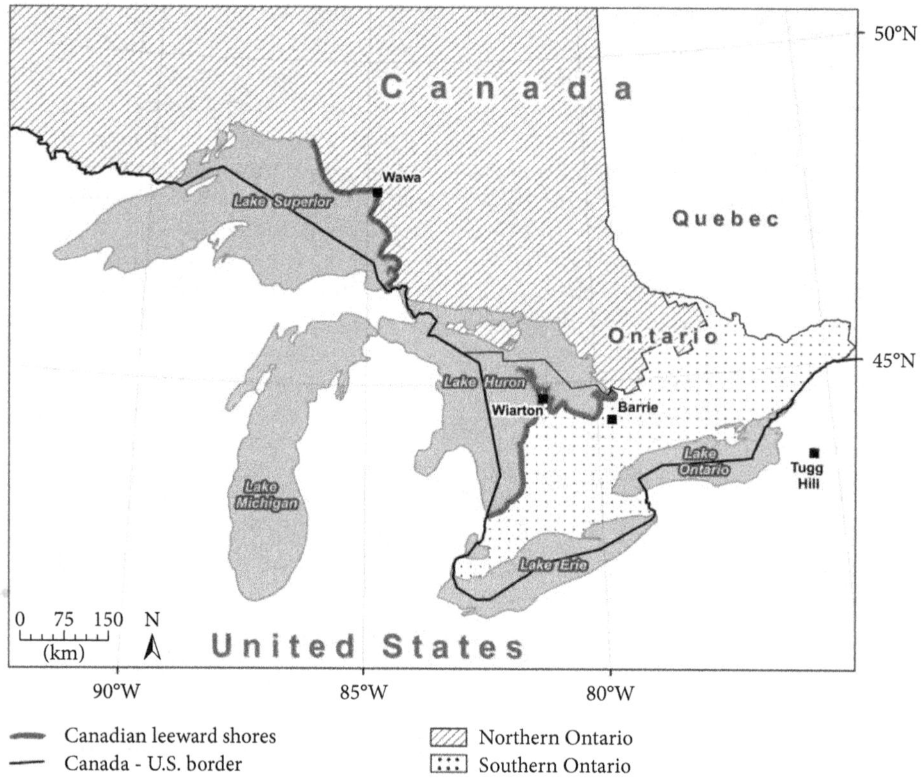

Figure 1: Map of the Laurentian Great Lakes, depicting the geographic boundaries and Canadian leeward shores (snowbelts) of Lakes Superior and Huron analysed in this study.

Table 1: List of datasets with associated variables, sources, and temporal availability used in this study.

Dataset	Monthly temporal availability	Variables	Source
Daymet version 3	1980–2015	(i) Precipitation	https://daac.ornl.gov/DAYMET/guides/Daymet_V3_CFMosaics.html#revisions
NOAA coast watch	1995–2015	(i) LST	https://coastwatch.glerl.noaa.gov
NOAA Great Lakes ice atlas	1980–2015	(i) Maximum ice cover concentration	https://www.glerl.noaa.gov/data/ice/atlas/daily_ice_cover/daily_averages/dailyave.html https://www.glerl.noaa.gov/pubs/tech_reports/glerl-135/Appendix2/DailyLakeAverages/
Historical weather station observations	1980–2014	(i) Maximum and minimum temperature (ii) Maximum wind gust velocity	http://climate.weather.gc.ca/historical_data/search_historic_data_e.html

2.1.1. Daymet. Daymet (Version 3) provides meteorological data used in this study and constitutes a collection of software and algorithms that produce gridded estimates of daily weather observations, including daily maximum and minimum temperature and precipitation. Daymet inputs contain land mask and the North American subset of the National Aeronautics and Space Administration (NASA) Shuttle Radar Topography Mission (SRTM) near-global 30-arc second digital elevation model, Version 2.1. It also includes spatially referenced ground-based observations obtained from NOAA's National Centers for Environmental Information's Global Historical Climatology Network (GHCN), Version 3.22 [52, 53].

An interpolation method at each prediction point employs an iterative estimation of location density that uses spatial convolution of a truncated Gaussian filter to derive 1 km × 1 km grids of each meteorological parameter. In the algorithm, the search radius of stations is reduced in data-rich regions and increased in data-poor regions. The dataset was developed from the Environmental Sciences Division at the Oak Ridge National Laboratory [54, 55]. A detailed description of this dataset can be found at http://daymet.ornl.gov. Daily minimum and maximum temperature and precipitation over North America, for the years 1980 to 2015, were obtained online at http://daac.ornl.gov/cgi-bin/dsviewer.pl?ds_id=1328 and used to derive gridded daily snowfall.

It is also acknowledged that precipitation and temperature variables from Daymet V3 are the estimated values. Extensive cross-validation statistics for Daymet V3 data were computed by station-based daily observations and predictions over 2° × 2° tiles for North America [55]. The tiles provide information on the period-of-record mean absolute error (MAE) and bias statistics for input weather observations of maximum and minimum temperature and precipitation for each year, 1980–2015. A detailed validation summary can be found at https://daac.ornl.gov/DAYMET/guides/Daymet_V3_CrossVal.html#revisions. Daymet data are ideal for this study after considering other dataset options, such as the North American Regional Reanalysis (NARR). While NARR offers 3 hourly precipitation products, it comes at the cost of a coarser spatial resolution of 32 km. The 1 km daily Daymet products provide a high spatial and temporal resolution for producing and delineating highly localized snowsquall bands over the Ontario snowbelts.

2.1.2. Environment and Climate Change Canada Information Archive. Meteorological datasets for the cities of Wiarton and Wawa are provided by ECCC. The geographic coordinates and associated snowbelts are shown in Table 2. The observational data from ECCC's historical archive data can be found at the National Climate Data and Information Archive, http://climate.weather.gc.ca. Historical ground-based weather observation stations from ECCC provide maximum and minimum air temperature and are defined as the highest and lowest observed air temperature at the specific location. The direction of maximum wind gust for each month is also attained and defined as the direction of maximum wind gust from which the wind blows, with the value of 360° indicating winds advecting from true north. Values are only reported for gust speeds that exceed 29 km/h. The speed of maximum wind gust is also analysed for both sites. The gust is defined as the peak instantaneous or single reading from the weather station anemometer. The elapsed time corresponding to the duration of the gust is typically between three to five seconds. Apart from the atmospheric meteorological observational variable, lake variables are also analysed.

2.1.3. National Oceanic and Atmospheric Administration Ice Atlas and Coast Watch. NOAA provides lake-wide annual maximum ice cover concentration and monthly average LST for Lakes Superior and Huron. LST datasets can be found at https://coastwatch.glerl.noaa.gov. Ice chart data are a blend of observations from ships, shore stations, aircraft, and satellites to estimate ice cover data for the entire Great Lakes. The ice charts were digitized and made available at https://www.glerl.noaa.gov/data/ice/atlas/index.html.

2.2. Snowfall Derivation. Snowfall is derived from daily precipitation, P_1 (mm), and daily midpoint 2 m air temperature, T_{mid} (°C), based on the method used by [56]. Because Daymet does not provide daily mean 2 m air temperature, the midpoint value is used and calculated by taking the sum of the maximum and minimum daily temperature and dividing by 2, as shown in the following equation:

Table 2: Cities analysed in this study with corresponding geographic coordinates and Ontario snowbelt.

City	Weather station coordinates	Associated snowbelt
Wawa, Ontario	47.9, −84.78	Superior
Wiarton, Ontario	44.75, −81.11	Huron

$$T_{\mathrm{mid}} = \frac{T_{\max} + T_{\min}}{2}. \quad (1)$$

The T_{mid} is used to estimate the ratio (R), which denotes the daily precipitation falling as snowfall to that of the total daily precipitation, where $0 \leq R \leq 1$, equation (2) [57]. The following function was derived using a logistic curve and fitted to monthly data with a reported absolute deviation of 0.06 mm per month [58]:

$$R = \left[1.0 + 1.61 \cdot (1.35)^{T_{\mathrm{mid}}}\right]^{-1}. \quad (2)$$

Multiplying P_1 and R yields the water equivalent of the new daily snowfall, P_{n} (mm), as shown in the following equation:

$$P_{\mathrm{n}} = P_1 \cdot (R). \quad (3)$$

A density value is then computed as a function of T_{mid} (4). It is noted that when $T_{\mathrm{mid}} \leq -15$°C, the density of snowfall is assigned to 0.05 g·cm^{-3}:

$$\rho = 0.05 + 0.0017 \cdot \left[T_{\mathrm{mid}} + 15\right]^{1.5}. \quad (4)$$

Finally, the estimated height of new snowfall, H_{n} (cm) is given by

$$H_{\mathrm{n}} = \frac{0.1 \cdot P_n}{\rho}. \quad (5)$$

Additional uncertainty analysis was conducted by [56] to quantify uncertainties in derived snowfall. Considering that H_{n} is a function of R, P_1, T_{mid}, and ρ, first, annual uncertainty values were calculated for each of these variables. To calculate the initial uncertainty of precipitation and T_{mid}, two 2° × 2° tiles that contained Wawa, a city within the snowbelt of Lake Superior, and Wiarton, a city within Lake Huron-Georgian Bay's snowbelt, were selected from Daymet's cross-validation data resources. Both tiles provided the MAE of precipitation, which were converted to the error of standard deviation and used as the precipitation uncertainty. Furthermore, annual uncertainty measures of T_{mid} were calculated by computing the annual standard deviation between both Daymet and an independent dataset, NARR, for the cities of Wawa and Wiarton. The uncertainty estimate of density was given as 0.005, 10% of its initial value. The calculated annual uncertainties of R, P_1, T_{mid}, and ρ were then inputted into the derivative of the above equations to determine the annual sensitivity of H_{n} to ρ, H_{n} to P_1, R to T_{mid}, ρ to T_{mid}, H_{n} to R, H_{n} to T_{mid} (if $T_{\mathrm{mid}} \geq -15$°C), and H_{n} to T_{mid} (if $T_{\mathrm{mid}} < -15$°C).

Uncertainty analysis shows that the overall 36-year annual uncertainties in H_n relative to T_{mid} are 1.7 cm for Wiarton and 6.2 cm for Wawa. The sensitivity of H_n to P_1 is 0.03 cm for Wiarton and 0.12 cm for Wawa. The sensitivity of snowfall to R averages 1.4 cm for both Wiarton and Wawa. It is expected that the tile containing Wawa would have higher snowfall uncertainties than that of Wiarton because it is farther north and contains less numbers of weather observation stations. Furthermore, when predicting snowfall accumulation, meteorologists usually provide a 5 cm predictive range in snowfall, for example, 1–5 cm or 10–15 cm of snow. Thus, an uncertainty value of 1.7 cm for Wiarton and 6.2 cm for Wawa are within a reasonable uncertainty. These uncertainties are fairly low and provide confidence in the derived snowfall method used in this study. The derived gridded daily snowfall is computed over the GLB and is bounded by the coordinates 95° W, 75° W, 40° N, and 45° N. Over this region, the extremes in snowfall (intensity, frequency, and duration) for each month (November, December, January, February, and March) over the 36-year duration (1980 to 2015) are computed.

2.3. Defining Snowfall Extremes. Snowfall extreme analyses are evaluated in two folds: first by exploring the monthly spatiotemporal extremes in snowfall intensity, frequency, and duration over the 36-year period (from here on referred to as the 36-year extreme) and secondly by determining the trends in these extremes. Firstly, the 36-year extremes are explained. Snowfall intensity is defined as the rate of snowfall over time. The 99th percentile of daily snowfall accumulation is computed over all time steps, that is, all days between 1980 and 2015 for each individual month. The 99th percentile over the 36-year period is then spatially mapped to determine the extreme values of snowfall intensity and its location for each month of the cold season (November, December, January, February, and March). Based on each month's extreme intensity snowfall, a snowfall midpoint value is chosen that best represent the extreme snowfall threshold intensity (S_threshold) for the overall cold months.

The S_threshold value is then used to evaluate the 36-year extreme in snowfall frequency and duration. Frequency is defined as the number of snowfall days over a particular time period. In this study, the extreme snowfall frequency is determined by calculating the number of snowfall days that the given S_threshold is met or exceeded for each month over all time steps of the 36-year period. Monthly maps are generated to determine regions within the GLB of high and low extreme snowfall frequencies during the 36-year period. Snowfall duration is described as the length of time it snows at a particular intensity. The 36-year duration extreme is presented by mapping the maximum number of consecutive snowfall days, for which events equaled or exceeded the acquired S_threshold value over this time period. This procedure is repeated for each month of the cold season to determine spatial patterns in extreme snowfall duration throughout the season.

Secondly, the trends in 36-year extremes are calculated. Trends in extreme intensity are computed by applying a filter to extract daily snowfall events greater than or equal to the S_threshold value for each month and for each year. The filtered values are aggregated and then divided by the total number of snowfall days in the month, yielding the monthly average snowfall intensity. Note a snowfall day is defined as a full 24-hour day, for which snowfall exceeds 0 cm of snow. The procedure is repeated for each of the 36 years. The Mann–Kendall (MK) test is then applied to this time series to compute the trend in monthly extreme snowfall intensity. The trend is computed for each month of the cold season, and monthly spatiotemporal extreme intensity trends are produced. Trends in snowfall frequency are computed by counting the number of monthly extreme events (filtered values) for each year. The MK test is then applied over this time series to evaluate the spatiotemporal trends in monthly snowfall extremes. The trend in extreme snowfall duration is calculated by computing the MK test over each monthly time series. The time series comprises the monthly maximum number of consecutive snowfall days, for which events equaled or exceeded the S_threshold.

The MK test is applied to the extreme snowfall time series to determine whether there are monotonic upward or downward trends in intensity, frequency, and duration for each grid cell over the 36-year period. If grid cells do not have a complete 36-year time series of extreme values, then no trend analysis is computed for that grid cell, and it is assigned "not a number" and shaded gray. The MK test evaluates the slope of an estimated linear regression line nonparametrically and does not require the residual from a fitted regression line to be normally distributed, like that of a parametric linear regression [59, 60]. In this study, slope is calculated over the 36-year period for each grid cell in order to generate spatiotemporal trend maps. For the purpose of this study, the significance of the extreme snowfall trends is calculated at the 90% confidence level. This is because a higher confidence level may not be able to capture the highly episodic and disorganized spatial patterns of lake-induced trends seen in the results. The predictor variable trends are also estimated using the MK approach for the cities of Wawa and Wiarton Ontario.

3. Results

3.1. Snowfall Intensity Extreme Values. A predefined extreme intensity snowfall value is required in order to conduct extreme snowfall analyses. For this study, the S_threshold value is computed by calculating the 99th percentile of daily snowfall for each month over the 36-year period (Figures 2(a)–2(e)). Within the Canadian GLB domain, the leeward shores of Lake Superior and Lake Huron-Georgian Bay show the greatest value in snowfall intensity for all months. Along these shores, the lower snowfall intensity values range between 5 cm/day and 15 cm/day for the months of November and March (Figures 2(a) and 2(e)) and the higher values range between 15 cm/day and 30 cm/day for December and January (Figures 2(b) and 2(c)).

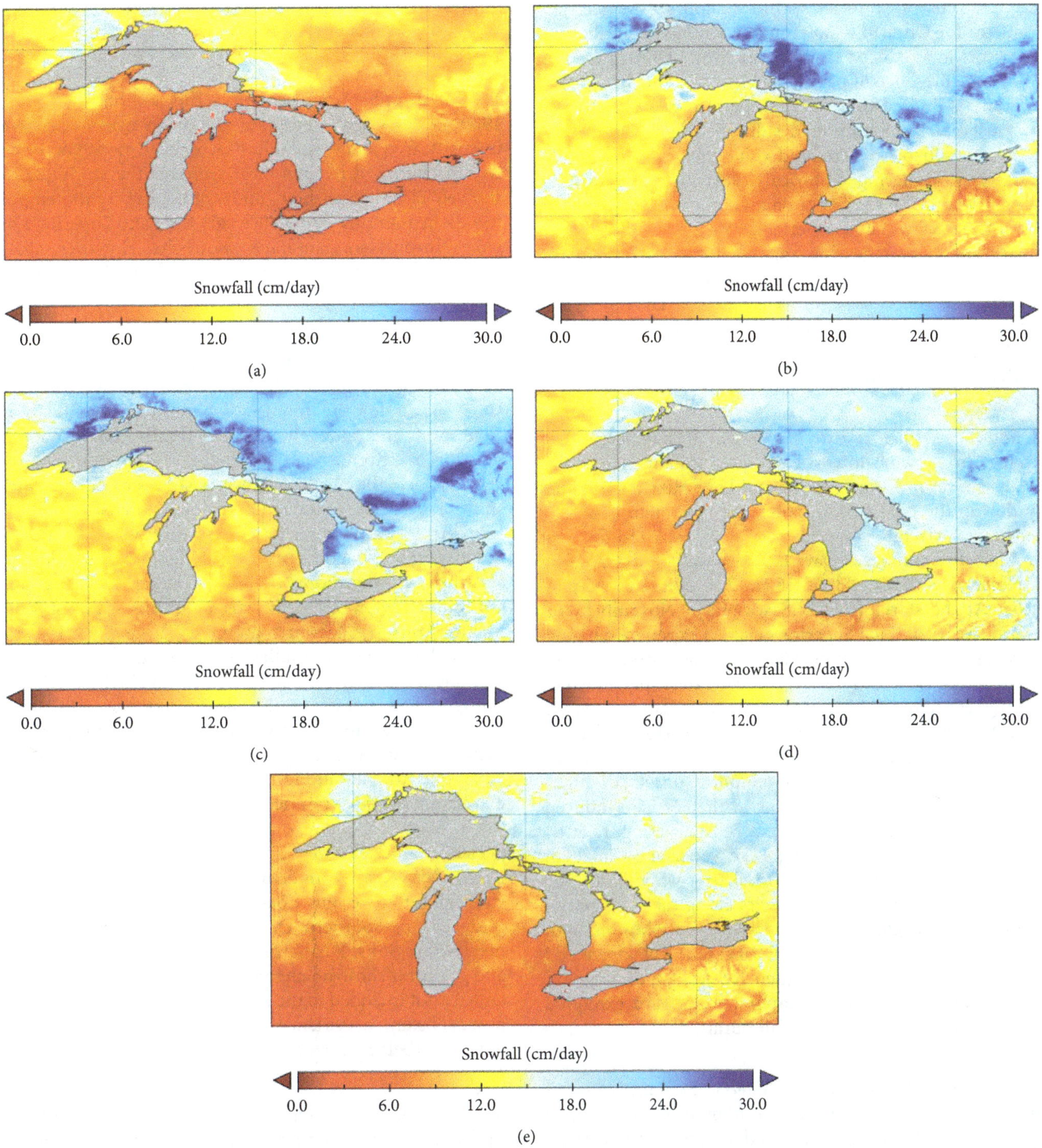

FIGURE 2: Extreme snowfall intensity for each month of the cold season: (a) November, (b) December, (c) January, (d) February, and (e) March, defined as the 99th percentile of daily snowfall between 1980 and 2015.

The S_threshold value of 15 cm/day of snowfall is adopted to evaluate extreme snowfall over the 36-year study period. It is noted that preliminary work also investigated smaller threshold values (5 cm/day and 10 cm/day) to determine whether there were any month-to-month spatiotemporal variations. However, results indicated similar spatiotemporal trends to that of 15 cm/day, but with smaller magnitudes of the extreme values. Due to these findings, only the 15 cm/day S_threshold values are plotted to reduce repetitiveness in the results.

Furthermore, the 15 cm/day S_threshold is ideal because it is the midpoint value among the cold months being analysed. This S_threshold is also in agreement with Environment and Climate Change Canada's warning criteria for Ontario. Environment Canada issues a snowfall warning for Ontario when the alerting snowfall parameter reaches or exceeds 15 cm of snowfall within 12 hours. The alerting parameter for Ontario's snowsquall warnings is similar but also factors in a reduced visibility criterion of 400 m with or without the presence of persistent blowing snow [61]. Since this study is limited to a daily temporal resolution, 15 cm/day is used as the S_threshold value. Furthermore, 15 cm/day is a reasonable S_threshold value. For example, when separately considering historical 1981 to 2015 monthly snowfall average for the city of Barrie, which lies within Lake Huron's snowbelt, Barrie experiences its highest snowfall in January with an average of 66 cm. This suggests an average of approximately only 2 cm/day with assumed continuous daily snowfall [62]. Therefore, a snowfall rate of 15 cm/day would be considered an extreme snowfall event of high intensity. This S_threshold is now applied to evaluate extremes in snowfall frequency and duration.

3.2. Snowfall Frequency Extreme Values. Results show that extremes in snowfall intensity are predominant along the leeward shores of Lake Superior and Lake Huron-Georgian Bay during the cold season. Next, the spatiotemporal frequency in extreme snowfall events is explored by determining the number of days that snowfall events equaled or exceeded S_threshold of 15 cm/day over the 36-year period. At the start of the LES season, in November, the leeward shores of Lake Superior begin to experience higher frequencies in extreme snowfall events compared to other regions of the GLB (Figure 3(a)). The greatest spatial extent with highest frequencies upward of 50 days is seen on the leeward shores of Lake Superior and Lake Huron for the months of December (Figure 3(b)) and January (Figure 3(c)). The spatial and temporal onsets of high-frequency snowfall behave similar to that of lake-induced snowfall spatiotemporal patterns. In February, (Figure 3(d)) the frequency is still high along Lake Superior's Canadian snowbelt, but becomes highly localized, while both the spatial extent and higher frequencies along the snowbelt of Lake Huron-Georgian Bay have decreased. By March, both the high-frequency values and spatial extent have decreased for both snowbelts with frequencies less than 10 days (Figure 3(e)). Similar spatiotemporal results are exhibited when examining extreme snowfall duration.

3.3. Snowfall Duration Extreme Values. Extreme snowfall duration plots show the maximum number of consecutive days when snowfall intensity equaled or exceeded 15 cm/day for the 36-year period. As early as November, highly localized areas along the leeward shores of Lake Superior experienced high duration of extreme snowfall events upwards of 3 days, while Lake Huron-Georgian Bay's snowbelt shows no maximum snowfall durations (Figure 4(a)). In December and January, most of Northeastern Ontario experienced extreme events greater than 2 days. However, the duration of extreme snowfall events greater than 5 days was spatially confined to the snowbelt regions. By February, the duration of extreme snowfall events has spatially decreased along the Lake Huron-Georgian Bay snowbelt, but continue to stay dominant along the leeward shores of Lake Superior. By March, both snowbelts show a spatially less extreme in snowfall duration (Figure 4(e)).

3.4. Trends in Snowfall Extremes. This section evaluates the 1980 to 2015 spatiotemporal trends in extreme snowfall intensity, frequency, and duration for each month of the cold season, with S_threshold of 15 cm/day. The study [62, 63] showed clear, consistent, and contiguous spatiotemporal trends in monthly snowfall totals for Lake Superior's and Lake Huron-Georgian Bay's snowbelts. However, spatiotemporal trends in monthly mean snowfall extremes are less coherent and less obvious to delineate.

3.4.1. November. Extreme snowfall intensity, frequency, and duration experience no significant changes over the Great Lakes Basin for the month of November. Although not significant, the snowfall intensity trends can provide indication of extreme snowfall evolution, shown in Figure 5. For instance, the northern leeward shores of Lake Superior show a decrease in extreme snowfall intensity, while its southern leeward shores present an increase, indicating the inconsistency of coherent spatial distribution along this snowbelt. Furthermore, there is a strong negative trend within the region of Northeastern Ontario, which is farther inland from the leeward shores of Lake Superior. This geographic location implies that the strong decrease in extreme snowfall trend may not be due to LES but perhaps to synoptic and extratropical scale snowstorms. The spatiotemporal trends for the month of November suggest that an increase in snowfall extremes in Northern Ontario is perhaps due to lake-induced processes, but a decrease in extratropical and synoptic scale-driven snowstorms.

In Southern Ontario, most of this region does not experience daily extreme snowfall greater than 15 cm/day in November. Recall from Figure 2, the monthly mean extreme snowfall for November does not exceed 12 cm/day for most grid cells in Southern Ontario and is indicated by gray pixelated regions. Furthermore, extreme snowfall frequency and duration show no changes for the month of November and are not included in the results. December results provide more substantive evidence of trends in snowfall extremes for both snowbelt regions.

3.4.2. December. December exhibits the most coherent significant spatiotemporal trends in snowfall extremes for all months of the cold season. In December, Lake Superior's snowbelt displays two distinct trends. The southern leeward shore reveals a significant decrease in extreme snowfall intensity, while the northern leeward shore shows a highly localized significant increase (Figure 6(a)). Both the northern and southern leeward shores exhibit a significant decrease in extreme snowfall frequency (Figure 6(b)). However, the northern shore displays a weaker negative trend. The duration also significantly decreases for the southern shore, with no change in extreme snowfall duration for the northern leeward shore of Lake Superior's snowbelt (Figure 6(c)). During this

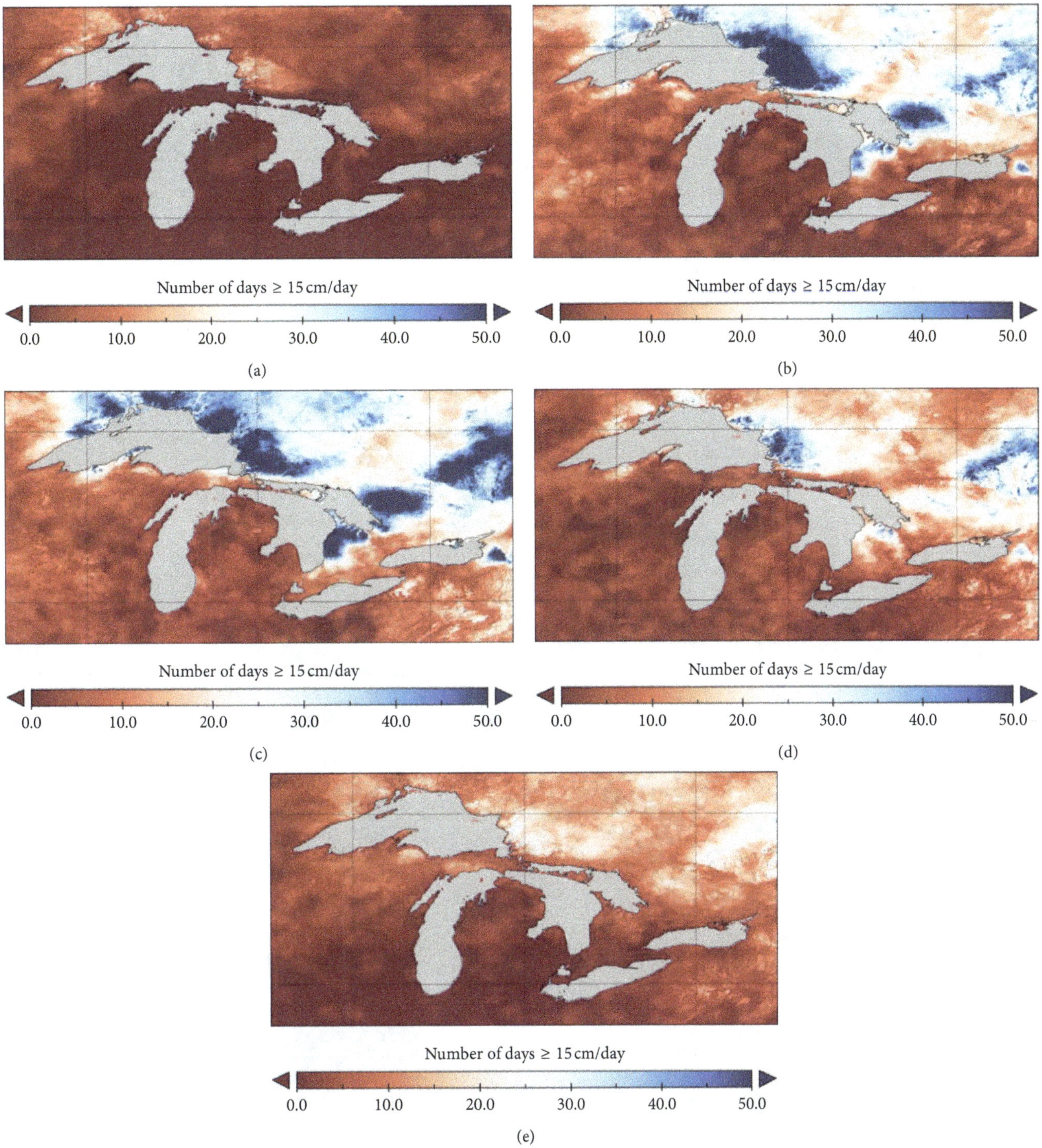

FIGURE 3: Extreme snowfall frequency for each month of the cold season: (a) November, (b) December, (c) January, (d) February, and (e) March. Extreme snowfall frequency is defined as the number of days when daily snowfall amounts equaled or exceeded S_threshold of 15 cm/day between 1980 and 2015.

month, the snowbelt of Lake Huron-Georgian Bay shows a highly localized decrease in snowfall intensity but no changes in extreme snowfall frequency and duration. It is evident that not only are spatiotemporal trends in snowfall extremes different for both snowbelts but also, within the same snowbelt, trends in snowfall extremes are not spatially coherent.

3.4.3. January. The month of January shows less spatial coherency than December for both snowbelt regions. The northern leeward shore of Superior shows a significant decrease in snowfall intensity (Figure 7(a)), frequency (Figure 7(b)), and duration (Figure 7(c)). Southern Ontario shows spatial incoherent negative trends in extreme snowfall intensity, while there are no changes in frequency and

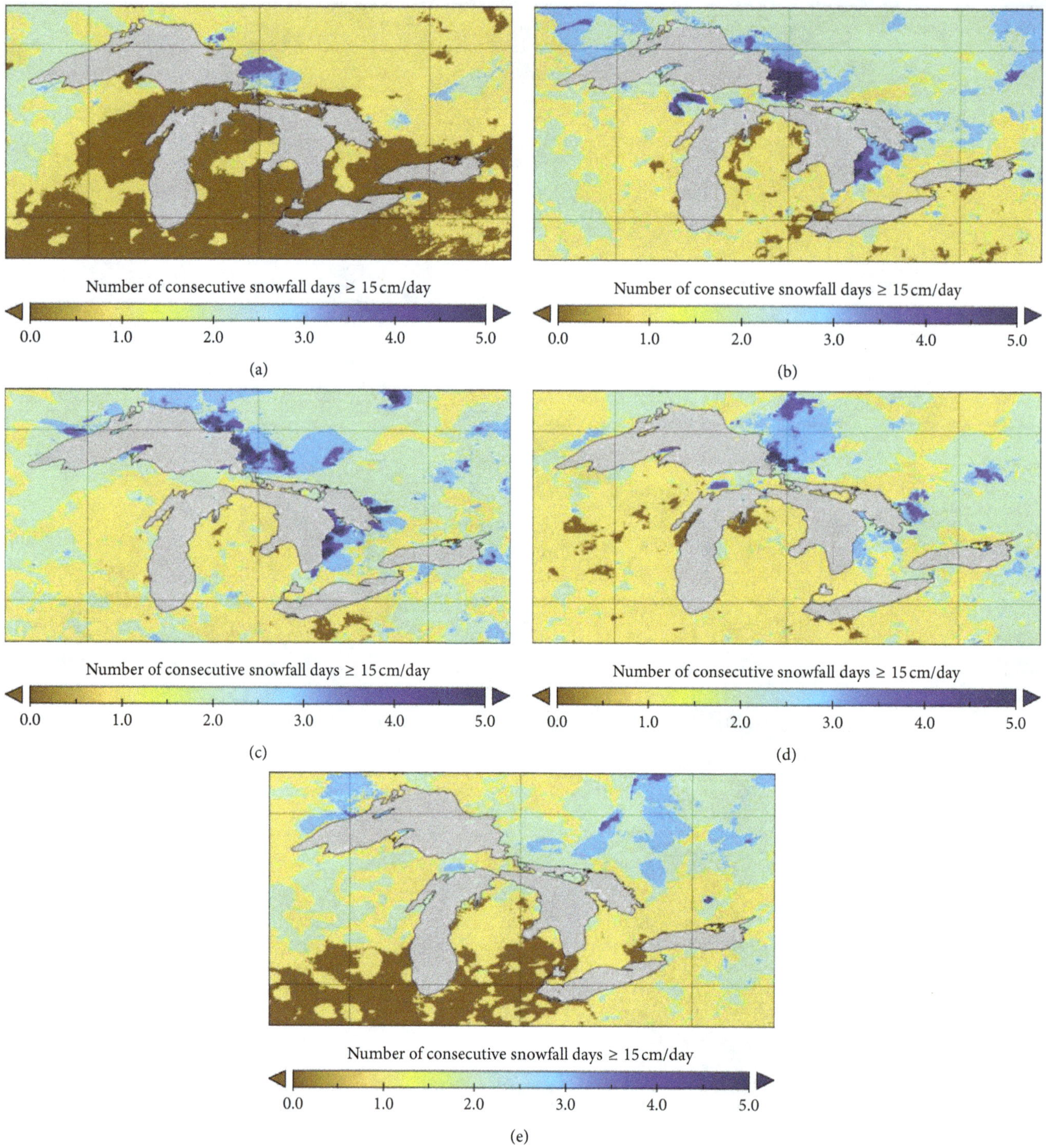

FIGURE 4: Extreme snowfall duration for each month of the cold season: (a) November, (b) December, (c) January, (d) February, and (e) March. Extreme snowfall duration is defined as the maximum number of consecutive days for which daily snowfall amounts equaled or exceeded S_threshold of 15 cm/day between 1980 and 2015.

duration. Compared to December, the spatiotemporal trends in extreme snowfall are highly localized and overall less spatially coherent.

3.4.4. February and March. The month of February shows incoherent spatial positive and negative trends in extreme snowfall intensity for both snowbelts (Figure 8(a)). Extreme snowfall frequency results show a highly localized decrease for the Superior snowbelt (Figure 8(b)), while there are no changes in extreme snowfall duration, results are not shown. There are also no changes in frequency and duration for Lake Huron-Georgian Bay's snowbelt.

March exhibits a significant decrease in extreme snowfall intensity along the leeward shores of Lake Superior and farther inland, outside of the snowbelt zone (Figure 9). The

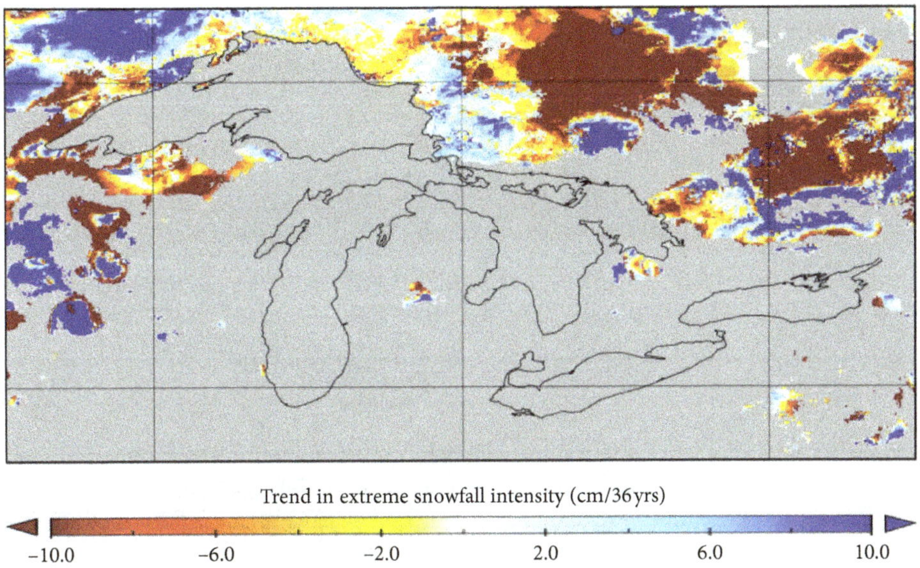

FIGURE 5: Trend in extreme snowfall intensity for the month of November. Gray pixels represent grid cells with no trend computed because the region contains some years with no extreme snowfall values.

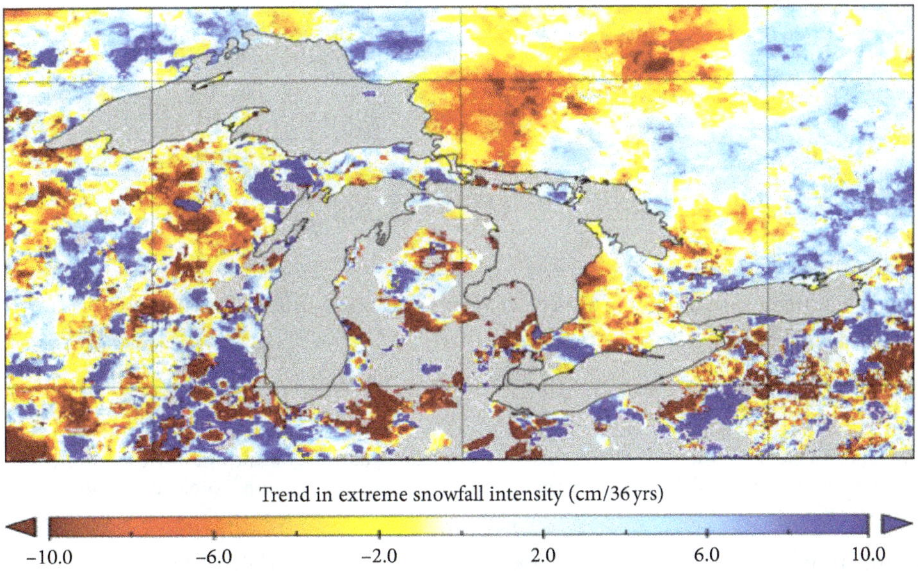

(a)

FIGURE 6: Continued.

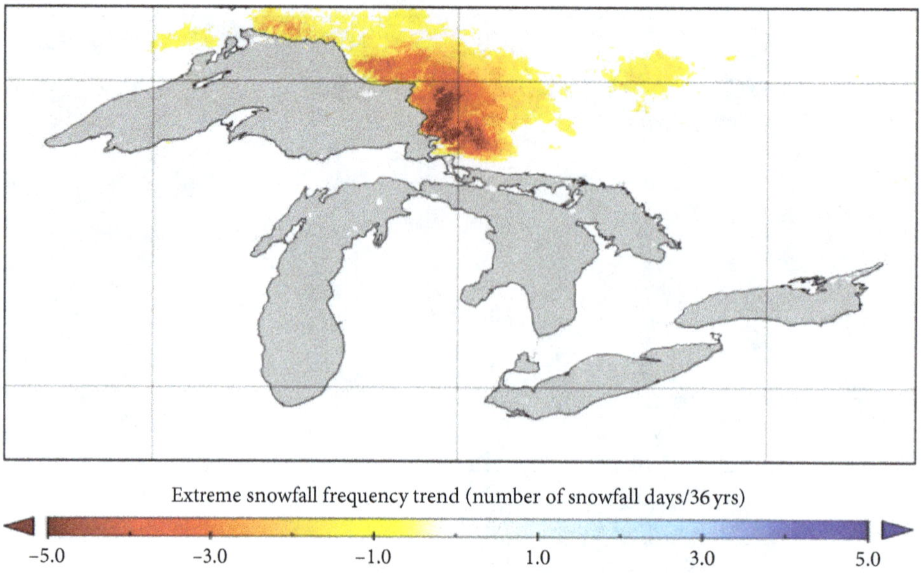

(b)

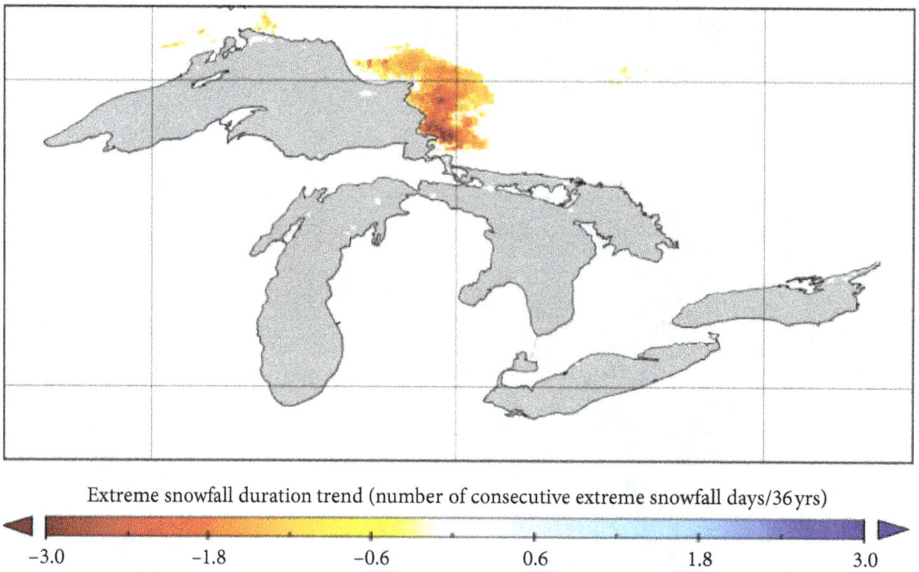

(c)

FIGURE 6: 1980 to 2015 trends in extreme snowfall (a) intensity, (b) frequency, and (c) duration, for the month of December. Gray pixels represent grid cells with no trend computed because the region contains some years with no extreme snowfall values.

geographic location and timing of this negative intensity trend suggest that the decrease in extreme snowfall intensity may not solely derive from LES but, perhaps as well from synoptic scale snowstorms. Frequency and duration in snowfall extremes exhibit no changes for this snowbelt, results are not shown. Lake Huron-Georgian Bay's snowbelt also shows negligible changes in snowfall extremes.

3.4.5. Predictor Variables. Lake-induced predictor variables are also analysed for each month of the cold season for the cities of Wawa (Table 3) and Wiarton (Table 4). Both locations and all months show a rise in maximum and minimum air temperature, with the exception of February (Wawa) and November (Wiarton). There is significant warming in maximum monthly air temperature for December (Wawa) and January (Wiarton). There is a significant warming in minimum monthly air temperature in both December and February (Wawa) and for the month of March (Wiarton).

Both cities do not show a significant change in direction of maximum wind gust. For Wawa, the trends in maximum wind gusts are positive for all months, except for March, when there is no change. For Wiarton, all trends are positive except for the month of December, which shows a negative trend. Furthermore, the speed of maximum wind gust shows no changes for the Wawa, except for a decrease occurring in November. In Wiarton, the speed of maximum wind gusts has decreased for each month of the cold season, with a significant decrease occurring in January.

LST has also warmed for Lakes Superior and Huron for all months of the cold season, with a significant increase

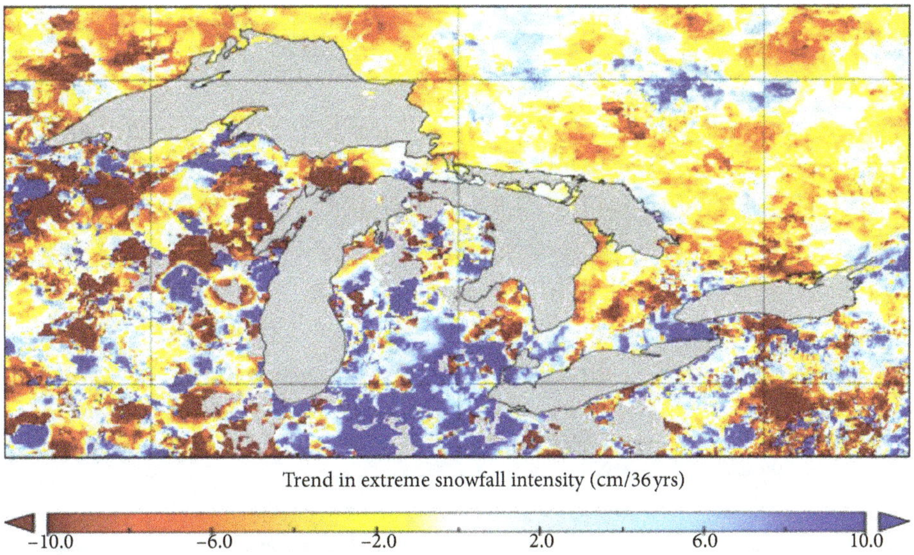

(a)

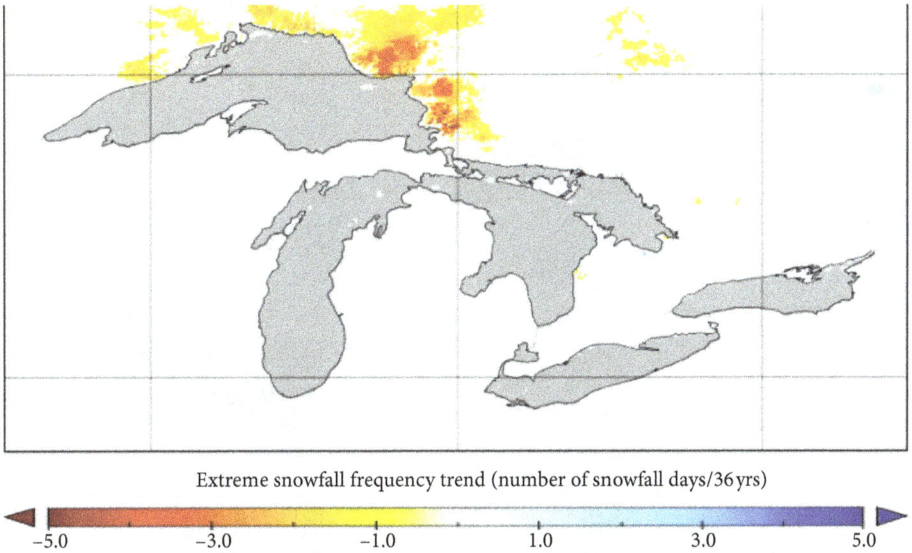

(b)

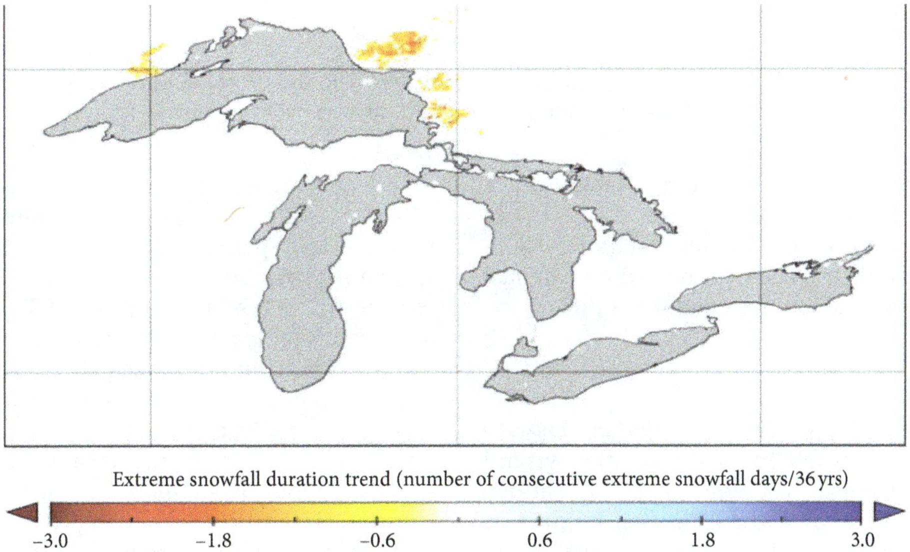

(c)

FIGURE 7: 1980 to 2015 trends in extreme snowfall (a) intensity, (b) frequency, and (c) duration, for the month of January. Gray pixels represent grid cells with no trend computed because region contains some years with no extreme snowfall values.

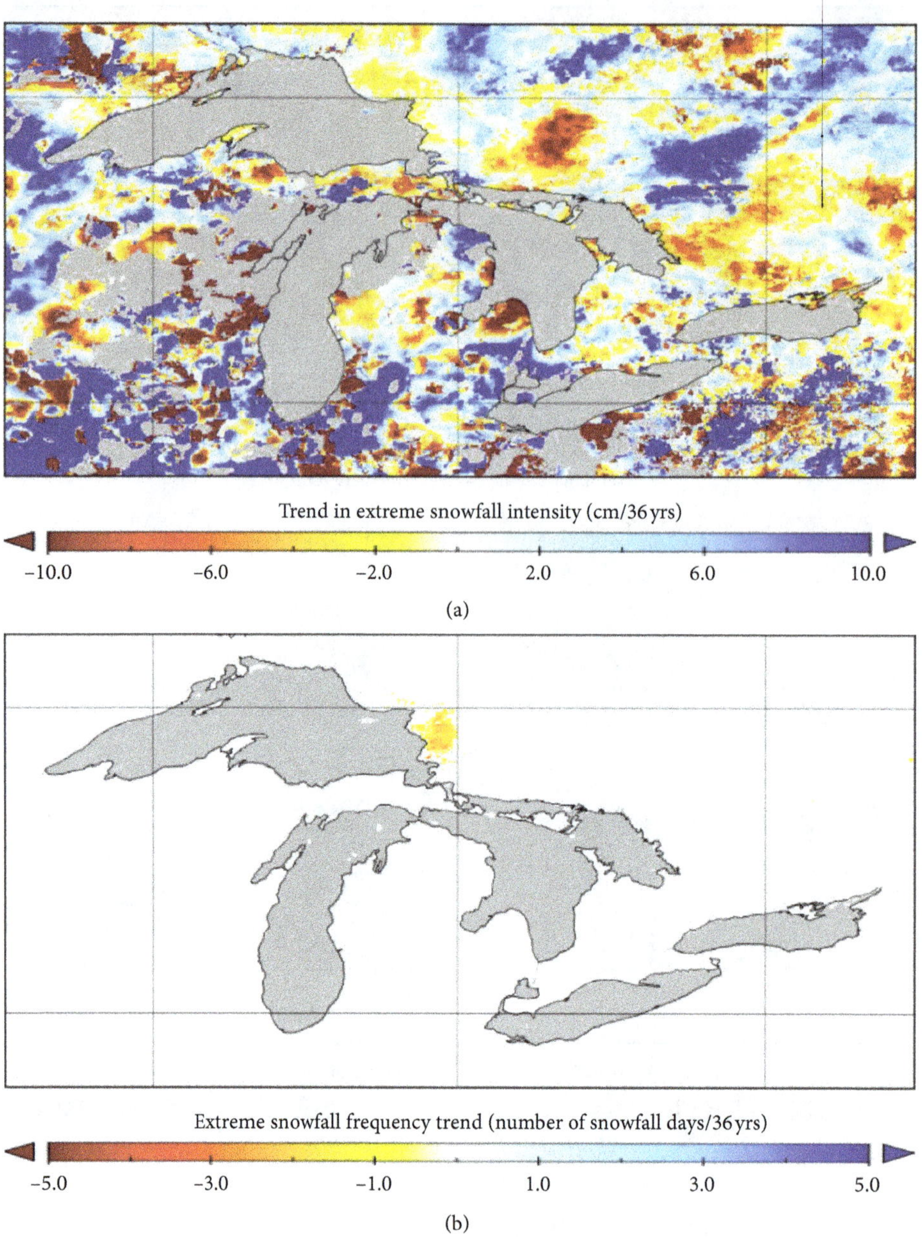

FIGURE 8: 1980 to 2015 trends in extreme snowfall (a) intensity and (b) frequency, for the month of February. Gray pixels represent grid cells with no trend computed because the region contains some years with no extreme snowfall values.

occurring for Lake Superior in January. Refer to Tables 3 and 4 for Sen slope values and significance. The final variable analysed is maximum annual ice cover concentration for both lakes. These lakes both show a decrease in maximum annual ice cover concentration, though not significant (Table 5).

4. Discussion

4.1. Snowfall Intensity, Frequency, and Duration Extreme Values. The 36-year snowfall intensity, frequency, and duration, as determined by the daily 15 cm/day S_threshold, are assessed over the Canadian domain of the GLB. The resultant behavioural patterns in the spatial and temporal snowfall extremes suggest that they are predominantly attributed to lake-induced snowfall rather than extratropical storms. This is because the snowfall extremes exhibit similar temporal and spatial patterns to that of LES. Spatially, these extremes are predominant along the Canadian leeward shores of Lakes Superior and Huron, where lake-induced snowfall occurs. Temporally, these resultant extremes in snowfall are dominant during the lake-effect months. According to [4], these snowbelt regions experience high annual snowfall totals, exceeding 250 cm during the autumn and winter months due to LES. These spatial patterns in extreme snowfall are also seen for all cold months in this study, with the highest extreme snowfall intensities occurring during December and January. This is in agreement with the study [64] that suggests that LES within the GLB is most intense during December and January when there is maximum open

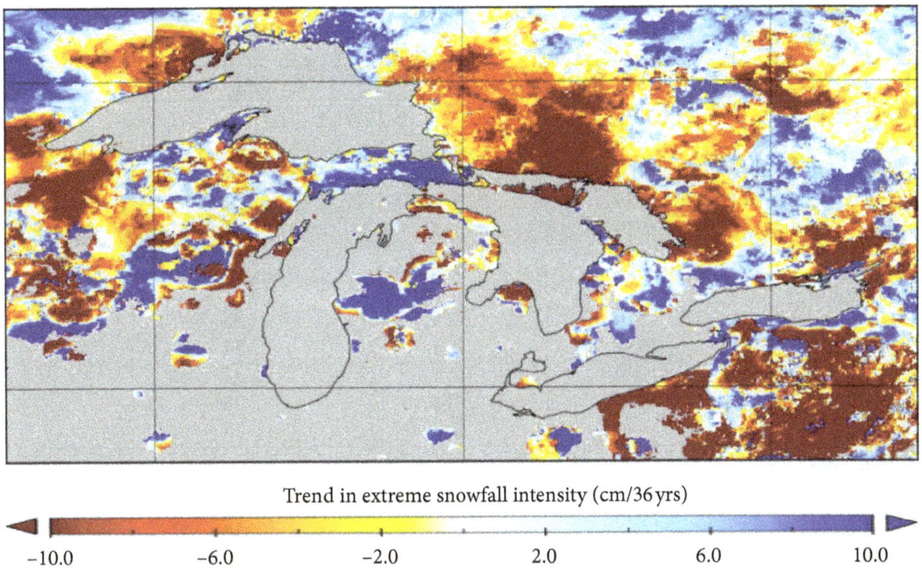

FIGURE 9: 1980 to 2015 trend in extreme snowfall intensity for the month of March. Gray pixels represent grid cells with no trend computed because the region contains some years with no extreme snowfall values.

TABLE 3: Predictor variable trend analysis for each month of the cold season with corresponding years used in Sen slope calculation for Wawa, Ontario.

Month	First year	Last year	Number of years	Sen slope and significance
Maximum monthly temperature				
November	1980	2013	34	0.044
December	1980	2013	34	0.089+
January	1980	2014	35	0.043
February	1981	2014	34	−0.024
March	1981	2014	34	0.038
Minimum monthly temperature				
November	1980	2013	34	0.067
December	1980	2013	34	0.212**
January	1980	2014	35	0.100
February	1981	2014	34	0.121**
March	1982	2014	33	0.060
Direction of maximum wind gust				
November	1980	2013	34	0.417
December	1980	2013	34	0.000
January	1980	2014	35	0.455
February	1980	2014	35	0.588
March	1980	2014	35	0.000
Speed of maximum wind gust				
November	1980	2013	34	−0.250
December	1980	2013	34	0.000
January	1980	2014	35	0.000
February	1980	2014	35	0.000
March	1980	2014	35	0.000
Lake surface temperature for superior				
November	1995	2015	21	0.040
December	1995	2015	21	0.022
January	1995	2015	21	0.049+
February	1995	2015	21	0.006
March	1995	2015	21	0.000

+, *, and ** represent 90%, 95%, and 99% confidence levels, respectively.

water and cold Arctic air masses that facilitate the exchange of moisture and heat between the lake and atmosphere.

Results show that snowfall extremes in February and March are lower than the other cold months. This is in agreement with the study [65] that suggests that only

TABLE 4: Predictor variable trend analysis for each month of the cold season with corresponding years used in Sen slope calculation for Wiarton, Ontario.

Month	First year	Last year	Number of years	Sen slope and significance
Maximum monthly temperature				
November	1980	2014	35	−0.032
December	1980	2013	34	0.062
January	1980	2014	35	0.125+
February	1980	2014	35	0.015
March	1980	2014	35	0.033
Minimum monthly temperature				
November	1980	2014	35	0.027
December	1980	2013	34	0.156
January	1980	2014	35	0.000
February	1980	2014	35	0.065
March	1980	2014	35	0.200*
Direction of maximum wind gust				
November	1980	2011	32	0.000
December	1980	2011	32	−0.889
January	1980	2012	33	0.000
February	1980	2012	33	1.181
March	1980	2012	33	0.513
Speed of maximum wind gust				
November	1980	2011	32	−0.472
December	1980	2011	32	−0.128
January	1980	2012	33	−0.725**
February	1980	2012	33	−0.077
March	1980	2012	33	−0.296
Lake surface temperature for Huron				
November	1995	2015	21	0.040
December	1995	2015	21	0.033
January	1995	2015	21	0.022
February	1995	2015	21	0.001
March	1995	2015	21	0.001

+, *, and ** represent 90%, 95%, and 99% confidence levels, respectively.

TABLE 5: Maximum annual ice cover trend analysis for Lakes Superior and Huron with corresponding years used in Sen slope calculation.

Lake	First year	Last year	Number of years	Sen slope and significance
Maximum annual ice cover				
Superior	1980	2015	36	−0.713
Huron	1980	2015	36	−0.453

approximately 14% of LES occurs downwind of Lakes Superior and Huron during February and March. This is attributed to the fact that ice cover is most extensive over the Great Lakes in February and March [66, 67]. As ice cover formation becomes more prominent, during the winter season, the dampening of the exchange in energy and moisture fluxes occurs, diminishing the production of snowfall [9]. Thus, lake ice is a regulator of LES [68], and the change in ice cover fraction can also influence the spatial distribution and extent of LES.

From a spatial perspective, when compared to the rest of the Canadian GLB, the leeward shores of Lakes Superior and Huron-Georgian Bay exhibit the highest snowfall extremes, with a decrease in the snowfall extremes farther inland. The spatial patterns of the extreme snowfall resemble that of lake-induced snowfall near the leeward shores of Lakes Superior and Huron. LES does not fall very far inland and is dependent on wind shear at upper levels to determine the intensity and organization of snow bands. LES can produce the heaviest amounts of snow within 50 to 150 km of the leeward shore. Most moisture supplied from the lake to the upper atmosphere is removed through precipitation within this distance [1, 69, 70].

Furthermore, in February, the spatial extent of extreme snowfall has substantially decreased for Lake Huron-Georgian Bay's snowbelt more than Lake Superior's. In November, the onset of extreme snowfall has started to develop for Lake Superior's snowbelt but not for Lake Huron's. These differences in extreme snowfall behaviours are attributed to the influence of lake-induced predictor variables, which are influenced by bathymetry and topography, giving additional credence that the resultant extremes in snowfall are mostly driven by lake-induced snowfall.

The Great Lakes have different geographic locations and lake bathymetry such as size, depth, and latitudinal extent that influence the onset of the unstable season and trigger

LES along each snowbelt region. The unstable season refers to the period when the lake is warmer than the ambient air temperature. Lake Superior, for example, is larger, deeper, and farther north than the other Great Lakes, and as a result, the warming season for Lake Superior begins relatively later in the spring and rarely ever attains the high LST values that are found for the lower Great Lakes. The prolonged unstable season over Lake Superior results in a longer LES season, lasting from mid November to early April [1], and is indicative of what is being observed in the current study. Lake depth is also a controlling factor when considering ice freeze-up and ice break-up dates at high latitudes [71, 72]. In shallow lakes, LST rises faster than deeper lakes, allowing for strong evaporation to occur earlier in the thaw season [73]. In the fall and winter, shallower lakes permit rapid and early cooling of LST [4]. These lakes also promote faster ice growth due to shorter thermal turnover rates, in the order of a week, which stores less heat. Thus, the spatiotemporal behaviours in snowfall extremes may vary among snowbelts because of lake bathymetry and other important factors.

Orographic lift is another factor that influences LES and can provide additional instability to an air parcel, affecting timing, and spatial distribution of precipitation [12]. Orographic lift can add approximately 13 to 20 cm of mean annual snowfall for every 30 m increase in elevation. It can also double the accumulation spawned from LES [27, 74]. The snowbelts of Lake Superior and the Tug Hill Plateau produce the highest annual snowfall precipitation although the Tug Hill Plateau is located farther south and experiences warmer air temperatures aloft than some of the other snowbelts [4]. The plateau rises 500 m above the leeward shores of Lake Ontario, and even with its modest altitude, it can influence the distribution of lake-effect precipitation off the long axis of the lake. Mean annual snowfall in this area can exceed 700 cm in Redfield, New York, on the western slope, which is more than twice the accumulation observed on the lowlands [12].

Similar behaviours in extreme snowfall found in this study, for both Lake Superior's snowbelt and the Tug Hill Plateau, further affirm that the extreme snowfall is produced by lake-induced processes and not solely by extratropical synoptic storms. If the extremes were mainly a product of large-scale frontal systems, then spatial distributions in snowfall extremes would follow climatologically persistent frontal boundaries and squall lines as opposed to snowbelt zones. The spatiotemporal snowfall extremes have been assessed for the 36-year duration; and the next step is to discuss whether there were any trends in these snowfall extremes.

4.2. Trends in Snowfall Extremes. Previous studies have shown an increase in the frequency of winter circulation patterns (between 1951 and 1982) such as 850 mb westerly winds and superadiabatic air temperature, which are both required for the development of LES and have given rise to an increase in LES events [3, 23, 25, 29, 64]. Comparatively, the last two decades of the 20th century and the first part of the 21st century show a significant decrease in monthly snowfall totals over the Canadian region of the Great Lakes Basin [62, 63]. Recent trends in extreme snowfall over the Canadian domain of the Great Lakes are not as apparent and show less coherent spatiotemporal patterns. While, overall, the results show a decrease in extreme snowfall intensity, frequency, and duration, they are spatially inconsistent for each month and fluctuate between the snowbelts and nonlake-effect zones.

Therefore, to further identify potential changes in extreme snowfall, lake-enhanced predictor variable trends are analysed for Wawa and Wiarton. Overall, lake-wide annual maximum ice cover has been decreasing for both lakes. These results are in agreement with [62, 63], who investigated trends in monthly averaged ice cover concentration from 1980 to 2015 and showed significant decrease for Lake Superior for the months of January and February, while Lake Huron showed a significant decrease in January. Trends in monthly ice cover do not exactly align with the timing of trends in extreme snowfall. While significant reduction in ice cover occurs in January for Lake Superior, the most significant reduction in extreme snowfall occurs in December (Figure 6). Decrease in annual maximum lake ice can be attributed to the lake-wide monthly LST warming trends.

Increase in LST and subsequent decrease in maximum ice cover can exhibit an increase in energy fluxes into the lower planetary boundary layer, which can be conducive to the development of lake-enhanced snowfall. However, extensive work in [62, 63] suggests that as air temperatures warm, air parcels can hold more moisture, extending the resonance time of water vapour in the air and delaying the saturation and precipitation of an air parcel along the immediate leeward shores of the Great Lakes.

Results from the current study display a similar idea. Snowfall extremes have been decreasing within the past 36 years along the leeward shores of Lake Superior and Lake Huron, despite the increase in LST and decrease in maximum annual ice cover. A potential reason for the decrease in extreme snowfall could be attributed to the increase in both maximum and minimum monthly air temperature. As the observed weather stations' maximum and minimum temperatures rise, the air temperature will become too warm for precipitation to fall as snow. Furthermore, an increase in both monthly maximum and minimum air temperatures suggest that the atmosphere is warming and, thereby, has the ability to retain moisture, as previously discussed.

The paper also presents trends in the direction and speed of maximum monthly wind gusts. Although there are no significant changes in direction of maximum wind gust, the cities of Wawa and Wiarton show an increase in direction of maximum wind gust for all months, with an exception to December for Wiarton. The predominant increase in these values suggests a clockwise shift in the origin location of the wind gusts. This clockwise shift is known as veering and can change the onset, location, frequency, intensity, and duration of lake-induced snowfall. For example, the Canadian leeward shores of Lake Superior and Huron are predominant snowbelt zones because of the prevalent cold and dry northwesterly winds that encounter maximum fetch across these lakes. However, if the origins of these wind directions begin to veer towards a northerly and easterly flow, the leeward snowbelt zones of Lake Superior will be predominant along the shores in Northern Michigan and Eastern Minnesota. The predominant snowbelt of Lake Huron would also shift to the shores along Eastern Michigan.

The speeds of the maximum wind gusts show no changes for all months, except for a decrease in November for the city of Wawa. All months of the cold season indicate a decrease in maximum wind gusts for the city of Wiarton, with a significant decrease in January. The speed of an air parcel will influence the fetch time over lakes. The fetch time of an air parcel will modify the air temperature over lakes and influence the development of lake-induced snowfall. A study by [75] indicates that it only takes approximately 10 minutes for an air parcel, at heights between one and 15 m above lakes, to undergo an increase in temperature modification. The longer the resonance time of an air parcel over lakes, the greater the increase in air temperature. If the air temperature warms too much (exceeds values above 0°C), this can decrease the production of lake-induced snowfall, as observed in this study.

This study has two apparent findings. Firstly, both Canadian snowbelt zones exhibit different spatiotemporal trends in snowfall extremes. For example, Lake Superior's Canadian snowbelt exhibits significant decreases in extreme snowfall intensity, frequency, and duration, predominantly, for the months of December and January, while there are no significant changes for Lake Huron's Canadian snowbelt. As discussed in the previous section, this is attributed to the geographic locations of the lakes, topography, lake bathymetry, and lake orientation that will influence fetch, ice cover, vertical temperature, and instability differently, thereby resulting in various spatial and temporal behaviours. It is therefore important to assess LES within each snowbelt separately, as the Laurentian Great Lakes and their locations have their own unique properties that influence LES and its ingredients such as lift, instability, and moisture.

Secondly, within each snowbelt, the spatiotemporal trends are not contiguously spatially coherent. This is evident from the resultant increase and decrease in extreme snowfall intensity seen along the northern and southern Canadian leeward shores of Lake Superior, respectively. The spatial variability within each snowbelt can result from local effects such as small-scale surface-atmosphere meteorological factors, which include localized ice cover fraction, shifts in vertical and horizontal wind velocity in the lower PBL, and storm tracks from synoptic scale systems, leading to lake-enhanced snowfall. Furthermore, changes in the frequency and intensity of cold air outbreaks can affect the vertical temperature gradient and oscillation patterns, both of which influence the production of LES. Assessing trends in Arctic air outbreak will be worth examining in future lake-effect and climate change studies.

5. Conclusion

The current study focused on examining the monthly cold season spatiotemporal extremes in snowfall intensity, frequency, and duration over the historical period of 1980 to 2015 for the Canadian GLB. The study has two folds: the first examines the 36-year period snowfall extremes, while the second explores the trends in snowfall extremes over the given period. In the first fold, extremes in snowfall intensity are derived by evaluating the 99th percentile of all daily snowfall for each cold month of the 36-year period over the GLB domain. Based on the spatiotemporal results, a snowfall intensity of 15 cm/day was found as the ideal threshold to evaluate extremes in snowfall. Counting the number of snowfall days that equaled or exceeded this S_threshold, the frequency of extreme snowfall days was determined. The maximum number of consecutive snowfall days that equaled or exceeded 15 cm/day represented the duration of extreme snowfall events. Spatiotemporal results show that the extremes in snowfall intensity, frequency, and duration are spatially coherent along the leeward shores of Lake Superior and Lake Huron-Georgian Bay. Results also indicate that snowfall extremes are most predominant in the months of December and January, suggesting that extremes in snowfall are attributed to lake-enhanced processes rather than driven solely by extratropical storms.

In the second fold, there are two apparent findings in the spatiotemporal trends of snowfall extremes. Firstly, individual snowbelts behave differently when assessing spatiotemporal trends. For instance, while portions of the Canadian leeward shores of Lake Superior show significant decreases in extreme snowfall, no changes are observed for Lake Huron's snowbelt. Secondly, within each snowbelt, there are spatial variability and inconsistent snowfall patterns. This is evident as Lake Superior's Canadian snowbelt shows a decrease in extreme snowfall intensity along the southern leeward shores and an increase along its northern leeward shores. The lakes' topographical, geographical, and bathymetric features influence these results, in addition to, mesoscale and synoptic scale shifts in lake-induced predictor variables.

Trends in lake-induced predictor variables were also analysed to provide potential explanations in the resultant negative trends in extreme snowfall intensity, frequency, and duration. The predominant variable influencing the decrease in extreme snowfall is suggested to be monthly maximum and minimum air temperature. This is because, in comparison with the other predictor variables, the maximum and minimum air temperature produces the most number of significant trends. The significant increases in these temperatures for both Wawa and Wiarton suggest that these variables can influence the decrease in snowfall, as previously discussed. The study [62, 63] noted that trends in monthly snowfall totals shifted from an increase in the first half of the 20th century to a decrease during the last part of the 20th and early 21st century. However, the current study suggests that not only are the totals in monthly snowfall decreasing along the Canadian leeward Shores of the Laurentian Great Lakes for the 1980 to 2015 period but also the extremes in snowfall intensity, frequency, and duration are also decreasing, albeit the spatial coherency is difficult to delineate.

Disclosure

This research was presented at the 2017 European Geosciences General Assembly, in Vienna, entitled *Climatological assessment of spatiotemporal trends in observational monthly snowfall totals and extremes over the Canadian Great Lakes Basin.*

Conflicts of Interest

The authors declare that they have no conflicts of interest.

Acknowledgments

The authors would like to thank Michele Thornton, from the Oak Ridge National Laboratory Distributed Active Archive Center (ORNL DAAC), for support with the Daymet product, as well as Nan Jiang, from Beijing Normal University, and Dr. Sofia Antonova from Alfred Wegener Institute, Potsdam, Germany, for their computational assistance. In addition, gratitude is expressed to Dr. Andrea Scott for her assistance with the uncertainty analyses and Mike Lackner for his technical mapping support, both from the University of Waterloo in Ontario, Canada. This research was conducted as part of the Canadian Network for Regional Climate and Weather Processes (CNRCWP) and funded by the Natural Sciences and Engineering Research Council (NSERC) of Canada.

References

[1] M. Notaro, A. Zarrin, S. Vavrus, and V. Bennington, "Simulation of heavy lake-effect snowstorms across the Great Lakes Basin by RegCM4: synoptic climatology and variability," *Monthly Weather Review*, vol. 141, no. 6, pp. 1990–2014, 2013.

[2] B. L. Wiggin, "Great snows of the Great Lakes," *Weatherwise*, vol. 3, no. 6, pp. 123–126, 1950.

[3] V. L. Eichenlaub, "Lake effect snowfall to the lee of the Great Lakes: its role in Michigan," *Bulletin of the American Meteorological Society*, vol. 51, no. 5, pp. 403–412, 1970.

[4] V. L. Eichenlaub, *Weather and Climate of the Great Lakes Region*, The University of Notre DamePress, Notre Dame, IN, USA, 1979, ISBN 0-268-01929-0.

[5] E. W. Holroyd III., "Lake effect cloud bands as seen from weather satellites," *Journal of the Atmospheric Sciences*, vol. 28, no. 7, pp. 1165–1170, 1971.

[6] K. Hozumi and C. Magono, "The cloud structure of convergent cloud bands over the Japan Sea in winter monsoon period," *Journal of the Meteorological Society of Japan*, vol. 62, no. 3, pp. 522–553, 1984.

[7] S. R. Pease, W. A. Lyons, C. S. Keen, and M. R. Hjelmfelt, "Mesoscale spiral vortex embedded within a Lake Michigan snow squall band: high resolution satellite observations and numerical model simulations," *Monthly Weather Review*, vol. 116, no. 6, pp. 1374–1380, 1988.

[8] D. J. Leathers and A. W. Ellis, "Relationships between synoptic weather type frequencies and snowfall trends in the lee of Lakes Erie and Ontario," in *Proceedings of 61st Annual Western Snow Conference*, pp. 325–330, Quebec City, Canada, June 1993.

[9] T. A. Niziol, W. R. Snyder, and J. S. Waldstreicher, "Winter weather forecasting throughout the Eastern United States, Part4: lake effect snow," *Weather and Forecasting*, vol. 10, no. 1, pp. 61–77, 1995.

[10] R. J. Ballentine, A. J. Stamm, E. F. Chermack, G. P. Byrd, and D. Schleede, "Mesoscale model simulation of the 4–5 January 1995 lake-effect snowstorm," *Weather and Forecasting*, vol. 13, no. 4, pp. 893–920, 1998.

[11] D. A. R. Kristovich and N. F. Laird, "Observations of widespread lake effect cloudiness: influence of lake temperature and upwind conditions," *Weather and Forecasting*, vol. 13, no. 3, pp. 811–821, 1998.

[12] L. S. Campbell, W. J. Steenburgh, P. G. Veals, T. W. Letcher, and J. R. Minder, "Lake effect mode and precipitation enhancement over the Tug Hill Plateau during OWELes IOP2b," *Monthly Weather Review*, vol. 144, no. 5, pp. 1729–1748, 2016.

[13] T. A. Niziol, "Operational forecasting of lake effect snow in western and central New York," *Weather and Forecasting*, vol. 2, no. 4, pp. 310–321, 1987.

[14] G. P. Byrd, R. A. Anstett, J. E. Heim, and D. M. Usinski, "Mobile sounding observations of lake-effect snowbands in western and central New York," *Monthly Weather Review*, vol. 119, no. 9, pp. 2323–2332, 1991.

[15] D. A. R. Kristovich, Neil F. Laird, and M. R. Hjelmfelt, "Convective evolution across Lake Michigan during a widespread lake-effect snow event," *Monthly Weather Review*, vol. 131, no. 4, pp. 643–655, 2003.

[16] J. J. Schroeder, D. A. R. Kristovich, and M. R. Hjelmfelt, "Boundary layer and microphysical influences of natural cloud seeding on a lake-effect snowstorm," *Monthly Weather Review*, vol. 134, no. 7, pp. 1842–1858, 2006.

[17] Z. J. Suriano and D. J. Leathers, "Twenty-first century snowfall projections within the eastern Great Lakes region: detecting the presence of a lake-induced snowfall signal in GCMs," *International Journal of Climatology*, vol. 36, no. 5, pp. 2200–2209, 2015.

[18] M. R. Hjelmfelt and R. R. Braham Jr., "Numerical simulation of the airflow over lake Michigan for a major lake effect snow event," *Monthly Weather Review*, vol. 111, no. 1, pp. 205–219, 1983.

[19] R. E. Passarelli Jr. and R. R. Braham Jr., "The role of the winter land breeze in the formation of Great Lake snowstorms," *Bulletin of the American Meteorological Society*, vol. 62, no. 4, pp. 482–492, 1981.

[20] N. F. Laird, "Observation of coexisting mesoscale lake-effect vortices over the western Great Lakes," *Monthly Weather Review*, vol. 127, no. 6, pp. 1137–1141, 1999.

[21] NOAA GLERL, *National Oceanic and Atmospheric Administration*, Great Lakes Environmental Research Laboratory, Ann Arbor, MI, USA, 2017, About our lakes, https://www.glerl.noaa.gov/education/ourlakes/intro.html.

[22] ECCC, "Environment and climate change Canada, Great lakes quickfacts," 2017, https://www.ec.gc.ca/grandslacsgreatlakes/default.asp?lang=En&n=B4E65F6F-1.

[23] D. C. Norton and S. J. Bolsenga, "Spatiotemporal trends in lake effect and continental snowfall in the Laurentian Great Lakes 1951–1980," *Journal of Climate*, vol. 6, no. 10, pp. 1943–1955, 1993.

[24] K. E. Kunkel, N. E. Wescott, and D. A. R. Kristovich, "Assessment of potential effects of climate change on heavy

lake-effect snowstorms near Lake Erie," *Journal of Great Lakes Research*, vol. 28, no. 4, pp. 521-536, 2002.

[25] A. W. Burnett, M. E. Kirby, H. T., W. P. Mullins, and Patterson, "Increasing Great Lakes-effect snowfall during the twentieth century: a regional response to global warming?," *Journal of Climate*, vol. 16, no. 21, pp. 3535–3542, 2003.

[26] D. A. R. Kristovich and M. L Spinar, "Diurnal variations in lake-effect precipitation near western Great Lakes," *Journal of Hydrometeorology*, vol. 6, no. 2, pp. 210-218, 2005.

[27] H. Hartmann, J. Livingstone, and M. G. Stapleton, "Seasonal forecast of local lake-effect snowfall: The case of Buffalo, U.S.A," *International Journal of Environmental Research*, vol. 7, no. 4, pp. 859-867, 2013.

[28] U. Cubasch and Coauthors, "Introduction," in *Climate Change 2013: The Physical Science Basis. Contribution of Working Group I to the Fifth Assessment Report of the Intergovernmental Panel on Climate Change*, Cambridge University Press, Cambridge, United Kingdom and New York, NY, USA, 2013, http://www.ipcc.ch/pdf/assessment-report/ar5/wg1/WG1AR5_Chapter01_FINAL.pdf.

[29] D. J. Leathers and A. W. Ellis, "Synoptic mechanisms associated with snowfall increases to the lee of Lakes Erie and Ontario," *International Journal of Climatology*, vol. 16, no. 10, pp. 1117-1135, 1996.

[30] D. A. R. Kristovich, G. S. Young, J. Verlinde et al., "The lake induced convection experiment and the snowband dynamics project," *Bulletin of the American Meteorological Society*, vol. 81, no. 3, pp. 519–542, 2000.

[31] K. E. Kunkel, M. Palecki, L. Ensor et al., "Trends in 20th Century U.S. snowfall using a quality-controlled data set," *Journal of Atmospheric and Oceanic Technology*, vol. 26, no. 1, pp. 33–44, 2009.

[32] J. J Hartnett, J. M. Collins, M. A. Baxter, and D. P. Chambers, "Spatiotemporal snowfall trends in Central New York," *Journal of Applied Meteorology and Climatology*, vol. 53, no. 12, pp. 2685–2697, 2014.

[33] G. M. Lackmann, "Analysis of a surprise Western New York snowstorm," *Weather and Forecasting*, vol. 16, no. 1, pp. 99–114, 2001.

[34] R. S. Hamilton, D. Zaff, and T. Nizol, *A Catastrophic Lake Effect Snow Storm Over Buffalo, NY October 12-14, 2006*, NOAA National Weather Service, Buffalo, NY, USA, 2006, http://ams.confex.com/ams/pdfpapers/124750.pdf.

[35] T. Maesaka, G. W. K. Moore, Q. Liu, and K. Tsuboki, "A simulation of a lake effect snowstorm with a cloud resolving numerical model," *Geophysical Research Letters*, vol. 33, no. 20, pp. 1–5, 2006.

[36] R. L. Lavoie, "A mesoscale numerical model of lake-effect storms," *Journal of Atmospheric Sciences*, vol. 29, no. 6, pp. 1025–1040, 1972.

[37] M. R. Hjelmfelt, "Numerical study of the influence of environmental conditions on lake-effect snowstorms over Lake Michigan," *Monthly Weather Review*, vol. 118, no. 1, pp. 138–150, 1990.

[38] T. T. Warner and N. L. Seaman, "A real-time, mesoscale numerical weather prediction system used for research, teaching, and public service at the Pennsylvania State University," *Bulletin of the American Meteorological Society*, vol. 71, no. 6, pp. 792–805, 1990.

[39] G. T. Bates, F. Giorgi, and S. W. Hostetler, "Toward the simulation of the effects of the Great Lakes on regional climate," *Monthly Weather Review*, vol. 121, no. 5, pp. 1373–1387, 1993.

[40] P. J. Sousounis and J. M. Fritsch, "Lake aggregate mesoscale disturbances. Part II: A case study of the effects on regional and synoptic-scale weather systems," *Bulletin of the American Meteorological Society*, vol. 75, no. 10, pp. 1793–1811, 1994.

[41] G. J. Tripoli, "Numerical study of the 10 January 1998 lake effect bands observed during lake-ice," *Journal of Atmospheric Sciences*, vol. 62, no. 9, pp. 3232–3249, 2005.

[42] J. J. Shi, W-K. Tao, T. Matsui et al., "WRF simulations of the 20–22 January 2007 snow events over eastern Canada: comparison with in situ and satellite observations," *Journal of Applied Meteorology and Climatology*, vol. 49, no. 11, pp. 2246–2266, 2010.

[43] D. Scott and D. Kaiser, "Variability and trends in United States snowfall over the last half century," in *Preprints, 15th Symposium on Global Climate Variations and Change*, Seattle, WA, USA, January 2004.

[44] T. R. Karl and R. W. Knight, "Secular trends of precipitation amount, frequency, and intensity in the United States," *Bulletin of the American Meteorological Society*, vol. 70, no. 2, pp. 231–241, 1998.

[45] J. P. Krasting, A. J. Broccoli, K. W. Dixon, and J. R. Lazante, "Future changes in Northern Hemisphere snowfall," *Journal of Climate*, vol. 26, no. 20, pp. 7813–7828, 2013.

[46] A. W. Ellis and J. J. Johnson, "Hydroclimatic analysis of snowfall trends associated with the North American Great Lakes," *Journal of Hydrometeorology*, vol. 5, no. 3, pp. 471–486, 2004.

[47] K. E. Kunkel, L. Ensor, M. Palecki et al., "A new look at lake-effect snowfall trends in the Laurentian Great Lakes using a temporally homogeneous data set," *Journal of Great Lakes Research*, vol. 35, no. 1, pp. 23–29, 2009.

[48] L. Bard and D. A. R. Kristovich, "Trend reversal in Lake Michigan contribution to snowfall," *Journal of Applied Meteorology and Climatology*, vol. 51, no. 11, pp. 2038–2046, 2012.

[49] X. Zhang, W. D. Hogg, and E. Mekis, "Spatial and temporal characteristics of heavy precipitation events over Canada," *Journal of Climate*, vol. 14, no. 9, pp. 1923–1936, 2001.

[50] K. E. Kunkel, T. R. Karl, H. Brooks et al., "Monitoring and understanding trends in extreme storms: state of knowledge," *Bulletin of the American Meteorological Society*, vol. 94, no. 4, pp. 499–514, 2013.

[51] P. A. Gorman, "Contrasting response of mean and extreme snowfall to climate change," *Nature*, vol. 512, no. 7515, pp. 416–418, 2014.

[52] M. J. Menne, I. Durre, R. S. Vose, B. E. Gleason, and T. G. Houston, "An overview of the global historical climatology network-daily database," *Journal of Atmospheric and Oceanic Technology*, vol. 29, no. 7, pp. 897–910, 2012.

[53] M. J. Menne, I. Durre, B. Korzeniewski et al., *Global Historical Climatology Network -Daily (GHCN-daily), Version 3.22*, NOAA National Climatic Data Center, 2012.

[54] P. E. Thornton, S. W. Running, and M. A. White, "Generating surfaces of daily meteorological variables over large regions of complex terrain," *Journal of Hydrology*, vol. 190, no. 3-4, pp. 214–251, 1997.

[55] P. E. Thornton, M. M. Thornton, B. W. Mayer et al., *Daymet: Daily Surface Weather Data on a 1-km Grid for North America, Version 3*, ORNL DAAC, Oak Ridge, TN, USA, 2016, http://dx.doi.org/10.3334/ORNLDAAC/1328.

[56] J. A. Baijnath-Rodino, C. R. Duguay, and E. LeDrew, "Climatological snowfall trends over the Canadian Snowbelts of the Laurentian Great Lakes Basin," *International Journal of Climatology*, vol. 38, no. 10, pp. 3942–3962, 2018.

[57] D. R. Legates and C. J. Willmott, "Mean seasonal and spatial variability in gage-corrected global precipitation," *International Journal of Climatology*, vol. 10, no. 2, pp. 111–127, 1990.
[58] M. A. Rawlins, C. J. Willmott, A. Shiklomanov et al., "Evaluation of trends in derived snowfall and rainfall across Eurasia and linkages with discharge to the Arctic Ocean," *Geophysical Research Letter*, vol. 33, no. 7, pp. 1–4, 2006.
[59] M. G. Kendall, *Rank Correlation Methods*, Charles Griffin, London, UK, 4th edition, 1975.
[60] R. O. Gilbert, *Statistical Methods for Environmental Pollution Monitoring*, Van Nostrand Reinhold Co, New York City, NY, USA, 1987, ISBN: 978-0-471-28878-7.
[61] ECCC, "Environment and climate change canada, public alerting criterion," https://www.canada.ca/en/environment-climate-change/services/types-weather-forecasts-use/public/criteria-alerts.html#snowFall.
[62] Current Results, Weather and science facts, Barrie snowfall totals and snowstorm averages, https://www.currentresults.com/Weather/Canada/Ontario/Places/barrie-snowfall-totals-snowstorm-averages.php.
[63] A. Janine, "Baijnath-Rodino," Ph.D. thesis, University of Waterloo, Waterloo, Canada, 2018.
[64] R. R. Braham Jr. and M. J. Dungey, "Quantitative estimates of the effect of Lake Michigan on snowfall," *Journal of Applied Meteorology and Climatology*, vol. 23, no. 6, pp. 940–949, 1984.
[65] M. Notaro, K. Holman, A. Zarrin, E. Fluck, S. Vavrus, and V. Bennington, "Influence of the Laurentian Great Lakes on regional Climate," *Journal of Climate*, vol. 26, no. 3, pp. 789–804, 2013.
[66] R. A. Assel, "An ice-cover climatology for Lake Erie and Lake Superior for the winter seasons 1897–98 to 1982–83," *International Journal of Climatology*, vol. 10, no. 7, pp. 731–748, 1990.
[67] R. A. Assel, "Great Lakes ice cover," in *Potential Climate Change Effects on Great Lakes Hydrodynamics and Water Quality*, D. C. L. Lam and W. M. Schertzer, Eds., pp. 1–21, American Society of Civil Engineers, Reston, VA, USA, 1999.
[68] L. C. Brown and C. R. Duguay, "Response and role of ice cover in lake climate interactions," *Progress in Physical Geography*, vol. 34, no. 5, pp. 671–704, 2010.
[69] K. F. Dewey, "An objective forecast method developed for Lake Ontario induced snowfall systems," *Journal of Applied Meteorology*, vol. 18, no. 6, pp. 787–793, 1979.
[70] R. M. Rauber, J. E. Walsh, and D. J. Charlevoix, *Severe & Hazardous Weather-An Introduction to High Impact Meteorology*, Kendall/Hunt Publishing Company, Dubuque, LA, USA, Second edition, 2005.
[71] C. R. Duguay, G. M. Flato, M. O. Jeffries, P. Ménard, K. Morris, and W. R. Rouse, "Ice covers variability on shallow lakes at high latitudes: model simulation and observations," *Hydrological Process*, vol. 17, no. 17, pp. 3465–3483, 2003.
[72] C. R. Duguay, T. D. Prowse, B. R. Bonsal, R. D. Brown, M. P. Lacroix, and P. Ménard, "Recent trends in Canadian lake ice cover," *Hydrological Process*, vol. 20, no. 4, pp. 781–801, 2006.
[73] W. R. Rouse, C. Oswald, J. Binyamin et al., "Role of northern lakes in a regional energy balance," *Journal of Hydrometeorology*, vol. 6, no. 3, pp. 291–305, 2005.
[74] J. D. Hill, "Snow squalls in the lee of Lakes Erie and Ontario," NOAA Tech. Memo. NWS ER-43, [NTIS COM-00959.], 1971.
[75] D. W. Phillips, "Modification of surface air over lake Ontario in winter," *Monthly Weather Review*, vol. 100, no. 9, pp. 662–670.

5

Spatiotemporal Variability of Arctic Soil Moisture Detected from High-Resolution RADARSAT-2 SAR Data

Adam Collingwood,[1,2] François Charbonneau,[3] Chen Shang,[1] and Paul Treitz[1]

[1]*Department of Geography and Planning, Queen's University, Kingston, ON, Canada K7L 3N6*
[2]*Waterton Lakes National Park, Parks Canada, Box 200, Waterton Park, AB, Canada T0K 2M0*
[3]*Canada Centre for Mapping and Earth Observation, Natural Resources Canada, 560 Rochester Street, Ottawa, ON, Canada K1A 0E4*

Correspondence should be addressed to Paul Treitz; paul.treitz@queensu.ca

Academic Editor: Jifu Yin

Various methods are used to determine soil moisture information from synthetic aperture radar (SAR) data, but none specific to High Arctic regions and their unique physical characteristics. This research presents a method for determining, at high spatial and temporal resolutions, surface soil moisture and its changes through time in the Canadian High Arctic. An artificial neural network (ANN) is implemented using input variables derived from RADARSAT-2 SAR data and previously modelled surface roughness information. The model is applied to SAR data collected at various incidence angles and acquisition dates across two study sites on Melville Island, Nunavut. The model results in absolute soil moisture errors of approximately 15% ($r^2 = 0.46$) for the primary study sites and 12% ($r^2 = 0.26$) for the verification study area. The ANN model is accurate for modelling (i) the spatial distribution of soil moisture and (ii) the changes in moisture through time across the study areas, two characteristics that are very important for inputs to hydrologic or climate models. In addition, the models appear to be scalable when applied at coarser spatial resolutions, showing potential for large-area mapping or modelling.

1. Introduction

The estimation or modelling of biogeophysical variables such as soil moisture in the Arctic is an important step towards understanding Arctic energy fluxes, the effects of changing climate, and hydrological processes and patterns. For example, the spatial patterns of carbon dioxide (CO_2) fluxes are heavily influenced by the spatial patterns of soil moisture in high-latitude ecosystems [1–7]. Areas in the Arctic with high soil moisture content are also thought to be less responsive to climate warming [3, 5, 8], thus making soil moisture an important variable in global climate-change models. Current and future climate variations in the Arctic will influence the spatial distribution of soil water content, which can affect vegetation, active layer depths, and formation of wetlands over large time scales [9–12]. A more comprehensive understanding of the spatial distribution of soil moisture across the tundra landscape will allow for more accurate and precise predictions of CO_2 flux as ecosystems adapt to climate change [6, 13]. Further, hydrologic variables are very important controls on Arctic geomorphology and ecosystem dynamics [14].

With traditional soil moisture measurement techniques, such as direct sampling (weighing wet soil versus dry soil) and time-domain reflectometry (TDR) measurements, the soil moisture values obtained represent point locations only and are usually averaged or extrapolated over large areas [15]. Spatial variation in soil moisture levels is an important consideration at a number of spatial scales. At subcatchment and finer scales in particular, the spatial distribution of soil moisture becomes as (or more) important than the absolute value of the soil moisture [16]. The demand for spatially distributed soil moisture is clear, but point measurement data are often insufficient, that is, due to soil heterogeneity, land use, and topography. In short, soil moisture may be very different in space and time from one point to another [17–19].

Information gathered from synthetic aperture radar (SAR) is ideal for the spatial estimation or modelling of soil moisture, at a range of different spatial scales [20–22]. In many cases, these data can be gathered instantaneously over large areas, day or night, in any weather, which is a tremendous advantage over traditional techniques. SAR data are especially useful for medium- and large-scale analysis of soil moisture levels [16, 23, 24]. However, there has been limited research on SAR-modelled soil moisture conducted in the Arctic [25–28], or in natural (nonagricultural) environments in general. Jagdhuber et al. [29] examined the utility of multitemporal space-borne C-band SAR polarimetry for estimating soil freezing and thawing states and found that the mean scattering alpha angle and polarimetric entropy were strongly correlated with the soil freezing and thawing states. A more recent study by Högström and Bartsch [28] identified the fraction of water bodies as a source of error in coarse-scale SAR-retrieved soil moisture due to variations in water surface roughness. However, high-resolution multi-incidence angle SAR data from sensors such as RADARSAT-2 have not been examined to any significant degree for their utility in modelling surface moisture conditions in the High Arctic at fine spatial scales.

There are a number of methods for deriving soil moisture values from SAR backscatter, each possessing certain advantages and disadvantages. Regression-based empirical models, such as those of Dubois et al. [20], are often used to model soil moisture [30–33]. However, empirical models require large quantities of field data and are often very site specific [34]. Physical models, the most common of which is the integrated equation model (IEM) [35], invert backscatter from inputs including radar parameters, surface roughness, and vegetation cover to derive soil moisture values [34–40]. Physical models, like empirical models, are difficult to apply over large areas, and detailed information on parameters such as topography and soil type is required. Artificial neural networks (ANNs) are commonly used to model surface parameters from SAR data [18, 41] and show great promise in both simplifying the modelling process and increasing the accuracy of the results. ANNs have the capacity to "learn" complex, nonlinear patterns and generalize these patterns in noisy environments [42, 43]. This capacity to generalize means that ANNs can be effective in situations where data may be missing or imprecise. ANNs are also able to incorporate prior knowledge and physical constraints into the analysis, while making no assumptions about the statistical nature of the input data [44, 45]. This allows for the incorporation of disparate data from many remote sensing and ancillary sources, including other ANN model outputs. ANNs are superior to empirical models for generalizing results for application to new areas [46] and do not have the same parameterization problems and assumption requirements as physical models (e.g., IEM).

Given the spatial detail captured by RADARSAT-2 SAR data and the utility of ANNs for modelling environmental

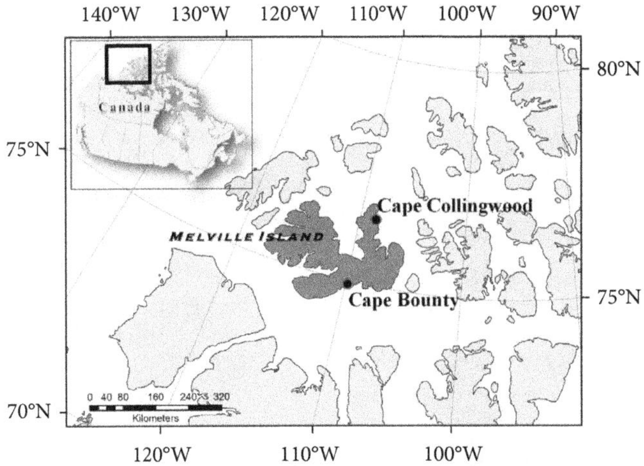

FIGURE 1: Study area location in the Canadian High Arctic.

variables, we designed a study to model soil moisture in the Canadian High Arctic. The objectives of the research presented here are to (i) examine the relationships that exist between SAR backscatter parameters and surface soil moisture, (ii) model those relationships using an ANN, and (iii) map the spatiotemporal dynamics of Arctic soil moisture using multitemporal RADARSAT-2 imagery acquired with varying incidence angles.

2. Methods

2.1. Site Description. The majority of the field work for this study was undertaken at the Cape Bounty Arctic Watershed Observatory (CBAWO), located on the southern coast of Melville Island, Nunavut, Canada (74.91°N, 109.44°W) (Figure 1) in 2009 and 2010. The CBAWO was used to develop, calibrate, and validate the ANN models. This High Arctic site is composed of two parallel watersheds, each covering approximately 15 km^2. The area is characterized by rolling topography of low to medium relief, with elevation varying between 5 m and 125 m above sea level. The site has been impacted by periods of glaciation, during which various tills have been deposited in the study region, including Bolduc, Dundas, and Winter Harbour tills [47]. Winter Harbour till is a thin (1-2 m) carbonate-rich till that contains many mafic and crystalline rock fragments and is draped over the other layers [47]. The Winter Harbour till also offlaps Holocene era fine-grained marine sediments, which are located between approximately 35 m and 90 m above sea level [47, 48]. These sediments are underlain by the Franklinian mobile belt, composed of Paleozoic sandstone, siltstone, and shale [48]. Vegetation in the area is extremely limited and rarely exceeds a few centimetres in height. Greater vegetation biomass is found in sedge and heath communities, while polar desert areas can be completely barren. Large areas of exposed, fractured bedrock are also present, and the entire area is underlain by permafrost, with an active layer of 0.5–1 m during the summer.

Additional field work was conducted in 2011 in the vicinity of Cape Collingwood on the Sabine Peninsula, located on northern Melville Island (approximately 76.53°N,

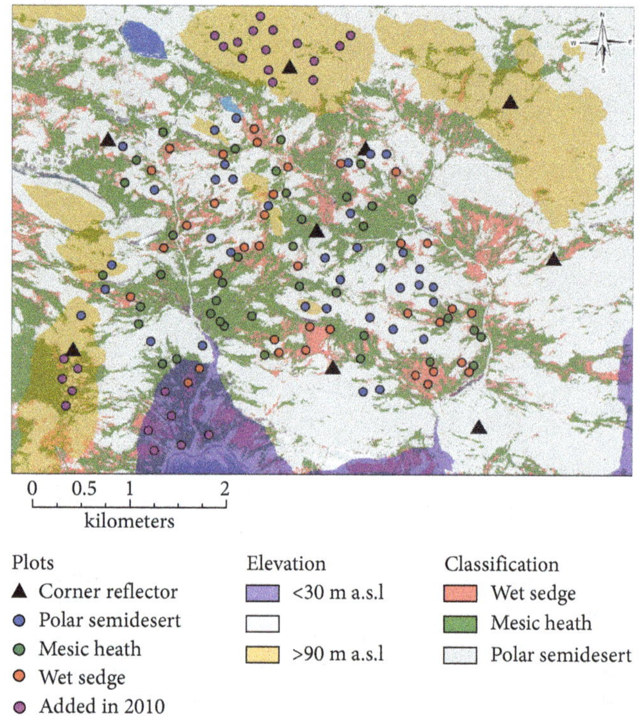

Figure 2: Unsupervised classification, elevation stratification, and plot locations at the Cape Bounty Arctic Watershed Observatory.

108.83°W) (Figure 1). Plots from this more northerly study site are representative of three different underlying bedrock types, all of which are different from the CBAWO. The Kanguk formation consists of a shale bedrock with surface materials of poorly sorted clay-silt mixed with a sand and gravel till; the Hassel formation has a sandstone bedrock with surface materials consisting of sand and silty sand; and the Christopher formation is a shale bedrock with a very fine, poorly drained silty clay surficial material [49]. These different surfaces will test the extension of the models developed at the CBAWO to other areas of the High Arctic. The topography of the area is characterized by low-to-moderate relief, with vegetation similar to the CBAWO and other locations in the High Arctic.

2.2. Field Methods. A digital elevation model (DEM) for the Cape Bounty study area was derived from a high-resolution GeoEye stereopair collected in August 2009. An unsupervised classification was also conducted on the GeoEye imagery, to distinguish between the three main vegetation communities in the region (polar desert, mesic heath, and wet sedge). The DEM and unsupervised classification results were then used to set up a stratified random sampling scheme across three elevations (<30 m, 30–90 m, and >90 m) and three vegetation classes (Figure 2). The number of samples across each vegetation class was determined on a relative basis by the spatial coverage of each class in the unsupervised classification. The elevation groupings were chosen to exploit the current knowledge of different till layers, with marine sediments thought to be present between approximately 35 m and 90 m above sea level, as explained previously. In 2009, 119 sample

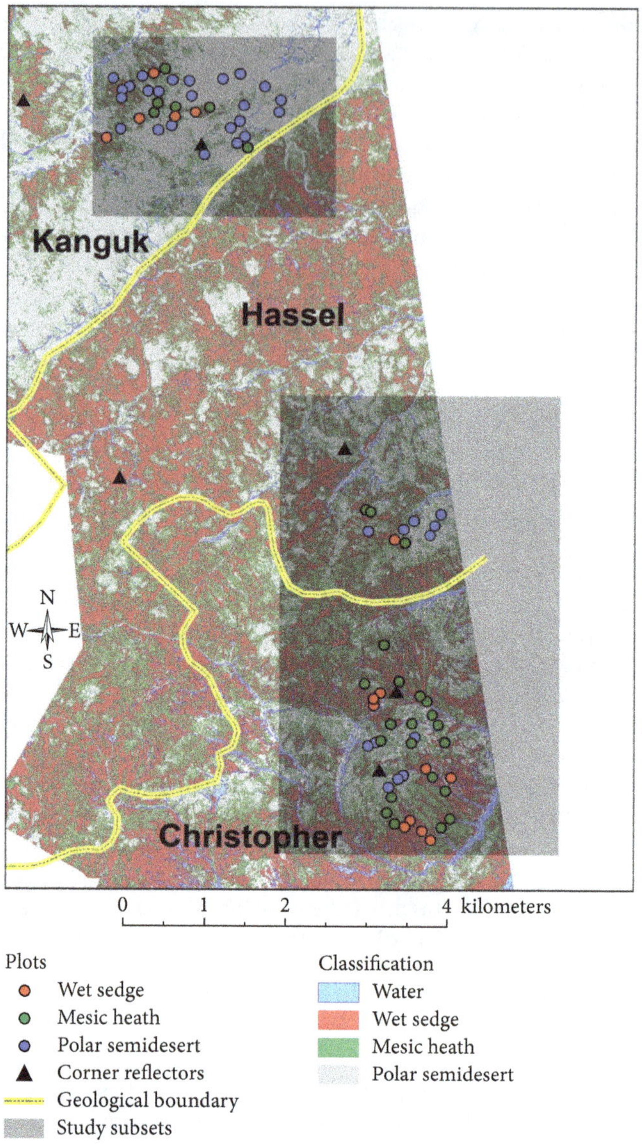

Figure 3: Unsupervised classification, study plots, and geological boundaries for the Cape Bounty Arctic Watershed Observatory.

locations were used, then expanded to 136 locations in 2010 to allow for increased coverage of the elevation classes. This stratified random sampling scheme was designed to obtain soil moisture measurements from areas with as many differences in backscatter-affecting variables as possible. Vegetation, surface roughness, and soil type, in addition to soil moisture, are the main components of a target that affect SAR backscatter response: surface roughness information could not be obtained a priori with the available optical imagery, however, so was not used as a stratification criterion. Similar methods were used at the Cape Collingwood locations, with the DEM and initial unsupervised vegetation classification being generated from a high-resolution 2011 Worldview-2 stereopair. A total of 79 sample plots were used at the Cape Collingwood location (Figure 3).

For both study locations, efforts were made to collect soil moisture measurements at each plot within three hours of the RADARSAT-2 acquisition, in order to minimize any daily fluctuations in soil moisture. For logistical reasons, this was sometimes expanded to ±6 hours. A meteorological station set up in the CBAWO gave precipitation information, which was used, along with in situ weather observations, to determine the timing and amount of precipitation with regard to overpass times and the temporal stability of the soil moisture measurements.

2.2.1. TDR Methodology. Soil moisture values were taken with a TDR (time-domain reflectometry) instrument, the Moisturepoint MP-917 from Environmental Sensors Inc. A TDR measures the apparent dielectric constant (K_a) of the soil by sending an electromagnetic pulse along a transmission line (i.e., metal probe) in a soil medium, and measuring the propagation velocity. The propagation velocity is influenced by the dielectric constant—higher dielectric values result in a slower velocity, when compared to the transmission velocity in a vacuum:

$$K_a = \left(\frac{c}{v}\right)^2, \quad (1)$$

where c is the speed of light and v is the propagation velocity.

The propagation velocity is determined by measuring the time it takes for a pulse to travel the distance to the end of the transmission line and back. Velocity can therefore be expressed as

$$v = \frac{2L}{t}, \quad (2)$$

where L is the linear distance travelled (probe length) and t is the measured travel time.

The values from the TDR were recorded as travel time, and so can be converted into the dielectric constant K_a by substituting (2) in (1):

$$K_a = \left(\frac{c}{(2L/t)}\right)^2, \quad (3)$$

where c is the speed of light (m/s), L is the length of the probe (mm), and t is the time delay (ns) measured with the TDR probe.

No site-specific calibration of the TDR readings was possible, so universal equations were applied. While not quite as accurate as a site-specific empirical regression could be, universal equations are very close in terms of overall accuracy and are an improvement over factory calibration settings [50]. The apparent dielectric constant was converted to % volumetric soil moisture (θ_v) using a 5°C three-phase model (4), that is, a linear approximation of the Topp universal equation [51, 52] for mineral soil and the Nagare equation [53] for organic soil (5).

$$\theta_v = 0.1209 * \sqrt{K_a} - 0.2032, \quad (4)$$

$$\theta_v = -0.0189 + (0.032 * K_a) - \left(0.000459 * \left(K_a^2\right)\right) + \left(0.0000027 * \left(K_a^3\right)\right). \quad (5)$$

While the Topp equation and its linear approximations have been shown to be accurate for a wide range of mineral soils [50–52], it is suggested very small bulk density values (such as in organic soils) can lead to large errors [54, 55], necessitating a separate equation (5) for low bulk density soils.

Soil moisture values were taken with custom 5 cm length probes, making the moisture values an integrated average of the top 5 cm of the soil. Three measurements per plot were taken in a triangular pattern around the plot center, approximately 2 m apart, with some measurements being taken in the vegetated areas, if present. This is important, as the vegetated areas hold different amounts of water than the mineral soil due to the different properties and bulk density of the heavily organic soils, and different equations are used to transform the TDR values to soil moisture values ((4) and (5)). The value for the plot is a weighted average of the organic and mineral soil moisture values based on an estimate of percent vegetation cover (organic soil) of the plot.

2.3. ANNs. The utility of ANNs in remote sensing (see review in [45]) and as a relatively simple way to invert radar backscatter to surface parameters is well documented [41, 56]. Surface parameter inversion from SAR backscatter data is usually carried out by some sort of linear or nonlinear regression approach [17] or through complex empirical inversion models such as IEM and others [57, 58]. ANNs, however, show great promise in both simplifying this procedure and increasing the accuracy of the results. For example, ANNs were compared to other models [41], that is, the Oh model [59] and the modified Dubois model (MDM) [60]. The ANN models outperformed the Oh model and MDM for surface roughness and soil moisture estimation. The selection of input predictor variables plays a critical role in the performance of ANNs, which may require different variable selection approaches typically adopted for parametric methods (i.e., due to the nonparametric nature of ANN) [61]. From a model specification perspective, the design of an ANN model is of crucial importance to the neural network learning the relationship between the input and output variables. More specifically, the optimal design (i.e., the number of hidden layers and the number of neurons in each layer) may vary across different applications, thus requiring prudent tuning on an individual basis.

However, even though ANNs do not operate based on assumptions regarding the statistical distribution of the input variables, the application of ANN is nontrivial due to the appropriate design and implementation needed to ensure its accuracy and robustness [45]. The performance of a trained ANN can be assessed using root mean square error (RMSE) between observed values and the values predicted by the network. Compared to some machine-learning algorithms that are less vulnerable to high variance, such as support vector machine (SVM), the size of the training sample can have a large impact on the performance of ANN [62]. ANNs learn the characteristics of the data from the sample values, not statistical derivatives. Hence, it is thought that very large numbers of training samples are required to optimally train the ANN, though excessive numbers of samples can slow the training process [45]. Various rules

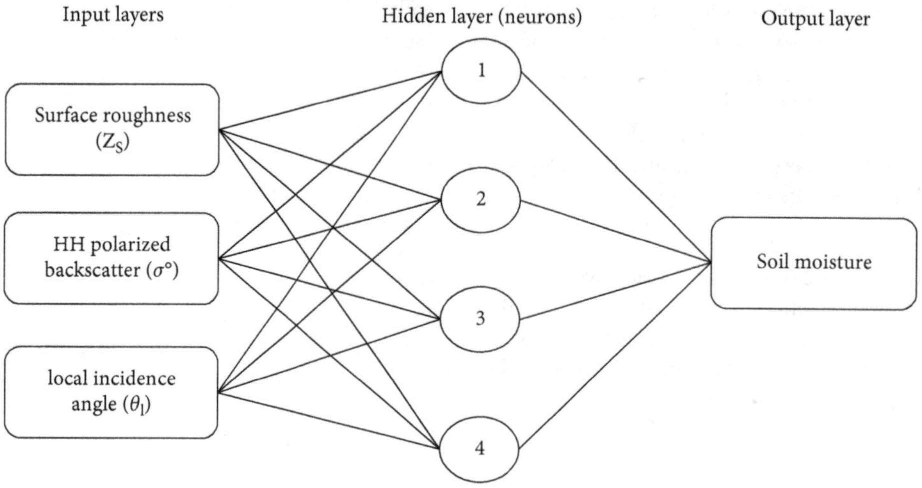

FIGURE 4: Schematic diagram of the spatially explicit input (and output) layers for the soil moisture ANN model.

have been proposed to determine the optimum number of training samples, for example, 5–10% of the image [63], or $30n(n+1)$ samples [64], with n being the number of input parameters to the ANN. However, these recommendations were for classification applications; that is, no suggestions have been made as to the number of training samples needed for the inversion of a continuous variable, though in this case it would seem that representativeness would be more important than the number of samples. If training samples were not fully representative of soil moisture variability, the predictive accuracy of the ANN model will likely deteriorate, resulting in less comprehensive understanding of the spatial variation of soil moisture over time.

The ANN models for this research were implemented in the MATLAB® software package (MathWorks, Natick, Massachusetts). This implementation is a two-layer (one hidden layer and one output layer) feed-forward network with sigmoid-function hidden neurons and linear-function output neurons, trained with the Levenberg–Marquardt back-propagation algorithm [65] (Figure 4). The input layers included surface roughness (Z_s), HH-polarized backscatter ($\sigma°$), and local incidence angle (θ_l). Input data were randomly separated into training (70% of the data), validation (15% of the data), and independent testing data (remaining 15%). Hundreds of models were trained, with each training run having slightly different starting weights and biases for each neuron, as well as randomly selected training data. Different numbers of hidden neurons were also examined, in a trial-and-error approach, until the best model of all the model runs could be selected. The final model for soil moisture had one hidden layer with four neurons (Figure 4).

2.4. Object-Based Image Analysis (OBIA) Methods. The models were built using object-based image analysis (OBIA), which involves pixels being grouped into objects based on the homogeneity of their spatial or spectral characteristics (termed segmentation). OBIA has a number of advantages over pixel-based methods, including reduced dependency on noise-filtering algorithms (especially with SAR data) and the objects being more natural representations of surface properties. OBIA is recognized as a very effective tool for analyzing geospatial and remotely sensed data [66]. With this method, different spatial scales can be investigated by changing the sizes of the image objects. The problem associated with OBIA is finding meaningful scales to investigate, that is, finding the scale of object segmentation that is objectively relevant based on the characteristics of the landscape being studied.

To solve this problem, Drăguţ et al. [67] have developed a tool that integrates with existing OBIA software (eCognition® 8.64) to estimate relevant scale parameters. The tool, ESP (estimation of scale parameter) plots values of local variance (LV) and the rate of change (ROC) of that variance between successive scale levels, that is, against scale levels at a set interval (Figure 5 for an example from the CBAWO). The "scale" of the object is a parameter in the eCognition software that determines the maximum allowed heterogeneity of the image objects, based on user-defined weightings of "color" (pixel values for various image bands) and "shape" (level of object compactness) [68]. In the ESP tool, LV increases with the increase in the scale parameter as the homogeneity of the objects increases; the highest values of LV relative to successive values indicate scales where objects have reached meaningful levels of organization in terms of the variation in their homogeneity. The ROC of the LV is the best way to show this and is a means of identifying how important the respective scale level is in structuring the information on objects' variability relative to the whole scene [67, 69].

The objects were generated from high-resolution optical imagery, 0.5 m pan-sharpened GeoEye-1 imagery for the CBAWO and 0.5 m pan-sharpened Worldview-2 imagery for the Cape Collingwood study site. In both cases, near-infrared, red, and green bands were used for the homogeneity criteria in order for the objects to be physically meaningful, in an ecological sense, on the ground. The Cape Collingwood site was split into two areas, north and south, to aid in processing times. Multiple scales were found to be meaningful (Figure 5), but only the largest (410) and smallest

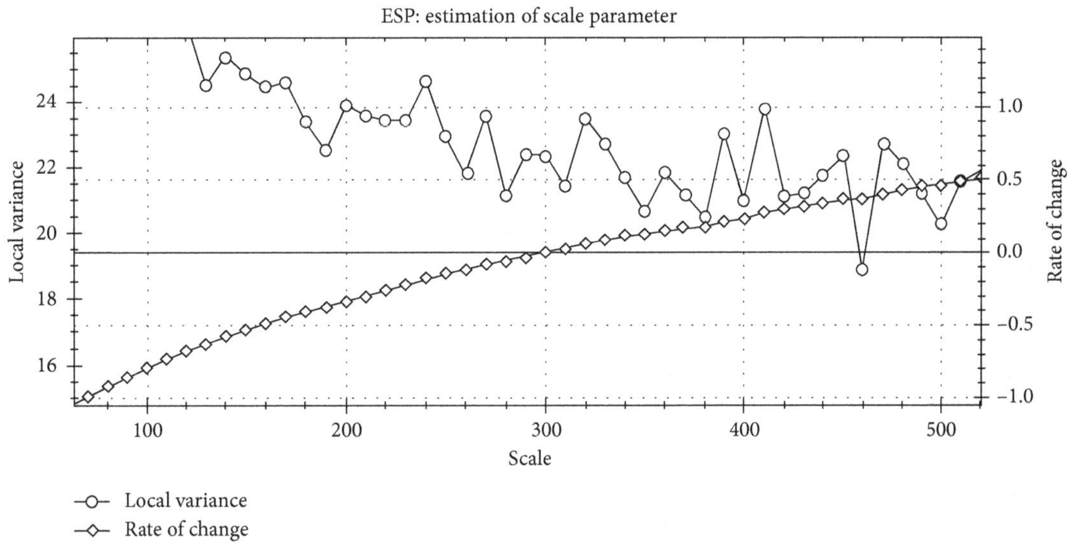

FIGURE 5: Local variance and rate of change values produced by the ESP tool for the Cape Bounty Arctic Watershed Observatory.

TABLE 1: RADARSAT-2 data used in the analysis.

RADARSAT-2 beam mode	Avg. incidence angle (°)	Polarization	Spatial resolution (m)	Acquisition date (local time)
U15	41.40	HH	3	01/07/2009
U4	33.00	HH	3	07/07/2010
U26	48.40	HH	3	09/07/2010
U75	25.75	HH	3	11/07/2010
U9	37.00	HH	3	17/07/2010
FQ2[a]	20.90	HH/VV/VH/HV	8	23/07/2010
U21	45.40	HH	3	23/07/2010
U75	25.75	HH	3	04/08/2010
U9	37.00	HH	3	10/08/2010
U8	36.25	HH	3	09/07/2011
U74	24.75	HH	3	10/07/2011
U13	40.00	HH	3	12/07/2011
U79	29.55	HH	3	13/07/2011
U4	33.00	HH	3	23/07/2011

[a]Pass was descending (all others were ascending).

(140) scale parameter values were used in further analysis. The scale values for the small objects at the Cape Collingwood site were slightly different, as the ESP tool estimates the scale parameter based on the local conditions. Scale values of 110 and 160 were used for the small objects for the north and south mapped portions, respectively, and 410 and 400 for the large objects. Small objects for both locations ranged between approximately 50 and 6000 m^2. The larger objects range from about 50 to 35000 m^2. These object sizes could have implications for the SAR spatial resolution requirements for surface roughness and soil moisture modelling.

2.5. SAR Methods. High spatial resolution fully polarimetric (FQ—fine-quad beam mode) and HH-polarized single polarization (U—ultra-fine beam mode) SAR data were collected at various incidence angles over the study areas during the summers of 2009–2011 (Table 1). The calibration of the soil moisture ANN relied on a single date of SAR data to model moisture for that specific date in time. Given that surface soil moisture content can change rapidly through time, only parameters from the date of interest, along with temporally invariant parameters, such as surface roughness, were used to develop the ANN model. The scenes available for different dates were also from different beam modes, with different incidence angles. To examine the capability of the soil moisture ANN model to generalize over other acquisition conditions, it was calibrated with data from one beam mode (one date) and applied to different beam modes from different dates, some with very different average incidence angles. For this reason, the local incidence angle of each image object was also included as a model input variable.

The ultra-fine mode RADARSAT-2 data were orthorectified before further analysis, while the fine-quad data were analyzed in slant range to preserve polarimetric information [70] and were only orthorectified to extract the plot data for each variable after they were calculated for the scene. Nine corner reflectors spaced around the CBAWO, and five around the Cape Collingwood study area, were used to assist with geometric correction. Since an object-based approach was implemented in the modelling framework, speckle reduction was handled through image averaging at the object level.

2.6. Soil Moisture Models. The relationship between SAR backscatter and soil moisture must be separated from other factors that influence the SAR signal, including surface roughness [71]. Hence, the output of a surface roughness ANN model [72, 73] was included as input to the soil moisture model. More specifically, this surface roughness ANN model was developed using multiangular and polarimetric RADARSAT-2 imagery and field data in an object-based framework for the Cape Bounty and Cape Collingwood study site. To investigate the impact of different spatial scales on surface roughness estimation, analyses were performed on two distinct image object sizes, and consistent performance was observed for both small and large image objects, with normalized RMSEs of approximately 15%. A variety of additional variables were derived and extracted from the SAR data [74], corresponding to the image objects that contained the field-measured soil moisture plots (Table 2). The texture variables were calculated on a per-pixel basis using an 11 × 11 pixel window for each RADARSAT-2 scene (i.e., HH backscatter intensity), before being averaged for each image object. The texture measures from each scene (and beam mode) were analyzed for correlation individually. Other variables include simple means of the pixel values for each image object.

Variables that were not directly correlated (Pearson $|r| < \sim 0.3$) with the field-measured soil moisture were removed from further analysis. Visual analysis of scatter plots helped to reveal possible nonlinear correlations that the Pearson correlation would not identify. Variables that were highly correlated (Spearman $|\rho| > \sim 0.7$) with other remaining variables were also removed, as multicollinearity can be problematic in the calibration of ANNs [77].

3. Results and Discussion

3.1. ANN Model. Of the SAR variables that were examined (Table 2), linear HH backscatter was selected for inclusion in the ANN, as it exhibited stronger correlation to soil moisture than VV and, most importantly, is the only polarization available in both FQ and U beam modes. High temporal coverage was an important consideration for the modelling effort, so variable selection was limited to allow the most possible acquisitions to be used. Reliance on polarimetric parameters, for example, would have (i) excluded the high volume of ultra-fine beam mode data, (ii) limited the model to FQ data, and (iii) reduced the temporal resolution of the analysis. A number of polarimetric variables were examined for completeness, such as Cloude–Pottier decomposition variables [78], intensity ratio, and phase difference, none of which showed strong correlation to soil moisture, something that has been noted in other studies [79].

The variables that were used in the final ANN model were (i) the surface roughness (Z_S) values modelled by the surface roughness ANN [72, 73], (ii) the HH-polarized

Table 2: Variables generated from SAR data.

Variable	Description
Homogeneity[a]	A measure of local homogeneity
Contrast[a]	A measure of local variation
Correlation[a]	A measure of the linear dependency of grey levels of neighbouring pixels
Mean[a]	Arithmetic mean of all pixel values
SD[a]	Standard deviation of pixel values
VI/VA/VL/U[b]	A normalized log measure of texture
HH	σ^0 intensity of HH polarized backscatter

[a]Haralick et al., [75]. [b]Oliver and Quegan, [76].

backscatter (σ°), and (iii) the local incidence angle that corresponds to the HH backscatter used, all averaged across each image object. Of these three variables, the surface roughness is time invariant, so the same values were used across each date for which the model was applied. The σ° values were taken from the scene that corresponds to the date of interest, and the local incidence angle values derived from that same scene. The model was created using the July 11, 2010 U75 CBAWO data, then applied to the other dates for the CBAWO and Cape Collingwood datasets, with the σ° and local incidence angle inputs changing accordingly. A total of 15 plots were not included in the model—five plots had no surface roughness data, and ten were classified as outliers based on the Mahalanobis distances when the inputs were compared to the field-measured moisture. The model was applied to both the small and large image objects at each study location. The results of the ANN modelling are presented in Tables 3–5.

3.2. Cape Bounty Arctic Watershed Observatory. The results for the 2009 and 2010 CBAWO data (Table 3) demonstrate that the best results tend to occur with the steepest incidence angles, that is, the U75 beam mode. Of course, the model was created using the U75 beam mode, but this is also expected due to the nature of the relationship between moisture-affected SAR backscatter and local incidence angles, that is, low incidence angles have been shown to be more accurate for soil moisture prediction [80] since the roughness signal is minimized, leaving moisture as the dominant control on the backscatter. The inclusion of surface roughness and incidence angle information in the ANN allows these differences to be accounted for across different beam modes, but the nature of the relationships between these variables cannot be fully modelled. The modelled soil moisture error is approximately 15% across the various dates and beam modes. Soil moisture values at the CBAWO cover nearly the full range of 0 to 1 m^3/m^3 volumetric water content. When the ANN was applied to the large image objects, the results were similar to the small image objects. The r^2 values are slightly lower, but the RMSE values are comparable.

The ANN soil moisture output, apart from the July 11 U75 data, which were used to create the ANN, is not at the same scale as the field-measured moisture. This difference in scale is because the local incidence angle and σ° backscatter are dependent on the beam mode used. The relationships are still linear, but the slope and intercept of the regression line

TABLE 3: r^2 and RMSE values of the ANN applied to the different scenes/dates at the CBAWO in 2010.

Date	Beam mode	Avg. inc. angle	Sm. obj. r^2	Sm. obj. RMSE	Lrg. obj. r^2	Lrg. obj. RMSE
1-Jul-2009	U15	43.3	0.15	0.171	—	—
7-Jul-2010	U4	33	0.09	0.210	0.03	0.217
9-Jul-2010	U26	48.4	0.29	0.159	0.18	0.149
11-Jul-2010[a]	U75	25.75	0.75	0.086	0.40	0.118
17-Jul-2010	U9	37	0.39	0.151	0.35	0.121
23-Jul-2010	U21	45.4	0.37	0.173	0.29	0.179
23-Jul-2010[b]	FQ2	20.9	0.31	0.181	—	—
4-Aug-2010	U75	25.75	0.44	0.153	0.32	0.167
10-Aug-2010	U9	37	0.39	0.165	0.32	0.162

Data are presented for both small and large image objects. [a]ANN created using this date. [b]Fine-quad data, descending pass (all others ascending).

TABLE 4: r^2 and RMSE values of the ANN applied to the different scenes/dates at Cape Collingwood in 2011.

Date	Beam mode	Sm. obj. r^2	Sm. obj. RMSE	Lrg. obj. r^2	Lrg. obj. RMSE
09-Jul-2011[a]	U8	0.13	0.142	0.21	0.136
10-Jul-2011[a]	U74	0.14	0.193	0.27	0.178
12-Jul-2011[b]	U13	0.16	0.163	0.28	0.134
13-Jul-2011[b]	U79	0.20	0.148	0.05	0.135
23-Jul-2011[b]	U4	0.04	0.150	0.05	0.147

[a]Covers northern subset of study area (Kanguk formation). [b]Covers southern subset of study area (Hassel and Christopher formations). Data are presented for both small and large image objects.

TABLE 5: Results of the ANN applied to the different scenes/dates at Cape Collingwood in 2011, after removal of outlier moss plots.

Date	Beam mode	Sm. obj. r^2	Sm. obj. RMSE	Lrg. obj. r^2	Lrg. obj. RMSE
09-Jul-2011	U8	$0.20^{\uparrow}$	$0.122^{\downarrow}$	$0.32^{\uparrow}$	$0.113^{\downarrow}$
10-Jul-2011	U74	$0.20^{\uparrow}$	$0.172^{\downarrow}$	$0.42^{\uparrow}$	$0.147^{\downarrow}$
12-Jul-2011	U13	$0.19^{\uparrow}$	$0.116^{\downarrow}$	$0.22^{\downarrow}$	$0.119^{\downarrow}$
13-Jul-2011	U79	$0.27^{\downarrow}$	$0.106^{\downarrow}$	$0.03^{\downarrow}$	$0.126^{\downarrow}$
23-Jul-2011	U4	$0.05^{\uparrow}$	$0.151^{\uparrow}$	$0.05^{\leftrightarrow}$	$0.147^{\leftrightarrow}$

Results are presented for both small and large image objects. ↑ = increase from Table 4 values. ↓ = decrease from Table 4 values. ↔ = no change.

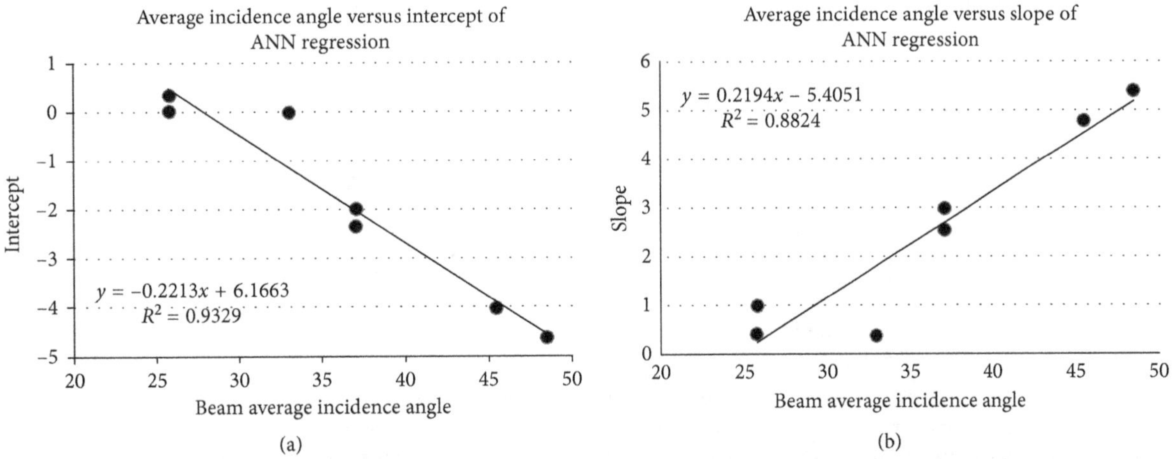

FIGURE 6: Beam mode incidence angle correlations to slope and intercept values calculated for ANN results.

that relates the ANN output to the field measurements vary by beam mode. However, the differences in the slopes and intercepts correspond closely to the average incidence angle of the beam mode (Figure 6). This relationship can therefore be used to determine the appropriate scaling needed for the ANN output for any given beam mode, even those not available for this study. The r^2 values of the slope and intercept equations are 0.88 and 0.93, respectively, giving high confidence in this relationship between beam average incidence angle and slope/intercept values for the ANN output. The results of the ANN for each different beam mode can be combined once the output scales have been matched

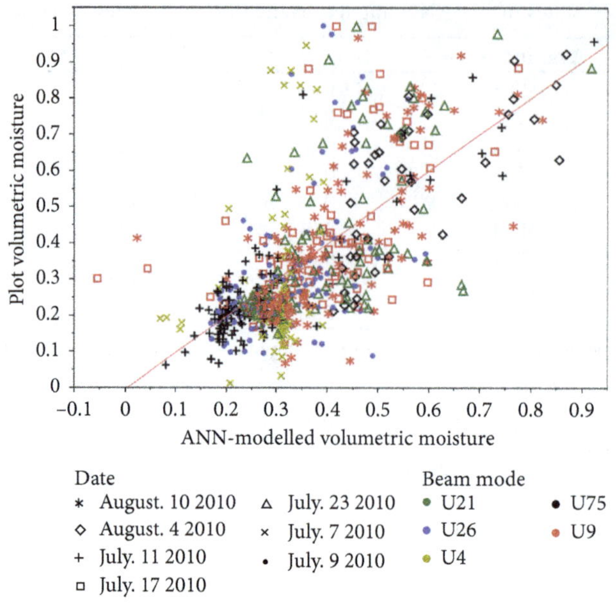

FIGURE 7: Combined ANN output for each 2010 date, after scaling the results.

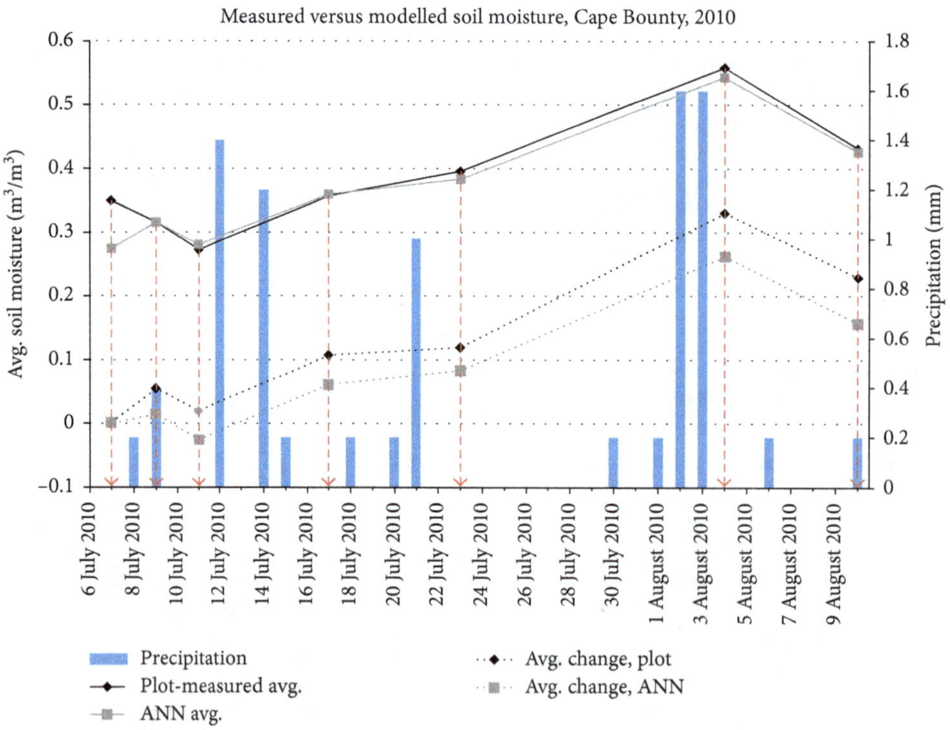

FIGURE 8: Precipitation and its effects on average soil moisture at the Cape Bounty Arctic Watershed Observatory, 2010. Arrows indicate RADARSAT-2 overpass dates. Plot-measured and ANN avg. values are average soil moisture values measured in the field at each date and the ANN output for that date, respectively. Avg. change values are differences in moisture between each date from a 0 start value (at July 7).

(Figure 7). The r^2 and RMSE values for the combined results are 0.46 and 0.155, respectively.

Once the ANN output is in the same scale as the field-measured soil moisture, then the absolute values can be compared directly. One of the strengths of this methodology is the ability to model soil moisture at different times, and to therefore have a temporal record of soil moisture. The mean soil moisture values across all the plots were compared for each date for which imagery was available and the differences between the field-measured and ANN-modelled moisture values analyzed (Figure 8). The average moisture values given by the ANN model are very similar to the field-measured soil moisture values across all dates (July 7 is the exception). The average soil moisture is affected by the precipitation across the

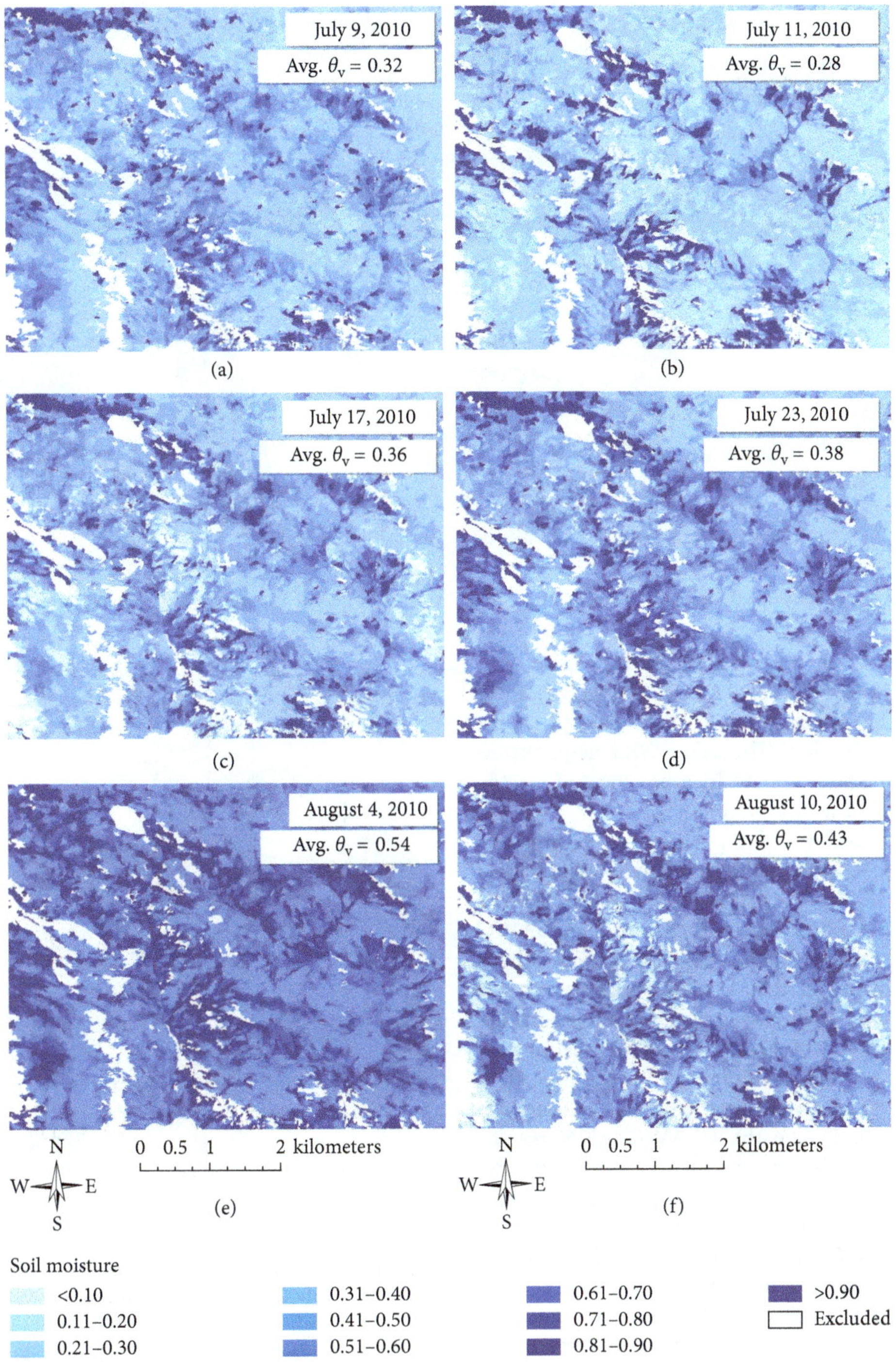

FIGURE 9: Spatiotemporal distribution of θ_v across 6 dates in 2010 at the Cape Bounty Arctic Watershed Observatory. Average θ_v values are the ANN-modelled averages for the field-measured plots.

area, and this is also reflected in the data. Precipitation data as measured by the CBAWO meteorological station is correlated to increases in average soil moisture, with soil moisture decreasing during periods without precipitation (Figure 8).

The pattern of average soil moisture increases and decreases with weather conditions, a pattern confirmed when the entire study area is modelled for each date. The ANN was applied to every image object for each date, excluding those objects that did not have surface roughness data [72, 73]. Mapping the spatial distribution of soil moisture at such high spatial resolutions is another major strength of the approach taken for this research. The change in moisture across both temporal and spatial scales is readily apparent (Figure 9). Comparing the small to the large image objects reveals the spatial consistency in modelled soil moisture across object sizes (Figure 10).

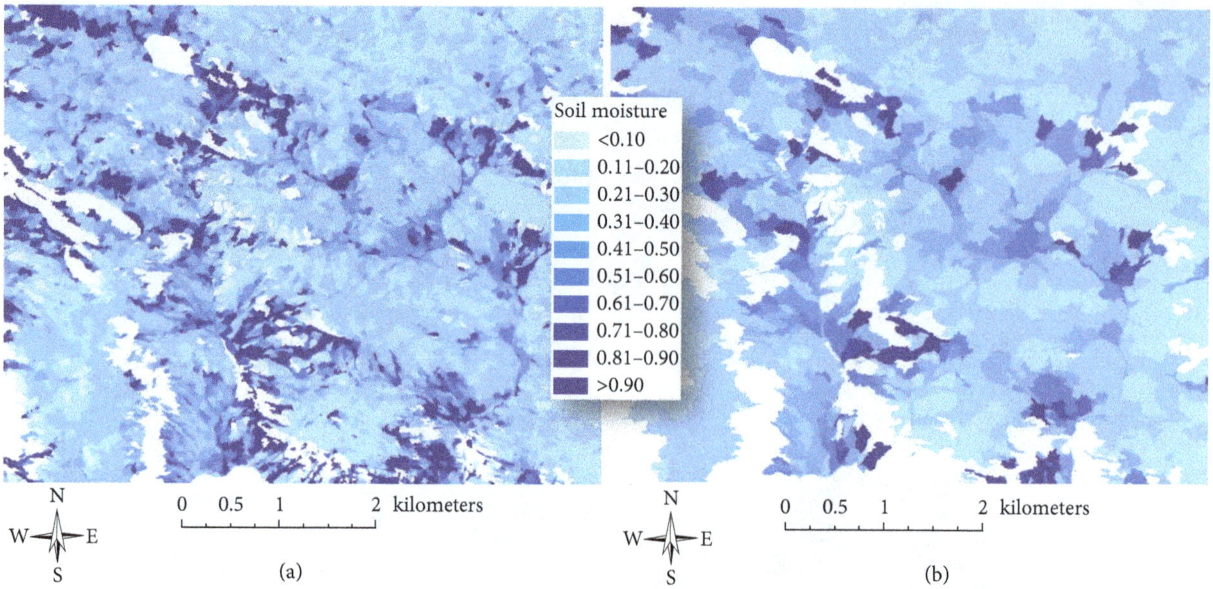

FIGURE 10: Comparison of ANN-modelled soil moisture for small (a) and large (b) image objects. Values shown are for July 11, 2010.

3.3. Cape Collingwood. In the analysis of the Cape Collingwood data, the July 9 (U8) and the July 10 (U74), 2011 acquisitions cover the northern study area, located in the Kanguk geological formation. The other 2011 acquisitions (Table 2) cover the southern study area, located in the Christopher and Hassel formations.

The results of the ANN model applied to the 2011 imagery are presented in Table 4. The r^2 values are not as strong as for the CBAWO results, though the RMSE values are similar. The total range of moisture values is smaller than at the CBAWO (θ_v of 0.1–0.8 for the northern subset, 0.2–1.0 for the southern subset), giving rise to higher RMSE values on a percentage basis. The July 23 scene resulted in a very poor relationship, although none are particularly strong. It was very wet on July 23, with over 16 mm of rain having fallen over the preceding few days (precipitation measured manually on location with rain gauge) (Figure 8). Unlike the CBAWO ANN results, there is less evidence for incidence angle impacting the accuracy of the results for Cape Collingwood. Beam modes with a small incidence angle, such as U79, have stronger results, while others, such as U74, are no better than the large incidence angle beam mode results (U8, U13).

When the results are scaled to the field-measured soil moisture values and combined (same methodology as for the CBAWO data), the model does not seem to apply as well to Cape Collingwood (Figure 11), though the RMSE values are acceptable. Without the July 23 data included, the combined r^2 and RMSE values are 0.26 and 0.122, respectively. Some of the error in the ANN-modelled moisture values for Cape Collingwood can be attributed to a few plots that are characterized by heavy moss cover, which can have unpredictable effects on SAR backscatter (Figure 12). When these plots are removed (1 plot from North subset, 3 plots from South subset), the results are improved in nearly every case (Table 5).

The absolute values of soil moisture, as has been noted earlier, are not as important as the spatiotemporal results, that is, the modelling of the relative moisture values across

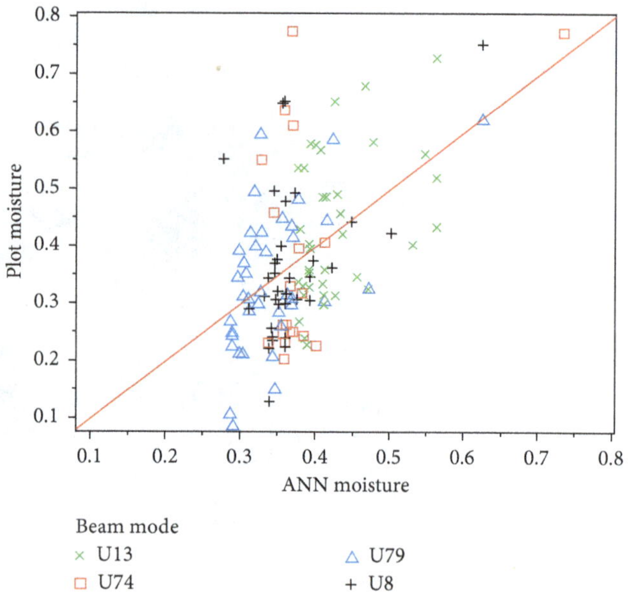

FIGURE 11: Combined ANN results versus plot-measured soil moisture. July 23 data are not included.

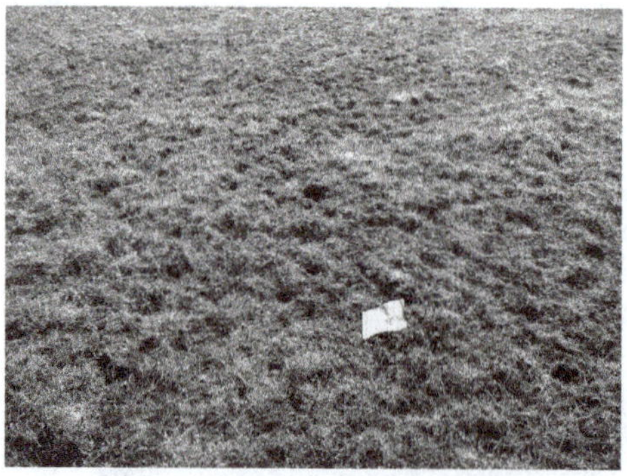

FIGURE 12: Cape Collingwood outlier plot, showing the heavy moss cover.

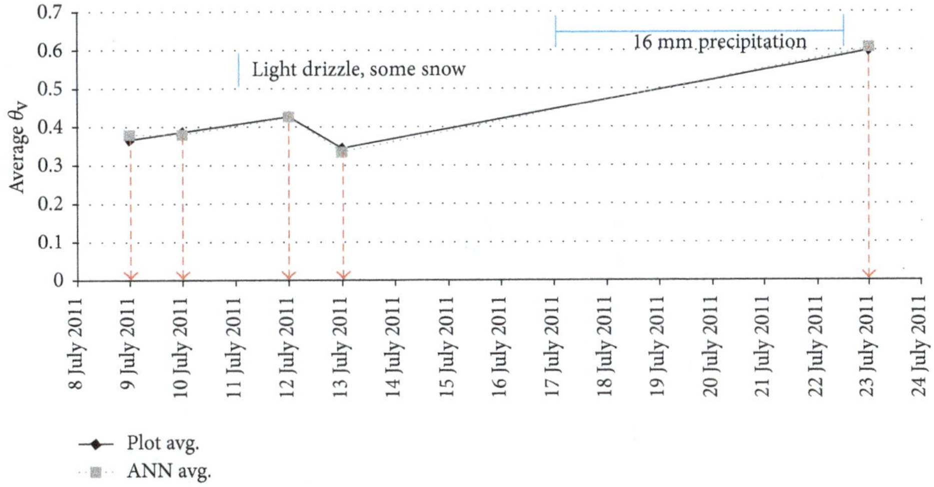

FIGURE 13: Comparison of field-measured average soil moisture to ANN-modelled average soil moisture. Values are within 0.015 θ_v of each other for each date. Arrows indicate RADARSAT-2 data acquisition dates. Timing and duration of precipitation events are also indicated.

space and through time [16]. The Cape Collingwood data again show good agreement with mean values of θ_v (Figure 13), the field-measured and scaled ANN-modelled values being within 0.015 θ_v of each other. The similar mean values indicate that the model is robust with respect to the relative differences between dates and across large numbers of image objects. The ANN results were applied to all image objects in the study area, and the resulting maps clearly indicate stream channels and the wet slopes surrounding them (Figure 14). Aside from the quantitative results already given, this demonstrates in a qualitative sense that the model is working as intended.

3.4. Sources of Error. The relationship between the absolute values of θ_v as determined by the ANN and field measurements is clearly not strong in all cases. With the aforementioned caveats about the relevance of absolute θ_v values in the context of this research in mind, it is still useful to point out some of the underlying causes of the nature of these relationships. The data used as input to the ANN have errors associated with them, specifically with regard to the surface roughness values [72, 73]. We know that even small inaccuracies in roughness parameterization can lead to large errors in soil moisture estimation [81, 82]. In addition to the errors inherent in the surface roughness ANN output [72, 73], there is also the inherent error in the difference between surface-measured roughness and SAR-perceived roughness.

To further the idea of inherent errors in field measurements, there is the problem of integrating soil moisture values across depth. The TDR measurements of θ_v were averaged over the top five cm of the soil surface, and there were often noticeable (qualitatively) moisture gradients between the top of the soil and at 5 cm depth. The SAR backscatter is not necessarily representing the soil to the same depth, a characteristic of the data that depends on θ_v—the lower the soil moisture, the farther the SAR energy penetrates the soil [83], up to a maximum of approximately five cm with RADARSAT-2 (equivalent to the TDR-measured depth). This is a problem that has been known for some time [84], yet is not readily resolved.

Soil moisture changes can also change the form of SAR scattering behaviour. It is known from previous work in the area [27] that saturated or very high soil moisture conditions cause a reduction in the strength of the relationship between SAR backscatter (and other SAR-derived variables) and surface soil moisture. Dobson and Ulaby [85] demonstrated how SAR backscatter reflection changes from diffuse to specular as the soil moisture nears saturation, reducing the backscatter to values associated with lower soil moisture; this is thought to be the reason for the underestimation of high soil moisture values in the ANN model presented here. Under these conditions (i.e., saturated soil moisture), the relationship between the dielectric constant and SAR backscatter is reduced or eliminated; hence, the modelling of soil moisture is confounded within the ANN. This response is also affected by the local incidence angle, with incidence angles closer to nadir being affected less than those at shallower angles. Other studies have revealed that low incidence angles (closer to nadir) are better for soil moisture modelling with SAR data [80], as these angles minimize the backscatter contribution from surface roughness, in addition to minimizing the effects of increased specular reflection due to near-saturated moisture conditions. With the dataset presented in this research, evidence in support of this idea can be seen by analyzing high and low incidence beam modes from the same date (and therefore similar soil moisture conditions). In this case, both a U21 scene (average incidence angle = 45.4°) and an FQ2 scene (average incidence angle = 20.9°) were acquired for the CBAWO on July 23, 2010. When the ANN model output for these two scenes are compared (Figure 15), it is clear that the smaller incidence angle FQ2 scene is modelling higher soil moisture values ($\theta_v > 0.60$) more accurately than the large incidence angle U21 data. Out of 20 field-measured values where $\theta_v > 0.60$, only four of those values are modelled above the 0.60 level for the U21 data, compared to 12 above that level for

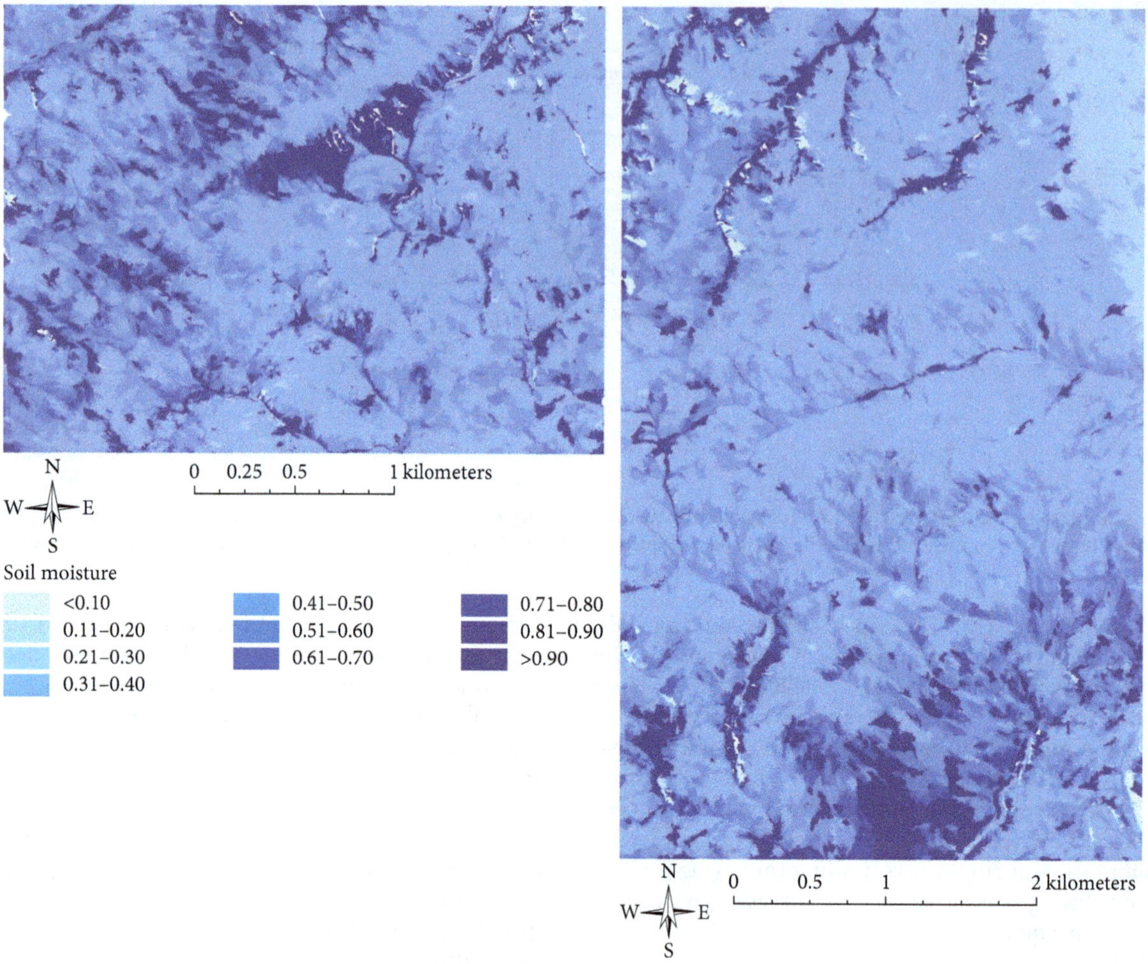

FIGURE 14: ANN-modelled soil moisture (θ_v) values for the Cape Collingwood study area. The North subset is on the left, and the South subset is on the right. Values are from July 1, 2011.

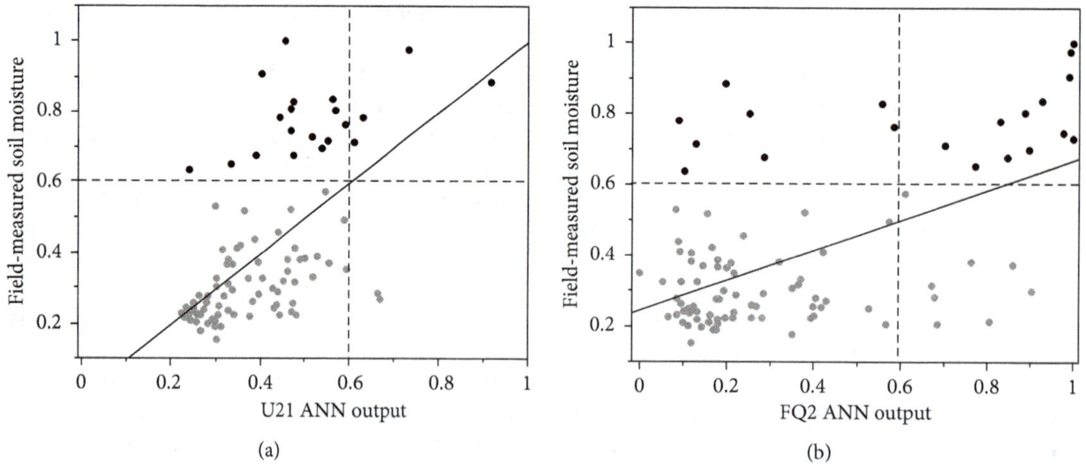

FIGURE 15: Comparison of incidence angle effects when modelling high soil moisture ($\theta_v > 0.60$) values at the Cape Bounty Arctic Watershed Observatory, July 23, 2010. Here, the U21 and FQ2 scenes have average incidence angles of 45.4° and 20.9°, respectively. Field-measured soil moisture values where volumetric water content was greater than 0.60 (i.e., $\theta_v > 0.60$) are represented in black, whereas values less than 0.60 (i.e., $\theta_v < 0.60$) are presented in grey. Dashed lines indicate where $\theta_v = 0.60$ for both field-measured and ANN-modelled soil moisture values.

the FQ2 data. The overall r^2 is lower for the FQ2 data, however, which could be a result of the comparatively larger spatial resolution or the fact that it is a descending pass (the model was created using ascending pass data). As a result, it is unclear as to why smaller incidence angles do not appear to perform as well when modelling low soil moisture conditions.

4. Conclusions

The soil moisture ANN was implemented using both SAR data (HH-polarized backscatter and local incidence angle) and previously modelled surface roughness values [72, 73]. The model was created from a single date and beam mode (July 11, 2010, U75) and then applied to multiple dates and beam modes, a methodology necessary to model soil moisture through time. The model was applied to the CBAWO, where it was calibrated, and the Cape Collingwood study area, where it was validated under different biophysical and geological conditions.

The 0.155 RMSE average that resulted from the application of the ANN to the different dates at the CBAWO, with the very large soil moisture range present in these environments ($\theta_v = 0$ to 1), is equivalent to other studies at southern latitudes with a θ_v RMSE of 0.05 (a commonly cited value), but maximum soil moisture values of $\theta_v = 0.40$. The absolute moisture values (Figure 8) are not as important as the spatiotemporal distribution of the soil moisture however (i.e., relative differences) (Figure 9), and the strong results of this metric indicate that the model tends to be robust with different input beam modes and across different physical spaces. Similar results were found at the validation site (i.e., Cape Collingwood), with an average r^2 of 0.26 and RMSE of 0.122, though again the spatiotemporal modelling of the moisture was much better (Figure 13).

The best results appear to occur when the model is applied using smaller incidence angle beam modes, which reduce the effects of surface roughness and specular scattering from areas of very high soil moisture. There are, however, very strong relationships between the slope ($r^2 = 0.88$) and intercept ($r^2 = 0.93$) of the regression used to match the ANN output for the different beam modes to the scale of soil moisture measured on the ground. Using these relationships, it is possible to use any beam mode (and associated incidence angle) with the model, as the output can be converted to the proper scale using the slope and intercept equations. However, it is acknowledged that smaller incidence angles may not be as suited to modelling low soil moisture, although the factors for this remain unclear.

When the model is applied to larger image objects, the results are very similar to the smaller image objects, giving rise to the potential to apply the ANN across larger areas (i.e., up-scaling the model output). This adaptability, along with the strong spatiotemporal modelling results, suggests that the ANN model presented in this research is capable of providing useful information for hydrological and climatic model assimilation, in addition to being used as input in other models (hazard susceptibility, vegetation, etc.) and for mapping soil moisture distribution at a high spatial resolution across watersheds.

Conflicts of Interest

The authors declare no conflicts of interest with respect to the publication of this paper.

Acknowledgments

This study received RADARSAT-2 data through the Science and Operational Applications Research–Education initiative (SOAR-E) program of the Canadian Space Agency and the Canada Centre for Remote Sensing. Funding was provided by a variety of sources, including NSERC Alexander Graham Bell Canada Graduate Scholarship (CGS D) (Collingwood), NSERC Discovery Grant (Grant no. RGPIN/03822, Treitz), NSERC Strategic Grant (Grant no. STPGP/380977, Lafrenière), Government of Canada Program for International Polar Year (Grant no. 2006-CC053; Lamoureux), NCE-ArcticNet (Grant no. NCE/12000; Lafrenière and Lamoureux), the Northern Scientific Training Program (NSTP), and the Polar Continental Shelf Project (PCSP).

References

[1] M. S. Torn and F. S. Chapin, "Environmental and biotic controls over methane flux from arctic tundra," *Chemosphere*, vol. 26, no. 1–4, pp. 357–368, 1993.

[2] L. Illeris, A. Michelsen, and S. Jonasson, "Soil plus root respiration and microbial biomass following water, nitrogen, and phosphorus application at a high arctic semi desert," *Biogeochemistry*, vol. 65, no. 1, pp. 15–29, 2003.

[3] J. M. Welker, J. T. Fahnestock, G. H. R. Henry, K. W. O'Dea, and R. A. Chimner, "CO_2 exchange in three Canadian High Arctic ecosystems: response to long-term experimental warming," *Global Change Biology*, vol. 10, no. 12, pp. 1981–1995, 2004.

[4] S. Sjögersten, R. van der Wal, and S. J. Woodin, "Small-scale hydrological variation determines landscape CO_2 fluxes in the high Arctic," *Biogeochemistry*, vol. 80, no. 3, pp. 205–216, 2006.

[5] S. F. Oberbauer, C. E. Tweedie, J. M. Welker et al., "Tundra CO_2 fluxes in response to experimental warming across latitudinal and moisture gradients," *Ecological Monographs*, vol. 77, no. 2, pp. 221–238, 2007.

[6] J. Dagg and P. Lafleur, "Vegetation community, foliar nitrogen, and temperature effects on tundra CO_2 exchange across a soil moisture gradient," *Arctic, Antarctic, and Alpine Research*, vol. 43, no. 2, pp. 189–197, 2011.

[7] M. Reichstein, M. Bahn, P. Ciais et al., "Climate extremes and the carbon cycle," *Nature*, vol. 500, no. 7462, pp. 287–295, 2013.

[8] R. G. Taylor, B. Scanlon, P. Döll et al., "Ground water and climate change," *Nature Climate Change*, vol. 3, no. 4, pp. 322–329, 2013.

[9] M.-K. Woo and Z. Xia, "Suprapermafrost groundwater seepage in gravelly terrain, Resolute, NWT, Canada," *Permafrost and Periglacial Processes*, vol. 6, no. 1, pp. 57–72, 1995.

[10] M. Woo and Z. Xia, "Effects of hydrology on the thermal conditions of the active layer," *Hydrology Research*, vol. 27, no. 1–2, pp. 129–142, 1996.

[11] K. M. Hinkel, F. Paetzold, F. E. Nelson, and J. G. Bockheim, "Patterns of soil temperature and moisture in the active layer and upper permafrost at Barrow, Alaska: 1993–1999," *Global and Planetary Change*, vol. 29, no. 3, pp. 293–309, 2001.

[12] S. M. Natali, E. A. G. Schuur, and R. L. Rubin, "Increased plant productivity in Alaskan tundra as a result of experimental warming of soil and permafrost," *Journal of Ecology*, vol. 100, no. 2, pp. 488–498, 2012.

[13] K. E. O. Todd-Brown, J. T. Randerson, F. Hopkins et al., "Changes in soil organic carbon storage predicted by Earth system models during the 21st century," *Biogeosciences*, vol. 11, no. 8, pp. 2341–2356, 2014.

[14] R. P. Daanen, D. Misra, and H. Epstein, "Active-layer hydrology in nonsorted circle ecosystems of the arctic tundra," *Vadose Zone Journal*, vol. 6, no. 4, pp. 694–704, 2007.

[15] P. Dobriyal, A. Qureshi, R. Badola, and S. A. Hussain, "A review of the methods available for estimating soil moisture and its implications for water resource management," *Journal of Hydrology*, vol. 458, pp. 110–117, 2012.

[16] K. C. Kornelsen and P. Coulibaly, "Advances in soil moisture retrieval from synthetic aperture radar and hydrological applications," *Journal of Hydrology*, vol. 476, pp. 460–489, 2013.

[17] S. S. Haider, S. Said, U. C. Kothyari, and M. K. Arora, "Soil moisture estimation using ERS 2 SAR data: a case study in the Solani River catchment/estimation de l'humidité du sol grâce à des données ERS-2 SAR: étude de cas dans le bassin de la rivière Solani," *Hydrological Sciences Journal*, vol. 49, no. 2, 2004.

[18] S. Said, U. C. Kothyari, and M. K. Arora, "ANN-based soil moisture retrieval over bare and vegetated areas using ERS-2 SAR data," *Journal of Hydrologic Engineering*, vol. 13, no. 6, pp. 461–475, 2008.

[19] L. Brocca, T. Tullo, F. Melone, T. Moramarco, and R. Morbidelli, "Catchment scale soil moisture spatial–temporal variability," *Journal of Hydrology*, vol. 422-423, pp. 63–75, 2012.

[20] P. C. Dubois, J. Van Zyl, and T. Engman, "Measuring soil moisture with imaging radars," *IEEE Transactions on Geoscience and Remote Sensing*, vol. 33, no. 4, pp. 915–926, 1995.

[21] D. Entekhabi, E. G. Njoku, P. E. O'Neill et al., "The soil moisture active passive (SMAP) mission," *Proceedings of the IEEE*, vol. 98, no. 5, pp. 704–716, 2010.

[22] F. De Zan, A. Parizzi, P. Prats-Iraola, and P. López-Dekker, "A SAR interferometric model for soil moisture," *IEEE Transactions on Geoscience and Remote Sensing*, vol. 52, no. 1, pp. 418–425, 2014.

[23] B. W. Barrett, E. Dwyer, and P. Whelan, "Soil moisture retrieval from active spaceborne microwave observations: an evaluation of current techniques," *Remote Sensing*, vol. 1, no. 3, pp. 210–242, 2009.

[24] A. Moreira, G. Krieger, I. Hajnsek et al., "Tandem-L: a highly innovative bistatic SAR mission for global observation of dynamic processes on the Earth's surface," *IEEE Geoscience and Remote Sensing Magazine*, vol. 3, no. 2, pp. 8–23, 2015.

[25] D. L. Kane, L. D. Hinzman, H. Yu, and D. J. Goering, "The use of SAR satellite imagery to measure active layer moisture contents in Arctic Alaska," *Hydrology Research*, vol. 27, no. 1-2, pp. 25–38, 1996.

[26] N. G. Meade, L. D. Hinzman, and D. L. Kane, "Spatial estimation of soil moisture using synthetic aperture radar in Alaska," *Advances in Space Research*, vol. 24, no. 7, pp. 935–940, 1999.

[27] J. Wall, A. Collingwood, and P. Treitz, "Monitoring surface moisture state in the Canadian high Arctic using synthetic aperture radar (SAR)," *Canadian Journal of Remote Sensing*, vol. 36, no. 1, pp. S124–S134, 2010.

[28] E. Högström and A. Bartsch, "Impact of backscatter variations over water bodies on coarse-scale radar retrieved soil moisture and the potential of correcting with meteorological data," *IEEE Transactions on Geoscience and Remote Sensing*, vol. 55, no. 1, pp. 3–13, 2017.

[29] T. Jagdhuber, J. Stockamp, I. Hajnsek, and R. Ludwig, "Identification of soil freezing and thawing states using SAR polarimetry at C-Band," *Remote Sensing*, vol. 6, no. 3, pp. 2008–2023, 2014.

[30] E. E. Sano, A. R. Huete, D. Troufleau, M. S. Moran, and A. Vidal, "Relation between ERS-1 synthetic aperture radar data and measurements of surface roughness and moisture content of rocky soils in a semiarid rangeland," *Water Resources Research*, vol. 34, no. 6, pp. 1491–1498, 1998.

[31] M. Shoshany, T. Svoray, P. J. Curran, G. M. Foody, and A. Perevolotsky, "The relationship between ERS-2 SAR backscatter and soil moisture: generalization from a humid to semi-arid transect," *International Journal of Remote Sensing*, vol. 21, no. 11, pp. 2337–2343, 2000.

[32] H. Lievens, N. E. C. Verhoest, E. De Keyser et al., "Effective roughness modelling as a tool for soil moisture retrieval from C-and L-band SAR," *Hydrology and Earth System Sciences*, vol. 15, no. 1, pp. 151–162, 2011.

[33] K. Millard and M. Richardson, "Quantifying the relative contributions of vegetation and soil moisture conditions to polarimetric C-Band SAR response in a temperate peatland," *Remote Sensing of Environment*, vol. 206, pp. 123–138, 2018.

[34] D. P. Thoma, M. S. Moran, R. Bryant et al., "Comparison of four models to determine surface soil moisture from C-band radar imagery in a sparsely vegetated semiarid landscape," *Water Resources Research*, vol. 42, no. 1, 2006.

[35] A. K. Fung, Z. Li, and K.-S. Chen, "Backscattering from a randomly rough dielectric surface," *IEEE Transactions on Geoscience and Remote Sensing*, vol. 30, no. 2, pp. 356–369, 1992.

[36] N. Baghdadi, S. Gaultier, and C. King, "Retrieving surface roughness and soil moisture from synthetic aperture radar (SAR) data using neural networks," *Canadian Journal of Remote Sensing*, vol. 28, no. 5, pp. 701–711, 2002.

[37] Z. Li, X. Ren, X. Li, and L. Wang, "Soil moisture mapping with C band multi-polarization SAR imagery," *Polarization*, vol. 900, no. 2, p. 35, 2005.

[38] M. B. Charlton and K. White, "Sensitivity of radar backscatter to desert surface roughness," *International Journal of Remote Sensing*, vol. 27, no. 8, pp. 1641–1659, 2006.

[39] S.-B. Kim, M. Moghaddam, L. Tsang, M. Burgin, X. Xu, and E. G. Njoku, "Models of L-band radar backscattering coefficients over global terrain for soil moisture retrieval," *IEEE Transactions on Geoscience and Remote Sensing*, vol. 52, no. 2, pp. 1381–1396, 2014.

[40] S.-B. Kim, J. J. van Zyl, J. T. Johnson et al., "Surface soil moisture retrieval using the l-band synthetic aperture radar onboard the soil moisture active–passive satellite and evaluation at core validation sites," *IEEE Transactions on Geoscience and Remote Sensing*, vol. 55, no. 4, pp. 1897–1914, 2017.

[41] M. R. Sahebi, F. Bonn, and G. B. Bénié, "Neural networks for the inversion of soil surface parameters from synthetic aperture radar satellite data," *Canadian Journal of Civil Engineering*, vol. 31, no. 1, pp. 95–108, 2004.

[42] S. Gopal and C. Woodcock, "Remote sensing of forest change using artificial neural networks," *IEEE Transactions on Geoscience and Remote Sensing*, vol. 34, no. 2, pp. 398–404, 1996.

[43] B. Pradhan and S. Lee, "Landslide susceptibility assessment and factor effect analysis: backpropagation artificial neural networks and their comparison with frequency ratio and bivariate logistic regression modelling," *Environmental Modelling and Software*, vol. 25, no. 6, pp. 747–759, 2010.

[44] F. Del Frate and L.-F. Wang, "Sunflower biomass estimation using a scattering model and a neural network algorithm," *International Journal of Remote Sensing*, vol. 22, no. 7, pp. 1235-1244, 2001.

[45] J. F. Mas and J. J. Flores, "The application of artificial neural networks to the analysis of remotely sensed data," *International Journal of Remote Sensing*, vol. 29, no. 3, pp. 617–663, 2008.

[46] S. Paloscia, P. Pampaloni, S. Pettinato, and E. Santi, "A comparison of algorithms for retrieving soil moisture from ENVISAT/ASAR images," *IEEE Transactions on Geoscience and Remote Sensing*, vol. 46, no. 10, pp. 3274–3284, 2008.

[47] D. A. Hodgson and J.-S. Vincent, "A 10,000 yr B.P. extensive ice shelf over Viscount Melville Sound, Arctic Canada," *Quaternary Research*, vol. 22, no. 1, pp. 18–30, 1984.

[48] P. Lajeunesse and M. A. Hanson, "Field observations of recent transgression on northern and eastern Melville Island, western Canadian Arctic Archipelago," *Geomorphology*, vol. 101, no. 4, pp. 618–630, 2008.

[49] D. M. Barnett, S. A. Edlund, L. A. Dredge, D. C. Thomas, and L. S. Prevett, *Terrain Classification and Evaluation, Eastern Melville Island, NWT*, Geological Survey of Canada Open File 252, p. 1318, 1975.

[50] G. C. Heathman, P. J. Starks, and M. A. Brown, "Time domain reflectometry field calibration in the Little Washita River Experimental Watershed," *Soil Science Society of America Journal*, vol. 67, no. 1, pp. 52–61, 2003.

[51] G. C. Topp, J. L. Davis, and A. P. Annan, "Electromagnetic determination of soil water content: measurements in coaxial transmission lines," *Water Resources Research*, vol. 16, no. 3, pp. 574–582, 1980.

[52] C. Yu, A. W. Warrick, and M. H. Conklin, "Derived functions of time domain reflectometry for soil moisture measurement," *Water Resources Research*, vol. 35, no. 6, pp. 1789–1796, 1999.

[53] R. M. Nagare, R. A. Schincariol, W. L. Quinton, and M. Hayashi, "Laboratory calibration of time domain reflectometry to determine moisture content in undisturbed peat samples," *European Journal of Soil Science*, vol. 62, no. 4, pp. 505–515, 2011.

[54] J. Stein and D. L. Kane, "Monitoring the unfrozen water content of soil and snow using time domain reflectometry," *Water Resources Research*, vol. 19, no. 6, pp. 1573–1584, 1983.

[55] M. A. Malicki, R. Plagge, and C. H. Roth, "Improving the calibration of dielectric TDR soil moisture determination taking into account the solid soil," *European Journal of Soil Science*, vol. 47, no. 3, pp. 357–366, 1996.

[56] F. Del Frate and D. Solimini, "On neural network algorithms for retrieving forest biomass from SAR data," *IEEE Transactions on Geoscience and Remote Sensing*, vol. 42, no. 1, pp. 24–34, 2004.

[57] R. Leconte, F. Brissette, M. Galarneau, and J. Rousselle, "Mapping near-surface soil moisture with RADARSAT-1 synthetic aperture radar data," *Water Resources Research*, vol. 40, no. 1, 2004.

[58] Y. Oh, "Quantitative retrieval of soil moisture content and surface roughness from multipolarized radar observations of bare soil surfaces," *IEEE Transactions on Geoscience and Remote Sensing*, vol. 42, no. 3, pp. 596–601, 2004.

[59] Y. Oh, K. Sarabandi, and F. T. Ulaby, "An empirical model and an inversion technique for radar scattering from bare soil surfaces," *IEEE Transactions on Geoscience and Remote Sensing*, vol. 30, no. 2, pp. 370–381, 1992.

[60] M. R. Sahebi and J. Angles, "An inversion method based on multi-angular approaches for estimating bare soil surface parameters from RADARSAT-1," *Hydrology and Earth System Sciences*, vol. 14, no. 11, pp. 2355–2366, 2010.

[61] G. J. Bowden, G. C. Dandy, and H. R. Maier, "Input determination for neural network models in water resources applications. Part 1—background and methodology," *Journal of Hydrology*, vol. 301, no. 1–4, pp. 75–92, 2005.

[62] G. Mountrakis, J. Im, and C. Ogole, "Support vector machines in remote sensing: a review," *ISPRS Journal of Photogrammetry and Remote Sensing*, vol. 66, no. 3, pp. 247–259, 2011.

[63] X. Zhuang, B. A. Engel, D. F. Lozano-Garcia, R. N. Fernandez, and C. J. Johannsen, "Optimization of training data required for neuro-classification," *Remote Sensing*, vol. 15, no. 16, pp. 3271–3277, 1994.

[64] D. R. Hush, "Classification with neural networks: a performance analysis," in *Proceedings of the IEEE International Conference on Systems Engineering*, pp. 277–280, Wrocław, Poland, 1989.

[65] M. T. Hagan and M. B. Menhaj, "Training feedforward networks with the Marquardt algorithm," *IEEE Transactions on Neural Networks*, vol. 5, no. 6, pp. 989–993, 1994.

[66] T. Blaschke, "Object based image analysis for remote sensing," *ISPRS Journal of Photogrammetry and Remote Sensing*, vol. 65, no. 1, pp. 2–16, 2010.

[67] L. Drăguţ, D. Tiede, and S. R. Levick, "ESP: a tool to estimate scale parameter for multiresolution image segmentation of remotely sensed data," *International Journal of Geographical Information Science*, vol. 24, no. 6, pp. 859–871, 2010.

[68] E. Cognition, *eCognition Developer (8.64. 0) User Guide*, Trimble Germany GmbH, München, Germany, 2010.

[69] L. Drăgut, T. Schauppenlehner, A. Muhar, J. Strobl, and T. Blaschke, "Optimization of scale and parametrization for terrain segmentation: an application to soil-landscape modeling," *Computers and Geosciences*, vol. 35, no. 9, pp. 1875–1883, 2009.

[70] J. Álvarez-Mozos, A. Larrañaga, M. González-Audicana, and J. Casali, "On the influence of surface roughness on RADARSAT-2 polarimetric observations," in *Proceedings of the 4th International Workshop on Science and Applications of SAR Polarimetry and Polarimetric Interferometry*, pp. 26–30, Frascati, Italy, January 2009.

[71] N. E. C. Verhoest, H. Lievens, W. Wagner, J. Álvarez-Mozos, M. S. Moran, and F. Mattia, "On the soil roughness parameterization problem in soil moisture retrieval of bare surfaces from synthetic aperture radar," *Sensors*, vol. 8, no. 7, pp. 4213–4248, 2008.

[72] A. Collingwood, P. Treitz, and F. Charbonneau, "Surface roughness estimation from RADARSAT-2 data in a high Arctic environment," *International Journal of Applied Earth Observation and Geoinformation*, vol. 27, pp. 70–80, 2014.

[73] A. W. Collingwood, *Modeling Biophysical Variables in the Canadian High Arctic using Synthetic Aperture Radar Data*, Queen's University, Kingston, ON, Canada, 2014.

[74] T. Lakhankar, H. Ghedira, M. Temimi, M. Sengupta, R. Khanbilvardi, and R. Blake, "Non-parametric methods for soil moisture retrieval from satellite remote sensing data," *Remote Sensing*, vol. 1, no. 1, pp. 3–21, 2009.

[75] R. M. Haralick, K. Shanmugam, and I. Dinstein, "Textural features for image classification," *IEEE Transactions on Systems, Man, and Cybernetics*, vol. 3, no. 6, pp. 610–621, 1973.

[76] C. Oliver and S. Quegan, *Understanding Synthetic Aperture Radar Images*, Artech House, Boston, MA, USA, 1998.

[77] Y. Wang, F. Wang, J. Huang, X. Wang, and Z. Liu, "Validation of artificial neural network techniques in the estimation of nitrogen concentration in rape using canopy hyperspectral reflectance data," *International Journal of Remote Sensing*, vol. 30, no. 17, pp. 4493–4505, 2009.

[78] S. R. Cloude and E. Pottier, "An entropy based classification scheme for land applications of polarimetric SAR," *IEEE Transactions on Geoscience and Remote Sensing*, vol. 35, no. 1, pp. 68–78, 1997.

[79] J. R. Adams, A. A. Berg, H. McNairn, and A. Merzouki, "Sensitivity of C-band SAR polarimetric variables to unvegetated agricultural fields," *Canadian Journal of Remote Sensing*, vol. 39, no. 1, pp. 1–16, 2013.

[80] F. T. Ulaby and P. P. Batlivala, "Optimum radar parameters for mapping soil moisture," *IEEE Transactions on Geoscience Electronics*, vol. 14, no. 2, pp. 81–93, 1976.

[81] K. J. Tansey and A. C. Millington, "Investigating the potential for soil moisture and surface roughness monitoring in drylands using ERS SAR data," *International Journal of Remote Sensing*, vol. 22, no. 11, pp. 2129–2149, 2001.

[82] J. Alvarez-Mozos, J. Casali, M. González-Audicana, and N. E. C. Verhoest, "Assessment of the operational applicability of RADARSAT-1 data for surface soil moisture estimation," *IEEE Transactions on Geoscience and Remote Sensing*, vol. 44, no. 4, pp. 913–924, 2006.

[83] F. T. Ulaby, P. C. Dubois, and J. Van Zyl, "Radar mapping of surface soil moisture," *Journal of Hydrology*, vol. 184, no. 1-2, pp. 57–84, 1996.

[84] F. T. Ulaby, P. P. Batlivala, and M. C. Dobson, "Microwave backscatter dependence on surface roughness, soil moisture, and soil texture: part I-bare soil," *IEEE Transactions on Geoscience Electronics*, vol. 16, no. 4, pp. 286–295, 1978.

[85] M. C. Dobson and F. Ulaby, "Microwave backscatter dependence on surface roughness, soil moisture, and soil texture: part III-soil tension," *IEEE Transactions on Geoscience and Remote Sensing*, vol. GE-19, no. 1, pp. 51–61, 1981.

6

A Multiple Kernel Learning Approach for Air Quality Prediction

Hong Zheng[1], **Haibin Li**[1], **Xingjian Lu**[1,2] **and Tong Ruan**[1]

[1]*Information Engineering and Computer Science College, East China University of Science and Technology, Shanghai 200237, China*
[2]*Smart City Collaborative Innovation Center, Shanghai Jiao Tong University, Shanghai 200240, China*

Correspondence should be addressed to Hong Zheng; zhenghong@ecust.edu.cn

Academic Editor: Ilan Levy

Air quality prediction is an important research issue due to the increasing impact of air pollution on the urban environment. However, existing methods often fail to forecast high-polluting air conditions, which is precisely what should be highlighted. In this paper, a novel multiple kernel learning (MKL) model that embodies the characteristics of ensemble learning, kernel learning, and representative learning is proposed to forecast the near future air quality (AQ). The centered alignment approach is used for learning kernels, and a boosting approach is used to determine the proper number of kernels. To demonstrate the performance of the proposed MKL model, its performance is compared to that of classical autoregressive integrated moving average (ARIMA) model; widely used parametric models like random forest (RF) and support vector machine (SVM); popular neural network models like multiple layer perceptron (MLP); and long short-term memory neural network. Datasets acquired from a coastal city Hong Kong and an inland city Beijing are used to train and validate all the models. Experiments show that the MKL model outperforms the other models. Moreover, the MKL model has better forecast ability for high health risk category AQ.

1. Introduction

With the development of the economy and society all over the world, most metropolitan cities are experiencing elevated concentrations of ground-level air pollutants, especially in fast developing countries like India and China. Exposure to air pollution can affect everyone, but it can be particularly harmful to people with a heart disease or a lung condition, elderly people, and children. Studies show that long-term exposure to fine particulate air pollution or traffic-related air pollution is associated with environmental-cause mortality, even at concentration ranges well below the standard annual mean limit value [1, 2]. Therefore, building an early warning system, which provides precise forecast and also alerts health alarm to local inhabitants will provide valuable information to protect humans from damage by air pollution.

Currently, three major approaches are used to forecast real-time air quality: simple empirical approaches, advanced physically based approaches, and machine learning approaches.

Simple empirical approaches like persistence method and climatology method are based on assumptions or hypothesis; that is, thresholds of forecasted meteorological variables can indicate future pollution level [3]. They are computationally fast but have low accuracy and are primarily used as references by other methods. Advanced physically based approaches like chemical transport models (CTMs) simulate the formation and accumulation of air pollutants by a solution of the conservation equations and transformation relationships among the mass of various chemical species and physical states. They can provide valuable insights for understanding pollutant diffusion mechanisms. But they are computationally expensive, demanding reliable meteorological predictions, and highly relevant to a high level of expertise [4].

Machine learning methods are computationally fast and cost-effective and can provide promising prediction accuracy. Various machine learning methods have been applied to predict the air quality. Widely used methods include classical autoregressive moving average (ARMA) methods like the autoregressive integrated moving average

(ARIMA) [5], support vector machine (SVM) methods like the support vector classifier (SVC) [6, 7], ensemble methods like the random forest (RF) [8, 9], artificial neural network (ANN) methods like the multiple layer perceptron (MLP) [10, 11], and deep learning methods like the long short-term memory neural network (LSTM NN) [12, 13].

Among the models mentioned above, ARIMA is a time series model and is often used as a baseline model. The performance of the SVM model is often hinged on the appropriate choice of the kernel. A kernel in SVM introduces nonlinearity into the problem by mapping new input data implicitly into a Hilbert space where it may then be linearly separable [14]. Neural network models, especially deep neural networks, can automatically learn the representations from raw data, but it takes a long time and a large volume of data to train a well-behaved network.

Multiple kernel learning (MKL) is proposed as an alternative to cross validation, feature selection, metric learning, and ensemble methods. MKL refers to using multiple kernels instead of a single one; most of the algorithms which make use of the kernel tricks can take the advantage of MKL, such as SVM and kernel ridge regression (KRR). In MKL, feature combination and classifier training are done simultaneously, and different data formats can be used in the same formulation. In addition, the inherent kernel trick of combining linear kernels and nonlinear kernels in MKL makes it more promising in solving fusing information problems. There is a significant amount of work in the literature for combining multiple kernels [15, 16]. Various applications indicate that performance gains can be achieved by linear and nonlinear kernel combinations using MKL methods [17–19].

In this paper, a novel multiple kernel learning-based air quality prediction approach that can inherently capture the characteristics of the heterogeneous time, meteorology, and air pollutant data is proposed. Real datasets from a coastal city Hong Kong and an inland city Beijing are used to demonstrate the effectiveness the proposed approach. Comprehensive comparison experiments with ARIMA, RF, SVCs, MLP, and LSTM are conducted. Though some of the algorithms can automatically learn the representative features of the data, pretraining featuring engineering is still necessary and will significantly affect the models' performance. In addition, hyperparameter tuning is critical for all the parametric models. Therefore, in this paper, special attention is paid to the feature engineering and parameter tuning process. The methodologies applied to Hong Kong and Beijing datasets are similar. Therefore, Hong Kong is used for demonstration in most of the paper. The main contributions of this paper are as follows:

(1) A multiple kernel learning approach is introduced into the domain of air quality prediction for the first time. Multiscale predictions over the next 1, 3, 6, 9, and 12 hours' air quality of an inland city Beijing and a coastal city Hong Kong are presented.

(2) The proposed method can effectively capture the air quality features from the hybrid time, meteorology, and air pollutant data. The experimental results demonstrated the advantages of this approach over some of the widely used models, especially in the prediction of severe air pollution conditions.

The rest of the paper is organized as follows: Section 2 presents the methodology of the multiple kernel learning algorithm; data preparation is introduced in Section 3; in Section 4, extensive experimentation results and necessary discussions are presented; and Section 5 concludes this paper.

2. Methodology

While classical kernel-based classifiers such as SVCs are based on a single kernel, in practice, it is often desirable to base classifiers on combinations of multiple kernels since data points typically can be due to multiple heterogeneous sources. A kernel implicitly represents a notion of similarity for the data, and different kernels will accommodate different nonlinear mappings, and MKL provides a way to combine different ideas of similarity. Using a specific kernel may be a source of bias, and MKL provides a way to select optimal kernels and parameters from a larger set of kernels. In the air quality prediction case, the source data are coming from different modalities. Therefore, in the paper, instead of using just a single kernel which is usually more suitable for the homogeneous data source, multiple kernels are combined, and the classical and empirically successful support vector classifier is used as the base learner. The detailed introduction of the kernel support vector machine is given in Appendix A. In this section, the multiple kernel learning approach is described first, and then, the centered alignment method is introduced for learning kernels.

2.1. Multiple Kernel Learning. MKL is conceptually similar to single kernel learning. In other words, single kernel leaning is a special case of MKL. In MKL, the final kernel is learnt as a combination (linear or nonlinear) of many base kernels from the data itself:

$$\kappa_\eta(x_i, x_j) = f_\eta\left(\left\{\kappa_m\left(x_i^m, x_j^m\right)\right\}_{m=1}^{P} \middle| \eta\right), \tag{1}$$

where f_η: $\mathbb{R}^P \rightarrow \mathbb{R}$ is the combination function, κ_m is the kernel function, m is the dimensionality of the corresponding feature representation, and η parameterizes the combination function.

It is also possible to integrate η into the kernel functions where it is optimized during training.

$$\kappa_\eta(x_i, x_j) = f_\eta\left(\left\{\kappa_m\left(x_i^m, x_j^m \middle| \eta\right)\right\}_{m=1}^{P}\right). \tag{2}$$

Most of the existing MKL algorithms fall into the first category and try to combine predefined kernels in an optimal way. Commonly used kernels are linear, polynomial, radial basis function (RBF), and sigmoid.

Input:	dataset: $(\mathbf{x}^{(1)}, y^{(1)}), \ldots, (\mathbf{x}^{(n)}, y^{(n)})$, n samples
Output:	decision function of MKSVC
Start	First, get the kernel coefficients by optimizing the single kernel-base learners $(\kappa_m(x_i, x_i))$
	Second, get the weight of each kernel by the centered kernel alignment algorithm (η)
	Third, get the number of kernels by boosting approach (P)
	Fourth, get the combined optimized kernel $\kappa_\eta(x_i, x_j) = \sum_{m=1}^{P} \eta_m \kappa_m(x_i, x_i)$
	Then, use SVC as the base learner and optimize it with a general optimizing algorithm
	Return $f_m(x) = \sum_{i=1}^{N} \alpha_i y_i \kappa_\eta(x_i, x_j) + b$
Stop	

Algorithm 1: MKSVC.

$$\kappa(x, x^i) = \begin{cases} (x^T \cdot x^i) & \text{linear} \\ (x^T \cdot x^i + 1)^d & \text{polynomial} \\ \exp(-\gamma \|x - x^i\|^2) & \text{RBF} \\ \tanh(\gamma x \cdot x^i) & \text{sigmoidal.} \end{cases} \quad (3)$$

The kernels can be combined in different ways, and each has its own combination parameter characteristics. Generally, linear combination methods are used, and they fall into two basic categories: unweighted sum (i.e., using sum or mean of the kernels as the combined kernel) and weighted sum. In the weighted sum case, the combination function is linearly parameterized:

$$\kappa_\eta(x_i, x_j) = f_\eta\left(\left\{\kappa_m(x_i^m, x_j^m)\right\}_{m=1}^{P} \middle| \eta\right) = \sum_{m=1}^{P} \eta_m \kappa_m(x_i^m, x_j^m), \quad (4)$$

where η denotes the kernel weights. Different versions of this approach differ in the way they put restrictions on η: the linear sum has arbitrary real value η_m and the conic sum requires η_m to be positive, while η sums to 1 for the convex sum.

The conic sum and convex sum are special cases of the linear sum, but the former two are used more often because the relative importance of the combined kernels can be extracted by looking at the kernel weights. Furthermore, the kernel weights of the conic and convex sum correspond to scaling the feature spaces when they are nonnegative [20].

In this paper, the conic sum restriction used as the convex sum is a special case of the conic sum. The resulting decision function of the multiple kernel support vector classifier (MKSVC) is defined as

$$f(x) = \sum_{i=1}^{N} \alpha_i y_i \sum_{m=1}^{P} \eta_m \kappa_m(x_i^m, x_j^m) + b, \quad \text{subject to} \quad \eta \in \mathbb{R}_+^P. \quad (5)$$

There are four important parameters: the number of kernels (P), the inner kernel coefficients of each kernel, features to use for each kernel (x_i^m), and the weight (η_m) of each kernel. In this paper, the inner kernel coefficients are obtained by optimizing the single kernel-based learners. η is obtained by the centered alignment approach proposed in [32]. P is obtained through the boosting approach by iteratively adding a new kernel until the performance stops improving (the kernels are added based on the weights learned by the centered alignment approach, kernel with higher weight first). As with the features used by each kernel, for simplicity, the canonical multiple kernel learning approach is used, namely, one kernel combination for all feature representations. The pseudo code of the MKSVC is described in Algorithm 1.

2.2. Centered Alignment Method for Learning Kernels. Centered alignment is used as a similarity measure between kernels or kernel matrices. Given p kernels matrices $\mathrm{K}_1, \mathrm{K}_2, \ldots, \mathrm{K}_p$, centered kernel alignment learns a linear combination of kernels resulting in a combined kernel matrix:

$$\mathrm{K}_{c\mu} = \sum_{q=1}^{p} \mu_q \mathrm{K}_{cq}, \quad (6)$$

where p is the number of kernels, μ_q is the centered kernel weight, and K_{cq} is the centered kernel:

$$\mathrm{K}_{cq} = \left(\mathbf{I} - \frac{11^T}{m}\right)\mathrm{K}_q\left(\mathbf{I} - \frac{11^T}{m}\right), \quad (7)$$

where $\mathbf{I}$ is the identity matrix, $1 \in \mathbb{R}^{m\times 1}$ denotes the vector with all entries equal to one, and K_q is the original kernel matrix.

The alignment between two kernel functions K and K′ is defined by

$$\widehat{\rho}(\mathrm{K}, \mathrm{K}') = \frac{\langle \mathrm{K}_c, \mathrm{K}'_c \rangle_F}{\|\mathrm{K}_c\|_F \|\mathrm{K}'_c\|_F}, \quad (8)$$

where K_c and K'_c are the centered kernels of K and K′ and $\langle \cdot, \cdot \rangle_F$ denotes the Frobenius product and $\|\cdot\|_F$ the Frobenius norm defined by

$$\forall \mathbf{A}, \mathbf{B} \in \mathbb{R}^{m\times m}, \quad \langle \mathbf{A}, \mathbf{B} \rangle_F = \mathrm{Tr}\left[\mathbf{A}^T\mathbf{B}\right], \quad \|\mathbf{A}\|_F = \sqrt{\langle \mathbf{A}, \mathbf{A}_F \rangle_F}, \quad (9)$$

and $\widehat{\rho}(\mathrm{K}, \mathrm{K}') \in [0, 1]$ by definition.

Using the independent alignment-based algorithm proposed in [32], the alignment between each kernel matrix K_q and the target K_Y ($\mathrm{K}_Y = \mathbf{y}\mathbf{y}^T$, $\mathbf{y}$ is the labels) can be computed independently by using the training samples and the centered kernel weight can be chosen proportional to that alignment. Thus, the resulting kernel matrix is defined by

$$K_\mu \propto \sum_{q=1}^{p} \hat{\rho}\left(K_q, K_Y\right) K_q = \frac{1}{\|K_Y\|_F} \sum_{q=1}^{p} \frac{\langle K_q, K_Y \rangle_F}{\|K_q\|_F} K_q. \quad (10)$$

3. Data Preparation

In this paper, two datasets are used: one is from Hong Kong, a coastal city, whose air condition is relatively good, and the other is from an inland city, Beijing, whose air condition is relatively poor. Dataset of HK contains two years' hourly meteorology data and pollutant data between 1 February 2013 and 31 January 2015 collected from HK's Sha Tin air quality monitoring station [21] and weather forecast station [22]. Dataset of Beijing contains five years' hourly PM2.5 data and meteorology data between 1 January 2010 and 31 December 2014 collected from UCI machine learning repository [23].

3.1. Prediction Target and Performance Metric

3.1.1. Prediction Target. The prediction targets in this paper are the air quality health index (AQHI) in Hong Kong and the PM2.5 individual air quality level (IAQL) in Beijing. AQHI and IAQL are scales designed to help understand the impact of air quality on health. Unlike air quality concentrations, these air quality indices provide the public with advice on how to protect their health during air quality levels associated with low, moderate, high, and very high health risks. They also provide advice on how to improve air quality by proposing behavioral change to reduce the environmental footprint [24, 25].

For any given hour, the AQHI is calculated from the sum of the percentage excess risk of daily hospital admissions attributing to the 3-hour moving average concentrations of four criteria air pollutants: ozone (O_3), nitrogen dioxide (NO_2), sulphur dioxide (SO_2), and particulate matter (PM) (respirable suspended particulates (RSP or PM10) or fine suspended particulates (FSP or PM2.5), whichever poses a higher health risk).

The IAQL is classified based on the individual air quality index (IAQI) which is calculated according to a formula published by China' Ministry of Environmental Protection (MEP) [26]. The highest IAQI among pollutants SO_2, NO_2, O_3, carbon monoxide (CO), PM2.5, and PM10 at a given time is called the primary or dominant pollutant and is chosen for the overall AQI value. In China, PM2.5 is the primary pollutant most of the time; therefore, its IAQI is usually the overall AQI.

The detailed information of calculating AQHI and IAQI is given in Appendix B. These indices are health protection tools used to make decisions to reduce short-term exposure to air pollution by adjusting activity levels during increased levels of air pollution. Table 1 shows the health risks with corresponding air quality classifications.

Table 1: Air quality classifications and health risk.

Health risk	Low	Moderate	High	Very high	Serious
Hong Kong (AQHI)	1–3	4–6	7	8–10	10+
Beijing (IAQL)	1–2	3	4	5	6

Table 2: Air pollutant samples.

Date	Hour	Station	FSP	NO_2	NO_x	O_3	RSP	SO_2
1/1/2014	1	SHATIN	91	131	266	N.A.	114	18
1/1/2014	2	SHATIN	88	124	262	3	110	14
1/1/2014	3	SHATIN	86	114	225	2	107	13
1/1/2014	4	SHATIN	85	107	197	3	104	15

3.1.2. Performance Metric. In this paper, accuracy, mean square error (mse), weighted precision (wp), weighted recall (wr), and weighted f1-score (wf) are used to evaluate the effectiveness of all the algorithms. The precision (P) is calculated by the formula $TP/(TP + FP)$ where TP is the number of correct predictions and FP is the number of incorrect predictions. Recall (R) is the proportion of instances classified as a given class divided by the actual total in that class. F1-score is a harmonic average of precision and recall [27].

For accuracy and mse,

$$\text{accuracy}(y, \hat{y}) = \frac{1}{n_{\text{samples}}} \sum_{i=0}^{n_{\text{samples}}} 1(y_i = \hat{y}_i),$$
$$\text{mse}(y, \hat{y}) = \frac{1}{n_{\text{samples}}} \sum_{i=0}^{n_{\text{samples}}} (y_i - \hat{y}_i)^2, \quad (11)$$

where $\hat{y}_i$ is the predicted value of the ith sample and y_i is the corresponding true value.

For wp, wr, and wf

$$\text{wp} = \frac{1}{\sum_{l\in L}|\hat{y}_l|} \sum_{l\in L} |\hat{y}_l| P(y_l, \hat{y}_l),$$
$$\text{wr} = \frac{1}{\sum_{l\in L}|\hat{y}_l|} \sum_{l\in L} |\hat{y}_l| R(y_l, \hat{y}_l), \quad (12)$$
$$\text{wf} = \frac{1}{\sum_{l\in L}|\hat{y}_l|} \sum_{l\in L} |\hat{y}_l| F_1(y_l, \hat{y}_l),$$

where y is the set of predicted (sample, classes) pairs, $\hat{y}$ is the set of true (sample, classes) pairs, L is the set of classes, and y_l is the subset of y with classes l; similarly, $\hat{y}_l$ is the subset of $\hat{y}$. $P(y_l, \hat{y}_l) = |y_l \cap \hat{y}_l|/|y_l|$ and $R(y_l, \hat{y}_l) = |y_l \cap \hat{y}_l|/|\hat{y}_l|$ (conventions vary on handling $\hat{y}_l = \phi$; this implementation uses $R(y_l, \hat{y}_l) = 0$, and similar for $P(y_l, \hat{y}_l)$). $F_1(y_l, \hat{y}_l) = (2 \times (P \times R))/(P + R)$.

3.2. Featured Data. Take dataset of HK for example. Following air pollutant data features are contained: FSP, NO_2, NO_x, O_3, RSP, and SO_2 (unit of measurement of all the air pollutants is $\mu g/m^3$). Air pollutant data samples are shown in Table 2.

TABLE 3: Meteorological samples.

Local time in Sha Tin	T	P0	P1	δP	H	WD	WP	dew
29.01.2015 02:00	15.2	763.4	764.5	1.0	73	Wind blowing from the east	3	10.3
29.01.2015 01:00	15.6	763.9	765.1	0.4	77	Calm, no wind	0	11.5

T, air temperature (degrees Celsius) at 2 meters height above the Earth's surface; P0, atmospheric pressure at weather station level (millimeters of mercury); P1, atmospheric pressure reduced to mean sea level (millimeters of mercury); δP, pressure tendency, changes in atmospheric in the last three hours; H, relative humidity (%) at a height of 2 meters above the Earth's surface; WD, mean wind direction (compass points) at a height of 10–12 meters above the Earth's surface over the 10-minute period immediately preceding the observation; WP, mean wind speed at a height of 10–12 meters above the Earth's surface over the 10-minute period immediately preceding the observation (meters per second); dew, dew point at 2 meters height above the Earth's surface (degrees Celsius).

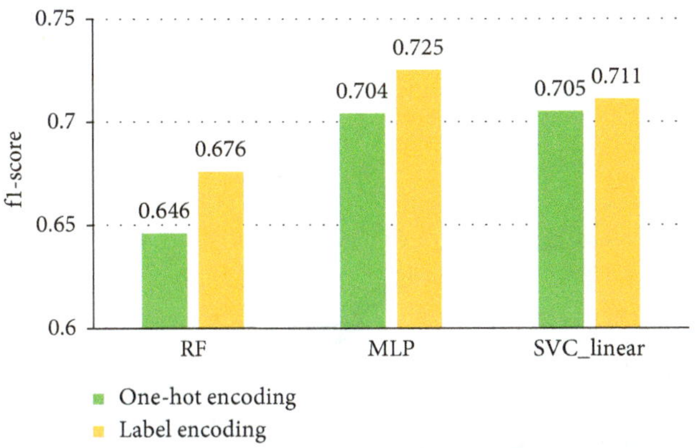

FIGURE 1: Comparison of one-hot encoding and label encoding over wind direction.

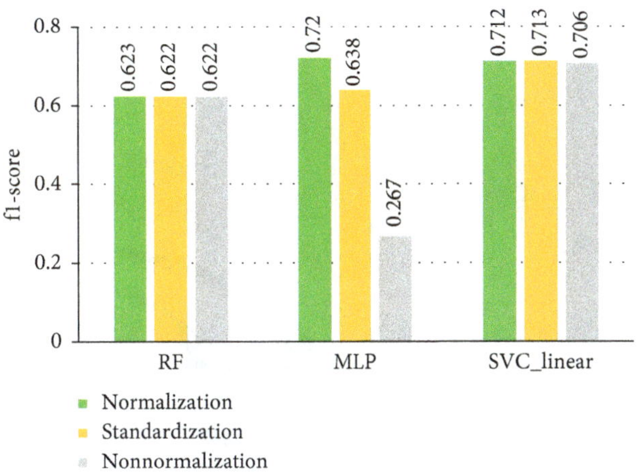

FIGURE 2: Comparison of with and without normalization.

Following meteorology data features are contained: T, P0, P1, δP, H, WD, WP, and dew. Meteorological samples are shown in Table 3.

Following time stamp features are contained: month, the day of the week (week), the day of the month (day), and the hour of the day (hour). There may be a yearly trend of the air quality, but we just have limited years of data, so "year" is not included in the feature set.

3.3. Feature Engineering

3.3.1. Feature Transformation

(1) Encoding Wind Direction. Among the data obtained, the wind direction is nonnumeric (i.e., "east," "east-southeast"). It has to be converted to numerical value so that the algorithms can make use of. One-hot encoding (e.g., "east" is encoded as [1,0,0,0,0,0,0,0,0,0,0,0,0,0,0,0,0,0]) and label encoding (e.g., "east" is encoded as 1, "south" is encoded as "2" etc.) were tried in this paper. Figure 1 shows the forecast performances of RF, MLP, and SVC_linear (SVC with linear kernel) algorithms when the wind direction was encoded by one-hot encoding and label encoding, respectively, and the parameters of the algorithms stayed unchanged. From the figure, it is obvious that label encoding is superior over one-hot encoding on the dataset. Therefore, in this paper, the wind direction was label encoded.

(2) Missing Data Imputation. Linear interpolation was used in the paper to interpolate the missing values in the two datasets.

$$V_t = \frac{V_s + (V_e - V_s)}{n+1}, \tag{13}$$

where V_t denotes the missing value at time t and n is the time gap between interval (V_s, V_e).

(3) Data Normalization. Normalization or standardization of either input or target variables tends to make the training process better behaved. Normalization scales the feature values in the range [0,1]:

$$V = \frac{V - V_{\min}}{V_{\max} - V_{\min}}. \tag{14}$$

Standardization transforms the feature values to have zero mean and unit variance:

$$V = \frac{V - \mu}{\sigma}. \tag{15}$$

To see whether normalization or standardization helps, both of them were tried and compared with the one without any processing. Again, RF, MLP, and SVC_linear were used as the validation algorithms. Results are shown in Figure 2. The figure shows that, generally, models benefit from normalization or standardization, especially for the neural network model. Normalization is slightly better than standardization. Therefore, in this paper, the data were normalized.

3.3.2. Feature Selection. Take Hong Kong for example. The source dataset contains 18 features, and they are as follows:

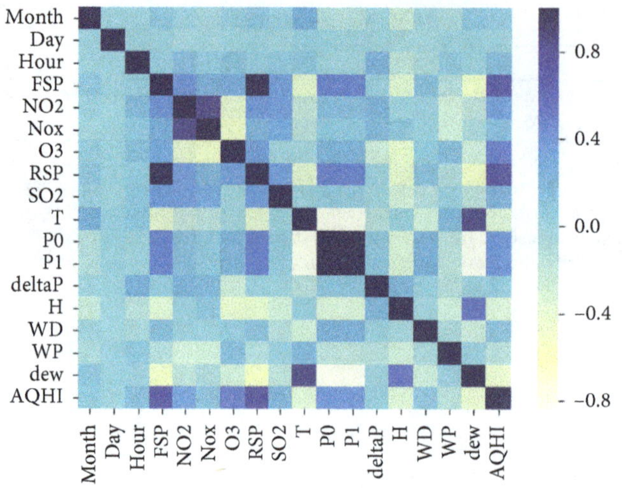

FIGURE 3: Spearman correlation between features.

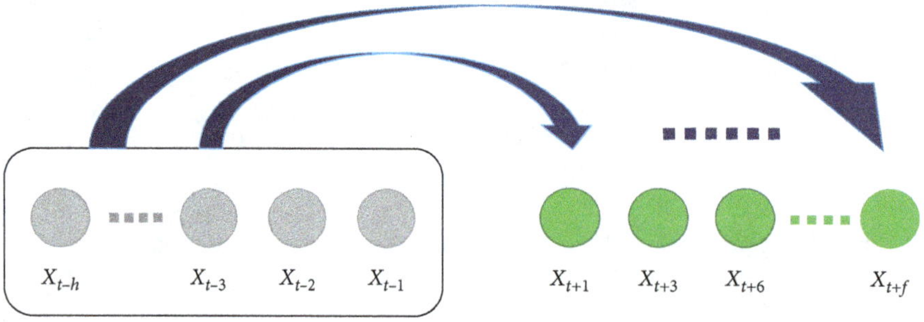

FIGURE 4: Multiscale predictor.

Meteorological (M) data features: <T, P0, P1, δP, H, WD, WP, dew>

Air pollutant (AP) data features: <FSP, NO_2, NO_x, O_3, RSP, SO_2>

Time data features: <month, week, day, hour>

The target is to forecast the near future AQHI. However, not all the features above are related to the AQHI, finding out the features which are correlated with the target would be beneficial. The historical pollutants and meteorology may impact the future air quality as the simple empirical approaches assume, finding out the influential historical time lag would be important as well.

(1) Feature Correlation Analysis. In this paper, Spearman's correlation analysis was used due to the possible nonlinear relationships between variables. Spearman's rank correlation coefficient measures the monotonic association between two variables and relies on the rank order of values [28]. The formula for Spearman's coefficient is

$$\rho_{\mathrm{rank}_x,\mathrm{rank}_y} = \frac{\mathrm{cov}\left(\mathrm{rank}_x, \mathrm{rank}_y\right)}{\sigma_{\mathrm{rank}_x}\sigma_{\mathrm{rank}_y}}, \tag{16}$$

where rank_x, rank_y are the ranked (sorted) values of variables x and y, $\mathrm{cov}(\cdot)$ is the covariance, and $\sigma(\cdot)$ is the standard deviation. Figure 3 shows the Spearman correlation coefficients between the features of HK dataset. Correlation scores go from −1 to 1. Perfect positive correlation is 1. Perfect negative correlation is −1. The figure shows that FSP, O_3, RSP, SO_2, P0, and P1 have strong positive correlations with the AQHI, while T, H, and dew have strong negative correlations with the AQHI. Cohen's standard [29] was used in this paper to select the correlated features. Features with association smaller than 0.30 are discarded. The picked features are as follows:

<FSP, NO_2, NO_x, O_3, RSP, SO_2, T, P0, P1, δP, H, dew, WP, WD, month, hour>

(2) Temporal Correlation Analysis. Intuitively, historical data from different periods have different effects on future time lags. More recent events have a stronger influence on the current status, while earlier events have a weaker influence. Denote current time as t, the historical time lag as h, and the future time lag as f, and then the prediction time is $t + f$ ($f = 1, 3, 6, 9, 12$) and the influential historical time is $t - h$ ($h = 1, 2, \ldots, n$). The multiscale prediction task is represented in Figure 4. In this paper, the LSTM NN model which is capable of learning long time series was used to select the appropriate influential historical time lag [30].

The network architecture of the LSTM model used in the paper is shown in Figure 5, which is the same as the LSTM-extended network proposed in [13]. The main input is the air pollutant data, and the auxiliary input is the time and meteorology data. There are two LSTM layers and one

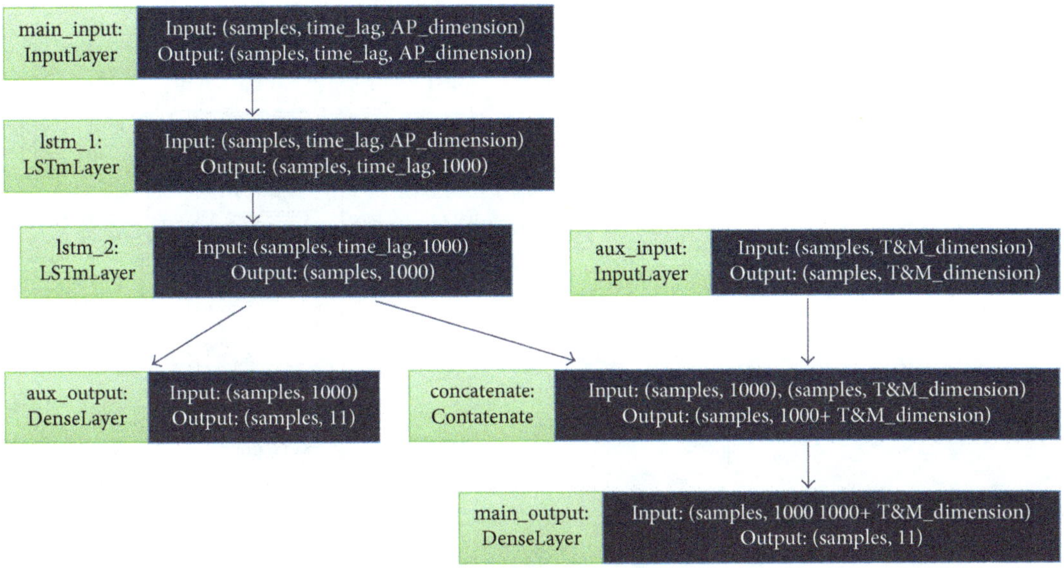

FIGURE 5: The LSTM network architecture used in this paper.

TABLE 4: Influences of different historical time lag over different future time lag.

f1			F-lag		
H-lag	1	3	6	9	12
0	0.659	0.543	0.452	0.409	0.388
1	0.647	0.488	0.483	0.446	0.408
2	0.603	0.573	0.443	0.417	0.395
3	0.606	0.528	0.467	0.434	0.431
4	0.642	0.465	0.439	0.452	0.417
5	0.706	0.654	0.511	0.454	0.413
6	0.739	0.504	0.574	0.486	0.423
7	0.711	0.700	0.682	0.508	0.436
8	0.736	0.705	0.713	0.447	0.481
9	**0.763**	0.721	**0.733**	**0.676**	0.512
10	0.674	**0.729**	0.693	0.557	0.495
11	0.630	0.702	0.632	0.631	0.541
12	0.728	0.656	0.584	0.607	**0.587**

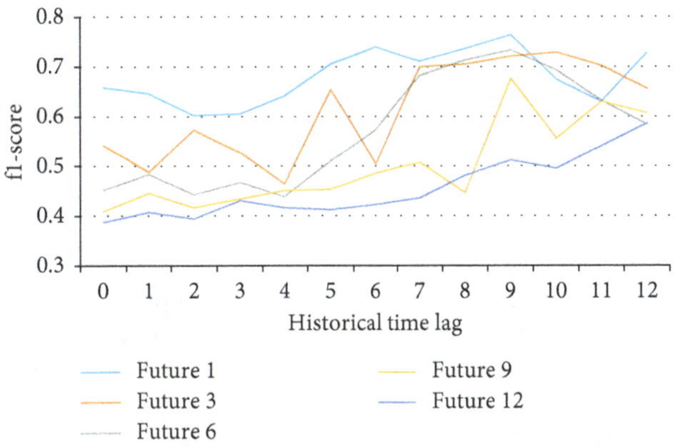

FIGURE 6: Influences of different historical time lag over different future time lag.

output layer which is a fully connected layer that has 11 neurons corresponding to the number of classes. The number of neurons in the LSTM layer has to be tuned. For simplicity, the number of neurons in each LSTM layer was set to an equivalent value chosen from a candidate set of {50, 100, 200, 500, 1000, 2000}. The most appropriate setting was chosen that yielded the best performance based on several comparative experiments. When the number of neurons in the LSTM was as 1000, the LSTM achieved the best performance. Therefore, in this paper, the number of neurons in the LSTM layers was set as 1000.

The future 1, 3, 6, 9, and 12 hours' AQHIs were predicted in this paper. With each future time lag, the influences of different historical time lags were examined. The results are given in Table 4. The evaluation metric is weighted f1-score (f1 in Table 4). The corresponding curve graph is given in Figure 6. The result shows that different future time lag (F-lag in Table 4) corresponds to slightly different optimal historical time lag (H-lag in Table 4). The general influential time of historical data for a specific future time's AQHI is around 9 hours.

Notably, the result shows that the prediction performances are poor for future time lag larger than 6, indicating that long-term prediction tasks are instinctively more difficult. Small-time lag cannot guarantee enough long-term memory inputs for the LSTM model, while large time lags permit an increased number of unrelated inputs, which increase the model's complexity and the difficulty of learning useful features. According to the above experiments, for simplicity, 9 was selected as the most appropriate influential historical time lag for different future time lag.

4. Results and Discussion

Algorithms used in the experiments are ARIMA, RF, MLP, SVC_linear (SVC with the liner kernel), SVC_rbf (SVC with the RBF kernel), SVC_sig (SVC with the sigmoid kernel), SVC_poly (SVC with the polynomial kernel), LSTM, and MKSVC. ARIMA was used as a baseline model, RF, MLP,

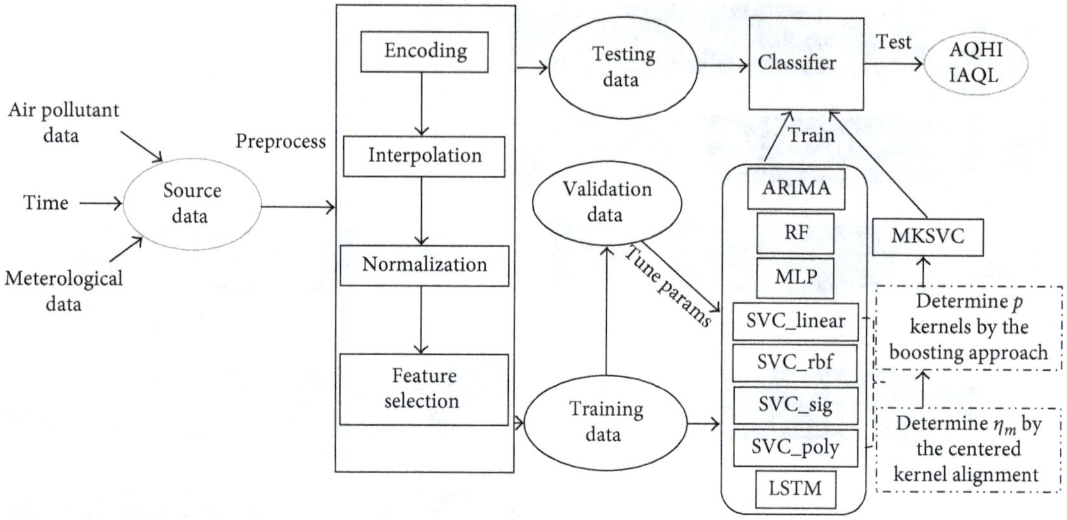

FIGURE 7: Experiment flow.

and SVC are widely used air quality forecast models, they were fine-tuned in this paper in order to make a fair comparison with MKSVC, and the LSTM in this paper has the same structure as the LSTM extended model proposed in [13]. Figure 7 shows the experimental flow. All algorithms were designed and tested with the same operation environment (Python 3.5.3, Windows 10, Intel® Core™ i7-5500U CPU @2.40 GHz, 16.0 GB RAM).

4.1. Parameter Optimization. Parameter optimization refers to the method of finding optimal parameters for a machine-learning algorithm. This is important since the performance of any machine learning algorithm depends to a huge extent on what the values of parameters are. For each prediction time lag, the parameters are different for each algorithm. It means an optimal model for each prediction task and each algorithm need to be tuned. The ways to get the parameters of MKSVC are detailed in Section 2 and Section 3.2.2 for LSTM. For the other algorithms, the parameter tuning process of the one-hour future time lag prediction task is presented in the following part, and the multiscale prediction tasks have identical fine-tuning processes.

First, the grid search interval of a parameter is narrowed by analyzing the influence curve of a single parameter on the training score and the validation score. For instance, by varying the kernel coefficient γ of the RBF kernel in SVC_rbf, the γ-score curve can be obtained as shown in Figure 8. The yellow line denotes the score over the training set. The purple line represents the score on the validation set, and the shadow represents the variance.

The figure shows that, at first, both the training and validation scores rise with the increase of γ. However, when γ reaches around 0.5, a further increase will result in the increase of the training score but the decrease of the validation score; it signifies that the model is getting overfitting. According to this influence curve, the grid search interval of γ in the next step can be narrowed between 0.0 and 1.0.

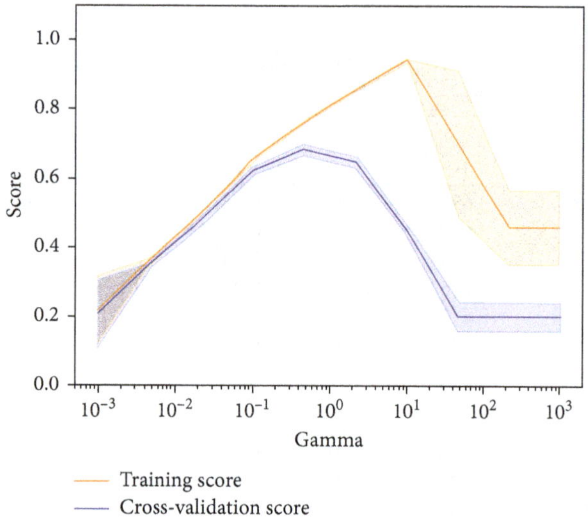

FIGURE 8: γ-score curve of SVC_rbf.

Based on the influence curve, the grid search intervals of the main parameters of the ARIMA, RF, MLP, and SVCs are shown in Table 5. RF, MLP, and SVCs used in this paper are implemented in scientific toolbox scikit-learn [31] and ARIMA implemented in statsmodels [32]. The unlisted parameters are set as default.

Then, a gird search with 5-fold cross validation was applied to find the optimum parameter. By exhaustively considering all parameter combinations in Table 5, the optimal parameter settings of the ARIMA, RFC, MLP, and SVCs are obtained as shown in Table 6. After getting the inner kernel coefficients of all the base kernels, the centered kernel alignment method described in Section 2.2. was used to get the optimal weight for each kernel.

4.2. Comparison. For HK, one year's data was used for training, and the other year's data was used for testing. For Beijing, the first two years' data was used for training, and

TABLE 5: Main parameters and their tuning range of the used algorithms.

Algorithm	Parameter	Algorithm	Parameter
ARIMA	p: [0,3], d: [0,10], q: [0,3]		
RF	n_estimators: [100, 1000; 50] max_depth: [10, 20; 1] max_features: [10, 30; 1] min_samples_split: [2,100; 1] min_samples_leaf: [1,100; 1]	SVC_linear	C: [100, 5000; 100]
		SVC_rbf	C: [100, 5000; 100] gamma: [0.0, 1.0; 0.01]
MLP	hidden_layer_sizes: {(50, 50), (100, 100), (10, 20, 10), (20, 40, 20)} activation: {'identity', 'logistic','tanh', 'relu'} solver: {'lbfgs', 'sgd', 'adam'}	SVC_sig	C: [100, 5000; 100] gamma: [0.0, 1.0; 0.01] coef0: [0, 1000; 50]
		SVC_poly	C: [100, 5000; 100] degree: {2,3} gamma: [0.0, 1.0; 0.01] coef0: [0, 1000; 50]

p: AR specification; d: integration order; q: MA specification; C: regularization coefficient in SVC; n_estimators: number of trees in the forest; max_depth: maximum depth of the tree; max_features: maximum number of features when looking for the best split; min_samples_split: the minimum number of samples required to split an internal node; min_samples_leaf: the minimum number of samples required to be at a leaf node; solver: algorithm used in the optimization problem; hidden_layer_sizes: hidden layer size; alpha: regularization term parameter in MLP; activation: activation function for the hidden layer; gamma: kernel coefficient for 'rbf', 'poly', and 'sigmoid'; degree: degree of the polynomial kernel function; coef0: independent term in kernel functions for 'poly' and 'sigmoid'; *[*a*, *b*; *c*] means within range [*a*, *b*], increase c every iteration; {} means set of values.

TABLE 6: The optimal parameter settings of the algorithms.

Algorithm	Parameter	Algorithm	Parameter
ARIMA	order (p,d,q): (2,0,2)		
RFC	n_estimators: 400 max_depth: 9 max_features: 11 min_samples_split: 95 min_samples_leaf: 71	SVC_linear	C: 400
		SVC_rbf	C: 300, gamma: 0.02
		SVC_sig	C: 100, gamma: 0.13, coef0: 400
MLP	hidden_layer_sizes: (20, 40, 20) activation: 'relu' solver: 'adam'	SVC_poly	C: 100, degree: 2, gamma: 0.04, coef0: 900
MKSVC	Kernel weights of linear, rbf, poly and sig kernels: (0.999, 0.212, 0,134, 0.00009)		

TABLE 7: Performance comparison for predicting the next hour's AQHI in HK.

	Accuracy	mse	wr	wf	wp
ARIMA	0.608	0.795	0.608	0.605	0.605
RF	0.782	0.279	0.782	0.779	0.782
MLP	0.908	0.101	0.908	0.908	0.909
SVC_linear	0.960	0.041	0.96	0.961	0.963
SVC_rbf	0.937	0.065	0.937	0.937	0.938
SVC_poly	0.959	0.042	0.959	0.959	0.961
SVC_sigmoid	0.267	4.996	0.267	0.113	0.071
LSTM	0.763	0.265	0.763	0.763	0.773
MKSVC	**0.972**	**0.030**	**0.972**	**0.971**	**0.972**

TABLE 8: Performance comparison for predicting the future 3 hour's AQHI in HK.

	Accuracy	Mse	wr	wf	wp
ARIMA	0.525	0.945	0.525	0.525	0.529
RF	0.782	0.275	0.782	0.778	0.781
MLP	0.938	0.09	0.938	0.936	0.935
SVC_linear	0.961	0.04	0.961	0.962	0.963
SVC_rbf	0.937	0.065	0.937	0.937	0.938
SVC_poly	0.954	0.047	0.954	0.955	0.956
SVC_sigmoid	0.267	4.994	0.267	0.113	0.071
LSTM	0.723	0.220	0.723	0.721	0.729
MKSVC	**0.974**	**0.028**	**0.974**	**0.975**	**0.974**

the other three year's data was used for testing. The comparisons of the predictions for the future 1, 3, 6, 9, and 12 hours are given below.

4.2.1. Predict the AQHI of Hong Kong. Tables 7–11 show the performances of the algorithms for forecasting the future 1, 3, 6, 9, and 12 hours' AQHI in Hong Kong. From the table, the following conclusions can be drawn:

(1) MKSVC performs best on all the three prediction tasks. SVC models with linear, RBF, and polynomial kernels perform better than other models except for the MKSVC. Sigmoid kernel SVC always makes the

Table 9: Performance comparison for predicting the future 6 hour's AQHI in HK.

	Accuracy	mse	wr	wf	wp
ARIMA	0.471	1.208	0.471	0.472	0.474
RF	0.785	0.27	0.785	0.781	0.783
MLP	0.942	0.086	0.942	0.939	0.938
SVC_linear	0.965	0.038	0.965	0.965	0.966
SVC_rbf	0.937	0.066	0.937	0.937	0.937
SVC_poly	0.959	0.043	0.959	0.960	0.960
SVC_sigmoid	0.267	4.992	0.267	0.113	0.071
LSTM	0.732	0.300	0.732	0.733	0.749
MKSVC	**0.976**	**0.028**	**0.976**	**0.976**	**0.976**

Table 10: Performance comparison for predicting the future 9 hour's AQHI in HK.

	Accuracy	mse	wr	wf	wp
ARIMA	0.467	2.020	0.467	0.433	0.436
RF	0.738	0.332	0.738	0.735	0.737
MLP	0.777	0.255	0.777	0.776	0.776
SVC_linear	0.799	0.231	0.799	0.798	0.799
SVC_rbf	0.787	0.244	0.787	0.786	0.786
SVC_poly	0.785	0.25	0.785	0.784	0.784
SVC_sigmoid	0.267	4.991	0.267	0.113	0.071
LSTM	0.681	0.393	0.692	0.676	0.645
MKSVC	**0.817**	**0.203**	**0.821**	**0.815**	**0.820**

Table 11: Performance comparison for predicting the future 12 hour's AQHI in HK.

	Accuracy	mse	wr	wf	wp
ARIMA	0.453	2.148	0.454	0.387	0.410
RF	0.559	0.806	0.559	0.554	0.551
MLP	0.597	0.66	0.597	0.598	0.603
SVC_linear	0.614	0.671	0.614	0.607	0.606
SVC_rbf	0.601	0.69	0.601	0.592	0.594
SVC_poly	0.59	0.716	0.590	0.585	0.583
SVC_sigmoid	0.267	4.992	0.267	0.113	0.071
LSTM	0.528	0.733	0.524	0.512	0.506
MKSVC	**0.630**	**0.609**	**0.641**	**0.629**	**0.633**

Table 12: Performance comparison for predicting the next hour's PM2.5 IAQL in Beijing.

	Accuracy	mse	wr	wf	wp
ARIMA	0.482	1.153	0.482	0.481	0.52
RF	0.472	1.956	0.472	0.442	0.443
MLP	0.486	1.686	0.486	0.466	0.465
SVC_linear	0.515	1.212	0.515	0.525	0.519
SVC_rbf	0.525	0.945	0.525	0.526	0.529
SVC_poly	0.520	1.033	0.520	0.520	0.521
SVC_sigmoid	0.391	3.999	0.391	0.219	0.153
LSTM	0.395	3.648	0.395	0.296	0.247
MKSVC	**0.605**	**0.806**	**0.605**	**0.605**	**0.620**

Table 13: Performance comparison for predicting the future 3 hour's PM2.5 IAQL in Beijing.

	Accuracy	mse	wr	wf	wp
ARIMA	0.471	1.208	0.471	0.472	0.474
RF	0.477	1.858	0.477	0.454	0.451
MLP	0.491	1.678	0.491	0.482	0.477
SVC_linear	0.444	2.363	0.444	0.37	0.336
SVC_rbf	0.496	1.641	0.496	0.469	0.471
SVC_poly	0.489	1.760	0.489	0.462	0.464
SVC_sigmoid	0.391	3.999	0.391	0.219	0.153
LSTM	0.391	3.999	0.391	0.219	0.153
MKSVC	**0.525**	**0.945**	**0.525**	**0.525**	**0.529**

Table 14: Performance comparison for predicting the future 6 hour's PM2.5 IAQL in Beijing.

	Accuracy	mse	wr	wf	wp
ARIMA	0.442	2.381	0.442	0.367	0.332
RF	0.468	2.024	0.468	0.433	0.437
MLP	0.493	1.67	0.493	0.463	0.462
SVC_linear	0.451	2.207	0.451	0.385	0.408
SVC_rbf	0.500	1.595	0.500	0.477	0.478
SVC_poly	0.490	1.844	0.490	0.435	0.441
SVC_sigmoid	0.391	3.999	0.391	0.219	0.153
LSTM	0.397	3.701	0.396	0.253	0.167
MKSVC	**0.513**	**1.275**	**0.513**	**0.520**	**0.519**

Table 15: Performance comparison for predicting the future 9 hour's PM2.5 IAQL in Beijing.

	Accuracy	mse	wr	wf	wp
ARIMA	0.410	1.208	0.471	0.472	0.474
RF	0.457	1.909	0.457	0.446	0.439
MLP	0.48	1.706	0.48	0.457	0.456
SVC_linear	0.45	2.236	0.45	0.385	0.424
SVC_rbf	0.492	1.746	0.492	0.452	0.456
SVC_poly	0.482	1.813	0.482	0.453	0.45
SVC_sigmoid	0.39	4.000	0.39	0.219	0.152
LSTM	0.390	4.000	0.391	0.217	0.151
MKSVC	**0.507**	**1.133**	**0.510**	**0.505**	**0.500**

Table 16: Performance comparison for predicting the future 12 hour's PM2.5 IAQL in Beijing.

	Accuracy	mse	wr	wf	wp
ARIMA	0.386	4.290	0.391	0.355	0.318
RF	0.456	1.933	0.456	0.442	0.433
MLP	0.477	1.663	0.477	0.457	0.451
SVC_linear	0.451	2.204	0.451	0.386	0.400
SVC_rbf	0.489	1.858	0.489	0.431	0.425
SVC_poly	0.478	1.851	0.478	0.452	0.449
SVC_sigmoid	0.390	4.001	0.39	0.219	0.152
LSTM	0.383	4.360	0.387	0.202	0.147
MKSVC	**0.501**	**1.536**	**0.500**	**0.498**	**0.491**

worst predictions which show that the sigmoid kernel is unable to capture the characters of the dataset.

(2) Time series models like ARIMA and LSTM fail to compete with the widely used parametric models like RF, MLP, and SVCs, and as the future time lag increases, the time series models' performances decrease, while the parametric models keep achieving very satisfying results.

(3) Among the well-performed SVC models, linear kernel model performs best, which demonstrates

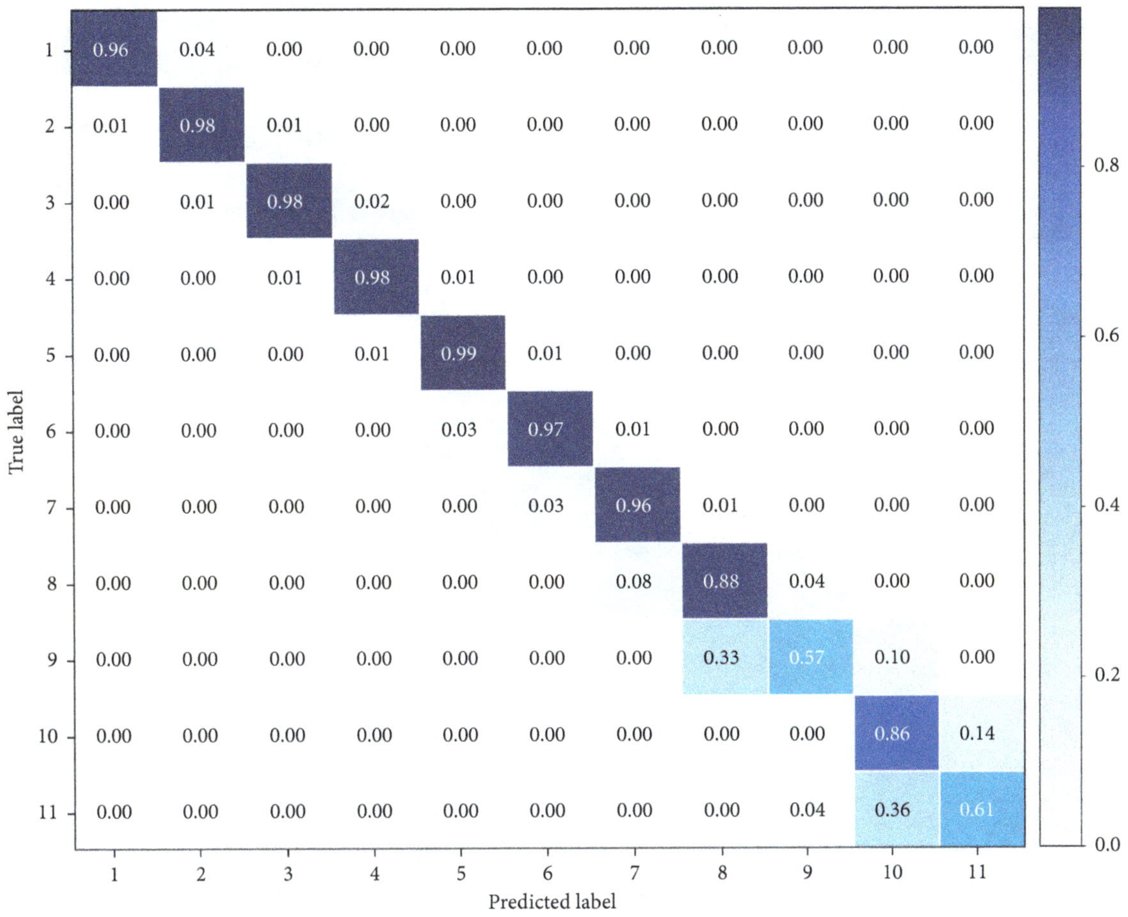

FIGURE 9: AQHI confusion matrix of MKSVC.

that the relation between the target and the input information has a lot of linear components, but there are also factors that influence the future air quality in a nonlinear way as the RBF and polynomial kernels also achieve promising performance.

(4) Models like MKSVC, MLP, and SVCs (except SVC_sigmoid) present very satisfying performance in the prediction for short-term air quality, larger than 90% of accuracy for the future 1, 3, and 6 hours. However, the performance for longer term predictions drops sharply from 0.976 of the 6 hour to 0.630 of the 12 hour (accuracy of MKSVC). It demonstrates that long-term air quality prediction is difficult.

4.2.2. Predict the PM2.5 IAQL of Beijing. Tables 12–16 show the performances of the algorithms for forecasting the next 1, 3, 6, 9, 12 hours' PM2.5 IAQL in Beijing. Similar conclusions can be drawn as that of HK, MKSVC is superior to other models, SVC_sigmoid and LSTM perform worst, SVCs behavior relatively better than other parametric models. But the overall performance of all the models on this dataset is much worse than that of HK. One possible reason is that there are fewer features in the Beijing dataset and the features in the dataset have a weaker correlation with the target. The other reason may be due to the generally worse air conditions in Beijing because higher polluting air conditions are harder to predict as demonstrated in the next part.

4.2.3. Comparison of Severe Air Pollution Prediction. Severe pollution prediction is a difficult task; however, it is critical as high-polluting air condition does way more damage to human health. Therefore, even a small improvement in the prediction of severe pollution is more meaningful than a large improvement in predicting good or less polluting air conditions.

As SVCs performed better than other algorithms except for the MKSVC, the best performing SVC was chosen to compare with the MKSVC in terms of forecasting severe air pollutions in the paper. AQHI greater than 6 is considered as severe pollution in HK. IAQL greater than 4 is considered as severe air pollution in Beijing. Figures 9 and 10 are the confusion matrixes of MKSVC and SVC_linear when predicting the next hour's AQHI of HK.

The x-axis denotes the predicted value, the y-axis denotes the true value, and the values on the diagonal of the matrix denote the probability of the correct prediction. The figures show that linear kernel SVC performs well in forecasting less polluting air conditions, so is the MKSVC. But MKSVC performs far better than linear kernel SVC when AQHI is larger than 8.

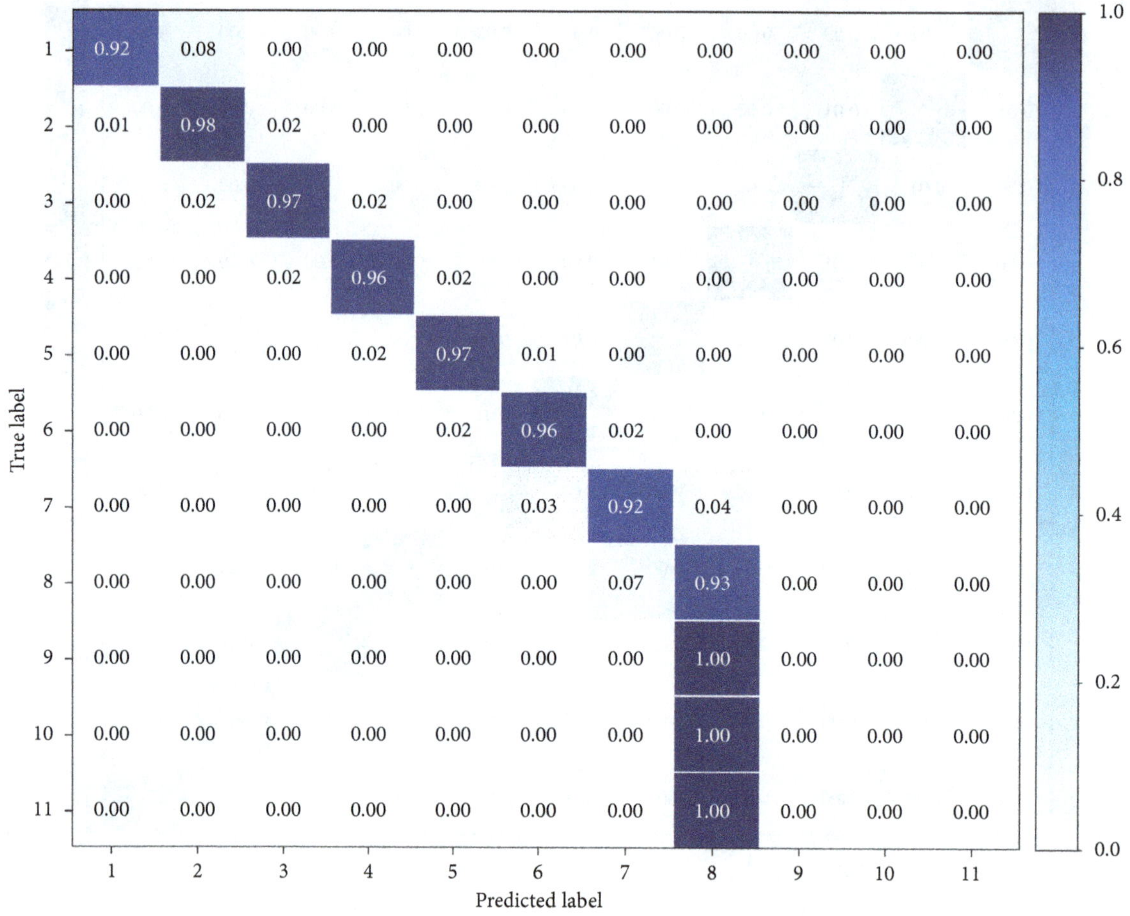

Figure 10: AQHI confusion matrix of SVC_linear.

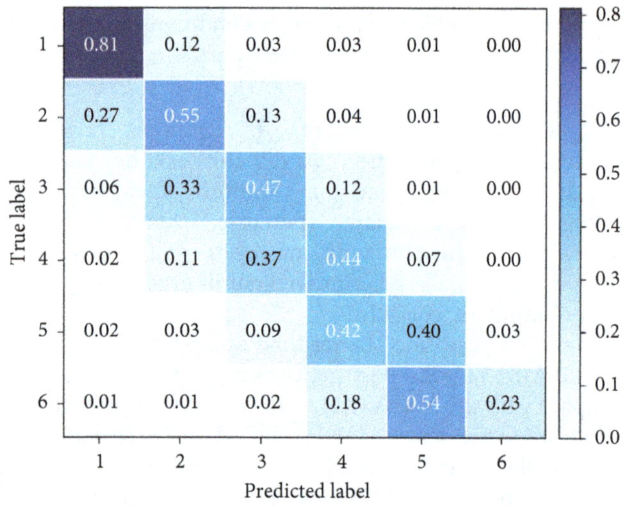

Figure 11: PM2.5 IAQL confusion matrix of MKSVC.

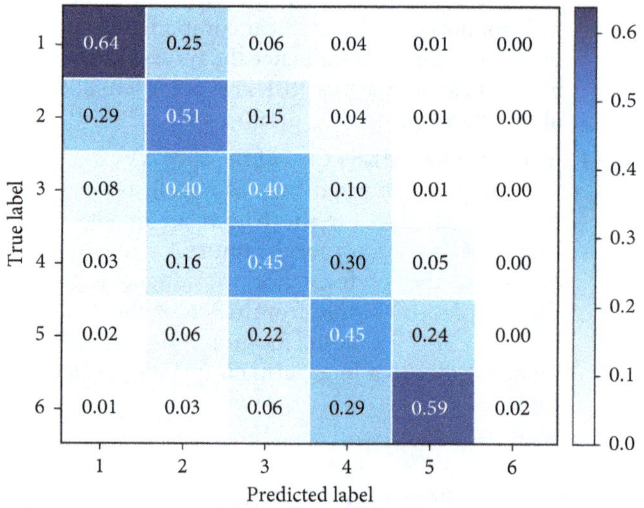

Figure 12: PM2.5 IAQL confusion matrix of SVC_rbf.

Figures 11 and 12 are the confusion matrixes of MKSVC and SVC_rbf when forecasting the next hour's PM2.5 IAQL of Beijing. The same conclusion can be drawn as that of HK. Generally, all the models make better prediction for light pollutions than severe ones due to the bias towards majority classes. It demonstrates that the task for severe air pollution prediction is challenging.

5. Conclusions

In this paper, a novel multiple kernel learning-based approach with SVC as the base learner was proposed for the near future's air quality prediction. It was the first time that multiple kernel learning method was applied to air quality forecasting. Special attention was given to the feature engineering process. MKSVC is capable of learning the optimal combination of different kernels with which information coming from multiple sources can be captured simultaneously. Extensive experiments were conducted to compare the performance of MKSVC with the baseline model ARIMA, widely used parametric air quality forecasting models RFC, MLP, and SVCs, and a deep recurrent neural network model LSTM. Historical air pollutant concentration data, meteorological data, and time stamp data of a coastal city Hong Kong and an inland city Beijing were used to validate the models. Based on the experiments, a number of conclusions can be drawn:

(1) The proposed MKSVC algorithm offers a better predictive ability than the other models.
(2) The proposed MKSVC algorithm is capable of forecasting severe air pollution much better than the other models.
(3) The widely used parametric models RF, MLP, and SVC exhibit better prediction performance than the time series models ARIMA and LSTM.
(4) Feature transformation and feature selection play a significant role in making better air quality forecasting.

As can be seen from the experiments, long-term prediction task is difficult, so is the task to predict severe air pollutions. Though the proposed multiple kernel learning-based approach demonstrated relatively good performance in terms of both long-term prediction and severe air pollution prediction, more sophisticated methods need to be explored in order to build a more comprehensive and effective air quality forecasting system.

Appendix

A. Kernel SVM

Given a dataset with training instances $\{\mathbf{x}_i, y_i\}$ $(i = 1, \ldots, N)$ where $\mathbf{x}_i$ is a vector in the input space $\mathbb{R}^D$ and y_i denotes the class index taking a value +1 or −1. SVM aims at minimizing an upper bound of the generalization error through maximizing the margin between the separating hyperplane and the data in the input space. In real-time problems, it is often not possible to determine an exact separating hyperplane dividing the data within the input space and also we might get a curved decision boundary in some cases. In such cases, the original input space can be mapped to a higher-dimensional feature space (Hilbert space) using nonlinear functions called feature functions ϕ: $\mathbb{R}^D \rightarrow \mathbb{R}^s$. The resulting discriminant function is

$$f(x) = \langle \mathbf{w}, \phi(\mathbf{x})\rangle + b. \tag{A.1}$$

The classifier can be trained by solving the following quadratic optimization problem:

$$\begin{aligned} &\text{minimize} \quad \frac{1}{2}\|\mathbf{w}\|_2^2 + C\sum_{i=1}^{N}\xi_i, \\ &\text{with respect to} \quad \mathbf{w} \in \mathbb{R}^s,\ \xi \in \mathbb{R}_+^N,\ b \in \mathbb{R}, \\ &\text{subject to} \quad y_i\left(\mathbf{w}^T\phi(\mathbf{x}_i) + b\right) \geq 1 - \xi_i \quad \forall_i, \end{aligned} \tag{A.2}$$

where $\mathbf{w}$ is the vector of weight coefficients, b is the bias term of the separating hyperplane, C is a predefined positive trade-off parameter between model simplicity and classification error, ξ represents parameters for handling nonseparable data. Instead of solving this optimization problem directly, the Lagrangian dual function enables us to obtain the following dual formulation:

$$\begin{aligned} &\text{maximize} \quad \sum_{i=1}^{N}\alpha_i - \frac{1}{2}\sum_{i=1}^{N}\sum_{j=1}^{N}\alpha_i\alpha_j y_i y_j \langle\phi(\mathbf{x}_i), \phi(\mathbf{x}_j)\rangle, \\ &\text{with respect to} \quad \alpha \in \mathbb{R}_+^N, \\ &\text{subject to} \quad \sum_{j=1}^{N}\alpha_i y_j = 0, \\ &\qquad C \geq \alpha_i \geq 0 \quad \forall_i, \end{aligned} \tag{A.3}$$

where α is the vector of dual variables corresponding to each separation constraint. Even though feature space is high dimensional, it could not be practically feasible to directly use the feature functions ϕ for classification of the hyperplane. So in such cases, nonlinear mapping induced by the feature functions is used for computation using special nonlinear functions called kernels.

$$\kappa(\mathbf{x}_i, \mathbf{x}_j) = \langle\phi(\mathbf{x}_i), \phi(\mathbf{x}_j)\rangle, \tag{A.4}$$

where κ: $\mathbb{R}^D \times \mathbb{R}^D \rightarrow \mathbb{R}$ is named the kernel function. By solving the above dual problem, we get $\mathbf{w} = \sum_{i=1}^{N}\alpha_i y_i \phi(\mathbf{x}_i)$, and the maximum margin separate hyperplane function can be rewritten as

$$f(\mathbf{x}) = \sum_{i=1}^{N}\alpha_i y_i \kappa(\mathbf{x}_i, \mathbf{x}) + b. \tag{A.5}$$

The multiclass support can be handled according to a one-versus-one or one-versus-rest scheme. The kernel trick allows SVMs to form nonlinear boundaries [14].

B. Calculation of AQHI in Hong Kong and IAQI in Mainland China

B.1. Calculation of AQHI. The AQHI of the current hour is calculated from the sum of the percentage added health risk (%AR) of daily hospital admissions attributable to the 3-hour moving average concentrations of four criteria air pollutants: ozone (O_3), nitrogen dioxide (NO_2), sulphur dioxide (SO_2), and particulate matter (PM) (respirable

Table 17: PM2.5 AQI of mainland China.

IAQI score	Description	PM2.5 concentration ($\mu g/m^3$)
0–50	Excellent	0–35
51–100	Good	35–75
101–150	Lightly polluted	75–115
151–200	Moderately polluted	115–150
201–300	Heavily polluted	150–250
301–500	Severely polluted	250–500

suspended particulates (RSP or PM10) or fine suspended particulates (FSP or PM2.5), whichever poses a higher health risk).

The %AR of each pollutant depends on its concentration and a risk factor which was derived from local health statistics and air pollution data. The %AR is then compared to a scale to obtain the appropriate banding of AQHI. The equations are as follows:

$$\%AR = \%AR(NO_2) + \%AR(SO_2) + \%AR(O_3) + \%AR(PM), \quad (B.1)$$

where %AR (PM) = %AR (PM10) or %AR (PM2.5), whichever is higher.

$$\begin{aligned} \%AR(NO_2) &= [\exp(\beta(NO_2) \times C(NO_2)) - 1] \times 100\%, \\ \%AR(SO_2) &= [\exp(\beta(SO_2) \times C(SO_2)) - 1] \times 100\%, \\ \%AR(O_3) &= [\exp(\beta(O_3) \times C(O_3)) - 1] \times 100\%, \\ \%AR(PM10) &= [\exp(\beta(PM10) \times C(PM10)) - 1] \times 100\%, \\ \%AR(PM2.5) &= [\exp(\beta(PM2.5) \times C(PM2.5)) - 1] \times 100\%, \end{aligned} \quad (B.2)$$

where %AR (NO_2), %AR (SO_2), %AR (O_3), %AR (PM), %AR (PM10), and %AR (PM2.5) are the added health risk of NO_2, SO_2, O_3, PM, PM10, and PM2.5 respectively;

C(NO_2), C(SO_2), C(O_3), C(PM10), and C(PM2.5) are the 3-hour moving average concentration of the respective pollutants in microgram per cubic meter ($\mu g/m^3$). $\beta(NO_2) = 0.0004462559$, $\beta(SO_2) = 0.0001393235$, $\beta(O_3) = 0.0005116328$, $\beta(PM10) = 0.0002821751$, and $\beta(PM2.5) = 0.0002180567$ are added health risk factors (technically known as regression coefficients) of the respective pollutants [24].

B.2. Calculation of IAQI. Each pollutant's individual AQI is called its IAQI. The highest IAQI among these six pollutants at a given time is called the primary or dominant pollutant and is chosen for the overall AQI value.

$$\begin{aligned} IAQI_p &= \frac{IAQI_{H_i} - IAQI_{L_o}}{BP_{H_i} - BP_{L_o}} (C_p - BP_{L_o}) + IAQI_{L_o}, \\ AQI &= \max\{IAQI_1, IAQI_2, IAQI_3, \ldots, IAQI_n\}, \end{aligned} \quad (B.3)$$

where C_p is mass concentration value of the air pollutant p, BP_{H_i} is the high value of the concentration limit which can be checked in the reference table from the paper [25], BP_{L_o} is the low value of the concentration limit which can be checked in the reference table from [25], $IAQI_{H_i}$ is the corresponding value of BP_{H_i} in the same reference table, and $IAQI_{L_o}$ is also the corresponding value of BP_{L_o} in the reference table. The detailed break down of China AQI for PM2.5 concentrations is shown in Table 17.

Conflicts of Interest

The authors declare no conflicts of interest.

Authors' Contributions

Hong Zheng is the group leader and she is responsible for the project management and in charge of revising this manuscript. Haibin Li is responsible for data analysis and planning and performing the experiments. Xingjian Lu and Tong Ruan provided valuable advice about the revised manuscript.

Acknowledgments

The authors are pleased to acknowledge the National Natural Science Foundation of China under Grant nos. 61103115 and 61103172; the National Natural Science Youth Foundation of China under Grant no. 61602175; the special fund for Software and Integrated Circuit Industry Development of Shanghai under Grant no. 150809; and the "Action Plan for Innovation on Science and Technology" Projects of Shanghai (Project no. 16511101000).

References

[1] Y. Wang, Y. Han, T. Zhu, W. Li, and H. Zhang, "A prospective study (SCOPE) comparing the cardiometabolic and respiratory effects of air pollution exposure on healthy and prediabetic individuals," *Science China Life Sciences*, vol. 60, no. 1, pp. 46–56, 2017.

[2] G. Cohen, I. Levy, Yuval et al., "Long-term exposure to traffic-related air pollution and cancer among survivors of myocardial infarction: a 20-year follow-up study," *European Journal of Preventive Cardiology*, vol. 24, no. 1, pp. 92–102, 2017.

[3] T. S. Dye, *Guidelines for Developing an Air Quality (Ozone and PM2. 5) Forecasting Program*, vol. 4, pp. 206-207, United States Environmental Protection Agency, Washington, DC, USA, 2013.

[4] Y. Zhang, M. Bocquet, V. Mallet, C. Seigneur, and A. Baklanov, "Real-time air quality forecasting, part I: history, techniques, and current status," *Atmospheric Environment*, vol. 60, pp. 632–655, 2012.

[5] U. Kumar and V. K. Jain, "ARIMA forecasting of ambient air pollutants (O_3, NO, NO_2 and CO)," *Stochastic Environmental Research and Risk Assessment*, vol. 24, no. 5, pp. 751–760, 2010.

[6] A. Saxena and S. Shekhawat, "Ambient air quality classification by grey wolf optimizer based support vector machine," *Journal of Environmental and Public Health*, vol. 2017, Article ID 3131083, 12 pages, 2017.

[7] C. M. Vong, W. F. Ip, P. K. Wong, and J. Y. Yang, "Short-term prediction of air pollution in Macau using support vector machines," *Journal of Control Science and Engineering*, vol. 2012, Article ID 518032, 11 pages, 2012.

[8] X. Hu, J. H. Belle, X. Meng et al., "Estimating PM2.5 concentrations in the conterminous United States using the random forest approach," *Environmental Science & Technology*, vol. 51, no. 12, pp. 6936–6944, 2017.

[9] R. Yu, Y. Yang, L. Yang, G. Han, and O. A. Move, "RAQ—a random forest approach for predicting air quality in urban sensing systems," *Sensors*, vol. 16, no. 1, p. 86, 2016.

[10] A. Russo, P. G. Lind, F. Raischel, R. Trigo, and M. Mendes, "Neural network forecast of daily pollution concentration using optimal meteorological data at synoptic and local scales," *Atmospheric Pollution Research*, vol. 6, no. 3, pp. 540–549, 2015.

[11] K. Karatzas, N. Katsifarakis, C. Orlowski, and A. Sarzyński, "Urban air quality forecasting: a regression and a classification approach," in *Proceedings of the Asian Conference on Intelligent Information and Database Systems*, vol. 2017, pp. 539–548, Springer, Kanazawa, Japan, April 2017.

[12] E. Pardo and N. Malpica, "Air quality forecasting in Madrid using long short-term memory networks," in *Proceedings of the International Work-Conference on the Interplay between Natural and Artificial Computation*, vol. 2017, pp. 232–239, Springer, Corunna, Spain, June 2017.

[13] X. Li, L. Peng, X. Yao et al., "Long short-term memory neural network for air pollutant concentration predictions: method development and evaluation," *Environmental Pollution*, vol. 231, pp. 997–1004, 2017.

[14] V. Vapnik, "The support vector method of function estimation," in *Nonlinear Modeling: Advanced Black-Box Techniques*, J. A. K. Suykens and J. P. L. Vandewalle, Eds., vol. 55, p. 86, Springer, New York City, NY, USA, 1998.

[15] C. Cortes, M. Mohri, and A. Rostamizadeh, "Algorithms for learning kernels based on centered alignment," *Journal of Machine Learning Research*, vol. 13, pp. 795–828, 2012.

[16] F. Aiolli and M. Donini, "EasyMKL: a scalable multiple kernel learning algorithm," *Neurocomputing*, vol. 169, pp. 215–224, 2015.

[17] S. Niazmardi, B. Demir, L. Bruzzone, A. Safari, and S. Homayouni, "Multiple kernel learning for remote sensing image classification," *IEEE Transactions on Geoscience and Remote Sensing*, vol. 56, no. 3, pp. 1425–1443, 2017.

[18] Y. Zhang, H.L. Yang, S. Prasad, E. Pasolli, J. Jung, and M. Crawford, "Ensemble multiple kernel active learning for classification of multisource remote sensing data," *IEEE Journal of Selected Topics in Applied Earth Observations and Remote Sensing*, vol. 8, no. 2, pp. 845–858, 2015.

[19] H. Wen, Y. Liu, I. Rekik et al., "Multi-modal multiple kernel learning for accurate identification of Tourette syndrome children," *Pattern Recognition*, vol. 63, pp. 601–611, 2017.

[20] M. Gönen and E Alpaydın, "Multiple kernel learning algorithms," *Journal of Machine Learning Research*, vol. 12, pp. 2211–2268, 2011.

[21] Aqhi.gov.hk, Environmental Protection Department, July 2017, http://epic.epd.gov.hk.

[22] RP5.ru: Weather for 243 Countries of the World, July 2017, http://rp5.ru.

[23] Uci.edu: Machine Learning Repository, Beijing PM2.5 Dataset, https://archive.ics.uci.edu.

[24] W. T. Wai, W. T. W. San, M. A. W. H. Shun et al., "A study of the air pollution index reporting system," *Statistical Modelling*, vol. 13, p. 15, 2012.

[25] F. Gao, "Evaluation of the Chinese new air quality index (GB3095-2012): based on comparison with the US AQI system and the WHO AQGs," Bachelor's thesis, Integrated Coastal Zone Management, Raseborg, Finland, 2013.

[26] Q. W. Yan, "Environmental protection department issued HJ633–2012 environmental air quality index (AQI) technical requirements (trial)," *CSG*, vol. 4, p. 49, 2012.

[27] D. M. Powers, "Evaluation: from precision, recall and F-measure to ROC, informedness, markedness and correlation," *Journal of Machine Learning Technologies*, vol. 2, no. 1, pp. 37–63, 2011.

[28] J. H. McDonald, *Handbook of Biological Statistics*, Sparky House Publishing, Baltimore, MD, USA, 2009.

[29] J. Cohen, P. Cohen, S. G. West et al., *Applied Multiple Regression/Correlation Analysis for the Behavioral Sciences*, Routledge, Abingdon, UK, 2013.

[30] S. Hochreiter and J. Schmidhuber, "Long short-term memory," *Neural Computation*, vol. 9, no. 8, pp. 1735–1780, 1997.

[31] F. Pedregosa, G. Varoquaux, A. Gramfort et al., "Scikit-learn: machine learning in Python," *Journal of Machine Learning Research*, vol. 12, pp. 2825–2830, 2011.

[32] S. Seabold and P. Josef, "Statsmodels: econometric and statistical modeling with Python," in *Proceedings of the 9th Python in Science Conference*, Austin, TX, USA, June 2010.

7

Analysis of SO_2 Pollution Changes of Beijing-Tianjin-Hebei Region over China based on OMI Observations from 2006 to 2017

Zhifang Wang,[1,2] Fengjie Zheng,[1] Wenhao Zhang,[1] and Shutao Wang[2]

[1]*Institute of Remote Sensing and Digital Earth, Chinese Academy of Sciences, Beijing 100101, China*
[2]*Department of Instrument Science & Engineering, Yanshan University, Qinhuangdao, Hebei 066004, China*

Correspondence should be addressed to Fengjie Zheng; zhengfj@radi.ac.cn

Academic Editor: Pedro Salvador

Sulfur dioxide (SO_2) in the planetary boundary layer (PBL) as a kind of gaseous pollutant has a strong effect regarding atmospheric environment, air quality, and climate change. As one of the most polluted regions in China, air quality in Beijing-Tianjin-Hebei (BTH) region has attracted more attention. This paper aims to study the characteristics of SO_2 distribution and variation over BTH. Spatial and temporal variations for a long term (2006–2017) over BTH derived from OMI PBL SO_2 products were discussed. The temporal trends confirm that the SO_2 loading falls from average 0.88 DU to 0.16 DU in the past 12 years. Two ascending fluctuations in 2007 and 2011 appeared to be closely related to the economic stimulus of each five-year plan (FYP). The spatial analysis indicates an imbalanced spatial distribution pattern, with higher SO_2 level in the southern BTH and lower in the northern. This is a result of both natural and human factors. Meanwhile, the SO_2 concentration demonstrates a decreasing trend with 14.92%, 28.57%, and 27.43% compared with 2006, during the events of 2008 Olympic Games, 2014 Asia-Pacific Economic Cooperation (APEC) summit, and 2015 Military Parade, respectively. The improvement indicates that the direct effect is attributed to a series of long-term and short-term control measures, which have been implemented by the government. The findings of this study are desirable to assist local policy makers in the BTH for drawing up control strategies regarding the mitigation of environmental pollution in the future.

1. Introduction

Sulfur dioxide (SO_2) is a short-lived gas primarily produced by volcanoes, power plants, refineries, metal smelting, and burning of fossil fuels. When SO_2 remains near the Earth's surface, it is toxic, causes acid rain, and degrades air quality. It forms sulfate aerosols that can alter cloud reflectivity and precipitation in the free troposphere [1, 2]. As a kind of important atmospheric pollutants, SO_2 critically affects the global environment,climate change, and public health. SO_2 has become one of the popular research topics in the past decades, to examine its changes over some of the world's most polluted regions [3–10].

China, with its incredible economic growth has been the focus of many studies during the previous decade because of its increasing sulfur dioxide emissions' contribution to the Earth's atmosphere [11–16]. The sources of SO_2 are both natural (volcanic) and anthropogenic emissions. Natural emissions include intentional biomass burnings and volcanic eruptions. Anthropogenic emissions are mainly due to fossil fuel burning (e.g., coal and oil), which accounts for more than 75% of global emissions [3]. Anthropogenic SO_2 emissions are predominantly in or slightly above the planetary boundary layer (PBL), impacting on regional variations of aerosol types [17].

Satellite measurements of trace gases have been widely used and been an essential way to provide global, consistent observations for detecting, monitoring, and quantifying the SO_2. The first space-based quantitative data on SO_2 mass of the El Chichon volcanic eruption in 1983 were obtained from Total Ozone Mapping Spectrometer (TOMS) on board Nimbus 7 [18]. Subsequently, anthropogenic SO_2 sources from power plants in eastern Europe [19, 20] and smelters in Peru and Russia [21] were demonstrated through detection of SO_2 emissions using Global Ozone Monitoring Experiment (GOME) measurements on the Earth Research

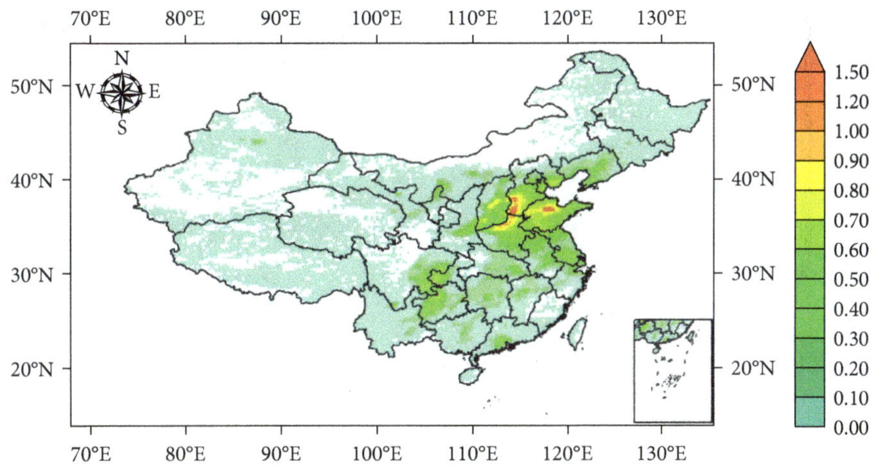

Figure 1: The average SO_2 spatial distribution (in DU) map (2006–2017) over China.

Satellite 2 (ERS-2). The tropospheric SO_2 were detected by the SCanning Imaging Absorption spectroMeter for Atmospheric CHartographY (SCIAMACHY) on board the ENVISAT [22] and the Global Ozone Monitoring Experiment-2 (GOME-2) instrument on MetOp-A [23]. The Ozone Monitoring Instrument (OMI) on NASA's Aura spacecraft enables to provide daily, nearly global measurements of ozone columns and aerosols, and the trace gases with the highest spatial resolution and the longest data record currently available [24, 25].

OMI data have been applied to assess the effect of pollutant transmission, analyze pollutant source contribution, evaluate pollutant emission inventory, observe regional pollution changes, and quantify the reduction of power plant emissions [14, 16, 17, 26–28], due to the higher spatial and temporal resolution. Three major air pollutants (NO_2, SO_2, and CO) in China before, during, and after the Olympic Games from Aura's Ozone Monitoring Instrument (OMI) and Terra's Measurements of Pollutants in the Troposphere (MOPITT) instrument have been measured [29]. Annual emissions by sector and fuel types calculated from satellite data show an increasing trend of SO_2 during 1996–2008 and decreasing thereafter in China [16]. Substantial changes in SO_2 emissions in the northern China for the period 2005-2008 were analyzed [14]. The spatiotemporal variation of SO_2 concentration during 2005–2008 over China from the planetary boundary layer (PBL) SO_2 column concentration retrieved from OMI has been analyzed [28]. Long-term SO_2 pollution changes over China or region have been observed through OMI observations [29]. A long-term trend of NO_2 and SO_2 levels (2005–2014) of the Henan province in China has been retrieved from the OMI [30]. In the past decades, China has adopted different policies for air quality control consistently, such as carbon reduction, energy saving, and other measures to reduce SO_2 emissions [31, 32]. The latest findings represent that large reductions in SO_2 are benefiting from the effective control policies in China [33, 34].

In this study, we analyze trend variation and distribution in SO_2 concentrations over the Beijing-Tianjin-Hebei (BTH) region observed by the OMI between January 2006 and December 2017. Figure 1 shows the multiyear average spatial distribution map of SO_2 based on OMI data (2006–2017) over China, which clearly shows the hotspots of SO_2 in the North China Plain. Meanwhile, it is noted that the BTH region is one of the most polluted regions in China. Accordingly, tight emission control arrangement of this area always adopted, the SO_2 emission of Hebei rank 3rd in 2013 went down to 5th with 17.4% rate of decline among all provinces in China, and Beijing and Tianjin declined 24.1% and 17.2%, respectively. Regional SO_2 time evolution and spatial distribution are discussed in the following. The datasets and monitoring area are presented in Section 2. The analysis and associated findings are described in Section 3. Finally, the main conclusions are summarized in Section 4.

2. Method Description

2.1. Data Sources. Data used in this study over the BTH are based on OMI SO_2 products with Dobson Units (DU, $1\,DU = 2.69 \times 10^{16}$ molecules/cm^2). The Ozone Monitoring Instrument (OMI) is a sun-synchronous polar orbiting Dutch/Finnish sensor on the AURA satellite launched on 15 July 2004. The science goals of OMI are directly related to these questions and focus on (1) measuring the ozone layer and its destroying trace gases BrO and OClO, (2) tropospheric pollution by ozone, nitrogen dioxide, tropospheric aerosols, SO_2, and formaldehyde, and (3) detection of species important for climate change such as aerosols, clouds, and ozone. The OMI measures the radiation backscattered by the Earth's atmosphere and surface over the entire wavelength range from 270 to 500 nm, with a spectral resolution of about 0.5 nm, and high spatial resolution ($13 \times 24\,km^2$), and daily global coverage [24]. OMI data have 4 processing grade products: Level-0, Level-1, Level-2, and Level-3. For this study, we used the PBL SO_2 vertical column density from the Level-3 0.25×0.25 degree gridded OMI/AURA SO_2 data product. The data used here with the time span of January 1, 2006 to December 31, 2017, were obtained from Giovanni interface (http://giovanni.gsfc.nasa.gov/giovanni/), derived from the NASA Goddard Earth Sciences Data Active Archive Center (GES DISC; http://disc.sci.gsfc.nasa.gov) [35, 36]. These Level-3 products have been widely used [15, 34, 37, 38]. The original OMI PBL SO_2

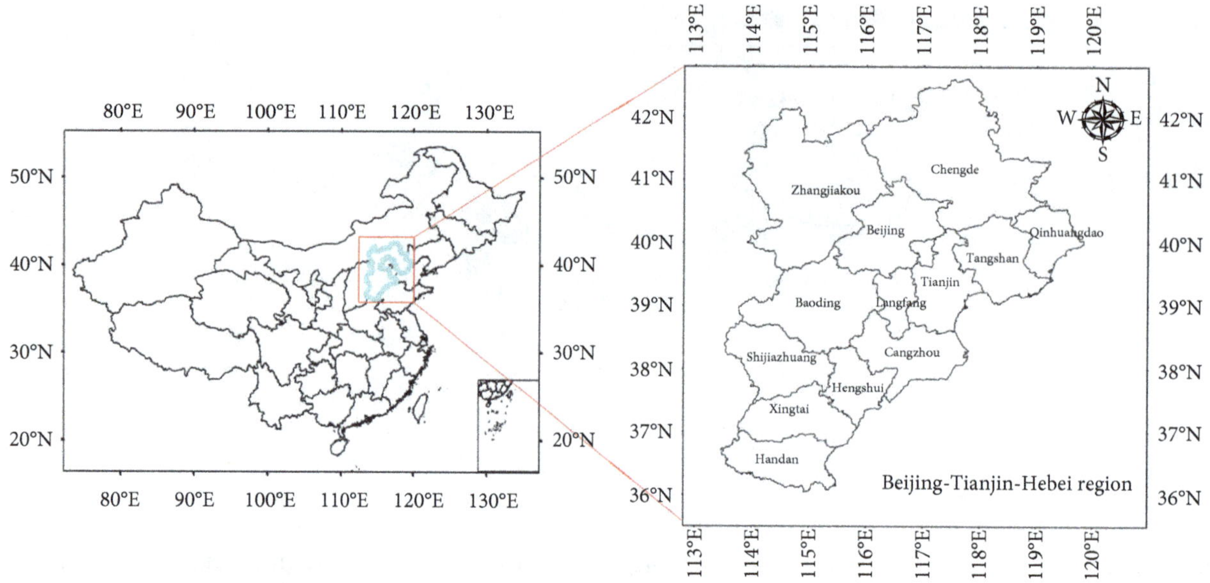

Figure 2: Location of the study area in this study. The right image represents the region of Beijing-Tianjin-Hebei.

Table 1: Relevant information for study areas (data based on 2016).

Name	Area (km^2)	Population (10^4)	Vehicles (10^4)	GDP (10^9)	Coal consumption (Mt)
Beijing	1.68	2173	547.44	25669	11.65
Tianjin	1.13	1562	273.69	17885	45.39
Hebei	18.77	7470	1245.89	32070	289.43

product employed the band residual difference (BRD) algorithm [39]. But this product has a high noise level and systematic artifacts that required empirical corrections [5, 6]. A new operational OMI PBL SO_2 product produced with the principal component analysis (PCA) algorithm was released [40, 41]. Validation of these two algorithmic products has been analyzed [42].

Daily satellite observations were retrieved with the given longitude and latitude for China and BTH region, as shown in Figure 2, to gain insight into the distribution of SO_2 columns in the BTH region. The data were gridded onto monthly 0.25×0.25 fields and then onto seasonal and yearly maps. The meteorological conditions used for analysis are from NCEP reanalysis data (https://www.esrl.noaa.gov/psd/).

2.2. Study Area. Our study area focuses on the Beijing-Tianjin-Hebei (BTH) region (Beijing, Tianjin, and Hebei integration), which is the most polluted industrialized regions in China (Figure 1). The BTH region is located in the northwest part of the North China Plain as shown in Figure 2 (36°05′–42°37′N, 113°11′–119°45′E), with a total area of 216,000 km^2 and more than 110 million residential population. The BTH region includes two municipalities (Beijing and Tianjin) and one province (Hebei) (Table 1), which contain thirteen cities: Beijing, Tianjin, Baoding, Langfang, Tangshan, Shijiazhuang, Xingtai, Handan, Cangzhou, Hengshui, Qinhuangdao, Chengde, and Zhangjiakou.

As one of the most economically vibrant regions in China, the BTH region covers only 2.3% of the Chinese territory but generates over 10% of the total national gross domestic product (GDP) in 2016 (National Bureau of Statistics of China (NBSC)) [43]. As the main high-tech and heavy industry base of China, there are mainly the automotive industry, electronic industry, machinery industry, iron industry, and steel industry. The SO_2 map (Figure 2) shows hotspots associated with the major coal-fired power plants and industrial activities. Figure 3 reveals that high sulfur coal-fired power plants are the major contributors to the SO_2 concentrations over the BTH region [44, 45]. OMI-derived spatial distribution shows generally good agreement with these main anthropogenic emission sources from burning sulfur-contaminated fossil fuels as Figure 3. The last decade has seen frequent occurrences of severe air pollution episodes (haze), and the high SO_2 loading observed certainly contributed to PM 2.5 problems, especially in winter. Atmospheric environment quality has attracted more and more attention related to air pollution prevention and control policy-making.

3. Results and Discussion

The temporal and spatial variations of SO_2 concentration in the region of Beijing-Tianjin-Hebei during the period of 2006–2017 were analyzed based on the satellite OMI data to characterize the variation of SO_2 columns.

3.1. Overall Temporal Trend of SO_2

3.1.1. SO_2 Decadal Change. Figure 4(a) clearly shows the annual average SO_2 concentration time series change trend over

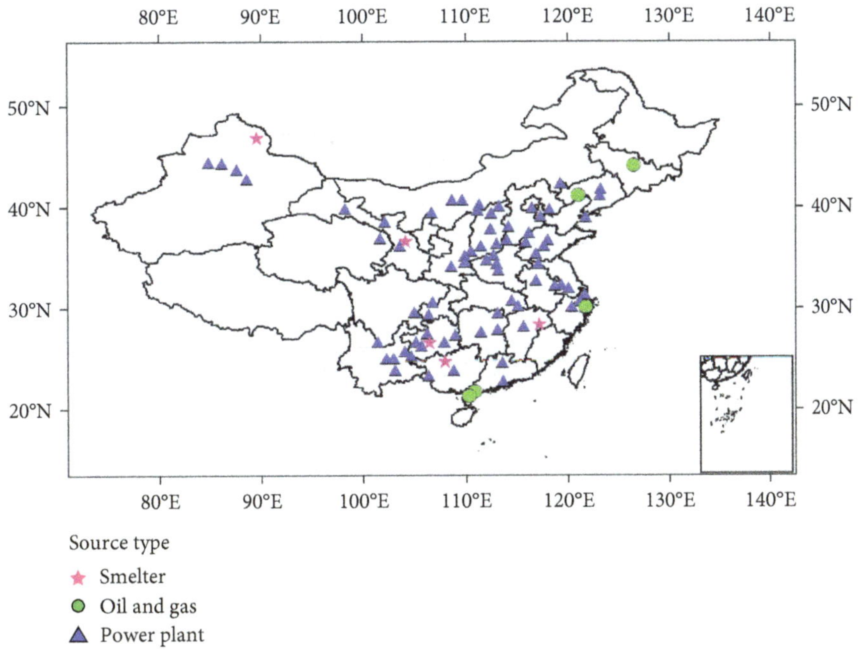

FIGURE 3: Geographic distribution of the major SO_2 sources in China.

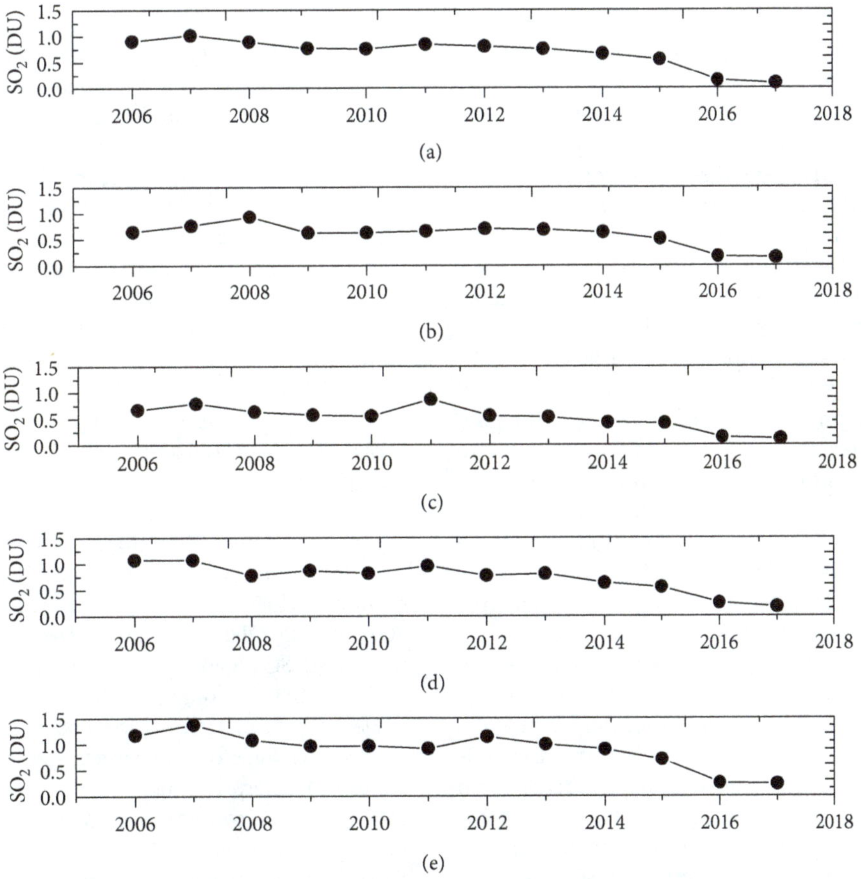

FIGURE 4: Long trend of SO_2 (DU) over Beijing-Tianjin-Hebei region. (a) Annual average, (b) spring average, (c) summer average, (d) autumn average, and (e) winter average.

BTH from 2006 to 2017. The SO_2 loading has decreased over the recent years without clear regularity, which is in line with the study results by others [41, 46]. The plot describes the irregular upward and downward rule. It is necessary to identify the specific period for the pollution attenuation, to a certain extent, related to governmental actions. It was found that SO_2 peak in

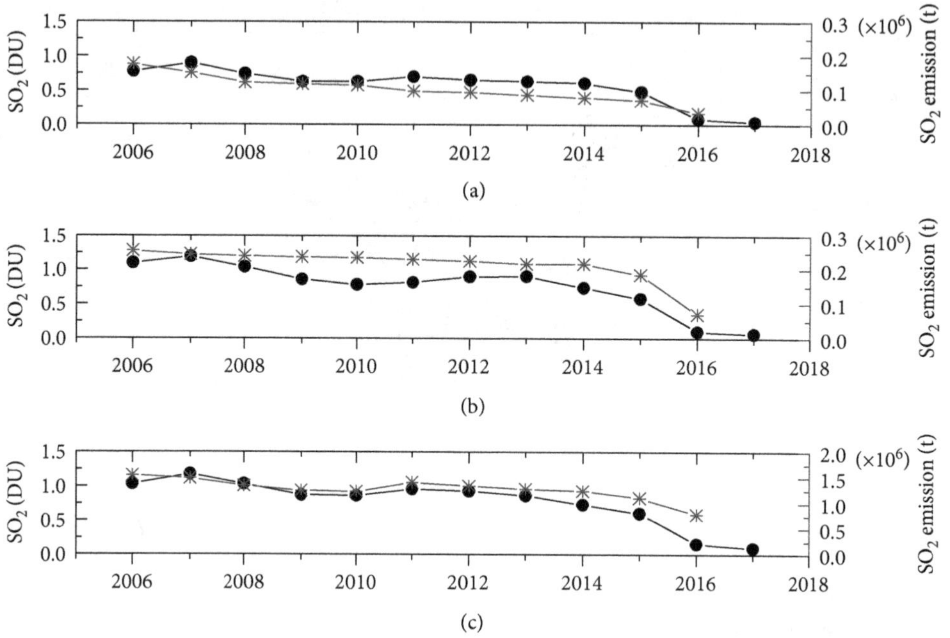

FIGURE 5: Annual average SO_2 column (DU) based on OMI (black circle dots) and SO_2 emission statistics (blue asterisks) in (a) Beijing, (b) Tianjin, and (c) Hebei province (statistical data for 2017 have not been published).

2007 with an upward of 14.96% compared to 2006, reflecting the total SO_2 emission in China had substantially increased from 2000 to 2007. The phenomenon has been largely driven by expansion in manufacturing industries and fueled by coal for the Chinese economic growth [14]. And then, an obvious downward trend appeared from 2007 to 2010, and the decline was 12.92% in 2007-2008 and 13.23% in 2008-2009, respectively. Afterwards from 2009 to 2010, the downward trend slowed down with the rate of decline 0.68%. The decrease was mainly due to China's 11th FYP requiring power plants to install FGD devices.

Nonetheless, there was a temporary rebound from 2010 to 2011 with 19.19% growth rate. The brief period of emission growth can probably be attributed to the government stimulus for resurgence of economy in response to the global financial crisis of 2007-2008. Subsequently, there was a sharp decrease with 37.48% reduction over the 7-year period during 2011–2017. Through the seasonal variation tendency (Figures 4(b)–4(e)), the increase in 2011 mostly comes from summer, accompanied by the industrial production slowdown in the latter half of 2011.

As an industrialized and populated region, the levels of BTH air pollution are determined by population density, economic activity, type of power generation and fuel used, and regulatory policies [47]. Here, we examine the SO_2 changes with the regional SO_2 emission data derived from national statistics (http://www.stats.gov.cn/tjsj/). It is interesting to find out the yearly emission trend that agrees well with the trend based on OMI retrieved SO_2 as mentioned earlier (Figure 5), along with the estimated SO_2 emissions data from power plants. Some mismatch is expected considering the different observation means and uncertainties with satellite and ground measurements.

Two fluctuations occur in the long trends; therefore, we broke down the data into two periods, 2006–2010 and 2011–2017, to verify the coherence with precursors' investigation [30, 48] as to the changes in SO_2 loadings. Figure 6 shows the 5-year mean SO_2 concentration distribution maps over the BTH region of China's 11th five-year plan (2006–2010) and second 7 years mean SO_2 meets China's 12th five-year plan (2011–2015) and China's new 13th five-year plan (2016–2017), respectively. The dramatic decrease in SO_2 loading (Figure 6) well illustrates the achievements and improvements due to a series of air pollution control policies. The average SO_2 was 0.87 DU for 2006–2010 (Figure 6(a)). The average concentration decreased to 0.72 DU for 2011–2015 and 0.23 DU for 2016-2017.

According to the above statistics (Figure 5), it is more clearly shown that the SO_2 loading over the BTH peaked in 2007, and then presented overall decreasing trend. This can be attributed to the emission control measures taken by the government. The sharp decline from 2007 to 2010 is closely related to the installation of flue gas desulfurization (FGD) and follow-up effects of strict pollution reduction measures implemented before the 2008 Beijing Olympic Games. The results also show that atmospheric SO_2 loadings in BTH have drastically decreased by 17.24% (2011–2015) and 75.29% (2016-2017) related to the average quantity in 11th five-year plan (2006–2010), due to more stricter emission reduction targets, such as flue gas desulfurization control of enterprises, new energy to replace polluting energy, 50% of privately owned vehicles were banned through an odd and even number system, and switching from coal to natural gas for heating. And it is noted that the SO_2 loadings have been with a short-lived upswing in the early of 11th and 12th five-year plan that may be caused by the government's economy stimulus.

3.1.2. Seasonal Variations. A more detailed temporal variation of SO_2 concentration during 2006–2017 in the BTH region is analyzed. As shown in Figure 7, the SO_2 seasonal

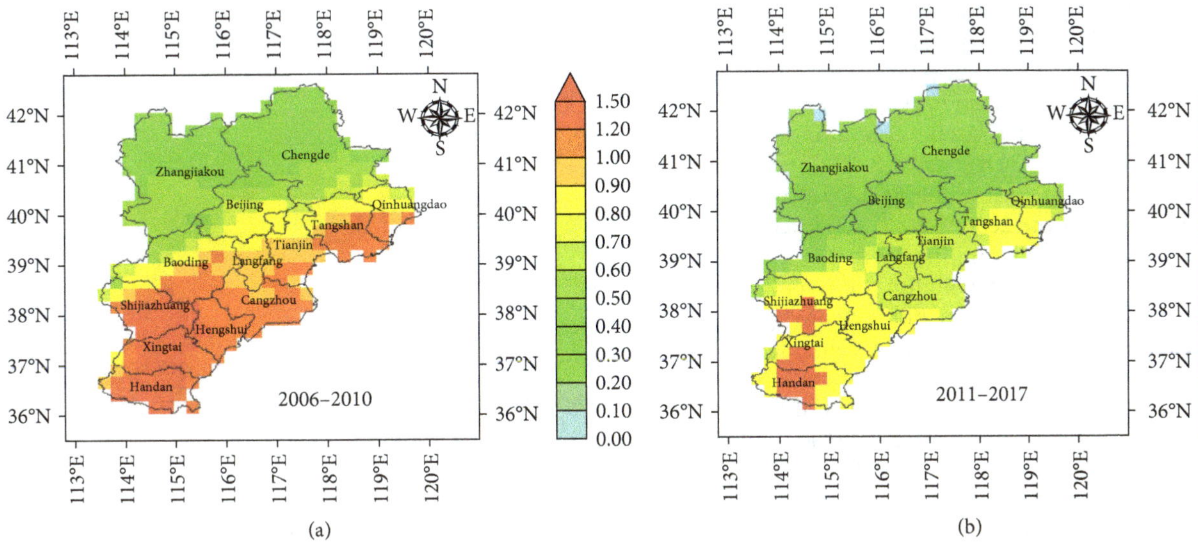

FIGURE 6: Average SO_2 column over the BTH region during 2006–2010 (a) and 2011–2017 (b).

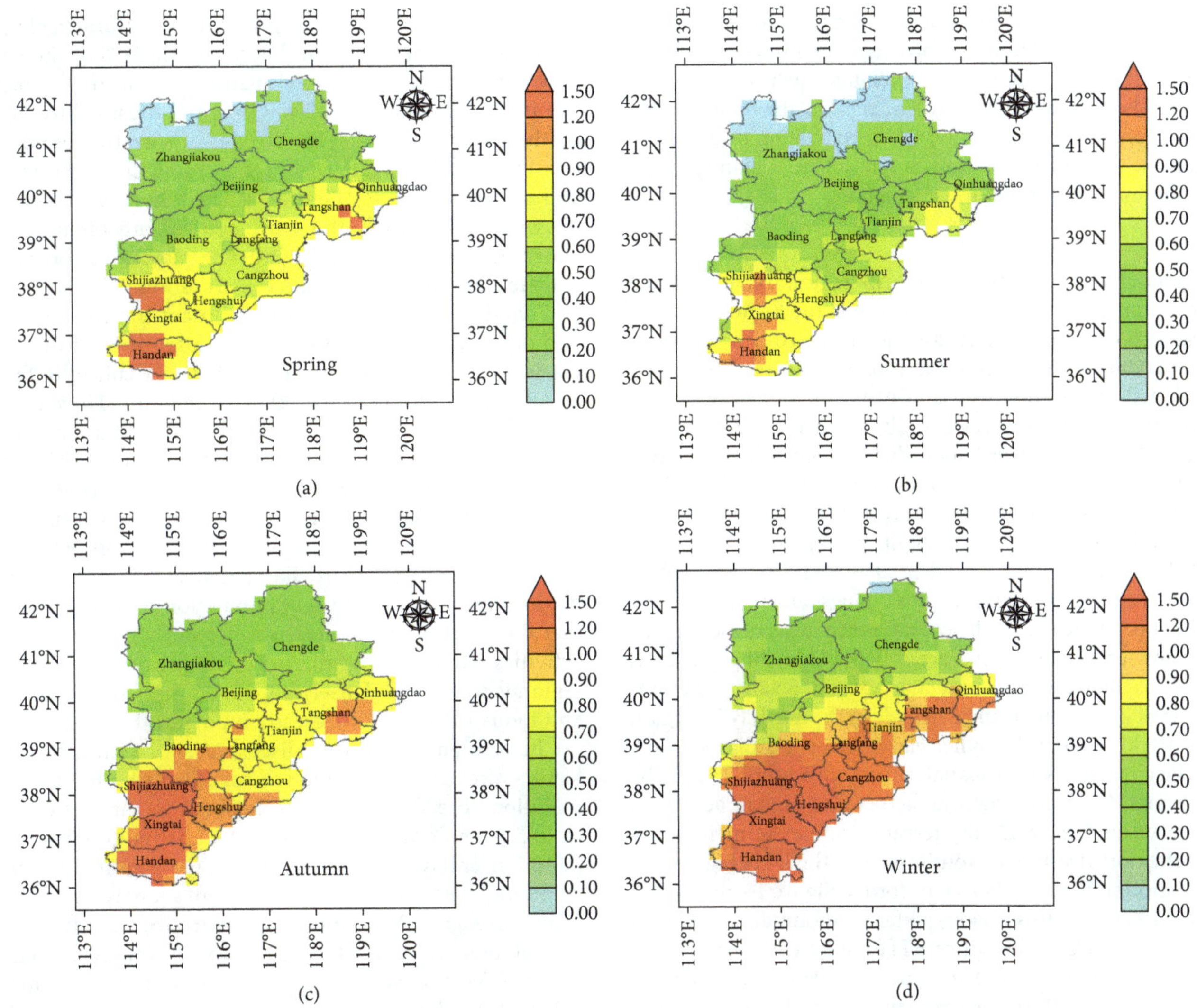

FIGURE 7: Seasonal average SO_2 over the Beijing-Tianjin-Hebei region (in DU), 2006–2017.

mean distribution has an obvious seasonal change sequence as winter (December, January, and February)>autumn (September, October, and November)>spring (March, April, and May)>summer (June, July, and August). The average SO_2 loading of four seasons (spring, summer, autumn, and winter) was 0.59 DU, 0.52 DU, 0.73 DU, and 0.89 DU, respectively. Figure 8 also shows seasonal variation by calculating the monthly SO_2 columns from 2006 to 2017. SO_2 levels are peak in winter and minimum in summer and they are mainly attributed to the difference in pollution diffusion, which is caused by anthropogenic activities and meteorological conditions. Coal heating is a major source accounting for high SO_2 levels in winter. The SO_2 rapid drawdown in 2016 and 2017 represents the fact that coal heating had been replaced by natural gas is successful in reducing sulfur dioxide emissions. In addition, the SO_2 in the PBL has short lifetimes during the warm season, on the short time scale of days to months, and meteorology plays an important role in regional air pollution [6]. The pollutants in the atmosphere have a dilution effect. The meteorological data of 2007, the highest loading over the past twelve years, were chosen as an example to interpret the seasonal differences (Figure 9). In summer, near-surface temperature is higher and air convection is stronger, besides relative humidity and amount of precipitation reaches the annual maximum. All these aforementioned basics can speed up the diffusion of atmospheric pollutants, which have given birth to short lifetimes of SO_2 loading. On the contrary, the temperature, relative humidity, and precipitation in winter go against the transformation from SO_2 to sulfate, making the SO_2 with the longest lifetime in winter [48].

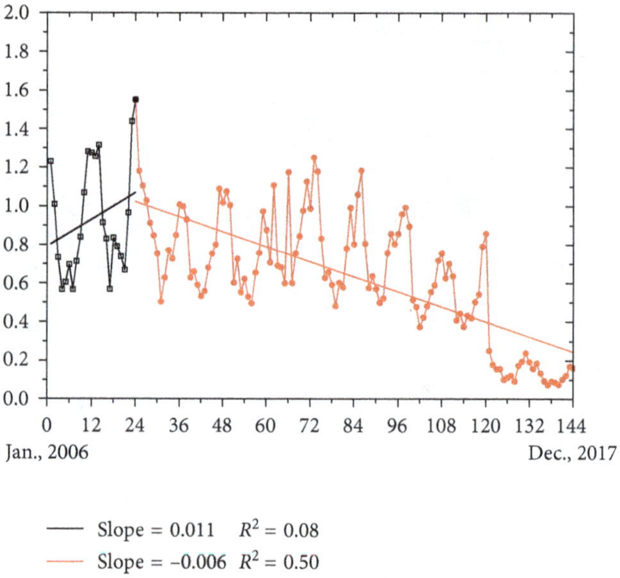

FIGURE 8: Long trend of SO_2 column over the BTH region during two periods, 2006–2007 and 2007–2017.

3.2. Spatial Distribution Pattern. In Figure 10, the spatial distribution of the average SO_2 column in the Beijing-Tianjin-Hebei region from 2006 to 2017 is presented. As shown in Figure 10, the high SO_2 concentration in BTH is significantly distributed in the southwest and eastern regions.

As for the thirteen cities, multiyear average SO_2 clearly presents the spatial distribution discrepancy. We set up two borderlines with SO_2 column amount (1.0 DU and 0.5 DU) to identify hotspots over BTH (Figure 11). The SO_2 concentrations of Handan, Xingtai, and Shijiazhuang in the southwest of BTH are more than 1.0 DU. Cities with SO_2 concentrations exceeding 0.5 DU include Tangshan, Tianjin, and Qinhuangdao located in the coastal beach areas of Bohai Bay, Hengshui, and Cangzhou adjacent to the Shandong province, as well as Baoding and Langfang around Beijing. The SO_2 loading of Zhangjiakou and Chengde in the north BTH region cities are lower than the other eleven cities less than 0.5 DU.

The characteristic of spatial distribution over BTH can be interpreted from both natural factors and human activities. As shown in Figure 12, the terrain declines semicircularly from the northwest to the southeast over the BTH region due to the mountains and plains landforms. Figure 13 also shows the landscapes in this region, plateaus, mountains, and hills account for 54% of the entire BTH region centering on the area of northwest. Surface pressures gradually decrease from eastern to the northwest over the mountains (Figure 14). The central and southeast plains account for 46% of the region. Forests and grasslands are representative natural ecological landscapes in the region and account for 38% of the study area, and they are mainly distributed in the Yanshan Mountains, the Taihang Mountains, and the northwest edge of the Inner Mongolia Plateau [49]. We also found that relative humidity is different in this region due to the terrain and surface coverage (Figure 9). The whole area is mostly above 30%, especially the northwest is with higher values above 50% in winter, and wind also plays a major part in contributing to the SO_2 spatial distribution. Wind direction and wind speeds determine the pollutant transmission path and diffusion velocity. The southern part of BTH underlies the leeward area which makes the SO_2 difficult to diffusion and dilution, leading to the SO_2 accumulation. However, the area of northern BTH located in the upwind accompanied with high wind speeds in favor of SO_2 diffusion as shown in Figure 14. From the point of human factors, coal, petrification, motor vehicles, and iron and steel industrial emissions are the major source of the BTH region, which has a close relationship with the spatial distribution of population. The spatial pattern of population over BTH exhibits that the southern plain area is more likely to cause human activities variance than the northern region. The southern region is so flat and densely populated that human activities and industrial emissions greatly influence the environment.

Besides natural factors discussed above, anthropogenic factors also take up significant role, contributing to heavy pollution. The BTH region is adjacent to Inner Mongolia, Shanxi, Shandong, and Henan, which is the top five major pollution emission provinces in China (Figure 12), where gathering the most heavily emission sources in China is shown in Figure 3. As shown in Figure 15, we calculate the annual SO_2 columns for each city of the BTH region. Handan, Xingtai, and Shijiazhuang as the emission hotspots will be served for further analysis. At first, the high SO_2 loadings in the southwest of BTH around the province of

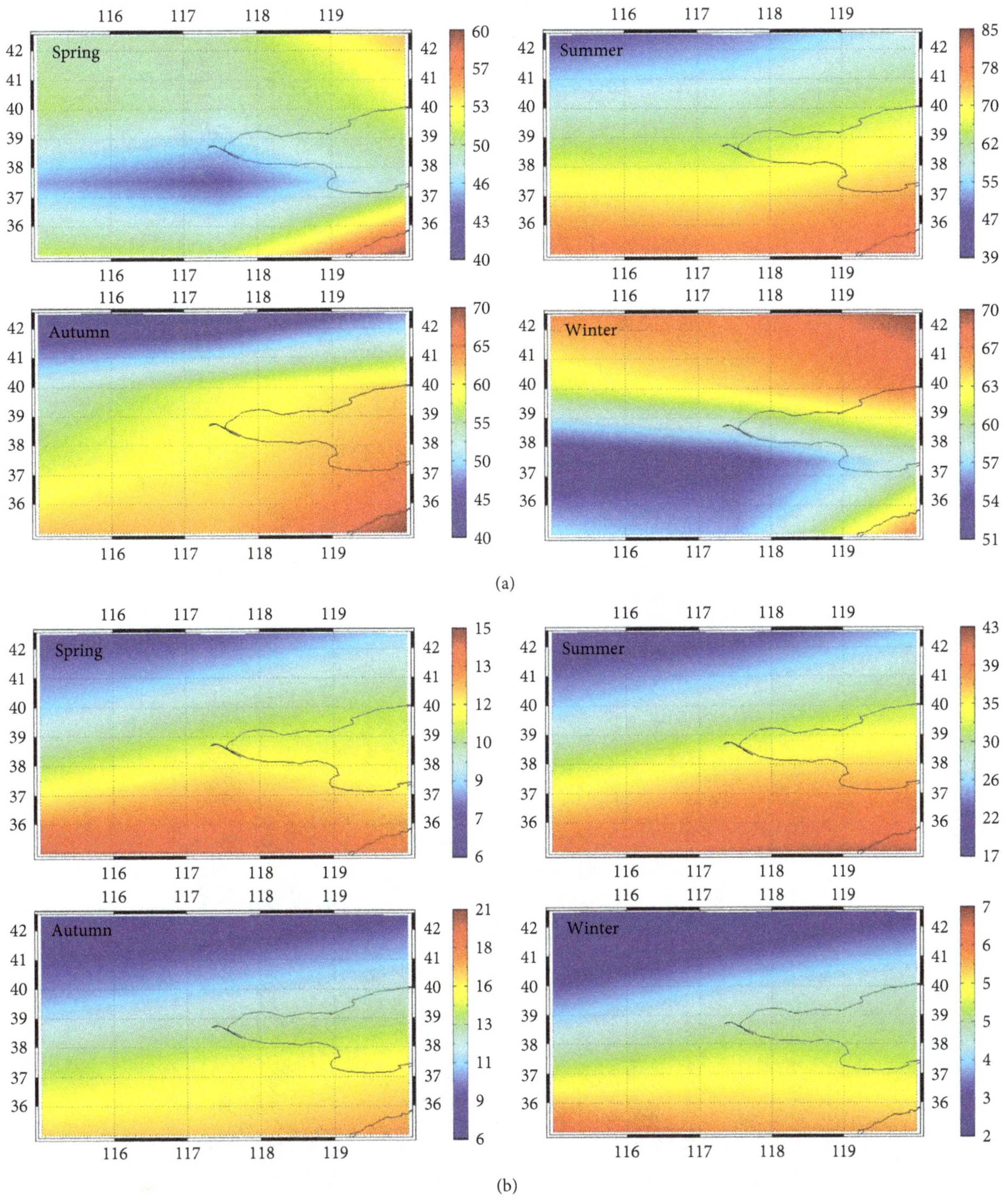

FIGURE 9: Seasonal mean of climatology variables. (a) Relative humidity (%) and (b) precipitable water content (kg/m^2).

Shanxi has nearly a hundred of coal-fired power plants that help contribute to the higher value. Secondly, the BTH region has the largest iron and steel industrial scale in China, among that Beijing, Tianjin, Shijiazhuang, Handan, and Tangshan occupies the majority. Meanwhile, these cities have been the most polluted in BTH region due to unfavourable factors, such as meteorological and geographic conditions and population density. Tangshan, the iron and steel producer of Hebei province, is the second most polluted city because of its high smoke SO_2 emission. Although population and enterprises of Beijing are high, the highest green coverage rate and strict pollution control measures make the pollution level relatively low in the BTH region. Zhangjiakou and Chengde had a low SO_2 column with the least amount of population density, industrial enterprises as well as the terrain advantage. The amplitude of each urban curves variation is able to reveal the degree of being influenced by natural and anthropogenic sources.

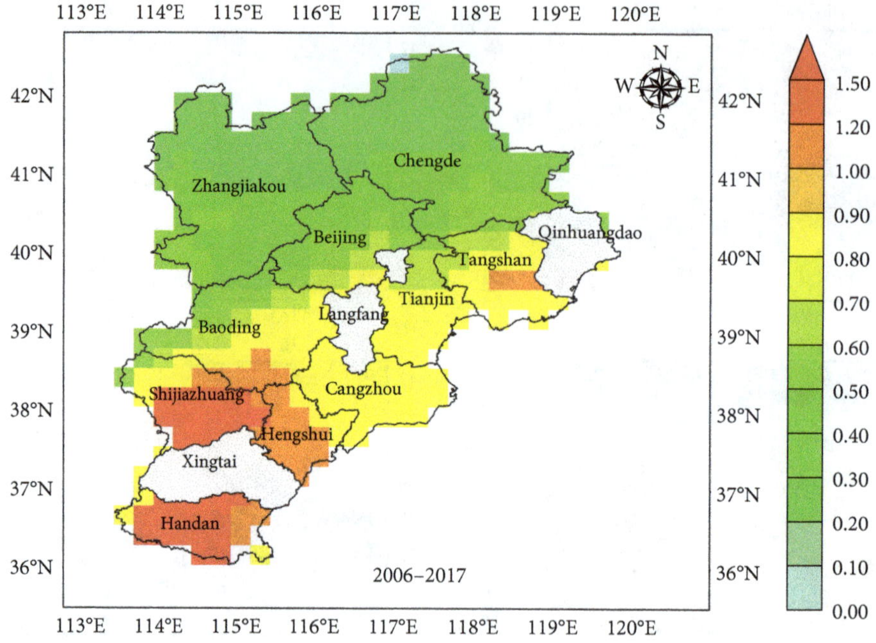

FIGURE 10: Multiyear average SO_2 spatial distribution map over the BTH region.

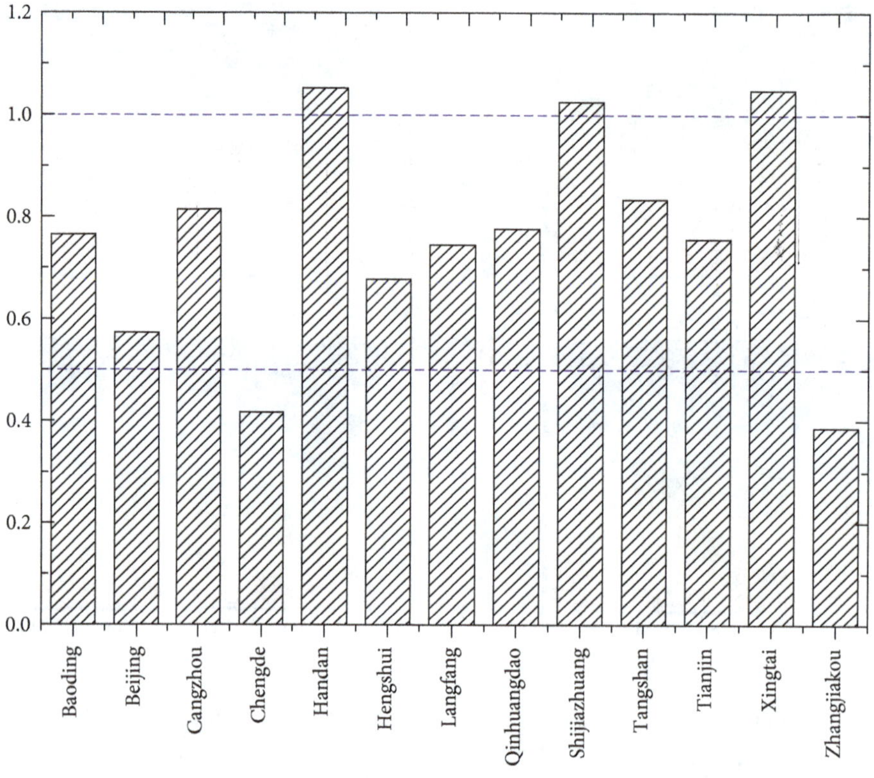

FIGURE 11: Average SO_2 column (in DU, twelve years) of thirteen cities in the BTH region.

3.3. Typical Pollution Control Cases. As analyzed in Sections 3.1 and 3.2, there is an overall large decreasing trend of SO_2 emission since 2011. The increasing/decreasing trend of SO_2 in this period has been discussed with decadal, seasonal, and district variability. The results suggest that long-term variations are attributed to stricter emission reduction. In particularly, three typical episodes (the Olympic Games in 2008, the Asia-Pacific Economic Cooperation (APEC) in 2014, and Military Parade in 2015) have been taken place in Beijing during 2006–2017. A series of strengthened emission reduction measures have played an important role in preventing and controlling air pollution, which significantly improved the air quality in Beijing and neighboring regions [50–54]. Therefore, we evaluate the variation (Figure 8) by

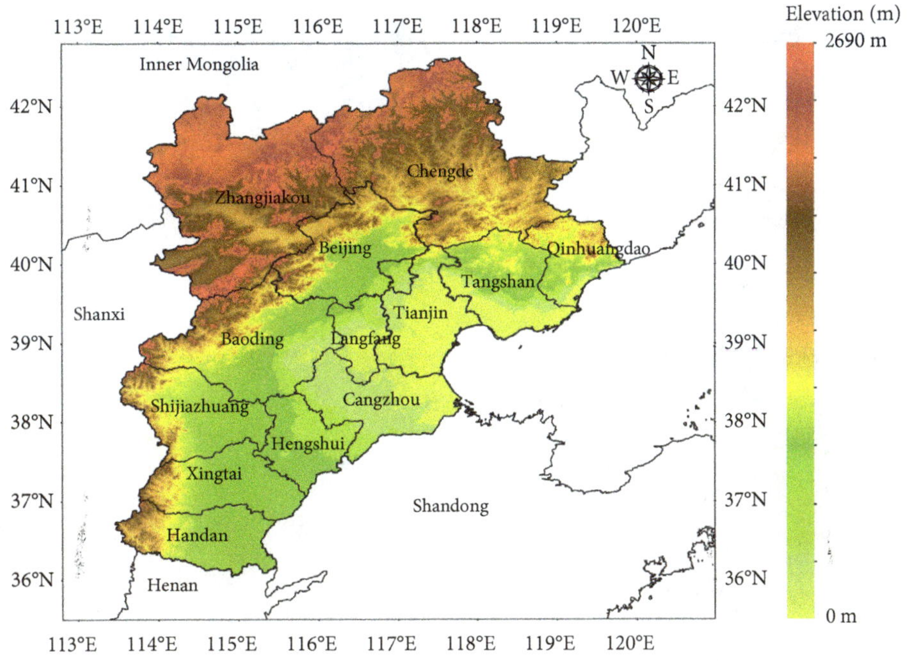

FIGURE 12: Topography of the Beijing-Tianjin-Hebei (BTH) region.

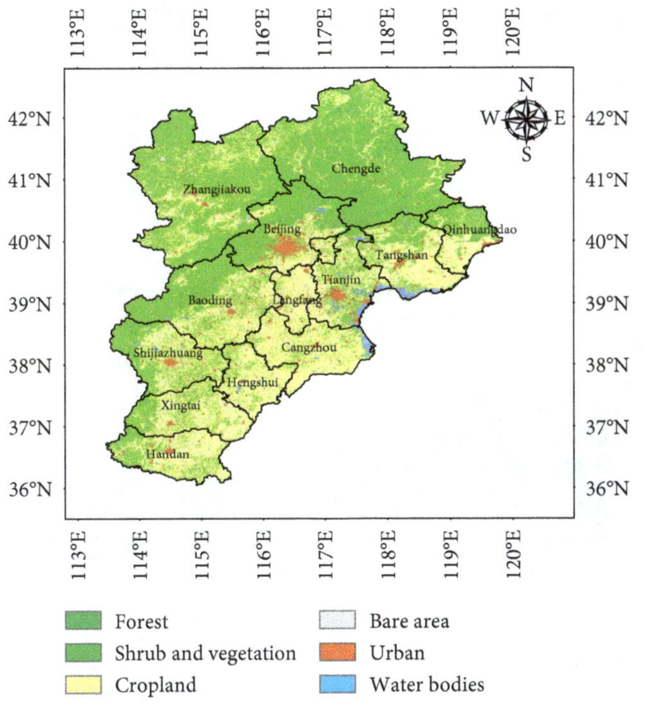

FIGURE 13: The main landscape type over the Beijing-Tianjin-Hebei (BTH) region.

tracking the changes of SO_2 to reveal the effective operations driven by environmental policy and measures. The 29th Olympic Games were held in Beijing, started on 8 August 2008, the 2014 Asia-Pacific Economic Cooperation (APEC) Economic Leaders' Meeting was held on November 10 and 11 in Beijing, and China Military Parade was held on 3 September 2015. Prior to the events, emission reduction measures were began to conduct. Monthly, SO_2 column concentration of other years from 2006 to 2017 corresponding to these periods have been compared with the basis of 2006.

As shown in Figure 16, the SO_2 average concentration for the period of Olympic Games decreased significantly to 14.92% and 7.76% compared to the neighboring years; the average SO_2 concentration in the APEC conference was significantly lower than that of the same periods in past ten years, and it declined 28.57% compared to the average of November in other years between 2006 and 2015; and the SO_2 concentration in September of the 2015 Military Parade, with 27.43% decline, reached the lowest value compared to ten years before. Standard deviations can reflect that more clearly. Besides, the standard deviations verify the previous analysis of seasonal characteristic and brief ascent in 2011.

During the three events, air pollution control policies have been reinforced. To achieve the goal of "green Olympic Games" [55], China has released a series of air pollution control policies to improve air quality. The government has implemented a series of long-term pollution reduction measures, such as coal-fired power plant in Beijing to install the desulfurization equipment, part of closure of small power plants in this area near Beijing during the Olympic Games, and about 94% of the small coal-fired boilers to use clean energy transformation. The government also implemented some short-term strategies; for example, from July 1st to September 20th of 2008, the vehicles with exhaust emissions that failed to meet the European No. 1 standard were all-day forbidden on the roads; from July 20th to September 20th, the odd/even license plate number rule was applied on personal vehicles in Beijing [12]; power generation facilities were run only 30% of the equipment to stop all construction activities; some heavy-polluting factories were closed during the Olympic Games; and some heavily polluting

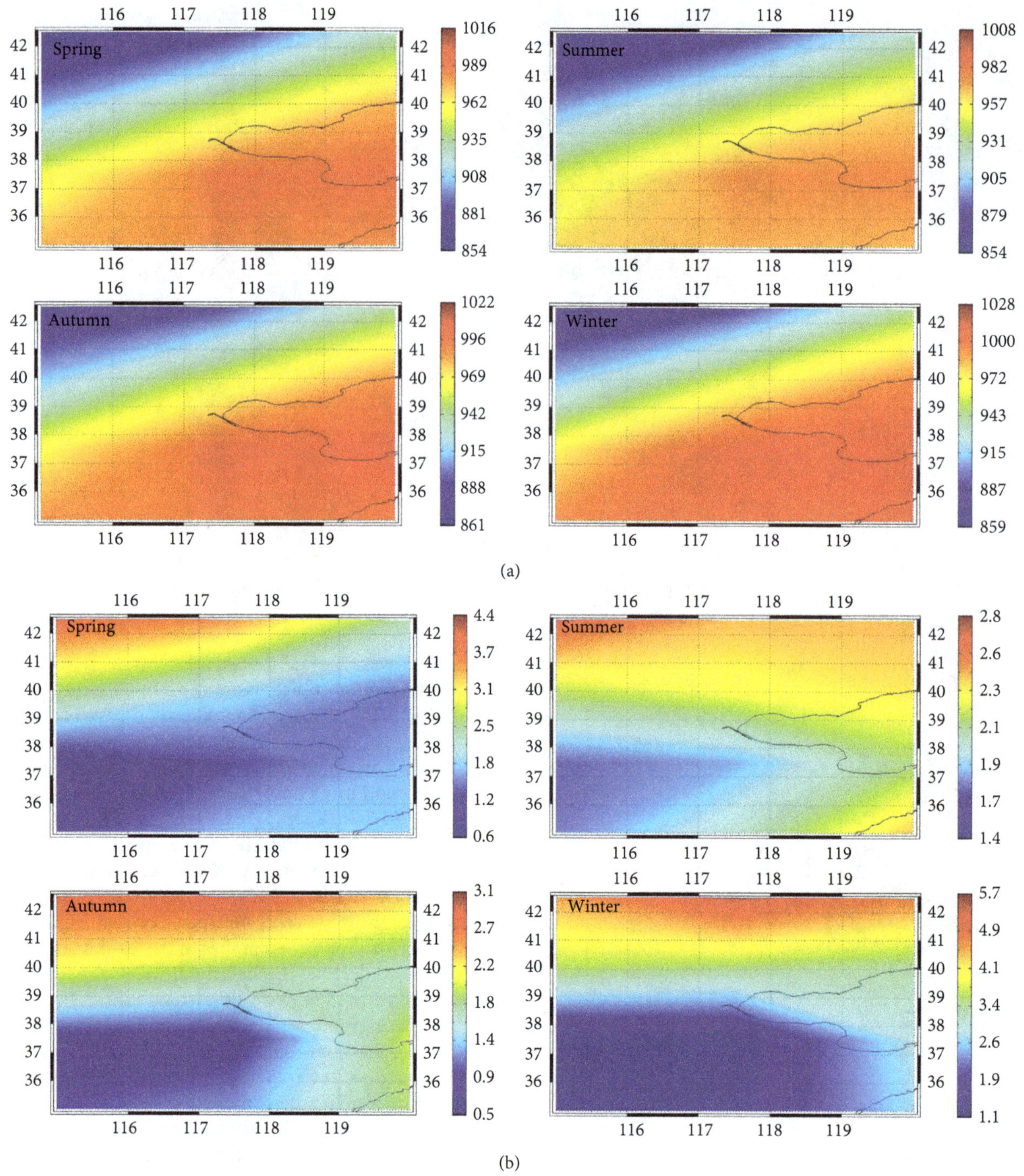

Figure 14: Seasonal mean of climatology variables. (a) Surface pressure (millibars) and (b) wind speed (m/s).

companies around Beijing city were closed [29, 31]. During the APEC meeting and the Military Parade, the government also implemented a series of measures to ensure the safety and environmental protection measures, and these measures are similar even more and stricter than the 2008 Olympic Games, as with Beijing, the surrounding six provinces also have taken similar measures [27, 56]. During the period from August 20th to September 3rd of 2015, Beijing adopted to strengthen urban transportation management to strictly limit the motor vehicle population [57].

Relevant conclusion comes out that any air pollution activities by human beings in the entire BTH region should be controlled with the most rigorous management. Beijing as China's political and administrative center has strict pollution control measures and high execution efficiency. Meanwhile, the environment policies to a certain extent affect the surrounding cities of Beijing. We focused on several affairs, such as the 2008 Olympic Games, 2014 APEC (Asia-Pacific Economic Cooperation), and 2015 Victory Day Military Parade, by means of analyzing the concentration of SO_2 during the

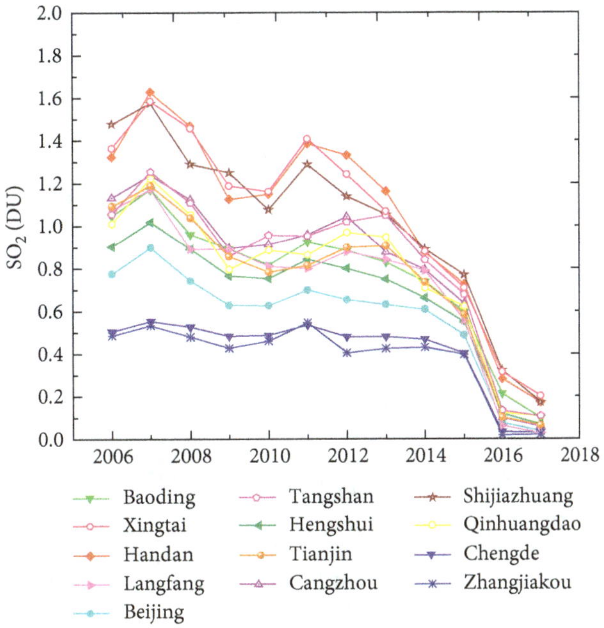

FIGURE 15: Annual mean curve of SO_2 column concentration of each city in the BTH region, 2006–2017.

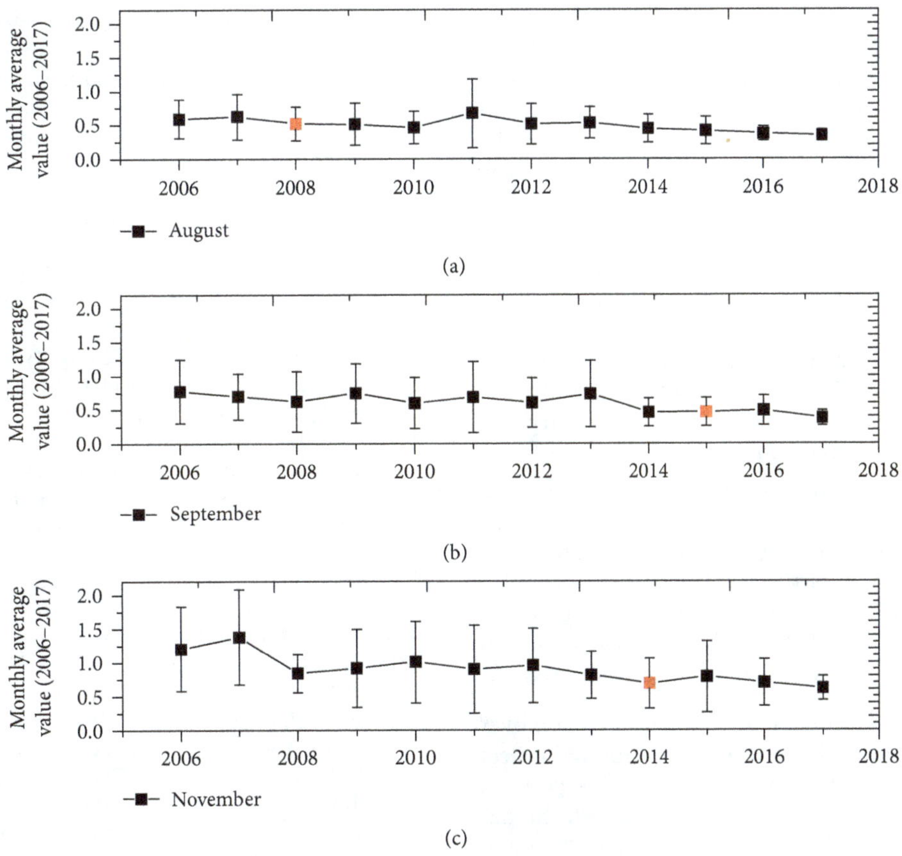

FIGURE 16: Monthly average SO_2 of August (a), September (b), and November (c) from 2006 to 2017. The black solid squares show relative changes based on 2006. The error bars express standard deviation. The red solid squares represent the months when events occur.

events to confirm the effectiveness of environmental management measures. The relative changes clearly demonstrate that these measures were effective in reducing SO_2 concentration during the periods of those affairs taken place.

4. Conclusions

In this study, the past 12 years (2006–2017), OMI observations have yielded profound insights into the spatial distribution and temporal trends in SO_2 emission over the Beijing-Tianjin-Hebei region. The SO_2 loading distribution has close correlation with their emission sources. Spatiotemporal variation characteristics over BTH can more clearly reflect the natural and anthropogenic emission sources, to provide references to air pollution prevention and control. The main conclusions are as follows:

(1) The temporal changes (2006–2017) over the BTH region exhibit the upward and downward trend consistent with the national trend in China. According to the dipping and heaving, we find that the gridded SO_2 data can be divided into two phases: the first (2006–2010) and second (2011–2017) period, to observe changes in SO_2 loadings presenting the Chinese government actions and policies and accomplishments in addressing air pollution. The SO_2 loadings have drastically decreased by more than 30% from the 2006–2010 period to the 2011–2017 period. SO_2 peaked in 2007 and the secondary peak was in 2011, to a certain extent, referring to the economic policy stimulus in the early of each FYP. Meanwhile, OMI observations show generally good agreement with independent SO_2 emission.

(2) The annual cycles of SO_2 show a pronounced seasonal pattern, with the highest values occurring in winter and the lowest values in summer. This seasonal variation can be explained mainly by the seasonality of emission strengths, lifetimes of these pollutants, and meteorological factors.

(3) Spatial distribution of BTH is also characterized, which is interesting to find that progressive changes from the northern to the southern regions are associated with the terrain and surface coverage, besides industrial pattern and population distribution. In the cities of BTH, we found that rapid SO_2 reductions generally correlate well with sharp reductions in industrial activity.

(4) China made great progress in pollution control with implementation of a series of major policies. The concentration of SO_2 during 2008 Olympic Games, 2014 APEC, and 2015 Military Parade have been compared to identical months in other years, and the result shows that concentration of SO_2 column was relatively low in these episodes because of reinforced emission controls.

These findings demonstrate that SO_2 concentration over BTH do not follow simple linear trends, but instead reflect a repercussion of environmental measures and political economic activities. In terms of temporal changes, the decreasing trend has been observed since 2011, mainly due to the government efforts to restrain emissions from the power and industrial sectors. On the other hand, spatial characteristics are regarding with natural and anthropogenic factors. One of the important goals in this work is to evaluate the effectiveness of a series of policy to control the pollution problems, and three typical time quantum status reflects that the national environmental pollution control measures have taken an evident effect. These findings can provide a basis for the development of environmental management measures during the Winter Olympics in 2022, also for the national environmental pollution prevention and control as a reference. The results clearly illustrate effectiveness of central governmental policies regarding emission mitigation of SO_2 and aid in important policy implications for the future reduction action plans to provide better air quality of China.

Conflicts of Interest

The authors declare that they have no conflicts of interest.

Acknowledgments

This study was supported by the National Natural Science Foundation of China (Grant no. 41501404) and Natural Science Foundation of Hebei Province under Project no. F2017203220.

References

[1] R. Wayne, *Chemistry of Atmospheres*, Oxford Science Publications, Oxford, UK, 1991.

[2] J. H. Seinfeld and S. N. Pandis, *Atmospheric Chemistry and Physics: From Air Pollution to Climate Change*, John Wiley & Sons, Hoboken, NJ, USA, 2nd edition, 2006.

[3] M. Chin, R. B. Rood, S. J. Lin et al., "Atmospheric sulfur cycle simulated in the global model GOCART: model description and global properties," *Journal of Geophysical Research: Atmospheres*, vol. 105, no. D20, pp. 24671–24687, 2000.

[4] B. Denby, I. Sundvor, M. Cassiani et al., "Spatial mapping of ozone and SO_2 trends in Europe," *Science of the Total Environment*, vol. 408, no. 20, pp. 4795–4806, 2010.

[5] V. E. Fioletov, C. A. Mclinden, N. Krotkov, M. D. Moran, and K. Yang, "Estimation of SO_2 emissions using OMI retrievals," *Geophysical Research Letters*, vol. 38, no. 21, 2011.

[6] C. Lee, R. V. Martin, A. Van Donkelaar et al., "SO_2 emissions and lifetimes: estimates from inverse modeling using in situ and global, space-based (SCIAMACHY and OMI) observations," *Journal of Geophysical Research Atmospheres*, vol. 116, no. D6, 2011.

[7] Z. Klimont, S. J. Smith, and J. Cofala, "The last decade of global anthropogenic sulfur dioxide: 2000–2011 emissions," *Environmental Research Letters*, vol. 8, no. 1, pp. 1880–1885, 2013.

[8] J. L. Hand, B. A. Schichtel, W. C. Malm et al., "Particulate sulfate ion concentration and SO_2 emission trends in the United States from the early 1990s through 2010," *Atmospheric Chemistry and Physics*, vol. 12, no. 21, pp. 10353–10365, 2012.

[9] N. Theys, I. De Smedt, J. Gent et al., "Sulfur dioxide vertical column DOAS retrievals from the Ozone Monitoring Instrument: global observations and comparison to ground-based and satellite data," *Journal of Geophysical Research: Atmospheres*, vol. 120, no. 6, pp. 2470–2491, 2015.

[10] H. He, K. Y. Vinnikov, C. Li et al., "Response of SO_2 and particulate air pollution to local and regional emission controls: a case study in Maryland," *Earth's Future*, vol. 4, no. 4, pp. 94–109, 2016.

[11] N. A. Krotkov, B. Mcclure, R. R. Dickerson et al., "Validation of SO_2 retrievals from the Ozone Monitoring Instrument over

NE China," *Journal of Geophysical Research: Atmospheres*, vol. 113, no. D16, pp. 16–40, 2008.

[12] C. Gao, H. Yin, N. Ai et al., "Historical analysis of SO_2 pollution control policies in China," *Environmental Management*, vol. 43, no. 3, pp. 447–457, 2009.

[13] H. He, C. Li, C. P. Loughner et al., "SO_2 over central China: measurements, numerical simulations and the tropospheric sulfur budget," *Journal of Geophysical Research: Atmospheres*, vol. 117, no. D16, pp. 812–819, 2012.

[14] C. Li, Q. Zhang, N. A. Krotkov et al., "Recent large reduction in sulfur dioxide emissions from Chinese power plants observed by the Ozone Monitoring Instrument," *Geophysical Research Letters*, vol. 37, no. 8, pp. 292–305, 2010.

[15] J. Lin, CP. Nielsen, Y. Zhao et al., "Recent changes in particulate air pollution over China observed from space and the ground: effectiveness of emission control," *Environmental Science & Technology*, vol. 44, no. 20, pp. 7771–7776, 2010.

[16] Z. Lu, D. G. Streets, Q. Zhang et al., "Sulfur dioxide emissions in China and sulfur trends in East Asia since 2000," *Atmospheric Chemistry and Physics*, vol. 10, no. 4, pp. 6311–6331, 2010.

[17] W. Zhang, H. Xu, F. Zheng et al., "Classifying aerosols based on fuzzy clustering and their optical and microphysical properties study in Beijing, China," *Advances in Meteorology*, vol. 2017, Article ID 4197652, 18 pages, 2017.

[18] A. Krueger, "Sighting of El chichón sulfur dioxide clouds with the nimbus 7 total ozone mapping spectrometer," *Science*, vol. 220, no. 4604, pp. 1377–1379, 1983.

[19] M. Eisinger and J. P. Burrows, "Tropospheric sulfur dioxide observed by the ERS-2 GOME instrument," *Geophysical Research Letters*, vol. 25, no. 22, pp. 4177–4180, 1998.

[20] J. P. Burrows, M. Weber, M. Buchwitz et al., "The global ozone monitoring experiment (GOME): mission concept and first scientific results," *Journal of Atmospheric Sciences*, vol. 56, no. 2, pp. 151–175, 1999.

[21] M. F. Khokhar, U. Platt, and T. Wagner, "Satellite observations of atmospheric SO_2 from volcanic eruptions," in *Proceedings of the 35th COSPAR Scientific Assembly*, Paris, France, June 2004.

[22] H. Bovensmann, J. P. Burrows, M. Buchwitz et al., "SCIAMACHY: mission objectives and measurement modes," *Journal of Atmospheric Sciences*, vol. 56, no. 2, pp. 127–150, 1999.

[23] J. Callies, E. Corpaccioli, M. Eisinger, A. Hahne, and A. Lefebvre, "GOME-2-Metop's second-generation sensor for operational ozone monitoring," *ESA Bulletin*, vol. 102, pp. 28–36, 2000.

[24] P. F. Levelt, G. H. J. V. D. Oord, M. R. Dobber et al., "The ozone monitoring instrument," *IEEE Transactions on Geoscience and Remote Sensing*, vol. 44, no. 5, pp. 1093–1101, 2006.

[25] W. Zhang, X. Gu, H. Xu, T. Yu, and F. Zheng, "Assessment of OMI near-UV aerosol optical depth over Central and East Asia," *Journal of Geophysical Research: Atmospheres*, vol. 121, no. 1, pp. 382–398, 2016.

[26] Q. Zhang, D. G. Streets, and K. He, "Satellite observations of recent power plant construction in Inner Mongolia, China," *Geophysical Research Letters*, vol. 36, no. 15, pp. 1–5, 2009.

[27] S. Wang, Q. Zhang, R. V. Martin et al., "Satellite measurements oversee China's sulfur dioxide emission reductions from coal-fired power plants," *Environmental Research Letters*, vol. 10, no. 11, article 114015, 2015.

[28] J. Jie, Z. Yong, G. Jay, and J. Jianjun, "Monitoring of SO_2 column concentration change over China form Aura OMI data," *International Journal of Remote Sensing*, vol. 33, no. 6, pp. 1934–1942, 2012.

[29] J. C. Witte, M. R. Schoeberl, A. R. Douglass et al., "Satellite observations of changes in air quality during the 2008 Beijing Olympics and Paralympics," *Geophysical Research Letters*, vol. 36, no. 17, pp. 37–44, 2009.

[30] Z. Leishi, L. C. Sheng, Z. Ruiqin, and C. Liangfu, "Spatial and temporal evaluation of long term trend (2005–2014) of OMI retrieved NO_2 and SO_2 concentrations in Henan Province, China," *Atmospheric Environment*, vol. 154, pp. 151–166, 2016.

[31] LT. Wang, C. Jang, Y. Zhang et al., "Assessment of air quality benefits from national air pollution control policies in China. Part I: background, emission scenarios and evaluation of meteorological predictions," *Atmospheric Environment*, vol. 44, no. 28, pp. 3449–3457, 2010.

[32] L. Feng and W. Liao, "Legislation, plans, and policies for prevention and control of air pollution in China: achievements, challenges, and improvements," *Journal of Cleaner Production*, vol. 112, pp. 1549–1558, 2017.

[33] C. Li, C. Mclinden, V. Fioletov et al., "India is overtaking China as the world's largest emitter of anthropogenic sulfur dioxide," *Scientific Reports*, vol. 7, no. 1, 2017.

[34] N. A. Krotkov, C. A. Mclinden, C. Li et al., "Aura OMI observations of regional SO_2 and NO_2 pollution changes from 2005 to 2014," *Atmospheric Chemistry and Physics*, vol. 15, no. 19, pp. 26555–26607, 2015.

[35] S. P. Ahmad, P. F. Levelt, P. K. Bhartia, E. Hilsenrath, G. W. Leppelmeier, and J. E. Johnson, "Atmospheric products from the ozone monitoring instrument (OMI)," *Earth Observing Systems VIII*, vol. 5151, pp. 619–630, 2003.

[36] NASA Goddard Earth Sciences, "Aura OMI sulphur dioxide data product-OMSO_2," April 2016, http://disc.sci.gsfc.nasa.gov/Aura/data-holdings/OMI/omso2_v003.shtml.

[37] A. I. Prados, G. Leptoukh, C. Lynnes et al., "Access, visualization, and interoperability of air quality remote sensing data sets via the Giovanni online tool," *IEEE Journal of Selected Topics in Applied Earth Observations and Remote Sensing*, vol. 3, no. 3, pp. 359–370, 2010.

[38] C. Mallik and S. Lal, "Seasonal characteristics of SO_2, NO_2, and CO emissions in and around the Indo-Gangetic Plain," *Environmental Monitoring and Assessment*, vol. 186, no. 2, pp. 1295–1310, 2014.

[39] N. A. Krotkov, S. A. Carn, A. J. Krueger et al., "Band residual difference algorithm for retrieval of SO/sub 2/from the aura ozone monitoring instrument (OMI)," *IEEE Transactions on Geoscience and Remote Sensing*, vol. 44, no. 5, pp. 1259–1266, 2006.

[40] C. Li, J. Joiner, N. A. Krotkov et al., "A fast and sensitive new satellite SO_2 retrieval algorithm based on principal component analysis: Application to the ozone monitoring instrument," *Geophysical Research Letters*, vol. 40, no. 23, pp. 6314–6318, 2013.

[41] C. Li, N. A. Krotkov, S. Carn et al., "New-generation NASA Aura Ozone Monitoring Instrument (OMI) volcanic SO_2 dataset: algorithm description, initial results, and continuation with the Suomi-NPP Ozone Mapping and Profiler Suite (OMPS)," *Atmospheric Measurement Techniques Discussions*, vol. 10, pp. 1–27, 2017.

[42] H. H. Yan, X. J. Li, X. Y. Zhang et al., "Comparison and validation of band residual difference algorithm and principal component analysis algorithm for retrievals of atmospheric SO_2 columns from satellite observations," *Acta Physica Sinica*, vol. 65, no. 8, 2016.

[43] NBS, *China Energy Statistical Yearbook*, National Bureau of Statistics of China, China Statistics Press, Beijing, China, 2016.

[44] V. E. Fioletov, C. A. Mclinden, N. Krotkov et al., "Lifetimes and emissions of SO_2 from point sources estimated from OMI," *Geophysical Research Letters*, vol. 42, no. 6, pp. 1969–1976, 2015.

[45] V. E. Fioletov, C. A. Mclinden, N. Krotkov et al., "A global catalogue of large SO_2 sources and emissions derived from the Ozone Monitoring Instrument," *Atmospheric Chemistry and Physics*, vol. 16, no. 18, pp. 11497–11519, 2016.
[46] N. A. Krotkov, C. A. Mclinden, C. Li et al., "Aura OMI observations of regional SO_2 and NO_2 pollution changes from 2005 to 2015," *Atmospheric Chemistry and Physics*, vol. 16, no. 7, pp. 4605–4629, 2016.
[47] X. Hualin, H. Yafen, and X. Xue, "Region using big data exploring the factors influencing ecological land change for China's Beijing-Tianjin-Hebei," *Journal of Cleaner Production*, vol. 142, no. 2, pp. 677–687, 2017.
[48] C. Calkins, C. Ge, J. Wang et al., "Effects of meteorological conditions on sulfur dioxide air pollution in the North China plain during winters of 2006–2015," *Atmospheric Environment*, vol. 147, pp. 296–309, 2016.
[49] P. Li, Y. Lv, C. Zhang et al., "Analysis and planning of ecological networks based on kernel density estimations for the Beijing-Tianjin-Hebei region in Northern China," *Sustainability*, vol. 8, no. 11, p. 1094, 2016.
[50] Y. Wang, J. Hao, M. B. McElroy et al., "Ozone air quality during the 2008 Beijing Olympics-effectiveness of emission restrictions," *Atmospheric Chemistry and Physics Discussions*, vol. 9, no. 2, pp. 9927–9959, 2009.
[51] Y. Gao and M. Zhang, "Sensitivity analysis of surface ozone to emission controls in Beijing and its neighboring area during the 2008 Olympic Games," *Journal of Environmental Sciences*, vol. 24, no. 1, pp. 50–61, 2012.
[52] Li Sheng a, K. Lu, X. Ma et al., "The air quality of Beijing-Tianjin-Hebei regions around the Asia-Pacific Economic Cooperation (APEC) meetings," *Atmospheric Pollution Research*, vol. 6, no. 6, pp. 1066–1072, 2015.
[53] G. Wang, S. Cheng, W. Wei et al., "Characteristics and emission-reduction measures evaluation of PM 2.5 during the two major events: APEC and Parade," *Science of the Total Environment*, vol. 595, pp. 81–92, 2017.
[54] Y. Zheng, H. Che, T. Zhao, X. Xia, K. Gui, and L. An, "Aerosol optical properties over Beijing during the World Athletics Championships and Victory Day Military Parade in August and September 2015," *Atmosphere*, vol. 7, no. 3, pp. 47–62, 2016.
[55] Beijing Organizing Committee for the Games of the XXIX Olympic Games (BOCOG), *Green Olympics in Beijing*, BOCOG, Beijing, China, 2005.
[56] F. Li, Z. Song, and W. Liu, "China's energy consumption under the global economic crisis: decomposition and sectoral analysis," *Energy Policy*, vol. 64, pp. 193–202, 2014.
[57] Beijing Municipal Environmental Protection Bureau, "Air pollution prevention and control action plan," 2014, http://www.gov.cn/zwgk/2013-09/12/content_2486773.htm.

8

Spatiotemporal Variation Characteristics of Vegetative PUE in China from 2000 to 2015

Haitao Xu,[1,2] Peng Hou,[3] Zhengwei He,[1,2] A. Duo,[4] and Bing Zhang[5]

[1]*State Key Laboratory of Geohazard Prevention and Geoenvironment Protection (Chengdu University of Technology), Chengdu 610059, China*
[2]*College of Earth Science, Chengdu University of Technology, Chengdu 610059, China*
[3]*State Environmental Protection Key Laboratory of Satellite Remote Sensing, Satellite Environment Center, Ministry of Environmental Protection of People's Republic of China, Beijing 100094, China*
[4]*Satellization Application Centre for Disaster Reduction of the Ministry of Civil Affairs, National Disaster Reduction Center of China, Beijing 100124, China*
[5]*College of Resource Environment and Tourism, Capital Normal University, Beijing 100048, China*

Correspondence should be addressed to Peng Hou; houpcy@163.com and Zhengwei He; hzw@cdut.edu.cn

Academic Editor: Hui Xu

Vegetative precipitation-use efficiency (PUE) is a key indicator for evaluating the dynamic response of vegetation productivity to the spatiotemporal variation in precipitation. It is also an important indicator for reflecting the relationship between the water and carbon cycles in a vegetation ecosystem. This paper uses data from MODIS Net Primary Production (NPP) and China's spatial interpolation data for precipitation from 2000 to 2015 to calculate the annual value, multiyear mean value, interannual standard deviation, and interannual linear trend of Chinese terrestrial vegetative PUE over the past 16 years. Based on seven major administrative regions, eleven vegetation types, and four climate zones, we analyzed the spatiotemporal variation characteristics of China's vegetative PUE. The research results are shown as follows: (1) China's vegetative PUE shows obvious spatial variation characteristics, and it is relatively stable interannually, with an overall slight increasing trend, especially in Northwest and Southwest China. The vegetative PUE is higher, and its stability is declined in Xinjiang, western Gansu, and the southern Tibetan valley. The vegetative PUE is lower, and its stability is increased in northeastern Tibet and southwestern Qinghai. An increasing trend in vegetative PUE is obvious at the edge of the Tarim Basin, in western Gansu, the southern Tibetan valley, and northwestern Yunnan. (2) There is a significant difference in the PUEs among different vegetation types. The average PUE of Broadleaf Forest is the highest, and the average PUE of Alpine Vegetation is the lowest. The stability of the PUE of Mixed Coniferous and Broadleaf Forest is declined, and the stability of the PUE of Alpine Vegetation is increased. The increasing speed of the PUE of Grass-forb Community is the fastest, and the decreasing speed of the PUE of Swamp is the fastest. (3) There is a significant difference in the PUEs among different vegetation types in the same climate zone, the difference in vegetative PUE in arid and semiarid regions is mainly affected by precipitation, and the difference in vegetative PUE in humid and semihumid regions is mainly affected by soil factors. The PUEs of the same vegetation type are significantly different among climate zones. The average PUE of Cultural Vegetation has the largest difference, the stability of the PUE of Steppe has the largest difference, and the increasing speed of the PUE of Swamp has the largest difference.

1. Introduction

The Earth's climate is strongly influenced by the characteristics of atmosphere and ground surface which is known as the ecological environment [1–3]. The human economic activities have been proved to have significant impact on the ecosystems. For example, Xu et al. [4, 5] make a comprehensive investigation on the spatial, temporal, and vertical distributions of dust over China, finding that airborne dust and its inducements like precipitation and vegetation are very important factors in the climate system and can be greatly influenced by the human economic activities.

Climate change has brought about profound effects on the patterns and functions of vegetation ecosystems. Vegetation has responded significantly to climate change through the exchange of energy, moisture, and material reaction [6, 7]. The interactions and mutual adaptations of ecosystems and the environment are embodied by the response relationship between global climate change and the vegetation ecosystem [8–10]. After its reform and opening up, China's ecological environment is worsening, leading to the degradation of vegetation, and vegetation degradation results from the interactions between climate change and human activities [11–13]. Climate change is the main reason for vegetative degradation, directly resulting in the decrease of vegetation, causing desertification. Land-surface vegetative change can also affect climate change. Human activities and climate change interact in complex ways [14, 15].

Climate change has obvious impacts on the spatiotemporal distribution of precipitation. Precipitation also greatly influences vegetative activity because fluctuations in interannual precipitation change vegetative biomass [16]. Vegetative biomass is usually measured by vegetative NPP, which refers to the total amount of organic dry matter produced from green plants in unit time and unit area. Vegetative NPP is an important characteristic of ecosystem function and structure and plays an important role in reducing atmospheric CO_2 content [17, 18]. Vegetative NPP is a major factor in determining carbon sources and sinks. It plays an important role in the global carbon cycle, reflecting the combined effects of climate change and human activities on terrestrial vegetation [19, 20]. Many scholars have performed many studies on the response relationship between vegetative NPP and climate change in China. This research has shown that precipitation is usually a key factor affecting the dynamic changes in ecosystem structure and function and is an important driving factor of the spatial distribution and interannual fluctuation of vegetative NPP [21–23]. PUE is a key indicator for exploring vegetative NPP response to precipitation change [17, 24]. PUE is the ratio of vegetative NPP to annual precipitation, which reflects the relationship between the photosynthetic production processes and the water consumption characteristics of vegetation. It is a key indicator for analyzing and evaluating vegetative productivity in response to the spatiotemporal variation characteristics of precipitation on a regional scale [25, 26]. It reflects the spatial variation characteristics of ecosystem water use along the climatic gradient [27].

Vegetative PUE is not only an important variable for linking the carbon and water cycles of vegetative ecosystems but also a way of regulating ecological populations and systems responding to climate change [28]. Changes in PUE are closely related to climatic zones, vegetation types, soil factors, and other elements. The difference in the PUEs among different vegetation types is affected by the biological community structure or biogeochemical factors [25, 27]. In past studies of China's vegetative PUE, the regions studied were almost entirely in the Qinghai-Tibet Plateau and the arid regions of Northwest China [21, 28], the vegetation types studied were usually selected from specific vegetation types such as Steppe and Desert, and the time series studied was short. As a result, there have been no reports on the study of vegetative PUE in the whole of China based on climatic zones and different vegetation types for more than 15 years [23–25]. Therefore, this paper begins at a national scale, using China's time-series remote-sensing NPP data and meteorological precipitation data from 2000 to 2015 to calculate the PUE of China's vegetation. Then, the paper analyzes changes and spatiotemporal patterns in Chinese terrestrial vegetative PUE from the most recent 16 years to assess the general characteristics of vegetative PUE in China, obtain the relationship between spatiotemporal patterns of vegetative PUE and vegetation types and climate zones, and discuss the main influencing factors of vegetative PUE. The results indicate that the spatiotemporal patterns of vegetative PUE can clearly show the response relationship between precipitation change and vegetative NPP, which will deepen our understanding of the formation process of vegetative productivity among different vegetation types and climatic zones. Therefore, the study of the relationship between ecosystem water and carbon cycles and climate change in China is of great theoretical and practical significance [17, 24].

2. Data and Methods

2.1. Method of Calculating Vegetative PUE. The concept of vegetative PUE is proposed based on vegetative water-use efficiency (WUE) [29, 30]. Usually, the ratio of vegetative NPP to precipitation is adopted for the simulated calculation of vegetative PUE [25]. Owing to the limitations of vegetative NPP data obtained by traditional ecological observation methods, some scholars have replaced vegetative NPP with the aboveground net primary productivity (ANPP) of vegetation when calculating the PUE based on measured data [25, 26]. Additionally, the normalized differential vegetation index (NDVI) has a significant linear correlation with vegetative NPP; thus, the NDVI can be used to replace vegetative NPP [21, 31]. In this paper, vegetative PUE is calculated based on the ratio of vegetative NPP to precipitation. Vegetative NPP data from 2000 to 2015 are obtained based on the terrestrial level-4 product of US Terra MODIS MOD17A3 (https://ladsweb.nascom.nasa.gov/search/) with a spatial resolution of 1 km.

2.2. Processing Method of Precipitation Data. Spatial processing methods of precipitation data in conventional ground meteorological stations mainly include inverse distance weight (IDW) tension, spline with tension, trend, ordinary kriging, and universal kriging. Many studies have analyzed and discussed the benefits and drawbacks of these methods [32–34]. Annual cumulative precipitation from 2000 to 2015 is calculated based on the daily precipitation of 833 standard meteorological stations in China, with data sourced from the China Meteorological Data Service Center (http://data.cma.cn/). First, the precipitation data are spatialized using IDW, spline, trend, ordinary kriging, universal kriging, and other methods based on measured data from 800 stations. Then, measured data from an additional 33 stations are processed using the normal population mean t-test method [34, 35]. The results show that the precipitation value obtained using the five interpolation methods is not

Table 1: Accuracy comparison of the five interpolation methods.

Interpolation method	Maximum (mm)	Minimum (mm)	Mean (mm)	Standard deviation (mm)	t-test	p
IDW	2686.0908	15.8132	567.8891	474.8156	0.2500	0.8043
Spline	8519.6455	−2830.9712	560.7704	487.1202	0.0273	0.9784
Trend	4136.6260	114.0831	577.2636	448.7920	0.3870	0.7014
Ordinary kriging	2613.2412	−28.6841	559.9145	476.0661	0.3520	0.7273
Universal kriging	2201.4834	−932.9639	560.7932	472.1631	0.2088	0.8360

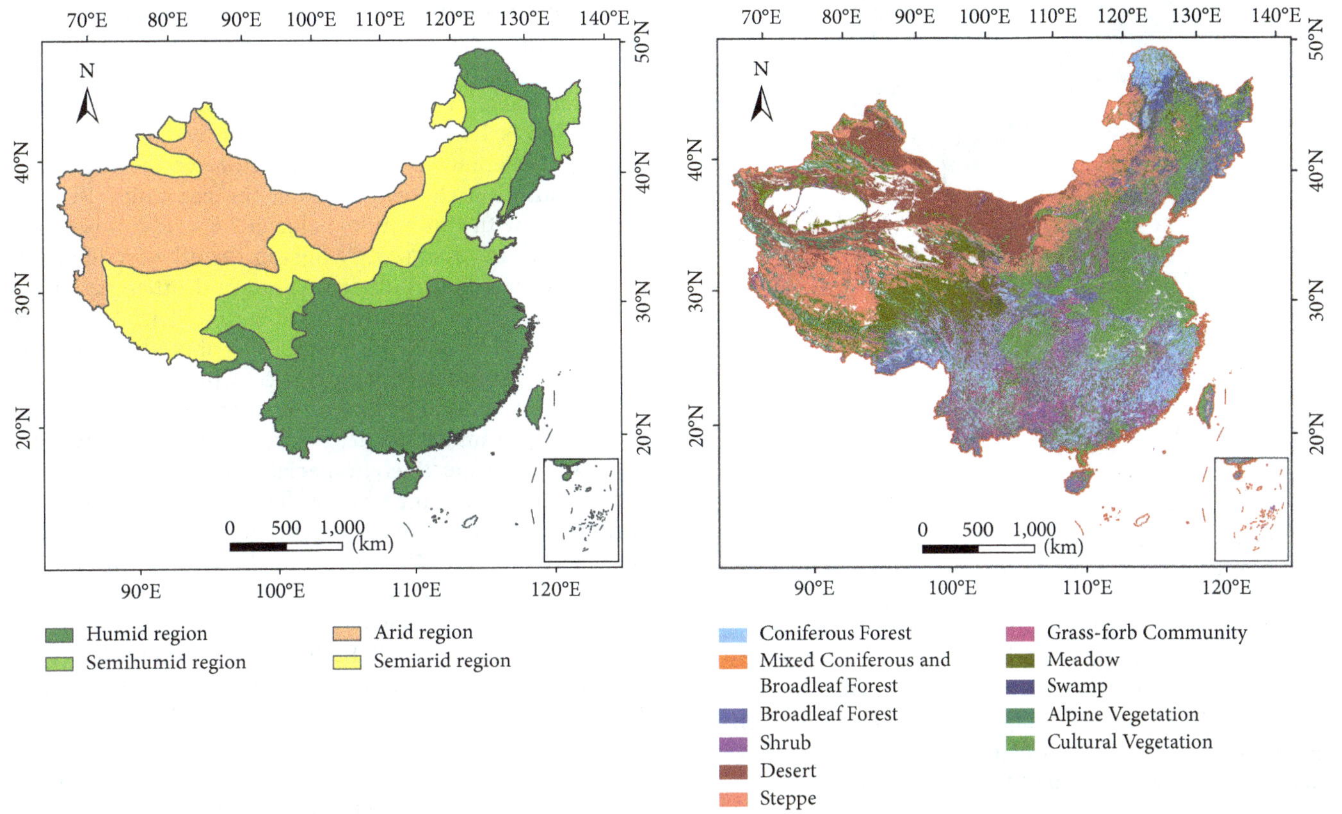

(a) (b)

Figure 1: The maps of (a) arid and humid climate zones and (b) vegetation types in China.

significantly different from the measured value at each station, as shown in Table 1. Because the precipitation values obtained using spline, trend, ordinary kriging, universal kriging, and other methods have problematic situations such as negative values and values that are too large or too small, these methods do not meet the calculation requirements of PUE. Therefore, this study uses the precipitation value obtained with the IDW interpolation method.

2.3. Spatiotemporal Variation Analysis Method of Vegetative PUE. China's climatic regionalization in 1981–2010, as proposed by Zheng et al. [36], is adopted to represent the classification diagram of China's arid and humid climate zones, as shown in Figure 1. The classification of vegetation types by regions mainly depends on the Vegetation Map of The People's Republic of China (1 : 1,000,000) compiled by the Chinese Vegetation Map Editorial Board of Chinese Academy of Sciences [37], in which vegetation types are divided into 11 categories: Coniferous Forest, Mixed Coniferous and Broadleaf Forest, Broadleaf Forest, Shrub, Desert, Steppe, Grass-forb Community, Meadow, Swamp, Alpine Vegetation, and Cultural Vegetation. In order to recognize the spatiotemporal variation feature of vegetative PUE, statistical indicators, including multiyear mean value, standard deviation, and linear trend, are calculated, and the comparative analysis method is used to find PUE difference of different statistical indicators of climatic regionalization and vegetation types.

2.4. Significance Testing of Linear Trend of Vegetative PUE. In statistics, the t-test method of the regression coefficient is usually used to test the statistical significance of linear trends, which is judged by calculating the rejection region or p value. The significance level is generally 0.05. If the calculated p value is less than 0.05, the regression coefficient is significant, and the linear trend is effective; otherwise, the

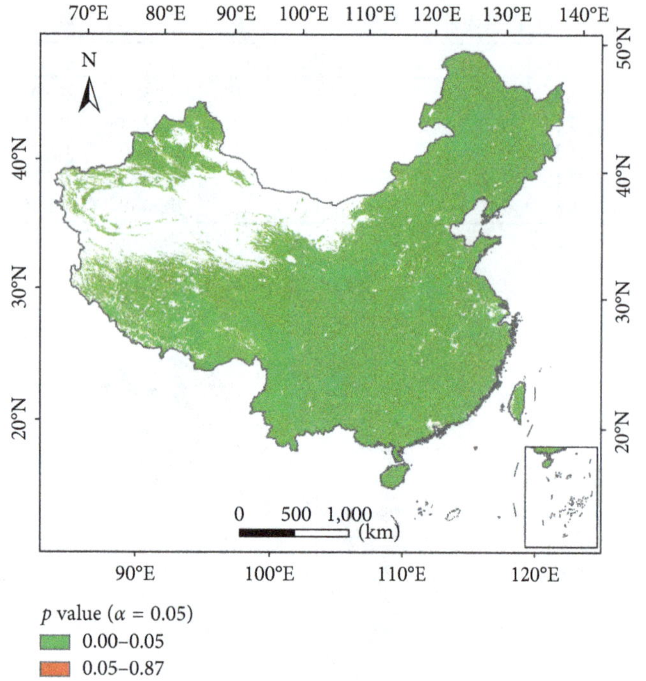

FIGURE 2: The p value of the linear trend of China's vegetative PUE.

regression coefficient is not significant and there is no linear relation [35]. This paper will also test the validity of the linear trend of vegetative PUE by a significance level of 0.05. The p value calculated is shown in Figure 2. It shows that the p value in green is less than 0.05, so the regression coefficient is significant, and there is a linear relation. However, the p value in red is greater than 0.05, which shows that there is no linear relation. In the following study, the analysis of linear trends of vegetative PUE will not include the pixels with linear trends that correspond with the red zone.

3. Results

In this section, the general characteristics of vegetative PUE are analyzed; then, there will be contrastive analysis on the difference in the PUEs among different vegetation types by studying Cultural Vegetation, Natural Vegetation, Woody Vegetation, and Herbaceous Vegetation; finally, there will be a comprehensive analysis on the difference in the PUEs among different vegetation types in climate zones and the difference of the same vegetation type's PUEs among climate zones.

3.1. General Characteristics of Vegetative PUE. The multiyear mean value of vegetative PUE has obvious spatial variation characteristics in China, as shown in Figure 3. The multiyear mean value of vegetative PUE in China ranges from 4.0×10^{-3} gC·m^{-2}·mm^{-1} to 8.597 gC·m^{-2}·mm^{-1}, with an average of 5.54×10^{-1} gC·m^{-2}·mm^{-1}, and with maximum in Northwest and minimum values in Central China. The vegetative PUE is higher in Northeast China, Northwest China, and North China, where the averages of multiyear mean values are 6.4×10^{-1} gC·m^{-2}·mm^{-1}, 6.23×10^{-1} gC·m^{-2}·mm^{-1}, and 5.89×10^{-1} gC·m^{-2}·mm^{-1}, respectively. These averages are significantly higher than the national average, especially in northeastern Jilin, southeastern Liaoning, most of Xinjiang and Gansu, northwestern Qinghai, northern Ningxia, western Inner Mongolia and other regions, and the southern Tibetan valley in Southwest China. The vegetative PUE in South China is the lowest, where the average of the multiyear mean value is 4.37×10^{-1} gC·m^{-2}·mm^{-1}; it is significantly lower than the national average, especially in northeastern Guangxi, northwestern Guangdong, and other regions.

China's vegetative PUE is stable interannually, but the spatial difference is relatively significant, as shown in Figure 3. It can be seen from the interannual standard variation that the mean value of the standard deviation of China's vegetative PUE is 1.173×10^{-1}, showing an overall trend of no significant interannual fluctuations. However, the interannual spatial variation characteristics of the vegetative PUE are significant, with a minimum value of 2.8×10^{-3}. It is mainly distributed in northern Xinjiang, southwest Qinghai, northeastern Tibet, southern Anhui and other regions in Northwest China, Southwest China, and East China; the maximum value is 7.734, mainly distributed in the edge area of the Tarim Basin in Northwest China. The vegetative PUE stability is declined in Northeast and Northwest China, where the interannual standard variations are 1.387×10^{-1} and 1.474×10^{-1}, respectively, especially in northeastern Jilin, southern Liaoning, the edge of the Tarim Basin, and other regions. The stability of the vegetative PUE in central China is increased, where the interannual standard variation is 8.8×10^{-2}, especially in the majority of Hunan, southern Hubei, and other regions.

China's vegetative PUE shows an overall slightly increasing trend, and the rate of increase in the mean values of 2011–2015 is 2.1% compared to those of 2000–2010. From the interannual trend of linear variation, it can be seen that the mean value of the increasing speed of China's vegetative PUE is 1.1×10^{-3}. However, China's vegetative PUE shows relatively obvious spatial variation characteristics, even decreasing trends in a few regions, accounting for 67.5% of the total land area of the country. The regions with the most obvious decreasing trends are mainly distributed in the western edge of the Tarim Basin, northern Ningxia, eastern Inner Mongolia, southwest of Heilongjiang, northwest of Jilin and other regions in Northwest China, North China, and Northeast China; the regions with the most obvious increasing trends are mainly distributed in the majority of Xinjiang, western Gansu, the southern Tibetan valley, northwest Yunnan and other regions in Northwest China, and Southwest China.

3.2. Variation Characteristics of Different Vegetation Types' PUEs

3.2.1. Variation Characteristics of Cultural and Natural Vegetative PUEs. The multiyear mean values of Cultural Vegetative and Natural Vegetative PUEs have obvious spatial variation characteristics, as shown in Figure 4. The average of the multiyear mean value of Cultural Vegetative PUE is 6.07×10^{-1} gC·m^{-2}·mm^{-1}, which is higher than that of China's vegetative PUE. The maximum value is 8.597 gC·m^{-2}·mm^{-1},

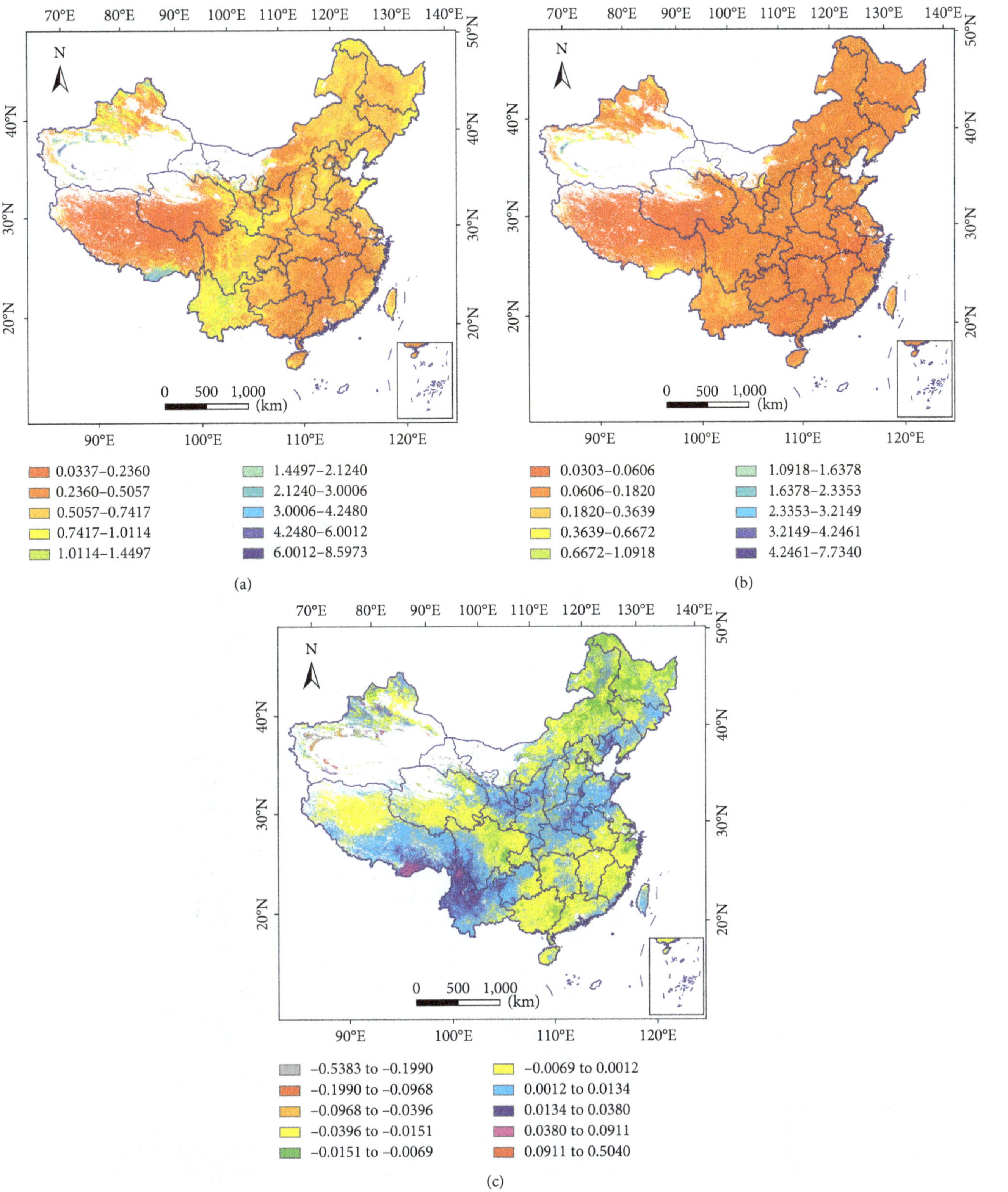

Figure 3: Continued.

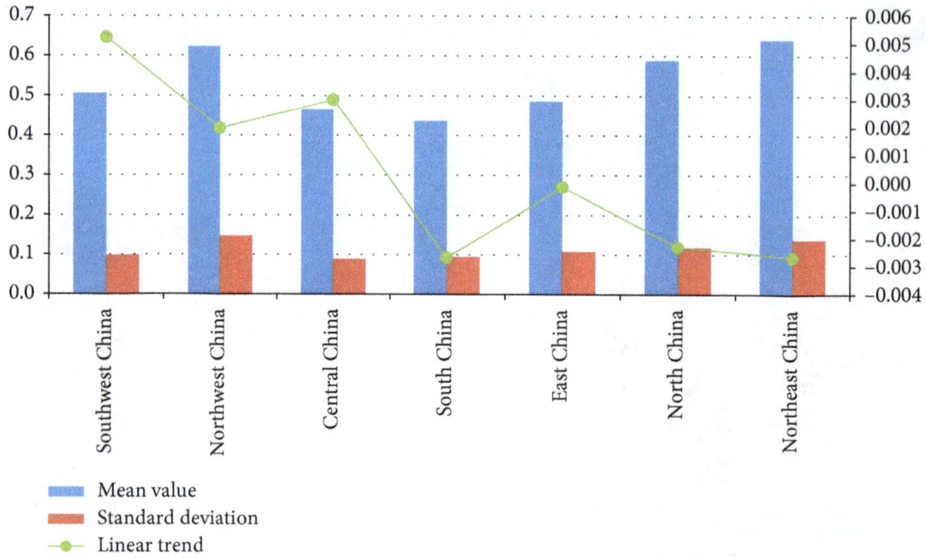

(d)

FIGURE 3: Temporal variation and spatial distribution of China's vegetative PUE from 2000 to 2015. (a) Spatial distribution of the multiyear mean value of China's vegetative PUE. (b) Standard deviation of China's vegetative PUE. (c) Linear trend of China's vegetative PUE. (d) Statistical central tendency of the mean value, standard deviation and linear trend of China's vegetative PUE.

and the minimum value is 3.4×10^{-2} gC·m^{-2}·mm^{-1}; this value is significantly higher than the national average, especially in the edge of the Tarim Basin, northwest of Gansu, the southern Tibetan valley, and other regions. The average of the multiyear mean value of Natural Vegetative PUE is 5.36×10^{-1} gC·m^{-2}·mm^{-1}, which is lower than that of China's vegetative PUE. The maximum value is 8.596 gC·m^{-2}·mm^{-1}, and the minimum value is 3.4×10^{-2} gC·m^{-2}·mm^{-1}; this value is significantly lower than the national average, especially in northeastern Tibet, southwestern Qinghai, and other regions.

The PUEs of Cultural Vegetation and Natural Vegetation are relatively stable interannually, but the spatial difference is relatively significant, as shown in Figure 4. It can be seen from the interannual standard deviation that the mean values of standard deviations of Cultural Vegetative and Natural Vegetative PUEs are 1.304×10^{-1} and 1.125×10^{-1}, respectively, showing an overall trend of no significant interannual fluctuations. However, the spatial variation characteristics of Cultural Vegetative and Natural Vegetative PUEs are significant. The maximum value of the standard deviation of Cultural Vegetative PUE is 6.274, which is mainly distributed in southwest Xinjiang, western Gansu, and other regions, and the minimum value is 2.46×10^{-2}, which is mainly distributed in the middle and lower reaches of the Yangtze River. The maximum value of the standard deviation of Natural Vegetative PUE is 7.734, which is mainly distributed in eastern and southern Xinjiang, northwest Qinghai and Inner Mongolia, and western Gansu, and other regions. The minimum value is 3.03×10^{-2}, which is mainly distributed in northeastern Tibet, southwestern Qinghai, and other regions.

Cultural and Natural Vegetative PUEs show an overall slightly increasing trend, as shown in Figure 4. From the interannual linear variation trend, it can be seen that the mean values of the increasing speeds of Cultural Vegetative and Natural Vegetative PUEs are all 1.1×10^{-3}, which is the same as that of China's vegetative PUE. However, the Cultural and Natural Vegetative PUEs show relatively obvious spatial variation characteristics and even decreasing trends in a few regions. The regions where Cultural Vegetative PUE shows a decreasing trend account for 21.9% of China's total land area; regions with the most obvious decreasing trends are mainly distributed in northwestern Xinjiang, western Gansu and Jilin, northern Ningxia, and other regions. The regions with the most obvious increasing trends of Cultural Vegetative PUE are mainly distributed in southeastern Xinjiang, the southern Tibetan valley, and other regions. The regions where Natural Vegetative PUE shows a decreasing trend account for 64.1% of China's total land area; the regions with the most obvious decreasing trend are mainly distributed in the edge of the Tarim Basin, northwestern Qinghai, eastern Inner Mongolia, western Heilongjiang, and other regions. The regions with the most obvious increasing trend of Natural Vegetative PUE are mainly distributed in the majority of Xinjiang, western Gansu, the southern Tibetan valley, northwestern Yunnan, and other regions.

3.2.2. Variation Characteristics of Different Natural Vegetation Types' PUEs. Natural Vegetation is divided into two main categories: Woody Vegetation (Coniferous Forest, Mixed Coniferous and Broadleaf Forest, Broadleaf Forest, and Shrub) and Herbaceous Vegetation (Desert, Steppe, Grass-forb Community, Meadow, Swamp, and Alpine Vegetation). The multiyear mean values of Woody Vegetative and Herbaceous Vegetative PUEs have obvious spatial variation characteristics, as shown in Figure 5. The average of the multiyear mean value of Woody Vegetative PUE is 6.22×10^{-1} gC·m^{-2}·mm^{-1}, which

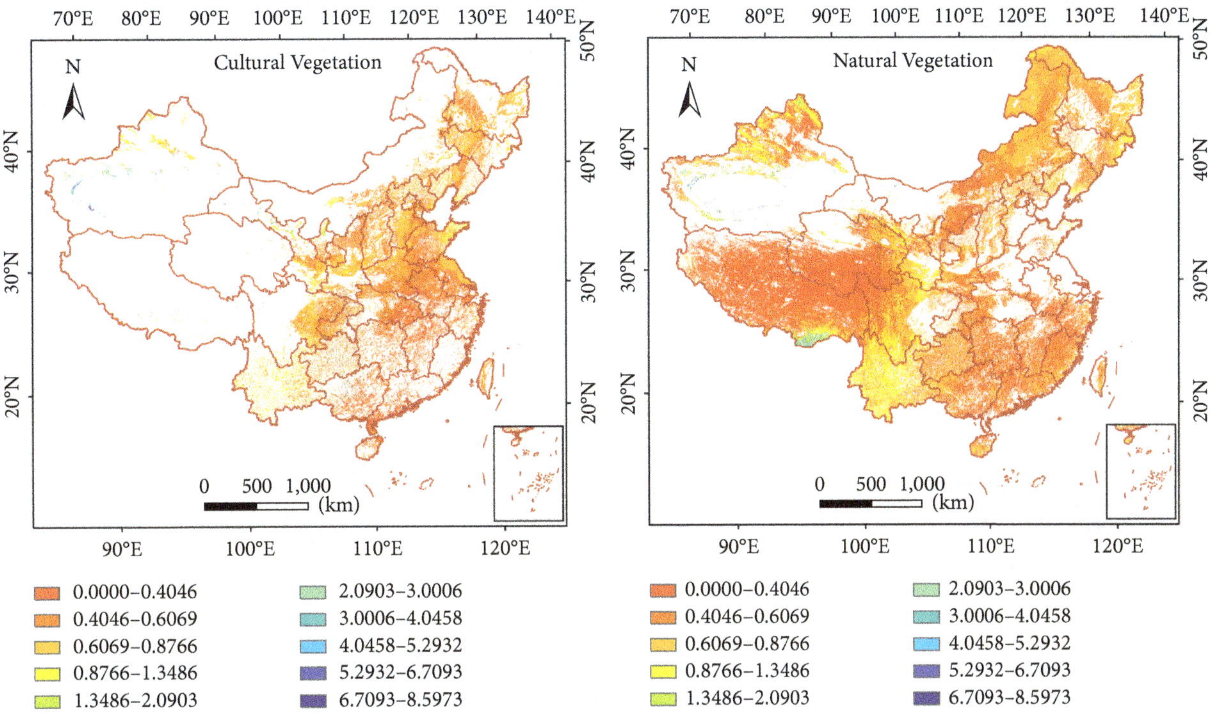

(a)

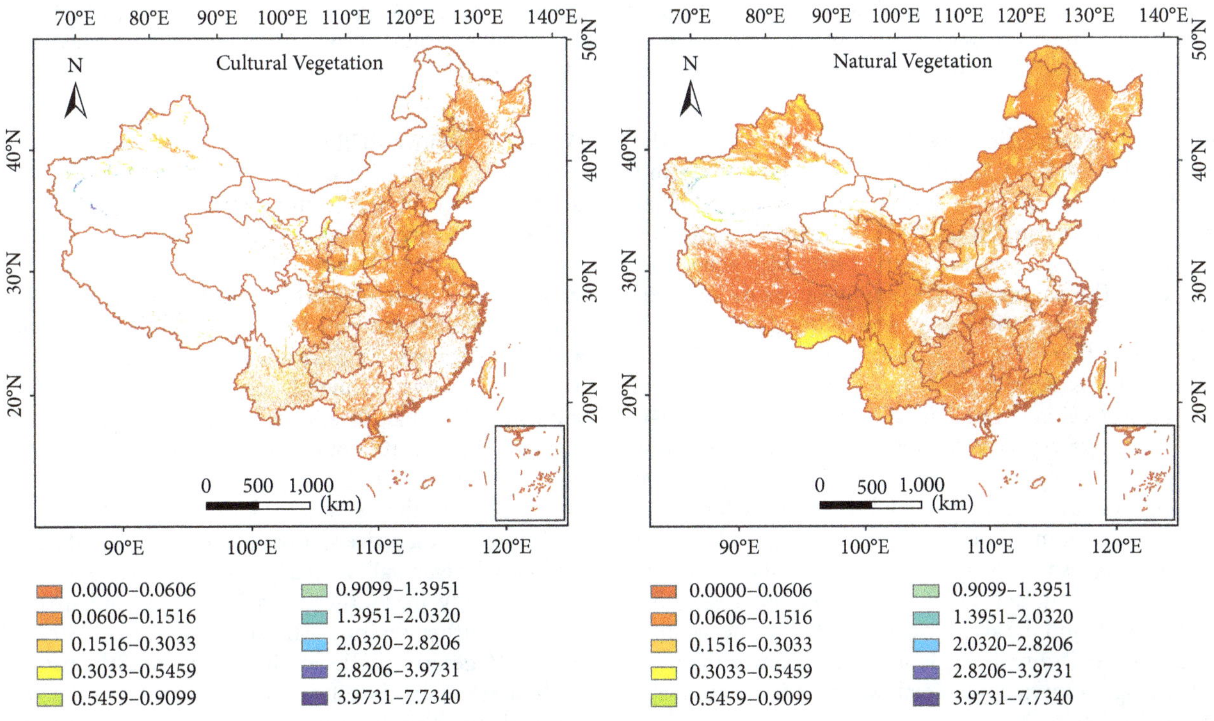

(b)

Figure 4: Continued.

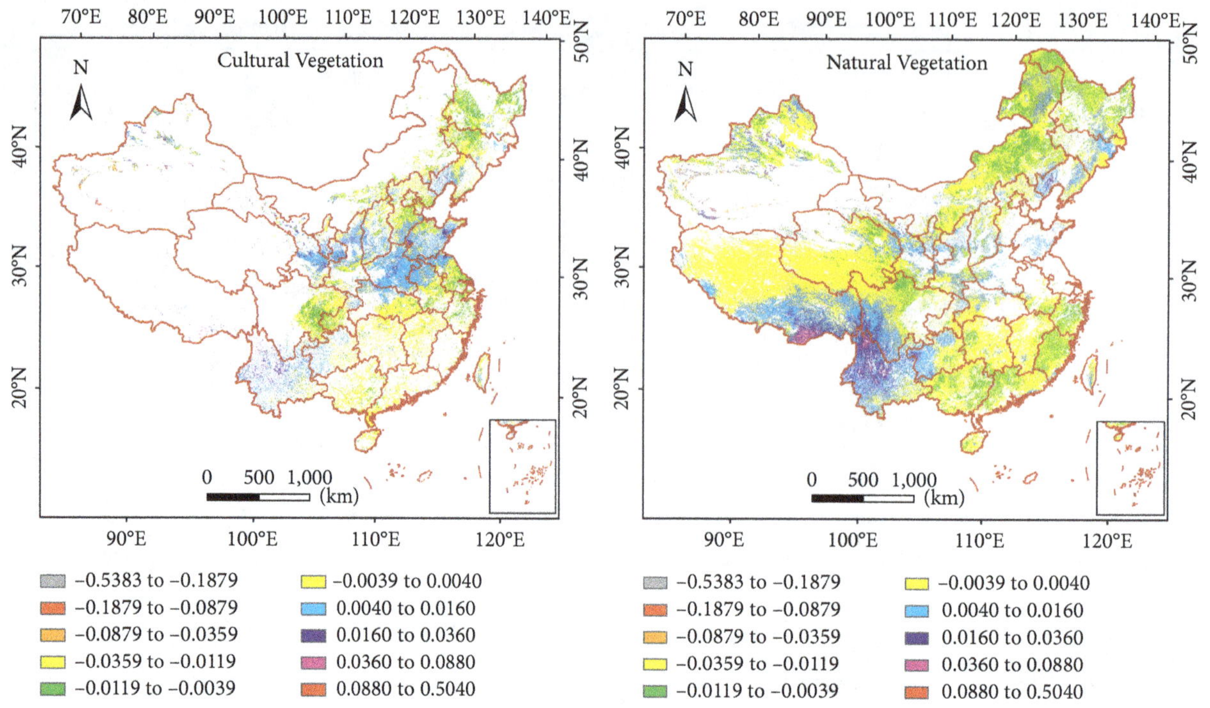

(c)

FIGURE 4: Temporal variation and spatial distribution of Cultural Vegetative and Natural Vegetative PUEs from 2000 to 2015. (a) Spatial distribution of the multiyear mean values of Cultural Vegetative and Natural Vegetative PUEs. (b) Standard deviations of Cultural Vegetative and Natural Vegetative PUEs. (c) Linear trends of Cultural Vegetative and Natural Vegetative PUEs.

is higher than that of China's vegetative PUE in China. The maximum value is 8.181 $gC{\cdot}m^{-2}{\cdot}mm^{-1}$ and the minimum value is 3.2×10^{-2} $gC{\cdot}m^{-2}{\cdot}mm^{-1}$; this is significantly higher than the national average, especially in northern Xinjiang, the southern Tibetan valley, northwestern Inner Mongolia, and other regions. The average of the multiyear mean value of Herbaceous Vegetative PUE is 4.66×10^{-1} $gC{\cdot}m^{-2}{\cdot}mm^{-1}$, which is lower than that of China's vegetative PUE; the maximum value is 8.596 $gC{\cdot}m^{-2}{\cdot}mm^{-1}$, and the minimum value is 3.4×10^{-2} $gC{\cdot}m^{-2}{\cdot}mm^{-1}$; this is significantly lower than the national average, especially in northeastern Tibet, southwest Qinghai, and other regions. The Broadleaf Forest's PUE is the highest, and the average of the multiyear mean values is 6.97×10^{-1} $gC{\cdot}m^{-2}{\cdot}mm^{-1}$; this is significantly higher than the national average, especially in the southern Tibetan valley, southwestern Yunnan, and other regions. Alpine Vegetative PUE is the lowest, and the average of the multiyear mean value is 1.87×10^{-1} $gC{\cdot}m^{-2}{\cdot}mm^{-1}$; this is significantly lower than the national average, especially in northeastern Tibet, southwest Qinghai, and other regions.

Woody Vegetative and Herbaceous Vegetative PUEs are relatively stable interannually, but the spatial difference is relatively significant, as shown in Figure 5. It can be seen from the interannual standard deviation that the mean values of the standard deviations of Woody Vegetative and Herbaceous Vegetative PUEs are 1.273×10^{-1} and 1.003×10^{-1}, respectively, showing an overall trend of no significant interannual fluctuations. However, the spatial variation characteristics of Woody Vegetative and Herbaceous Vegetative PUEs are significant. The maximum value of the standard deviation of Woody Vegetative PUE is 6.6877, which is mainly distributed in some regions in northwestern Inner Mongolia and the northern edge of the Tarim Basin; the minimum value is 2.62×10^{-2}, which is mainly distributed in most of Hunan, southern Anhui, and other regions. The maximum value of the standard deviation of Herbaceous Vegetative PUE is 7.734, which is mainly distributed in some regions in western Gansu and the edge of the Tarim Basin; the minimum value is 3.03×10^{-2}, which is mainly distributed in northeast Tibet, southwest Qinghai, locally in northern Xinjiang, and other regions. The stabilities of the Mixed Coniferous and Broadleaf Forest's and Broadleaf Forest's PUEs are declined, and the mean values of their standard deviations are 1.486×10^{-1} and 1.476×10^{-1}, respectively; this is especially significant in the southern Tibetan valley, locally at the northern edge of Tarim Basin, locally in southeastern Liaoning, and in northeast Jilin, and other regions. The stabilities of the Steppe's and Alpine Vegetative PUEs are increased, and the mean values of their standard deviations are 8.39×10^{-2} and 5.16×10^{-2}, respectively; this is especially significant in northeastern Tibet, western Qinghai, and other regions.

Woody Vegetative and Herbaceous Vegetative PUEs show an overall slightly increasing trend, as shown in Figure 5. From the interannual linear variation trend, it can be seen that the mean values of the increasing speeds of Woody Vegetative and Herbaceous Vegetative PUEs are 1.9×10^{-3} and 4×10^{-4}, respectively. However, Woody Vegetative and Herbaceous Vegetative PUEs show relatively obvious spatial variation characteristics and even decreasing trends in a few regions. Woody Vegetative PUE regions showing a decreasing trend

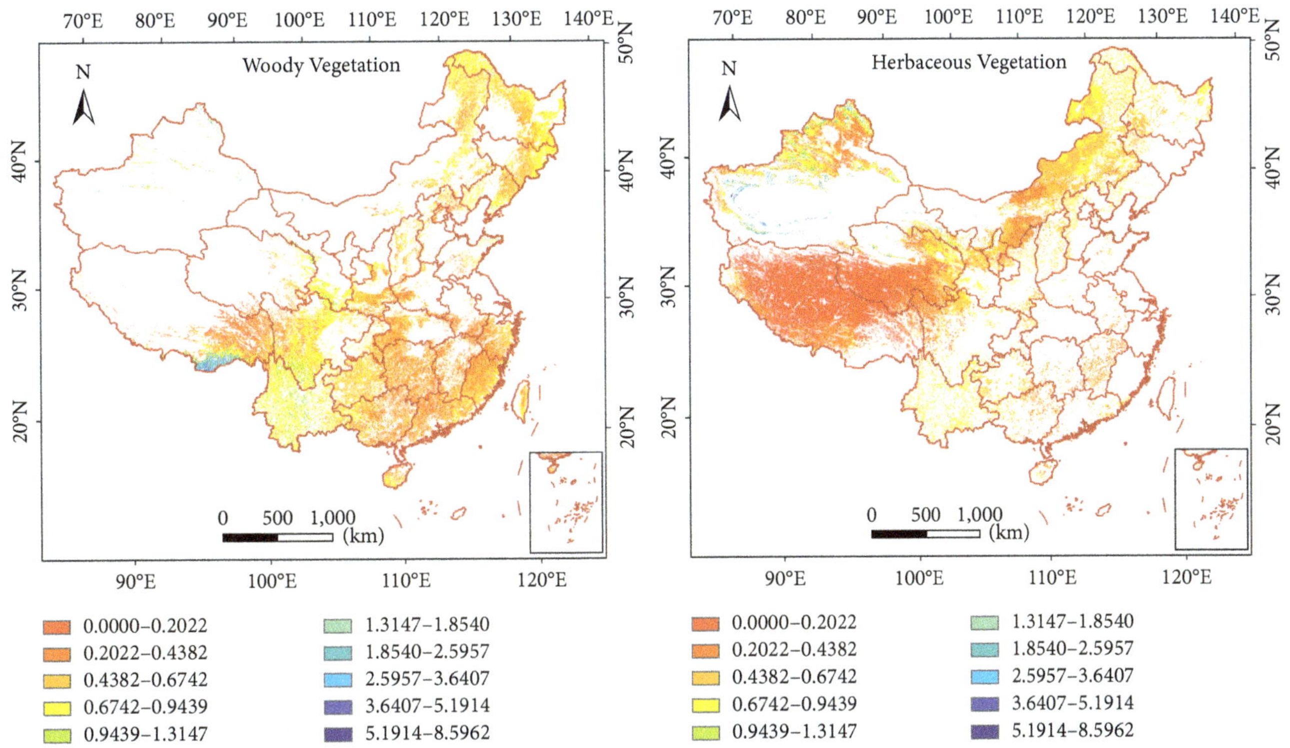

(a)

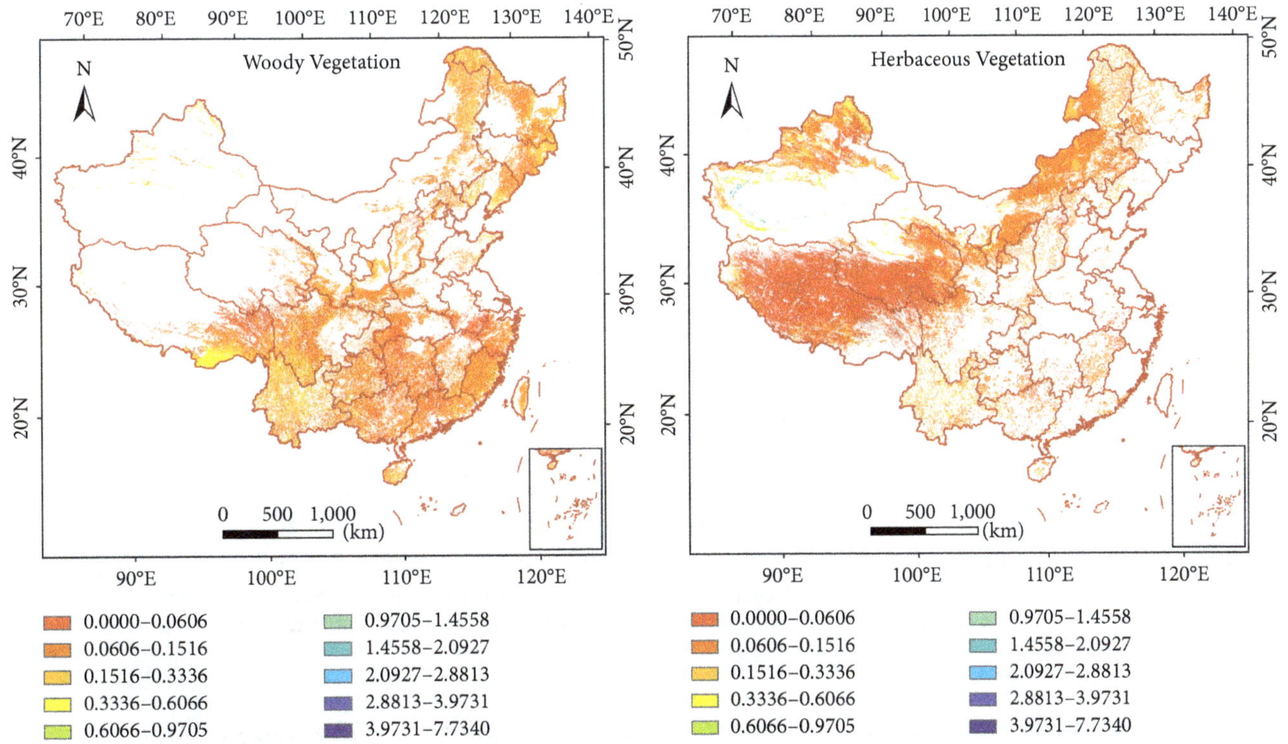

(b)

Figure 5: Continued.

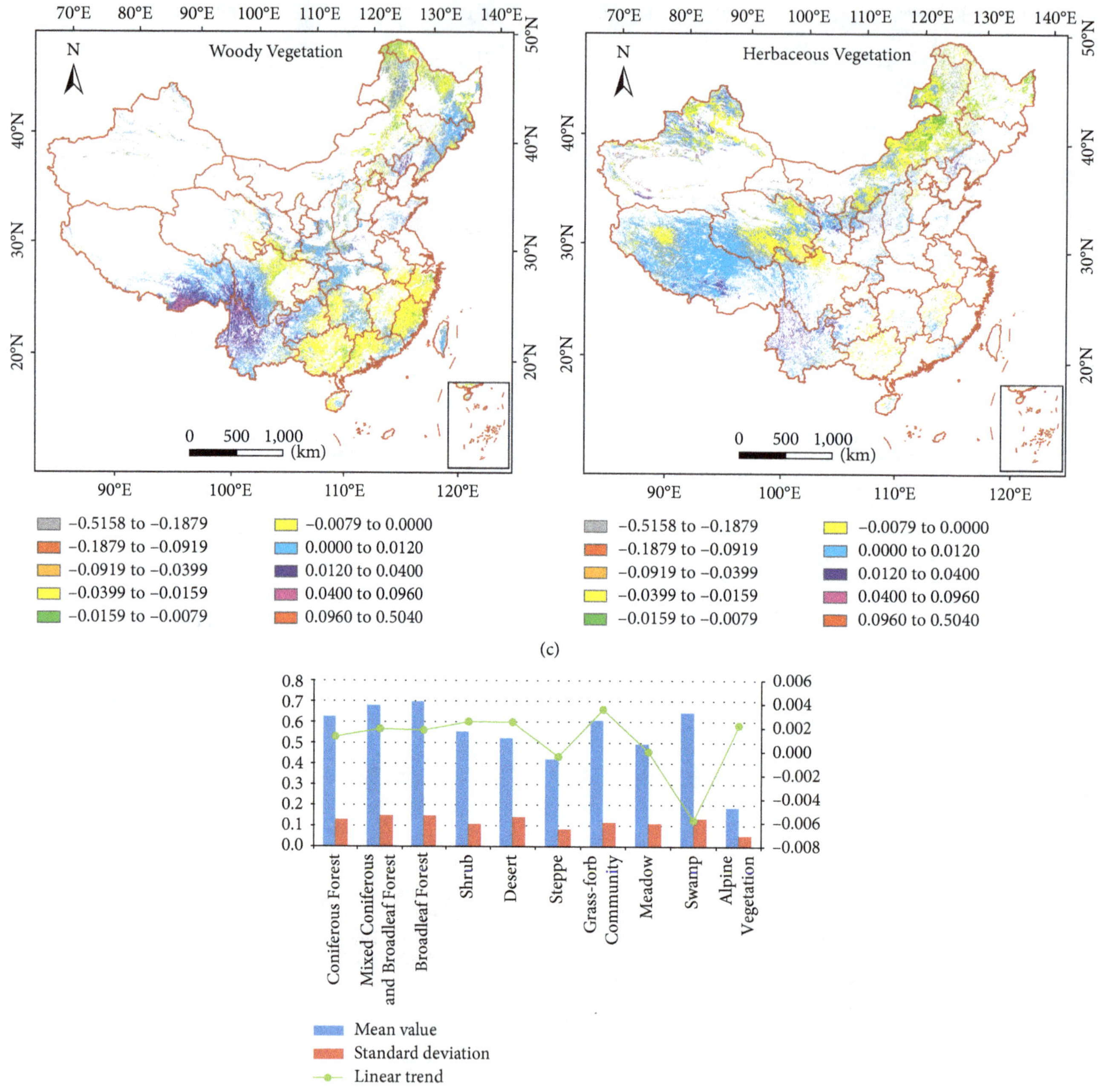

(d)

Figure 5: Temporal variation and spatial distribution of different Natural Vegetation types' PUEs. (a) Spatial distribution of the multiyear mean values of Woody Vegetative and Herbaceous Vegetative PUEs. (b) Standard deviations of Woody Vegetative and Herbaceous Vegetative PUEs. (c) Linear trends of Woody Vegetative and Herbaceous Vegetative PUEs. (d) Statistical central tendency of the mean values, standard deviations, and linear trends of different Natural Vegetation types' PUEs.

account for 24.5% of China's total land area; regions with the most obvious decreasing trends are mainly distributed in the northern edge of the Tarim Basin, northeast Inner Mongolia, northwest Heilongjiang, and other regions. Woody Vegetative PUE with the most obvious increasing trend is mainly distributed in the southern Tibetan valley, northwest of Yunnan, and other regions. Herbaceous Vegetative PUE regions showing a decreasing trend account for 42.2% of the total land area in China; regions with the most obvious decreasing trends are mainly distributed in the western edge of the Tarim Basin, northeastern Inner Mongolia, and most of Heilongjiang. A small number of Herbaceous Vegetative PUE regions with the most obvious increasing trend are mainly distributed in Xinjiang and western Gansu. The increasing speed of Grass-forb Community's PUE is the fastest, with a mean value of 3.6×10^{-3}; the increasing trend is most obvious in a small number of regions in northwestern Yunnan. The decreasing speed of Swamp's PUE is the fastest, with a mean value of -5.8×10^{-3}; the decreasing trend is most obvious in a small number of

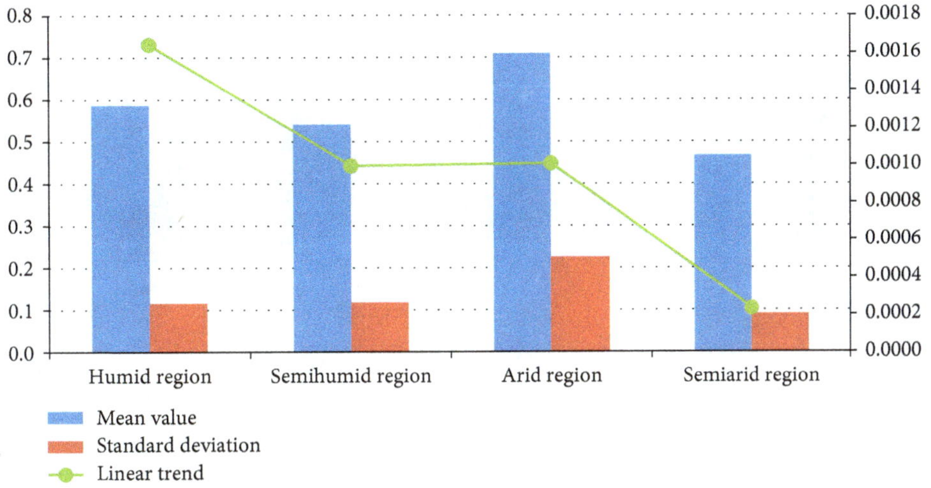

(a)

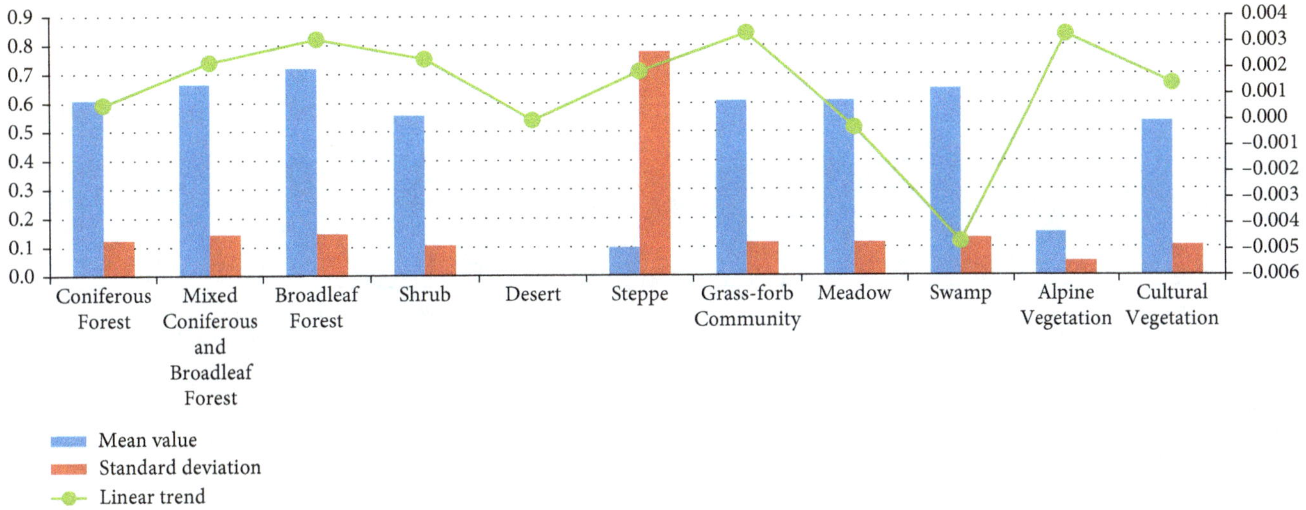

(b)

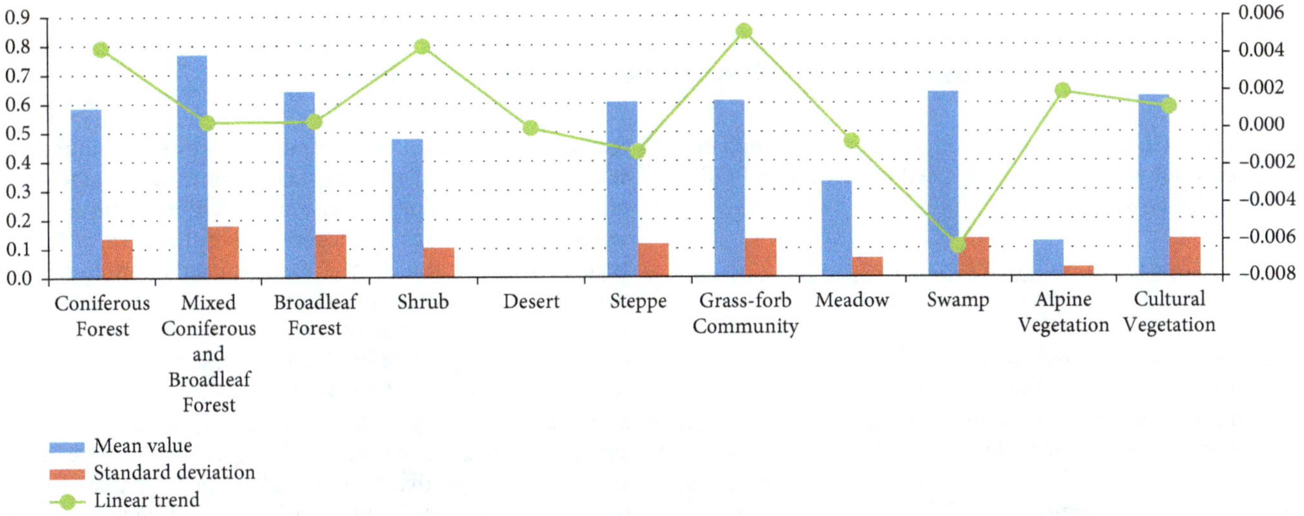

(c)

FIGURE 6: Continued.

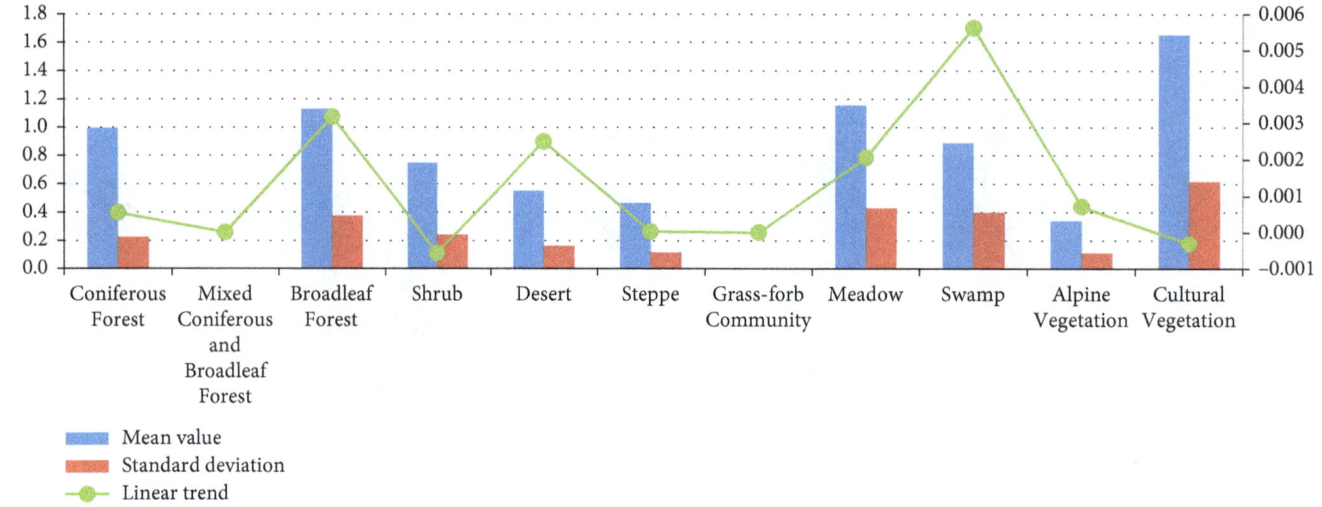

(d)

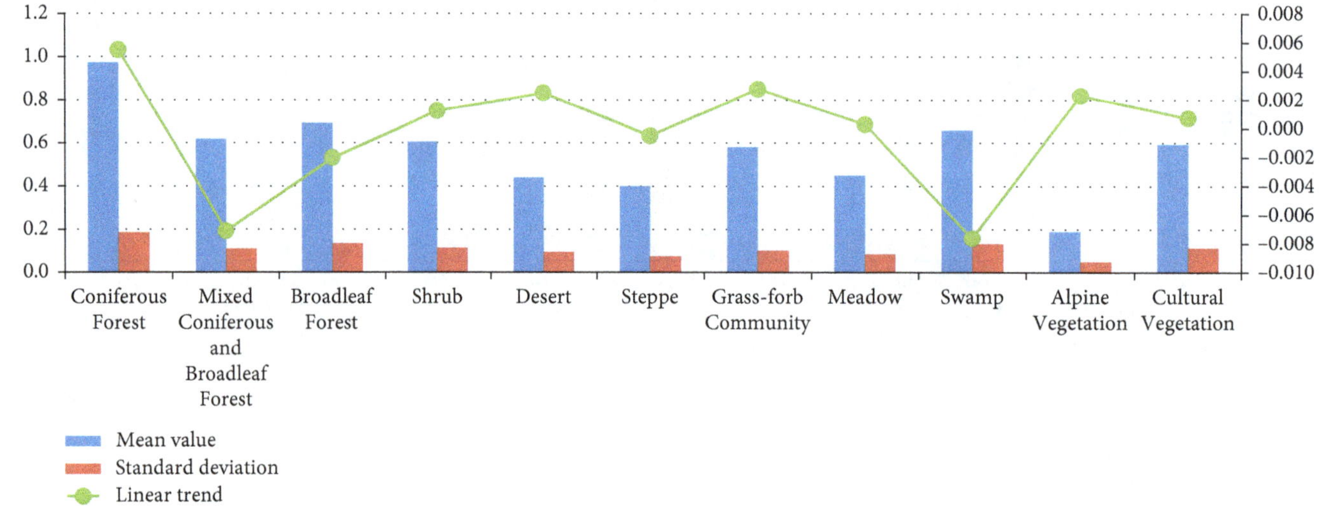

(e)

Figure 6: Statistical central tendency of the mean value, standard deviation, and linear trend of vegetative PUE in climate zones from 2000 to 2015: (a) climate zones; (b) humid regions; (c) semihumid regions; (d) arid regions; (e) semiarid regions.

regions in northeastern Inner Mongolia and Heilongjiang. The mean value of the increasing speed of Meadow's PUE is 0, tending to be stable.

3.3. Variation Characteristics of Climate Zones and Different Vegetation Types' PUEs

3.3.1. Variation Characteristics of Different Vegetation Types' PUEs in Climate Zones. The multiyear mean value of vegetative PUE in climate zones has obvious spatial variation characteristics, as shown in Figure 6. The averages of the multiyear mean values of vegetative PUE in humid, semihumid, arid, and semiarid regions are 5.86×10^{-1} gC·m^{-2}·mm^{-1}, 5.41×10^{-1} gC·m^{-2}·mm^{-1}, 7.09×10^{-1} gC·m^{-2}·mm^{-1}, and 4.67×10^{-1} gC·m^{-2}·mm^{-1}, respectively. The averages of the multiyear mean values of vegetative PUE in humid and arid regions are significantly higher than that of China's vegetative PUE, and they are significantly higher than the national average, especially along the edge of the Tarim Basin, south of Sichuan and Gansu, in central and western Yunnan, the southern Tibetan valley, and other regions. The averages of the multiyear mean values of vegetative PUE in semihumid and semiarid regions are significantly lower than that of China's vegetative PUE, and they are significantly lower than the national average, especially in northeastern Tibet, southwestern Qinghai, and other regions. The mean values of the standard deviations of vegetative PUE in humid, semihumid, arid, and semiarid regions are 1.155×10^{-1}, 1.162×10^{-1}, 2.237×10^{-1}, and 8.85×10^{-2}, respectively, showing an overall trend of no significant interannual fluctuations. However, interannual variation in vegetative PUE is significant in spatial. The stability of vegetative PUE in the arid region is declined, especially along the edge of the Tarim Basin. The stability of vegetative PUE in the semiarid region is increased, especially in northeastern Tibet, western Qinghai, and other regions. The vegetative PUE in climate zones shows an overall slightly increasing trend. From the interannual linear variation trend, it can be seen that the mean values of the increasing speed of vegetative PUE in humid, semihumid, arid, and semiarid regions are 1.6×10^{-3}, 1.0×10^{-3}, 1.0×10^{-3}, and 2.0×10^{-4}, respectively. However, the vegetative PUE shows relatively obvious spatial variation characteristics and even decreasing trends in a few regions. The regions in which vegetative PUEs have the most obvious lowering trends are mainly distributed in humid and semihumid regions. The regions in which vegetative PUEs have the most obvious increasing trends are mainly distributed in arid and semiarid regions.

In humid regions, as shown in Figure 6, the PUEs of Mixed Coniferous and Broadleaf Forest, Broadleaf Forest, and Swamp are higher, and the averages of their multiyear mean values are 6.64×10^{-1} gC·m^{-2}·mm^{-1}, 7.2×10^{-1} gC·m^{-2}·mm^{-1}, and 6.5×10^{-1} gC·m^{-2}·mm^{-1}, respectively; they are significantly higher than the national average, especially in the southern Tibetan valley, locally in northeastern Inner Mongolia, and other regions. Steppe's PUE is the lowest, and the average of the multiyear mean value is 9.9×10^{-2} gC·m^{-2}·mm^{-1}; this is significantly lower than the national average, especially in a small number of regions in southern Gansu. The stability of Steppe's PUE is declined, especially in a few regions of southern Gansu, and the mean value of the standard deviation is 7.775×10^{-1}. The stability of Alpine Vegetative PUE is increased, especially in a small number of regions in eastern Tibet, and the mean value of the standard deviation is 4.78×10^{-2}. The increasing speeds of Broadleaf Forest's, Grass-forb Community's, and Alpine Vegetative PUEs are faster, with mean values of 3.1×10^{-3}, 3.3×10^{-3}, and 3.3×10^{-3}, respectively; the increasing trend is most obvious in a small number of regions in the southern Tibetan valley and northwestern Yunnan. The decreasing speed of Swamp's PUE is the fastest, with a mean value of -4.7×10^{-3}; the decreasing trend is most obvious in a small number of regions in northeastern Inner Mongolia. No desert exists in humid regions.

In semihumid regions, as shown in Figure 6, the PUE of the Mixed Coniferous and Broadleaf Forest is the highest, with an average of the multiyear mean value of 7.69×10^{-1} gC·m^{-2}·mm^{-1}; this is significantly higher than the national average, especially in a small number of regions in eastern Heilongjiang and northeastern Jilin. The Alpine Vegetative PUE is the lowest, with an average of the multiyear mean value of 1.23×10^{-1} gC·m^{-2}·mm^{-1}; this is significantly lower than the national average, especially in a small number of regions in northeastern Tibet and southwestern Qinghai. The stability of the Mixed Coniferous and Broadleaf Forest's PUE is declined, especially in a small number of regions in northeastern Jilin, and the mean value of the standard deviation is 1.801×10^{-1}. The stabilities of the Meadow and Alpine Vegetative PUEs are increased, especially in southwestern Qinghai, locally in northeastern Tibet and other regions, with mean values of the standard variations of 6.41×10^{-2} and 3.19×10^{-2}, respectively. The increasing speed of the Grass-forb Community's PUE is the fastest, with a mean value of 5.2×10^{-3}; it is the most obvious increasing trend, especially in southwestern Liaoning. The decreasing speed of Swamp's PUE is the fastest, with a mean value of -6.4×10^{-3}; it is the most obvious decreasing trend, especially in a small number of regions in Heilongjiang and in northeastern Inner Mongolia. No desert exists in semihumid regions.

In arid regions, as shown in Figure 6, Cultural Vegetative PUE is the highest, with an average of the multiyear mean value of 1.652 gC·m^{-2}·mm^{-1}; this is significantly higher than the national average, especially in a small number of regions along the Tarim Basin's edge and in western Gansu. Alpine Vegetative PUE is the lowest, with an average of the multiyear mean value of 3.39×10^{-1} gC·m^{-2}·mm^{-1}; it is significantly lower than the national average, especially in a small number of regions in northern Tibet. The stability of Cultural Vegetative PUE is declined, especially in a small number of regions along the Tarim Basin's edge and in western Gansu, with a mean value of the standard deviation of 6.169×10^{-1}. The stability of Alpine Vegetative PUE is increased, especially in a small number of regions in northern Tibet, with a mean value of standard deviation of 1.101×10^{-1}. The increasing speed of Swamp's PUE is the fastest, with a mean value of 5.6×10^{-3}; this is the most obvious increasing trend, especially in a small number of regions in central Xinjiang. The increasing speeds of Shrub's and Cultural Vegetative PUEs are slower, with the mean values of -6×10^{-4} and -3×10^{-4}, respectively; they are the most obvious decreasing trends, especially in a small number of regions along the edge of the Tarim Basin. The mean value of the increasing speed of Steppe's PUE is 0, tending to be stable. No Mixed Coniferous and Broadleaf Forest or Grass-forb Community exists in arid regions.

In semiarid regions, as shown in Figure 6, Coniferous Forest's PUE is the highest, with an average of the multiyear mean value of 9.73×10^{-1} gC·m^{-2}·mm^{-1}; this is significantly higher than the national average, especially in a small number of regions in the southern Tibetan valley and in northern Xinjiang. Alpine Vegetative PUE is the lowest, with an average of the multiyear mean value of 1.9×10^{-1} gC·m^{-2}·mm^{-1}; this is significantly lower than the national average, especially in a small number of regions in northern Tibet and western Qinghai. The stability of Coniferous Forest's PUE is declined, with a mean value of standard deviation of 1.976×10^{-1}, especially in a small number of regions in the southern Tibetan valley and in northern Xinjiang. The stability of Alpine Vegetative PUE is increased, with a mean value of standard deviation of 4.82×10^{-2}, especially in a small number of regions in central Tibet and western Qinghai. The increasing speed of Coniferous Forest's PUE is the fastest, with a mean value of 5.5×10^{-3}; this is the most obvious increasing trend, especially in a small number of regions in the southern Tibetan valley. The lowering speeds of Mixed Coniferous and Broadleaf Forest's and Swamp's PUEs are faster, with the mean values of -7.1×10^{-3} and -7.6×10^{-3}, respectively; these are the most obvious decreasing trends, especially in a small number of regions in southwestern Heilongjiang and northwestern Jilin.

3.3.2. Variation Characteristics of the Same Vegetation Type's PUEs among Climate Zones. The averages of the multiyear mean values of the same vegetation type's PUEs among climate zones are significantly different, as shown in Figure 7. The difference in the average of Cultural

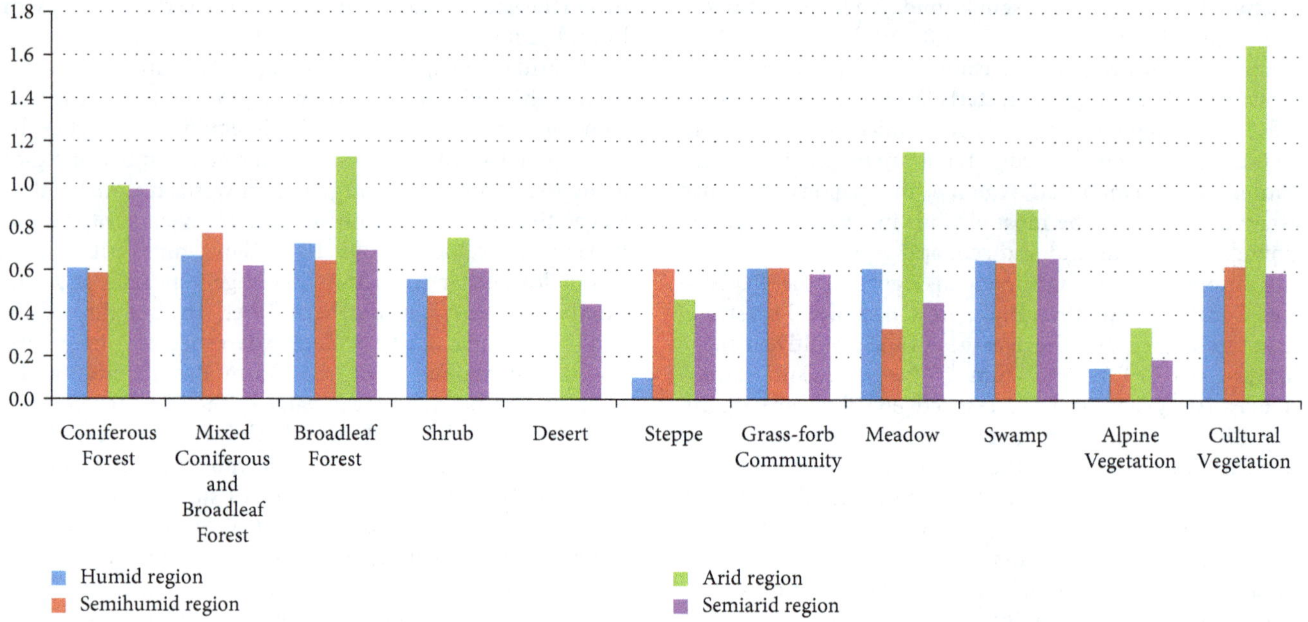

(a)

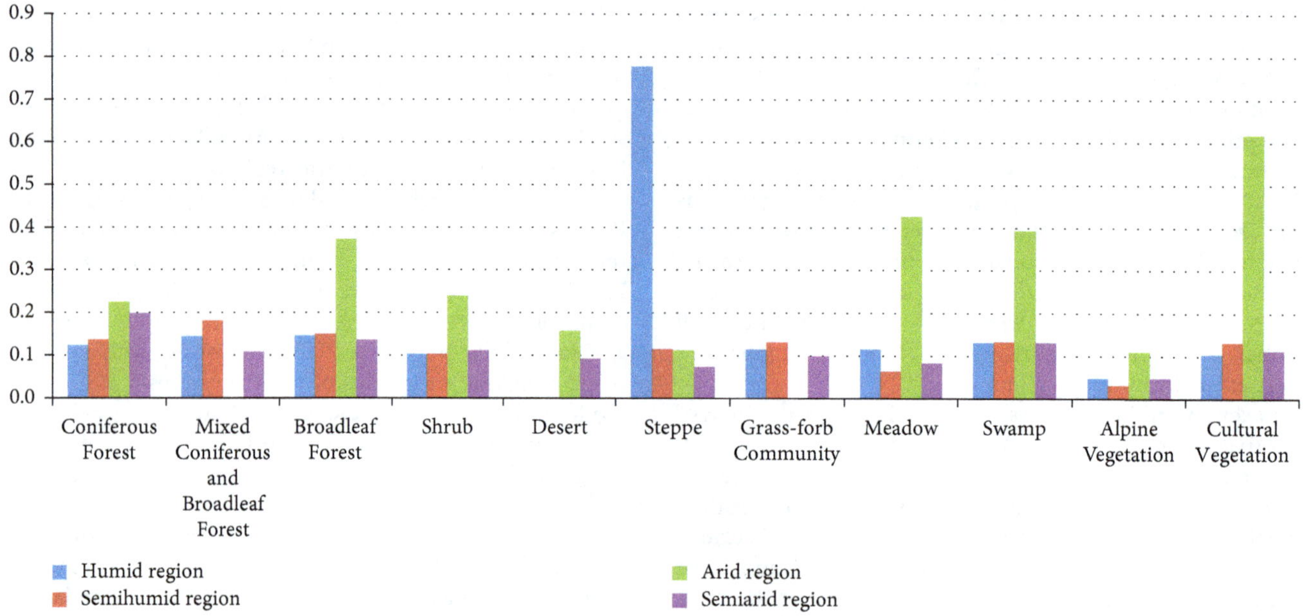

(b)

Figure 7: Continued.

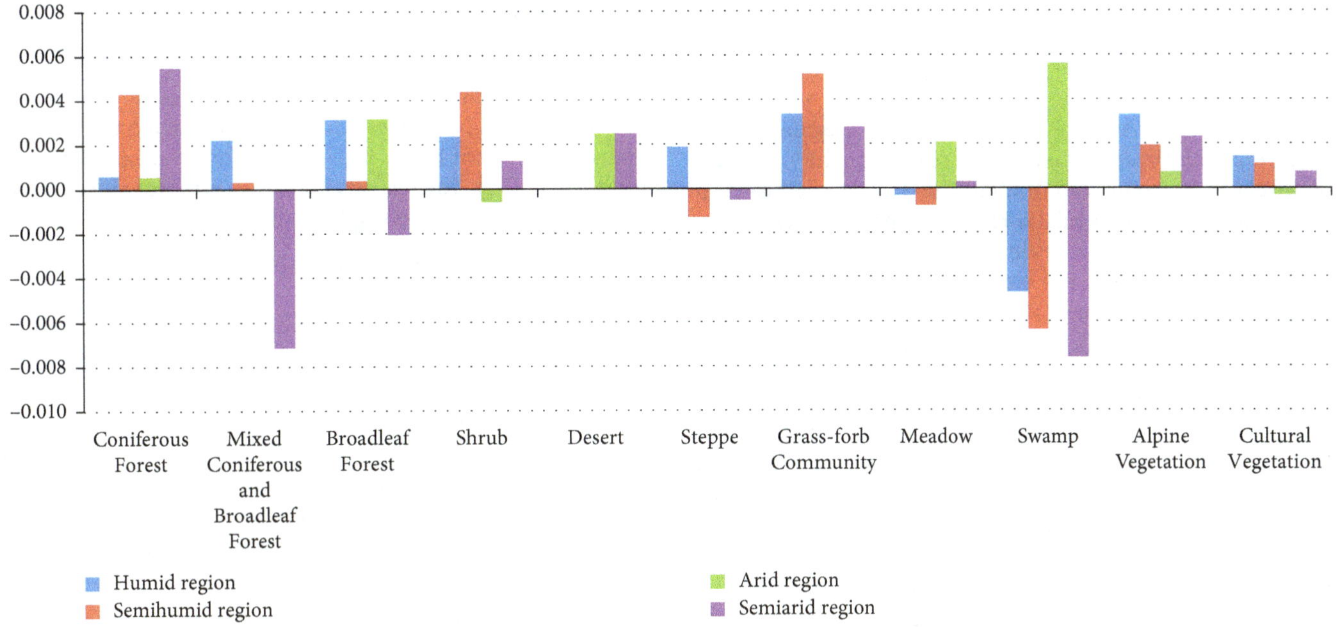

(c)

FIGURE 7: Statistical central tendency of the same vegetation type's PUEs among climate zones from 2000 to 2015: (a) mean values; (b) standard variations; (c) linear trends.

Vegetative PUE is the largest, with the averages of the multiyear mean values in humid, semihumid, arid, and semiarid regions are $5.35\times10^{-1}\,gC{\cdot}m^{-2}{\cdot}mm^{-1}$, $6.22\times10^{-1}\,gC{\cdot}m^{-2}{\cdot}mm^{-1}$, $1.652\,gC{\cdot}m^{-2}{\cdot}mm^{-1}$, and $5.93\times10^{-1}\,g\,C{\cdot}m^{-2}{\cdot}mm^{-1}$, respectively. The average of Cultural Vegetative PUE in arid regions is the highest, and it is significantly higher than the national average of Cultural Vegetation, especially in a small number of regions along the edge of the Tarim Basin and in western Gansu. The average of Cultural Vegetative PUE in humid regions is the lowest, and it is significantly lower than the national average of Cultural Vegetation, especially in eastern Guangxi, southern Guangdong, northern Jiangxi, and other regions. The difference in the average of Meadow's PUE is larger, and the averages of the multiyear mean values in humid, semihumid, arid, and semiarid regions are $6.08\times10^{-1}\,gC{\cdot}m^{-2}{\cdot}mm^{-1}$, $3.3\times10^{-1}\,gC{\cdot}m^{-2}{\cdot}mm^{-1}$, $1.152\,gC{\cdot}m^{-2}{\cdot}mm^{-1}$, and $4.52\times10^{-1}\,gC{\cdot}m^{-2}{\cdot}mm^{-1}$, respectively. The average of Meadow's PUE in arid regions is the highest, and it is significantly higher than the national average of Meadow, especially in a small number of regions along the edge of the Tarim Basin. The average of Meadow's PUE in semihumid regions is the lowest, and it is significantly lower than the national average, especially in a small number of regions in western Qinghai. The difference in the average of Grass-forb Community's PUE is the least, and the averages of the multiyear mean values in humid, semihumid, and semiarid regions are $6.08\times10^{-1}\,gC{\cdot}m^{-2}{\cdot}mm^{-1}$, $6.1\times10^{-1}\,gC{\cdot}m^{-2}{\cdot}mm^{-1}$, and $5.82\times10^{-1}\,gC{\cdot}m^{-2}{\cdot}mm^{-1}$, respectively. The average of Grass-forb Community's PUE in semihumid regions is the highest, and it is significantly higher than the national average of Grass-forb Community, especially in a small number of regions on the Shandong Peninsula, the Liaodong Peninsula, and in southwestern Liaoning. The average of Grass-forb Community's PUE in semiarid regions is the lowest, and it is significantly lower than the national average, especially in a small number of regions in western Shanxi. The difference in average of Mixed Coniferous and Broadleaf Forest's PUE is less, and the averages of the multiyear mean values in humid, semihumid, and semiarid regions are $6.64\times10^{-1}\,gC{\cdot}m^{-2}{\cdot}mm^{-1}$, $7.69\times10^{-1}\,gC{\cdot}m^{-2}{\cdot}mm^{-1}$, and $6.17\times10^{-1}\,gC{\cdot}m^{-2}{\cdot}mm^{-1}$, respectively. The average of Mixed Coniferous and Broadleaf Forest's PUE in semihumid regions is the highest, and it is significantly higher than the national average, especially in a small number of regions in eastern Heilongjiang and northeastern Jilin. The average of Mixed Coniferous and Broadleaf Forest's PUE in semiarid regions is the lowest, and it is significantly lower than the national average, especially in a small number of regions in southwestern Heilongjiang.

The interannual PUEs of the same vegetation type among climate zones is stable, but the spatial difference is relatively significant, as shown in Figure 7. The difference in stability of Steppe's PUE is the largest, and the mean values of standard deviation in humid, semihumid, arid, and semiarid regions are 7.775×10^{-1}, 1.151×10^{-1}, 1.132×10^{-1}, and 7.43×10^{-2}, respectively. The stability of Steppe's PUE in humid regions is declined, especially in a small number of regions in southern Gansu. The stability of Steppe's PUE in semiarid regions is increased, especially in northeastern Tibet, western Qinghai, and other regions. The difference in the stability of Cultural Vegetative PUE is larger, and the mean values of the standard deviation in humid, semihumid, arid, and semiarid regions are 1.042×10^{-1}, 1.317×10^{-1}, 6.169×10^{-1}, and 1.123×10^{-1}, respectively. The stability of Cultural Vegetative PUE in arid regions is declined, especially in a small number of regions along the edge of the Tarim Basin and in western Gansu. The stability of Cultural Vegetative PUE in humid regions is increased, especially in eastern Hubei, southern Anhui, and other regions. The difference in stability of Grass-forb Community's PUE is the least, and the mean values of the standard deviation in humid, semihumid, and semiarid regions are 1.147×10^{-1}, 1.318×10^{-1}, and 9.96×10^{-2}, respectively. The stability of Grass-forb Community's PUE in semihumid regions is declined, especially in a small number of regions on the Shandong Peninsula and in southern Liaoning. The stability of Grass-forb Community's PUE in semiarid regions is increased, especially in a small number of regions in northern Shanxi and western Hubei. The difference in stability of Alpine Vegetative PUE is less, and the mean values of standard deviation in humid, semihumid, arid, and semiarid regions are 4.78×10^{-2}, 3.19×10^{-2}, 1.101×10^{-1}, and 4.82×10^{-2}, respectively. The stability of Alpine Vegetative PUE in arid regions is declined, especially in a small number of regions in southern Xinjiang. The stability of Alpine Vegetative PUE in semihumid region is increased, especially in a small number of regions in northeastern Tibet and southwestern Qinghai.

The PUEs of the same vegetation type among climate zones with overall variation trends are significantly different, as shown in Figure 7. The difference in the increasing speed of Swamp's PUE is the largest, and the mean values of the increasing speeds in humid, semihumid, arid, and semiarid regions are -4.7×10^{-3}, -6.4×10^{-3}, 5.6×10^{-3}, and -7.6×10^{-3}, respectively. The increasing speed of Swamp's PUE in arid regions is the fastest, and it is the most obvious increasing trend, especially in a small number of regions in central Xinjiang. The decreasing speed of Swamp's PUE in semiarid regions is the fastest, and it is the most obvious decreasing trend, especially in a small number of regions in southwestern Heilongjiang and northwestern Jilin. The difference in the increasing speed of Mixed Coniferous and Broadleaf Forest's PUE is larger, and the mean values of the increasing speeds in humid, semihumid, and semiarid regions are 2.2×10^{-3}, 3×10^{-4}, and -7.1×10^{-3}, respectively. The increasing speed of Mixed Coniferous and Broadleaf Forest's PUE in humid regions is the fastest, and it is the most obvious increasing trend, especially in a small number of regions in northwestern Yunnan. The decreasing speed of Mixed Coniferous and Broadleaf Forest's PUE in semiarid regions is the fastest, and it is the most obvious decreasing trend, especially in a small number of regions in southwestern Heilongjiang. The difference in the increasing speed of Desert's PUE is the least; the mean values of the increasing speeds in arid and semiarid regions are both 2.5×10^{-3}; and it is the most obvious increasing trend, especially in a small number of regions in Xinjiang and western Gansu. The difference in the increasing speed of Cultural Vegetative PUE is less, and the mean values of the increasing speeds in humid, semihumid, arid, and semiarid regions are 1.4×10^{-3}, 1.1×10^{-3}, -3×10^{-4}, and 7×10^{-4}, respectively. The increasing speed of Cultural Vegetative PUE in humid regions is the fastest, and it is the most obvious increasing trend, especially in a small number of regions in the southern Tibetan valley and in northwestern Yunnan. The decreasing speed of Cultural Vegetative PUE in arid regions is the fastest, and it is the most obvious decreasing trend, especially in a small number of regions in western Xinjiang.

4. Discussion

4.1. Analysis of Extreme Values of Vegetative PUE. The extreme value regions of vegetative PUE are mainly distributed in arid areas in Northwest China, especially along the edge areas of the Tarim Basin. This is consistent with the findings of Mu et al. [38]. The reasons for these extreme values can be explained by the following three aspects:

(1) The Tarim Basin is surrounded by mountains such as the Tianshan, Kunlun, and Altun. Regional runoff is replenished by melting snow and ice from the mountains; thus, the vegetation shows good growth with limited precipitation. Terrain factor may be one of the reasons leading to higher vegetative PUE in the area [39].

(2) Because the Taklamakan Desert is located in the center of the Tarim Basin, the sand content of the soil surface around the desert is higher; thus, the infiltration rate of precipitation also increases, increasing the vegetative PUE [27, 38].

(3) In arid areas, the vegetative root system is well developed, with a lower canopy conductance being able to utilize the soil moisture in the lower layer; thus, the production volume of consumed water per unit of vegetation is higher [40]. In particular, some dominant plant communities have lower transpiration and higher photosynthetic rates in arid environments, so vegetative PUE may also be higher [41–43].

4.2. Influencing Factors on Spatial Variation in Vegetative PUE. The variation in vegetative PUE is closely related to climate zones, soil factors, and vegetation types [16, 27]. Due to the differences in precipitation, soil factors, and vegetation types in different climatic zones, in discussing the influencing factors on spatial variations in vegetative PUE, this study obtained the following four results:

(1) The difference in vegetative PUE may be affected by different vegetation types. The differences in biological community structure, photosynthetic efficiency, and fractional vegetation cover of different vegetation types can result in different PUEs among different vegetation types [44, 45].

(2) The difference in vegetative PUE may be affected by different climatic zones. For example, a significant spatial difference in the PUEs of the same vegetation type among climatic zones may be due to the distribution of climate conditions, such as heat and moisture, being different in different climatic zones, which can determine vegetative PUE as a zonal distribution [25, 46].

(3) The difference in vegetative PUE may be affected by precipitation, and water is an important factor that limits vegetative growth in arid areas [45]. For example, along the edge area of the Tarim Basin, the vegetation is replenished by melting snow and ice in the mountains, making the vegetative PUE in this area higher than that of other arid areas [39, 47].

(4) The difference in vegetative PUE may be affected by soil factors in humid areas. Generally, in humid regions, the soil moisture is in a saturation condition with sufficient precipitation, but the biological activity of the soil will be lower. In addition, a superfluous amount of precipitation will result in surface runoff, washing away key nutrient substances that are easily affected by eluviations, such as nitrogen and phosphorus, creating an indirect impact on vegetative growth [48]. In humid regions, improving soil fertility and permeability can be beneficial to vegetative PUE [16].

4.3. Limitations and Prospects of the Study. The meteorological stations used in this study are rare in the western China, with a generally uneven distribution that may affect the accuracy of the spatial interpolation of the precipitation data. In the future, the use of remote-sensing data to perform supplemental interpolation of spatial data based on DEM may be preferable to improve the spatial data quality of precipitation. The NPP directly uses the product of MODIS NPP because of the limitation of scale and the lack of measured data verification, which may result in errors when using NPP to calculate PUE. In the future research, we should use this model or improve the relevant model, carry out NPP simulation calculations, and use the measured data for verification, which may improve the accuracy of NPP. In further PUE studies, the NDVI can be used instead of NPP to enable PUE calculations, which may improve PUE quality and provide better data for further research on the spatiotemporal variations of PUE.

Due to the limitations of the research scale and data in this paper, it is impossible to deeply discuss the influencing factors for the evolution of the spatiotemporal patterns of vegetative PUE. As for the relationship between the evolution process of vegetative PUE and climatic change, altitude, biological characters of vegetation, soil, human activities, and so on, we can assess these factors from the following aspects: the driving relationship between changes in vegetative PUE and precipitation and temperature, the difference in vegetative PUE at different altitudes, the correlation between fractional vegetation cover (FVC), leaf area index (LAI) and spatial distribution and interannual fluctuations of vegetative PUE, and the impact of different soil types on spatial differences of vegetative PUE. In future research, other statistical models and analysis methods should be used to further improve the research effectiveness of PUE. In addition, the different influencing factors of vegetative PUE in arid and humid regions also need further in-depth study to gain important findings.

5. Conclusions

This paper starts from the national scale, using time-series MODIS NPP data and meteorological precipitation data from 2000 to 2015 in China to calculate the Chinese vegetative PUE from 2000 to 2015. Then, it analyzes the changes and spatiotemporal patterns of the Chinese terrestrial vegetative PUE in the most recent 16 years. The main conclusions are as follows:

(1) The multiyear mean value of China's vegetative PUE shows obvious spatial variation characteristics. It is relatively stable interannually, with an overall slightly increasing trend. The regions with extreme values, declined stability, and the most obvious decreasing trends of China's vegetative PUE all appear along the edge areas of the Tarim Basin.

(2) There is a significant difference in the PUEs among different vegetation types. Broadleaf Forest's PUE is the highest, and Alpine Vegetative PUE is the lowest. The stabilities of Mixed Coniferous and Broadleaf Forest's and Broadleaf Forest's PUEs are declined, and the stabilities of Steppe's and Alpine Vegetative PUEs are increased. The increasing speed of Grass-forb Community's PUE is the fastest, the decreasing speed of Swamp's PUE is the fastest, and the increasing speed of Meadow's PUE tends to be stable.

(3) There is a significant difference in the PUEs among different vegetation types in climate zones. No desert exists in humid and semihumid regions. No Mixed Coniferous and Broadleaf Forest or Grass-forb Community exists in arid regions. It includes all vegetation types in semiarid regions. The PUEs of the same vegetation type are significantly different among climate zones. The difference in the average of Cultural Vegetative PUE is the largest, and the difference in the average of Grass-forb Community's PUE is the least. The difference in the stability of Steppe's PUE is the largest, and the difference in the stability of Grass-forb Community's PUE is the least. The difference in the increasing speed of Swamp's PUE is the largest, and the difference in the increasing speed of Desert's PUE is the least.

Research on the spatiotemporal patterns of vegetative PUE in China will help us gain an in-depth understanding of the mechanism of vegetation response to global climate change, and we can more clearly recognize the formation process of the productivity of different vegetation types in different climatic zones. The spatiotemporal variation characteristics of vegetative PUE in arid regions, especially the possible influencing factors of vegetative PUE in extreme regions, can provide important references for many researchers of vegetative PUE in arid regions. A relatively comprehensive study of China's vegetative PUE, based on different vegetation types and different climate zones, can accumulate valuable information for other researchers of China's vegetative PUE in the future. Vegetative PUE has extensive application prospects in the assessment of vegetation degradation and regional water-carbon cycles. PUE has important practical and theoretical significance for the scientific study of China's ecological safety constructs, land-vegetation ecosystems, and reaction to global changes.

Data Availability

The vector data of the climate zones and vegetation types in China used to support the findings of this study are included within the article. The MODIS NPP data are downloaded from [NASA, https://ladsweb.nascom.nasa.gov/search/]. The meteorological precipitation data used to support the findings of this study were supplied by [the China Meteorological Data Service Center] under license and so cannot be made freely available. Requests for access to these data should be made to [the China Meteorological Data Service Center, http://data.cma.cn/].

Conflicts of Interest

The authors declare that there are no conflicts of interest regarding the publication of this paper.

Acknowledgments

This research was supported by National Key R&D Program of China (no. 2017YFC0506506 and no. 2016YFC0500206) and State Key Laboratory of Geohazard Prevention and Geoenvironment Protection Independent Research Project (no. SKLGP2017Z005).

References

[1] W. Zhang, H. Xu, and F. Zheng, "Aerosol optical depth retrieval over East Asia using Himawari-8/AHI data," *Remote Sensing*, vol. 10, no. 1, p. 137, 2018.
[2] W. Zhang, H. Xu, and F. Zheng, "Classifying aerosols based on fuzzy clustering and their optical and microphysical properties study in Beijing, China," *Advances in Meteorology*, vol. 2017, Article ID 4197652, 18 pages, 2017.
[3] R. Nemani, C. Keeling, H. Hashimoto et al., "Climate-driven increases in global terrestrial net primary production from 1982 to 1999," *Science*, vol. 30, no. 5625, pp. 1560–1563, 2003.
[4] H. Xu, F. Zheng, and W. Zhang, "Variability in dust observed over China using A-train CALIOP instrument," *Advances in Meteorology*, vol. 2016, Article ID 1246590, 11 pages, 2016.
[5] H. Xu, T. Cheng, D. Xie, J. Li, Y. Wu, and H. Chen, "Dust identification over arid and semiarid regions of Asia using AIRS thermal infrared channels," *Advances in Meteorology*, vol. 2014, Article ID 847432, 16 pages, 2014.
[6] X. F. Liu, X. F. Zhu, Y. Z. Pan et al., "Spatio-temporal changes in vegetation coverage in China during 1982–2012," *Acta Ecologica Sinica*, vol. 35, no. 16, pp. 5331-5332, 2015.
[7] J. Peuelas, T. Rutishauser, and I. Filella, "Phenology feedbacks on climate change," *Science*, vol. 324, no. 5929, pp. 887-888, 2009.
[8] L. Yu, K. R. Li, B. Tao et al., "Simulating and assessing the adaptability of geographic distribution of vegetation to climate change in China," *Progress in Geography*, vol. 29, no. 11, p. 1326, 2010.
[9] M. S. Zhao, C. B. Fu, X. D. Yan et al., "Study on the relation ship between different ecosystem sand climate in China using NOAA/AVHRR data," *Acta Geographica Sinica*, vol. 56, no. 3, p. 287, 2011.
[10] J. F. Mao, B. Wang, and Y. J. Dai, "Sensitivity of the carbon storage of potential vegetation to historical climate variability and CO_2 in continental China," *Advances in Atmospheric Sciences*, vol. 26, no. 1, pp. 87–100, 2009.
[11] J. Y. Fang, S. L. Piao, J. S. He et al., "China's vegetation activity has been increasing in the past 20 years," *Science in China (Series C)*, vol. 33, no. 6, p. 554, 2003.
[12] D. Q. Sun, J. X. Zhang, C. G. Zhu et al., "An assessment of China's ecological environment quality change and its spatial variation," *Acta Geographica Sinica*, vol. 67, p. 1599, 2012.
[13] Z. B. Xin, T. X. Xu, W. Zhang et al., "Effects of climate change and human activities on vegetation cover change in the Loess Plateau," *Science in China (Series D)*, vol. 37, no. 11, pp. 1504–1514, 2007.
[14] K. D. Arnab, N. R. Patel, S. K. Saha, and D. Dutta, "Desertification in western Rajasthan (India):an assessment using remote sensing derived rain-use efficiency and residual trend methods," *Natural Hazards*, vol. 86, pp. 297–313, 2017.
[15] S. X. Cao, G. C. Liu, and H. Ma, "Dynamic analysis of vegetation change in north China," *Acta Ecologica Sinica*, vol. 37, no. 15, pp. 2-3, 2017.
[16] Z. H. Gao, Z. Y. Li, G. D. Ding et al., "New approach for desertification assessment by remote sensing based upon rain use efficiency of vegetation," *Science of Soil and Water Conservation*, vol. 3, no. 2, pp. 37–41, 2005.
[17] G. Liu, R. Sun, Z. Q. Xiao et al., "Analysis of spatial and temporal variation of net primary productivity and climate controls in China from 2001 to 2014," *Acta Geographica Sinica*, vol. 35, no. 15, p. 2, 2017.
[18] J. Sun and W. P. Duan, "Effects of precipitation and temperature on net primary productivity and precipitation use efficiency across China's grasslands," *GIScience & Remote Sensing*, vol. 54, no. 5, pp. 881–897, 2017.
[19] Y. L. Zhang, W. Qi, C. P. Zhou et al., "Spatial and temporal variability in the net primary production of alpine grassland on the Tibetan Plateau since 1982," *Journal of Geographical Sciences*, vol. 24, no. 2, pp. 269–287, 2013.
[20] Q. Z. Yuan, S. H. Wu, E. F. Dai et al., "NPP vulnerability of China's potential vegetation to climate change in the past 50 years," *Acta Geographical Sinica*, vol. 71, no. 5, p. 798, 2016.
[21] S. J. Mu, K. X. Zhou, Y. Qi et al., "Spatio-temporal patterns of precipitation-use efficiency of vegetation and their controlling factors in Inner Mongolia," *Chinese Journal of Plant Ecology*, vol. 38, no. 1, pp. 1–16, 2014.
[22] H. H. N. Le, R. L. Bingham, and W. Skerbek, "Relationship between the variability of primary production and the variability of annual precipitation in world arid lands," *Journal of Arid Environments*, vol. 15, pp. 1–18, 1988.
[23] Y. F. Bai, X. G. Han, J. G. Wu, Z. Z. Chen, and L. H. Li, "Ecosystem stability and compensatory effects in the Inner Mongolia grassland," *Nature*, vol. 431, no. 7005, pp. 181–184, 2004.
[24] F. Huang and S. L. Xu, "Spatio-temporal variations of rain-use efficiency in the west of Songliao Plain, China," *Sustainability*, vol. 8, no. 4, p. 308, 2016.
[25] Y. F. Bai, J. G. Wu, X. Qi et al., "Primary production and rain use efficiency across a precipitation gradient on the Mongolia Plateau," *Ecology*, vol. 89, pp. 2140–2153, 2008.
[26] Z. M. Hu, G. R. Yu, J. W. Fan, H. P. Zhong, S. Q. Wang, and S. G. Li, "Precipitation-use efficiency along a 4500-km grassland transect," *Global Ecology and Biogeography*, vol. 19, no. 6, pp. 842–851, 2010.
[27] H. H. N. Le, "Rain use efficiency: a unifying concept in arid-land ecology," *Journal of Arid Environments*, vol. 7, pp. 213–247, 1984.
[28] H. Ye, J. B. Wang, M. Huang et al., "Spatial pattern of vegetation precipitation use efficiency and its response to precipitation and temperature on the Qinghai-Xizang Plateau of China," *Chinese Journal of Plant Ecology*, vol. 36, no. 12, p. 1237, 2012.
[29] F. G. Viets, "Fertilizers and efficient use of water," *Advances in Agronomy*, vol. 14, pp. 223–265, 1962.

[30] J. R. Wight and A. L. Black, "Range fertilization: plant response and water-use," *Journal of Range Management*, vol. 32, no. 5, pp. 345–349, 1979.
[31] H. X. Li, H. Y. Zhou, and X. H. Wei, "Analysis of the impact of human disturbance on vegetation based on RUE and NDVI: a case study in Northwest Guangxi, China," *Journal of Desert Research*, vol. 34, no. 3, pp. 928-929, 2014.
[32] F. Cai, G. R. Yu, and Q. L. Zhu, "Comparison of precisions between spatial methods of climatic factors: a case study on mean air temperature," *Resources Science*, vol. 27, no. 5, pp. 173–179, 2005.
[33] Y. J. Fan, X. X. Yu, H. X. Zhang et al., "Comparison between kirging interpolation method and inverse distance weighting tension for precipitation data analysis: taking Lijiang river basin as a study case," *Journal of China Hydrology*, vol. 34, no. 6, pp. 61-62, 2014.
[34] Z. Y. Liu, X. Zhang, and R. H. Fang, "Analysis of spatial interpolation methods to precipitation in Yulin based on DEM," *Journal of Northwest A & F University (Natural Science Edition)*, vol. 38, no. 7, pp. 228–231, 2010.
[35] Z. Sheng, S. Q. Xie, and C. Y. Pan, *Probability Theory and Mathematical Statistics*, China Higher Education Press, Beijing, China, 4nd edition, 2008.
[36] J. Y. Zheng, J. J. Bian, Q. S. Ge et al., "The climate regionalization in China for 1981–2010," *Chinese Science Bulletin*, vol. 58, no. 30, p. 3094, 2013.
[37] Chinese Vegetation Map Editorial Board of Chinese Academy of Sciences, *People's Republic of China Vegetation Map 1: 1000000*, China Geological Publishing House, Beijing, China, 2007.
[38] S. J. Mu, Y. L. You, C. Zhu, and Z. Kexin, "Spatio-temporal patterns of precipitation-use efficiency of grassland in Northwestern China," *Acta Ecologica Sinica*, vol. 37, no. 5, pp. 1–13, 2017.
[39] Y. P. Shen, "Tianshan in Central Asia is a hot spot for global climate change and water cycle changes," *Journal of Glaciology and Geocryology*, vol. 31, no. 4, pp. 780-781, 2009.
[40] E. G. Jobbágy and O. E. Sala, "Controls of grass and shrub aboveground production in the Patagonian steppe," *Ecological Applications*, vol. 10, no. 2, pp. 541–549, 2000.
[41] P. X. Su and Q. D. Yan, "Photosynthetic characteristics of C4 desert plants *Haloxylon ammodendron* and *Calligonum mongolicum* under different water conditions," *Acta Ecologica Sinica*, vol. 26, no. 1, pp. 75–82, 2006.
[42] J. R. Ehleringer, T. E. Cerling, and B. R. Helliker, "C4 photosynthesis, atmospheric CO2, and climate," *Oecologia*, vol. 112, no. 3, pp. 285–299, 1997.
[43] G. D. Farquhar, M. H. O'Leary, and J. A. Berry, "On the relationship between carbon isotope discrimination and the intercellular carbon-dioxide concentration in leaves," *Australian Journal of Plant Physiology*, vol. 9, no. 2, pp. 121–137, 1982.
[44] J. M. Paruelo, W. K. Lauenroth, I. C. Burke et al., "Grassland precipitation-use efficiency varies across a resource gradient," *Ecosystems*, vol. 2, no. 1, pp. 64–68, 1999.
[45] T. E. Huxman, M. D. Smith, P. A. Fay et al., "Convergence across biomes to a common rain-use efficiency," *Nature*, vol. 429, no. 6992, pp. 651–654, 2004.
[46] Y. H. Yang, J. Y. Fang, P. A. Fay, J. E. Bell, and C. Ji, "Rain use efficiency across a precipitation gradient on the Tibetan Plateau," *Geophysical Research Letters*, vol. 37, no. 15, pp. 78–82, 2010.
[47] S. A. Halse, M. D. Scanlon, J. S. Cocking, M. J. Smith, and W. R. Kay, "Factors affecting river health and its assessment over broad geographic ranges: the Western Australian experience," *Environmental Monitoring & Assessment*, vol. 134, no. 1–3, pp. 161–175, 2007.
[48] A. T. Austin and P. M. Vitousek, "Nutrient dynamics on a precipitation gradient in Hawai'i," *Oecologia*, vol. 113, no. 4, pp. 519–529, 1998.

9

Studies on the Climate Effects of Black Carbon Aerosols in China and their Sensitivity to Particle Size and Optical Parameters

Xingxing Ma, Hongnian Liu, Xueyuan Wang, and Zhen Peng

School of Atmospheric Sciences, Nanjing University, Nanjing 210023, China

Correspondence should be addressed to Hongnian Liu; liuhn@nju.edu.cn

Academic Editor: Roberto Fraile

In this paper, based on the principle of Mie scattering, we calculated the optical parameters of BC aerosols at different scales and then applied the new optical parameters to simulate the BC aerosols concentration distribution, radiative forcing, and their climate effects. We also compared the results of optical parameters of BC aerosols with homogeneous scales and analyzed the effect on climate. Compared with the conventional uniform-scheme optical parameterization, the concentrations of the first mode of BC aerosols simulated with the optical parameters that were recalculated based on the particle size are significantly higher, while the concentrations of the other modes and the total of BC aerosols are lower. In the respective of statistics, the changes of column burdens of BC in four modes are 0.085, −0.095, −0.089, −0.054 mg/m^2. The clear-sky TRF of BC are weakened in the value of 0.03 W/m^2 averaged over the domain, while the all-sky TRF of BC are enhanced of 0.06 W/m^2 in general. The warming effect of BC becomes weaker when using the new scheme by −0.04 K to −0.24 K. When using the new optical parameters scheme, the regional average surface concentrations of BC in four modes are 0.372, 0.264, 0.055 and 0.004 μg/m^3, respectively. Especially, the first and the second mode account for as large as 53% and 38%. The surface concentration and column burden of total BC are 0.69 μg/m^3 and 0.28 mg/m^2 can be dropped. The regional average direct RFs of BC at the top of the atmosphere are 0.49 W/m^2 under clear-sky and 0.36 W/m^2 under all-sky averaged over the domain. Over most areas of central China, North China, and East China, BC may increase the temperature in a range of 0.05~0.15 K, while over South China, BC shows cooling effect. In average, the precipitation variations caused by BC over East China, North China, South China, and Northeast China are −0.83, −0.05, −0.11, and −0.13 mm/d, respectively. As a whole, the variations of circulation, pressure, and temperature show a good correspondence.

1. Introduction

Atmospheric aerosols can affect the earth's climate through both direct and indirect effects and play important roles in atmospheric radiation and climate change. Among all of the anthropogenic aerosols, black carbon (BC) aerosols can effectively absorb solar radiation in the visible and infrared bands to heat the atmosphere and affect the climate and air quality [1, 2] and therefore play a unique and important role in the climate systems. The Intergovernmental Panel on Climate Change (IPCC) report [3] indicates that the average direct radiative forcing of global BC aerosols is 0.4 W·m^{-2}. BC aerosols make an important contribution to global warming [4] and have become the second major warming factor after CO_2 [2].

East Asia is an important source region of global aerosol emissions [5]. The anthropogenic aerosols in East Asia have increased, which has had a nonnegligible influence on the regional climate. The global atmosphere model CAM3 was used to study the influence of sulfate aerosols and BC aerosols on the East Asian summer monsoon, and the results indicate that sulfate aerosols reduce the temperature in most areas of China, decreasing precipitation in some regions and weakening the East Asian summer monsoon [6]. BC aerosols also weaken the East Asian summer monsoon, but their influence is more complicated. The relationship between BC aerosols and regional precipitation has been studied, and researchers suggested that the BC aerosol emissions in China are related to the droughts in southern China and the floods in northern China over the past several decades [7]. BC aerosols have a significant influence on the weakening of the wind velocity [8]. BC aerosols have a relatively heterogeneous spatiotemporal distribution, which makes it difficult to evaluate their influence on the regional climate. Coal combustion is the main anthropogenic source of BC aerosols [9]. Coal is the main component of the energy structure in China; therefore, several researchers have concluded that one quarter of the BC aerosols generated by global anthropogenic emissions comes from China [10, 11]. With the accelerated development in China in recent years, the emissions of BC aerosols have increased significantly, and BC aerosols in China have drawn broad international attention. China is located in the Asian monsoon area. The response of the climate in this region to the atmospheric compositions could be different from those in other regions, so it is necessary to use a regional model to study the radiative forcing and climate effect of BC aerosols in this region.

Aerosols affect atmospheric radiation through scattering and absorption, and their optical parameters are important for the calculation of radiation effects. In numerical simulations, the optical parameters of BC aerosols directly affect the calculated radiation effects. The radiative forcing effect of BC aerosols has received wide attention, and the analysis of their optical properties is especially important [12]. Many analyses of aerosol optical parameters have focused on dust and sulfates, and calculations of the optical properties BC aerosols are relatively limited. No analyses of the light extinction coefficient, absorption coefficient, and scattering phase function of BC aerosols have been conducted.

The climate forcing of carbon aerosols varies significantly with their size distribution [13, 14]. The average indirect radiative forcing effect of carbon aerosols at the top of the atmosphere ranges from −0.34 to 1.08 W/m^2, and its variation strongly depends on the assumed particle size distribution of the aerosols in the model [14]. The sensitive of BC aerosols climate effects on the particle size distribution is due to the two reasons: different assumptions of the particle size distribution will cause differences in the calculation of the BC aerosols' concentration distribution and then result in changes in the optical parameters of the BC aerosols. In many previous studies, the particle size distribution of BC aerosols adopted the homogeneous scale (median radius) method, and the optical parameters only varied with the wavelength. This method can cause relatively large errors and increase the uncertainty of the climate effects of BC aerosols.

Because the concentration of BC aerosols and the climate effect are relatively sensitive to the particle size, in previous studies, we adopted a size-resolved model of BC aerosols and studied the climate effect of BC aerosols based on the measured particle size distribution [15]. However, the optical parameters did not vary with the particle size, which may cause several errors. In this paper, we calculate the optical parameters of BC aerosols with different scales mainly through the principle of Mie scattering and further use the calculated optical parameterization scheme to simulate the size-resolved climate effects of BC aerosols. The conventional scale division scheme for the BC aerosols was used [15]. We compare the differences in the simulation results caused by the differences in the optical parameterizations of BC aerosols.

2. Calculation of the Optical Coefficient of Black Carbon Aerosols

In numerical models, the optical coefficient of aerosols directly affects the calculated radiation effect, and the radiation transfer scheme usually requires three input parameters: single scattering albedo (SSA), dissymmetry factor (DF), and mass absorption coefficient (MAC). The acquisition of these three optical parameters must be based on Mie scattering theory.

The SSA (ω_0) is the ratio between the aerosol scattering coefficient and the light extinction coefficient, and it measures the absorption strength of aerosols. It is an important optical property of aerosols. The MAC (K) measures the absorption properties of a unit aerosol mass, and the DF (g) describes the asymmetry of the forward and backward scatterings of aerosols. In numerical models, these parameters play important roles in calculating the radiative forcing of aerosols.

The three conditions for the numerical calculation are (1) the complex refractive index of BC aerosols in different wave bands, (2) the spectral distribution of the BC aerosols, and (3) the program of Mie scattering theory.

2.1. Complex Refractive Index. The real and imaginary parts of the complex refractive index of aerosol particles represent the characteristics of scattering and absorption, respectively, and the magnitudes of the absolute values determine the strengths of the scattering and absorption properties of aerosols, which have a direct influence on the radiation of aerosols. For a given scale spectrum of particles, the variation in the radiation characteristics of aerosols with the wavelength is determined by its complex refractive index. Figure 1 shows the values of complex refractive index of BC aerosols. These values of the complex refractive index used by Shettie are selected in this paper [16]. This data source has also been used to study the optical properties of BC aerosols by others [17]. The variation of the imaginary part of the complex refractive index of BC aerosols with the wavelength is relatively small, but its absolute value is usually greater than 0.4; therefore, BC aerosols mainly absorb the long radiation.

2.2. Particle Size Distribution. The particle size of the aerosols also has an important effect on the final radiation. In previous studies, the RIEMS2.0 model divided BC aerosols into four bins, which somewhat improves the simulation performance of the model. We use the same division and assign different optical parameters to BC aerosols of every particle size bin according to the Mie scattering theory.

Using the program of the Mie scattering model, we calculate the SSA, DF, and MAC of the BC aerosols in different bands.

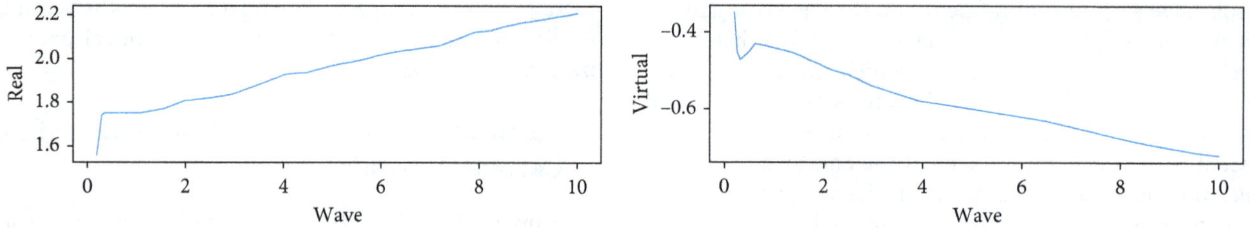

(a) (b)

FIGURE 1: Complex refractive index of BC aerosols in different wavelengths.

Because the radiation parameters of BC aerosols have an important relationship with the particle size distribution, the spectral distribution of the particles is very important to numerical calculations of the optical parameters. Winchester et al. [18] indicated that a lognormal distribution can accurately describe the spectral distribution of particles. Other study also adopted a lognormal distribution to describe the spectral distribution of aerosols [19]. In this chapter, we adopt this distribution for the calculations:

$$\frac{dN(r)}{d\ln r} = \frac{N_0}{\sigma\sqrt{2\pi}}\exp\left(-\frac{\ln\left(r/r_g\right)^2}{2\sigma^2}\right). \tag{1}$$

Based on the division of BC aerosol particle sizes, we consider BC with four bins (i.e., BC1, BC2, BC3, and BC4) with radii of 0.075, 0.205, 0.48, and 2.7 μm, respectively. In Equation (1), the values of r_g and N_0 are from the results of Levy et al. [4]. Based on the conversion between the median radius and the distribution radius of the number concentrations, we obtain the radius distribution of the number concentrations for four modes, r_g:

$$r_g = r_v \exp\left(-3\sigma^2\right). \tag{2}$$

In Equation (1), N_0 represents the particle number density of BC aerosols, and we can calculate different values of N_0 from r_g under the different modes. V_0 is the amplitude of the volume distribution. σ represents the standard deviation of the radius:

$$\begin{aligned} N_0 &= V_0\frac{3}{4\pi r_g^3}\exp\left(-\frac{9}{2}\sigma^2\right),\\ V_0 &= \int_0^\infty \frac{dV}{d\ln r}d\ln r. \end{aligned} \tag{3}$$

2.3. Mie Scattering Model. In using the Mie scattering theory to calculate the solution, we assume that the particles are spheres, and we can obtain the relevant optical characteristics (optical coefficient) based on the radius and complex refractive index of the particles. The Mie scattering is based on the Maxwell equation, and it assumes that the vector wave equation has separable solutions and is derived in the spherical coordinate system. Many studies have focused on numerical algorithms for Mie scattering, and the complete process is described by Liou, who used the Legendre function and Bessel function [20].

Figure 2 shows the numerical calculation results of the optical coefficients of BC in four particle sizes bins. BC1, BC2, BC3, and BC4 represent BC particles with a radius of 0.075, 0.205, 0.48 and 2.7 μm, respectively. BC5 represents the BC optical parameters in the conventional scheme.

As shown in the figure, BC aerosols in different modes show different absorption ability. BC5 in the conventional scheme shows the highest absorption while BC1, BC2, BC3, and BC4 show lower absorption in order.

The optical parameters of the BC aerosols used in the conventional model are taken from database of the University Corporation for Atmospheric Research (UCAR) (http://www.ccsm.ucar.edu/models/atm-cam/download/) and are not differentiated based on the particle size. BC5 represents the conventional settings of the optical parameters. A comparison of the results shows that the calculated optical parameters for the first mode of BC aerosols (particle size of 0.075 μm) are most similar to those of the conventional settings, and the calculated MAC is smaller than that of the conventional settings.

3. Design of the Numerical Scheme

In this paper, we use the RIEMS2.0 model, and the emission source data are taken from the MEIC-2010 emissions inventory, which has a spatial resolution of 0.25° × 0.25°. The time period of the numerical experiments is from January 1, 2010 to December 31, 2010.

We input the optical coefficients of the BC aerosols calculated for the four particle sizes bins into the RIEMS2.0 model and simulate the concentration, direct radiation, and the climate effect of the BC aerosols. In addition, we compare and analyze the results with those obtained by the conventional simulation scheme of BC aerosols with uniform-scale-optical parameters (hereinafter referred to as Scheme 1) [15]. In the following analysis, the scheme used in this paper is called the size-resolved scheme of optical parameters (referred to as Scheme 2), and the results show the differences between Schemes 2 and 1 simulation results. In this paper, we consider the direct radiative forcing effect of BC aerosols and its climate effect and compare them with the previously calculated direct radiative forcing and climate effects. The method used for the regional division is the same as those in Reference [15].

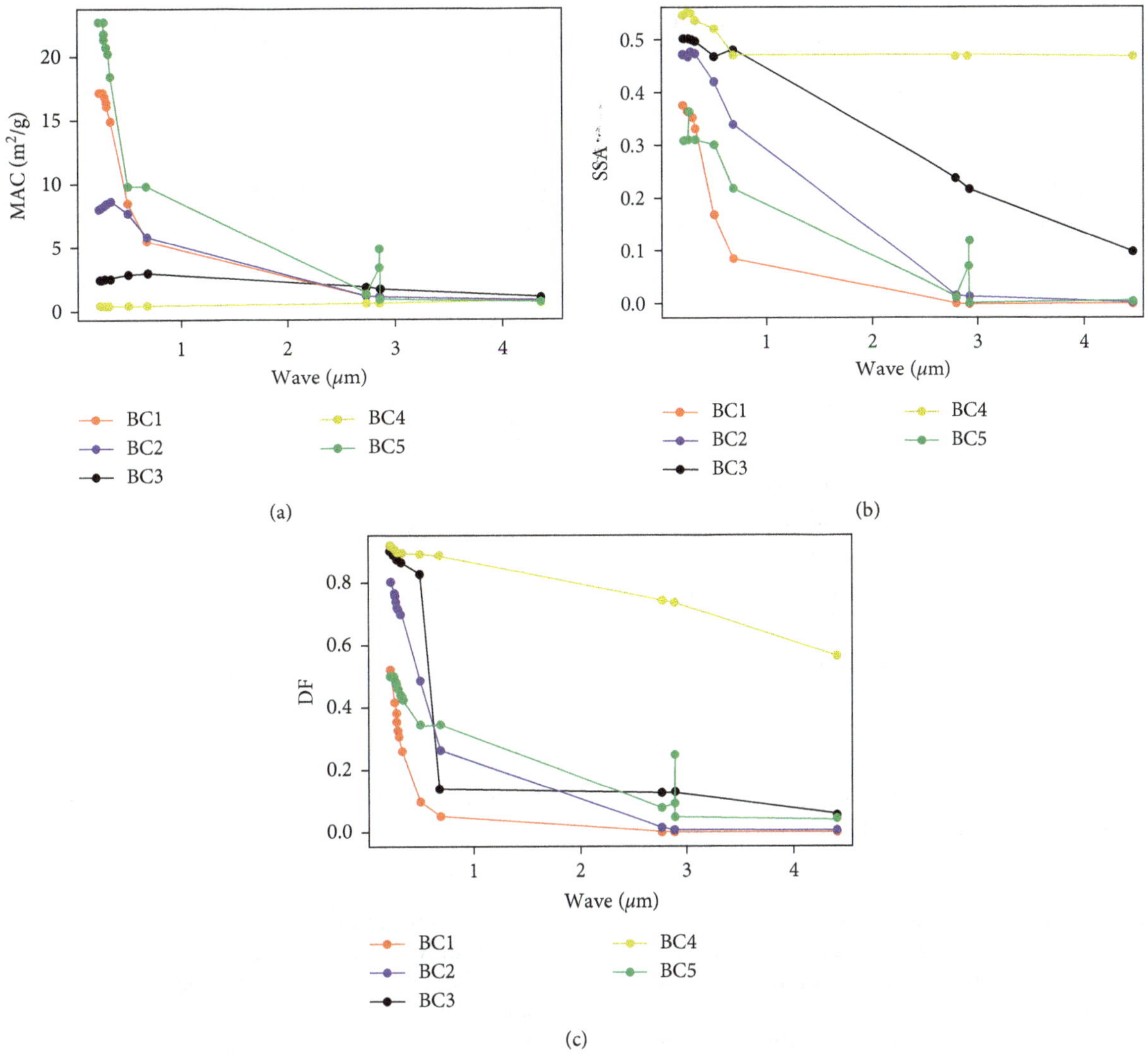

FIGURE 2: Variations of the optical properties of BC aerosols of different bins with the wavelength.

4. Results and Analysis

4.1. Difference between the Two Schemes of Optical Parameters on the Simulation of BC Aerosols. We use Schemes 1 and 2 to simulate the BC aerosols in China and find that there are differences in the concentrations, optical thicknesses, direct radiative forcing, and climate effects of the BC aerosols simulated using the two optical parameterization schemes.

The differences between the ground surface concentrations of BC aerosols simulated by the two schemes of optical parameters (Figure 3) show that the BC aerosol concentration of the first mode simulated by Scheme 2 is significantly higher, and the maximum difference is 3.30 $\mu g/m^3$. The ground surface concentrations of the other modes of BC aerosols are lower under Scheme 2. The total ground surface concentrations of BC aerosols are also lower under Scheme 2, and the range of differences is −0.2 to −2.8 $\mu g/m^3$. The differences in the regional average concentrations of the four modes of BC aerosols and in the total BC aerosol ground surface concentrations are 0.24, −0.16, −0.18, −0.10, and −0.21 $\mu g/m^3$, respectively.

The average variations in the column concentrations of the different modes of BC aerosols are 0.085, −0.095, −0.089, and −0.054 mg/m^2, respectively, and the total column concentrations of BC aerosols decrease by 0.08–0.56 mg/m^2. The distributions of the variations in the column concentrations of BC aerosols are similar to that of the ground surface concentrations (not shown). The differences in the BC aerosol concentrations occur due to the different optical parameters, which has the complicated feedback effect in the aerosol-radiation-meteorological field could generate different meteorological fields, and further affect the transport and diffusion of aerosols.

The optical thicknesses simulated by Scheme 2 are lower than those simulated by Scheme 1 and the differences range from −0.008 to −0.0064 (figure omitted). In the region of high BC aerosol optical thicknesses, the differences between the simulations are relatively large. For example, in the region near the Sichuan Basin and the North China Plain, the difference is approximately −0.0064; in the area with lower optical thicknesses, the differences are smaller. The regional

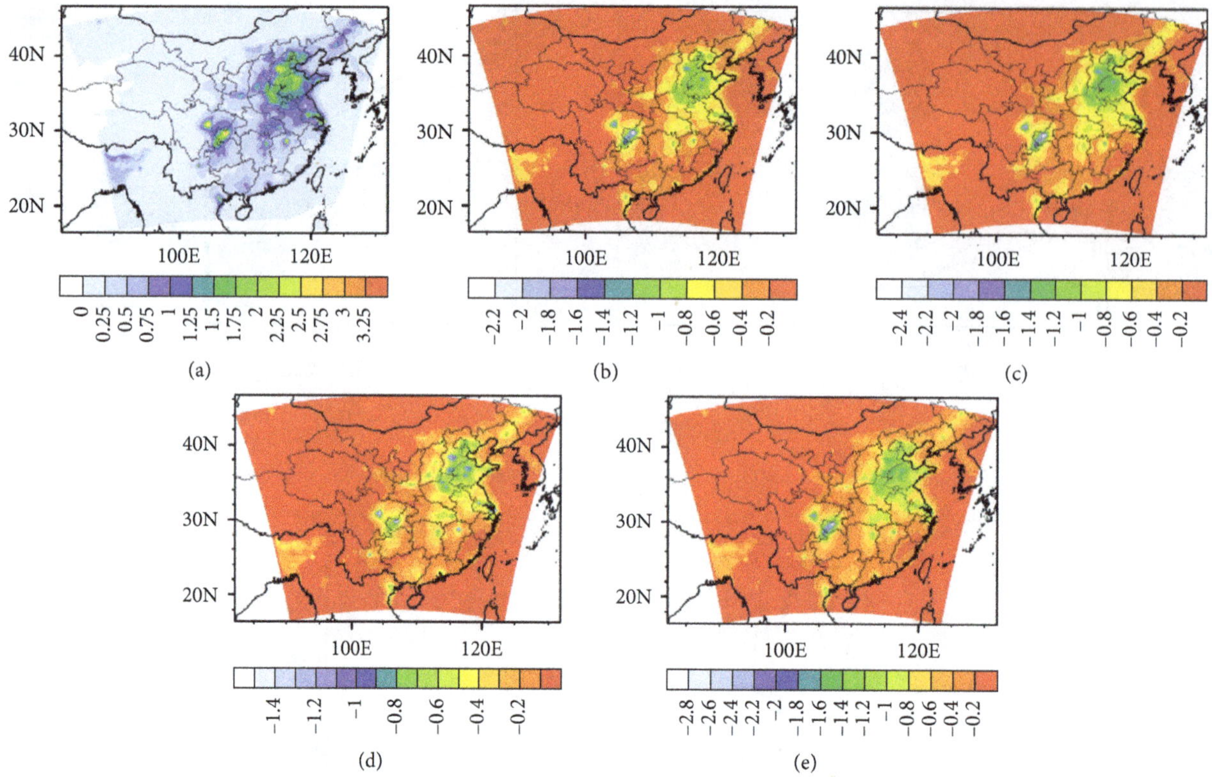

FIGURE 3: Differences between the four modes and the total ground surface concentrations of BC aerosols calculated by the two schemes ($\mu g/m^3$). (a) BC1. (b) BC2. (c) BC3. (d) BC4. (e) BC (total).

average difference in the BC aerosol optical depth (AOD) is −0.0019. The causes of the differences in optical thickness include the differences in the concentrations and the differences in the optical parameters.

Figure 4 shows the distributions of the differences in climate effects of BC aerosols simulated with two optical parameterization schemes. The direct radiative forcings under all-sky conditions simulated by Scheme 2 are higher than those from Scheme 1 with the range of 0.1–0.7 W/m^2 in most regions of central and eastern China but are slightly lower near the Sichuan Basin. The differences in clear-sky top-of-atmosphere radiative forcing (TRF) are negative in most areas. In the North China Plain and South China, the TRF changes by −0.2 to −0.8 W/m^2; near the Sichuan Basin, there is a region of strongly negative variations of TRF with a maximum value of −0.94 W/m^2, which is consistent with the variations in the BC aerosol optical depth. Over south of the Sichuan Basin, the clear-sky TRF of BC is also intensified, but the range of enhancement is smaller than that of the all-sky TRF, which is approximately 0.2–0.4 W/m^2. The regional average clear-sky TRF of BC aerosols is 0.034 W/m^2 lower, whereas the all-sky TRF is 0.06 W/m^2 higher. In general, in comparison with the change of the all-sky TRF, the variation in the clear-sky direct TRF of BC aerosols is more similar to the distributions of the variations in the concentration. Using the size-resolved optical parameters, the total concentration of BC aerosols, the simulated AOD, and TRF are lower. Because the calculation of the all-sky TRF considers the complicated influence of clouds, there are several differences between the changes in the all-sky TRF of BC aerosols and the changes in the other variables. Some research results also show that the effects of BC on clouds vary between regions and present much uncertainty [21].

The use of the principle of Mie scattering to calculate the size-resolved optical parameters causes a change in the direct climate effect of BC aerosols. The ground surface temperatures simulated by the two schemes increase in many regions, and there are several similarities. However, the extents of the warming regions simulated by Scheme 2 are smaller, and the magnitudes of the warming are lower in some regions, which is consistent with the weakening of the clear-sky TRF of BC in these regions. In Northeast China, the warming effect of BC is enhanced by 0.04 to 0.16 K. Figure 4(b) shows that this region also has significant increase in the clear-sky TRF of BC aerosols. In addition, in the Beijing-Tianjin-Hebei area and the Shandong Peninsula, the warming effect of BC is intensified by 0 to 0.16 K. BC simulated by the two schemes mostly have cooling effects in the arid and semiarid areas of western China. The cooling effects simulated using Scheme 2 are greater in some regions, and the cooling ranges from −0.04 to −0.24 K. The simulation of BC using Scheme 2 results in increased precipitation in South China, and the variations range from 0.4 to 2.4 mm/d. In some regions in the middle and lower reaches of the Yangtze River Basin, BC aerosols cause greater reductions in

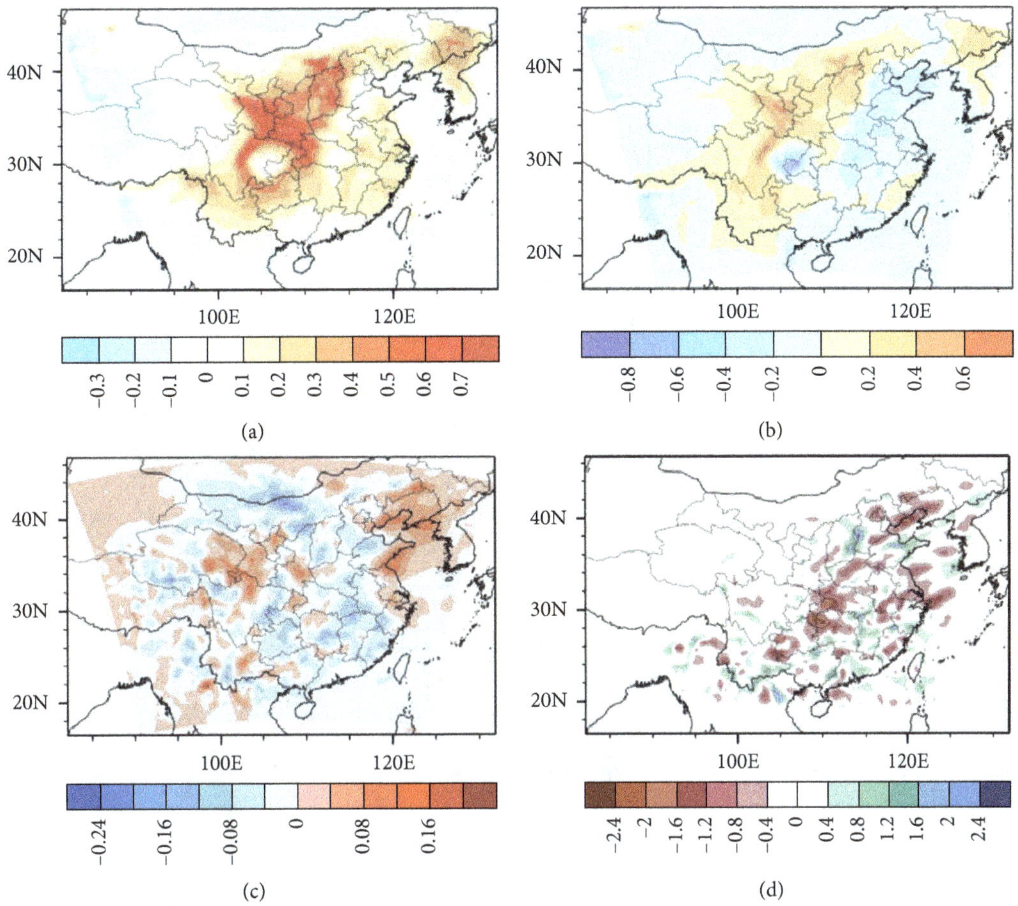

FIGURE 4: Differences between the climate effects of BC aerosols simulated by the two parameterization schemes. (a) All-sky top-of-atmosphere radiative forcing (TRF; W/m^2). (b) Clear-sky TRF (W/m^2). (c) Temperature (K). (d) Precipitation (mm/d).

precipitation, and the variations of precipitation range from −0.4 to −2.4 mm/d. In Northeast China and some regions in North China, BC aerosols simulated by Scheme 2 reduce the precipitation, whereas the results of Scheme 1 are the opposite. The climate effects of aerosols show much uncertainty due to many factors and so far the mechanism has not been formed [22]. Compared to the effects on temperature, the effects of BC on precipitation are more unclear by the research [23]. In general, BC aerosols show a complicated influence on precipitation, and the sensitivity of optical parameterization schemes makes it more difficult to analyze, which merits further studies.

4.2. The Climate Effect of BC Aerosols with the Size-Resolved Scheme of Optical Parameters. The results described above indicate that the radiative forcing and climate effect of BC aerosols are relatively sensitive to the optical parameters. If the dependence of the BC aerosols' optical parameters on the particle size is not considered in simulations of the climate effect of BC aerosols, errors may occur. In this section, we analyze the distribution of BC aerosol concentrations and its climate effect for the distribution scheme that considers the dependence of the aerosols' optical parameters on the particle size (Scheme 2).

4.2.1. Concentration Distribution of BC Aerosols. Figure 5 shows the spatial distributions of the annual average ground surface concentrations for the four modes and the total of BC aerosols simulated using Scheme 2. Figures 5(a)–5(d) show the results for the modes of BC aerosols with sizes of 0.075, 0.205, 0.48, and 2.7 μm, respectively, and Figure 5(e) shows the results for the total BC aerosols. The spatial distributions of BC aerosols with different particle sizes are similar, and the ground surface concentrations decrease from east to west, which is generally consistent with the emissions distribution of BC aerosols. In the North China Plain, Yangtze River Delta, and Sichuan Basin, which have dense industries and relatively high population densities, the BC aerosol concentrations are relatively high. Although the BC aerosol distributions of the four modes are similar, there are differences between the modes. The regional average ground surface concentrations of the four modes of BC aerosols are 0.372, 0.265, 0.055, and 0.004 μg/m^3, respectively. Before we adjust the optical properties of BC aerosols based on the particle size, the simulated proportions of the four modes are essentially consistent with their initial emissions proportions, whereas after recalculating the optical parameters according to Mie

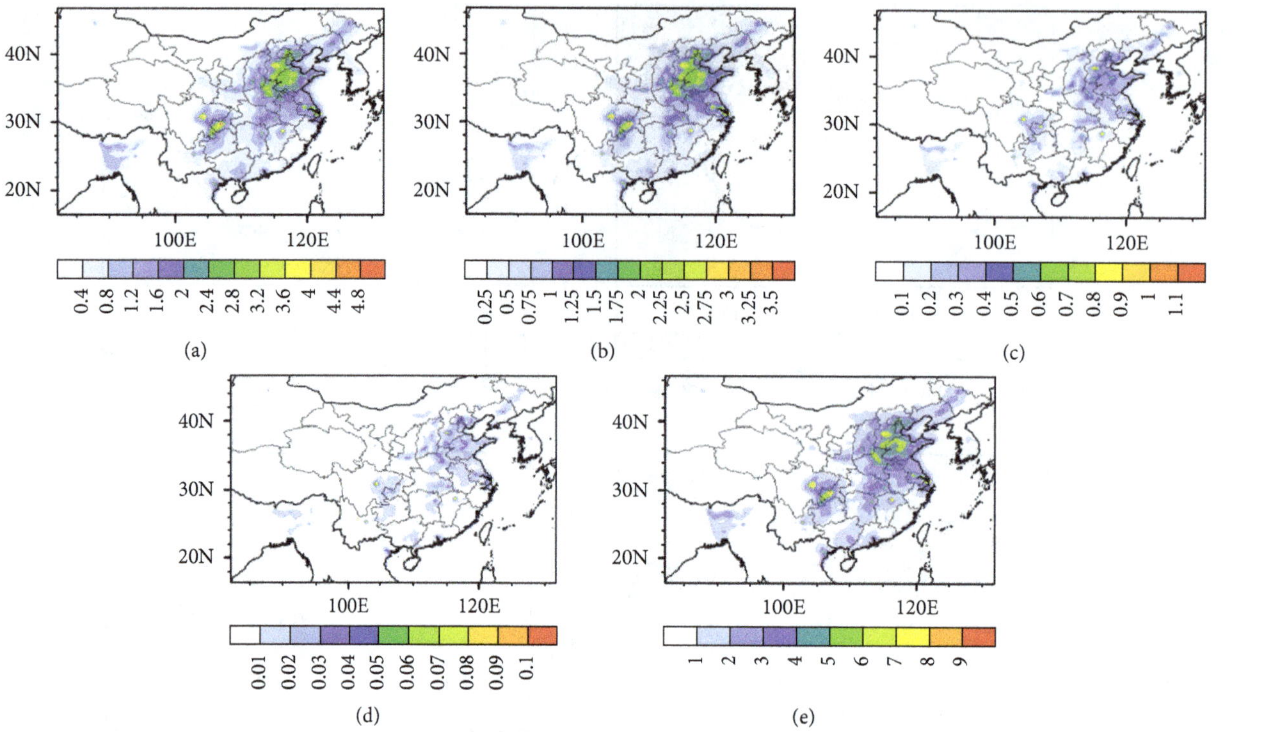

FIGURE 5: Distributions of the different modes and the total ground surface concentrations of BC aerosols ($\mu g/m^3$). (a) BC1. (b) BC2. (c) BC3. (d) BC4. (e) BC (total).

scattering, there are obvious differences in the proportions of the four modes. The proportions of the first and second modes, which have relatively small aerosol radii, are 53% and 38%, respectively, and the proportions of the third and fourth modes are very small, especially the fourth mode. The range of the total ground surface concentrations of BC aerosols in North China and East China is 1–6 $\mu g/m^3$ (Figure 5(e)). In the lower reaches of the Yellow River and the Yangtze River Basin, as well as the Sichuan Basin, there are regions of high BC aerosol ground surface concentrations, and the maximum concentration reaches 9.83 $\mu g/m^3$. The regional average total ground surface concentration of BC aerosols is 0.70 $\mu g/m^3$. The distribution of column concentrations of BC aerosols is similar to the distribution of ground surface concentrations (not shown). The regional average column concentration of the first mode of BC aerosols is the highest (0.15 mg/m^2) and that of the second mode is slightly lower (0.10 mg/m^2). The regional average total BC aerosol column concentration is 0.28 mg/m^2, and the maximum is 2.05 mg/m^2. In most other regions except for the arid and semiarid areas in western China, the range of BC aerosol column concentration is 0.20–2.0 mg/m^2.

The distributions of BC aerosol ground surface concentrations in this study are similar to the results of previous studies, but the concentrations in this study are smaller, which could be caused by the size-resolved scheme [24, 25]. Because many previous studies adopted a particle size distribution scheme with a constant scale and simply assumed the radius of BC aerosols to be approximately 0.1 μm [26–28]. They did not accurately consider the presence of BC aerosols with larger particle sizes, which caused the simulated BC aerosol concentrations to be higher than they actually are.

The wind field at the 850 hPa height during the spring (figure omitted) shows that the eastern part of the Indian Peninsula is mainly affected by westerly airflow, and the BC aerosols accumulate. In China, regions of high concentrations of BC aerosols are located in the Sichuan Basin and North China Plain. There are many mountains and hills in the Sichuan Basin, and the topography makes it difficult for pollutants to diffuse. The westerly airflow also causes the accumulation of BC aerosols in this region, which causes the concentration of BC aerosols to be very high. The high-concentration region in the North China Plain is mainly caused by industrial and anthropogenic emissions. In the summer, southern China and the Indian Peninsula are both affected by southwesterly airflows, and long narrow areas of high BC aerosol concentrations form in the Sichuan Basin and eastern China. Compared with the spring, the climate characteristics of northward and eastward transportation are obvious in southern China, and the column concentrations of BC aerosols increase in eastern China. Northwesterly airflows prevail in most regions north of the Yangtze River; therefore, the high-concentration area of BC aerosols in the North China Plain weakens. In the fall, the distribution of column BC aerosol concentrations is similar to those in the other seasons, but the concentrations are lower; southerly winds prevail in most of central and eastern China. In the winter, northwesterly winds prevail to the north of the Yangtze River, and easterly winds prevail to the south of the Yangtze River. The concentrations of BC aerosols in North China are higher than that in the fall.

To validate the effect of models on the simulation of BC aerosol concentrations, we compare the simulated BC aerosol concentrations with observations. The observed values in different regions are obtained from several studies [29–36]. The simulated and observed annual average concentrations of BC aerosols in different regions of China and the linear relationship between the data have been studied (figure omitted). The straight black line is the fit between the observed values and simulated values, and the different symbols represent different regions.

The results show that the model can simulate the ground surface concentrations of BC aerosols relatively well, and the correlation coefficient with the observations is 0.78. The linear relationship is given by the equation $y = 0.78x + 0.29$. In general, the simulated values are lower than the observed values, which is likely because the emission source we used does not consider straw combustion. The simulated BC aerosol concentrations of Zhuang et al. [37] are also lower than the observed values.

Figure 6 compares the monthly variations in the simulated BC aerosol ground surface concentrations with the observed values. The black dotted lines represent the simulated BC aerosol ground surface concentrations, the red dots represent the observations, and the red vertical line segments represent the standard deviations. The observation data are from the Atmospheric Composition Observation Network of the China Meteorological Administration and mainly include the observed concentrations of BC aerosols averaged over two years (2006 and 2007) at fourteen stations. In the areas of Longfengshan, Jinsha, Taiyangshan, Nanning, and Chengdu, the simulation results and observations are relatively consistent. In the winter, in Jinsha and Zhengzhou stations, the model-simulated BC aerosol concentrations are lower than the observations. At most stations, the observations are low in the summer and high in the winter. Except for the stations in Gucheng, Dalian, and Gaolanshan, the model simulates the monthly variations in the BC aerosol concentrations relatively well. In Xi'an, Dunhuang, and Lhasa, the simulated concentrations of BC aerosols are significantly lower than the observations, which is likely because the anthropogenic emission sources in western areas are difficult to estimate. This problem was also found in a similar study [25].

4.2.2. Optical Thickness Distribution and Radiative Forcing of BC Aerosols. Similar to the distribution of column concentrations, the regions of high BC aerosol optical thicknesses are located in North China and the Sichuan Basin (figure omitted). In most areas of North China and East China, the BC aerosol optical thicknesses range from 0.003 to 0.006. In the Sichuan Basin, the middle and lower reaches of the Yellow River and the Yangtze River Basin, and the Yangtze River Delta, the optical thicknesses of the BC aerosols are greater. In these regions, the industry is more well-developed, economic development has occurred more rapidly, and the population density is higher, which leading to higher emissions and therefore inducing greater BC aerosol concentrations. The regional average BC aerosol optical thickness is 0.0013, and the maximum is 0.0073. The model can simulate the spatial distribution of the BC aerosol optical thicknesses relatively well, and the distribution is similar to the results of previous studies [38, 39]. However, the simulated optical thicknesses of BC aerosols in this study are slightly lower than those in some early studies [40].

Figure 7 shows the distributions of the all-sky TRF and clear-sky TRF of BC aerosols. The distributions of the TRF are similar to the distributions of the BC aerosol emissions and concentrations. However, the radiative forcing is affected by the aerosol mass as well as by the vertical structure of the aerosols, optical properties, ground surface albedo, and clouds [41]. Therefore, there is a difference between the distribution of the BC aerosol radiative forcing and the concentrations. The all-sky TRFs of BC aerosols have a range of 0–2.0 W/m^2. The high values are located near the Sichuan Basin and in Northeast China, and the maximum TRF is 2.11 W/m^2. In the region between the Yangtze River and the Yellow River Basin, the all-sky TRFs of BC aerosols are relatively strong (1.4–1.8 W/m^2). The TRFs of BC aerosols under the clear-sky conditions are lower (0–1.8 W/m^2); the high values are located north of the Yellow River, and the maximum value is 1.98 W/m^2, which is similar to the distribution of the BC aerosol optical thicknesses. In comparison with the all-sky TRFs, the distribution of the clear-sky TRFs is more similar to the distribution of the optical thicknesses. This is because the all-sky TRF is also affected by the mass and cloud water content of the aerosols, so it is more complicated. The regional average clear-sky and all-sky TRFs of BC aerosols are 0.49 and 0.36 W/m^2, respectively, which are similar to the global average BC aerosol TRF (0.4 W/m^2) in the IPCC report [3]. In the same class of research areas, the regional average of the BC aerosol direct TRF estimated by Chang and Park [42] is 0.5 W/m^2, and the BC aerosol TRFs calculated by Wu et al. [43] range from 0.5 to 4.1 W/m^2. The regional average TRF of BC aerosols estimated by Li et al. [25] is 1.22 W/m^2, and the maximum is 5-6 W/m^2, whereas the average and maximum BC aerosol TRFs simulated by Zhuang et al. in 2010 [37] are 0.75 and 5.5 W/m^2, respectively. In Chang et al. [44], the regional average TRF of BC aerosols is 0.58 W/m^2. The results of this study are lower than those in these previous studies. This is because the size-resolved scheme assumes that a proportion of the BC aerosols is larger and has weaker optical properties. In the numerical model, the radiation effect of BC aerosols is weakened by a feedback effect between the concentration and other physical quantities. If we do not consider the particle size distribution of BC aerosols and their optical parameters, the climate effect of BC aerosols will likely be overestimated.

4.2.4. Influence on Temperature and Precipitation. BC aerosols change the energy equilibrium in the atmosphere through the absorption of solar radiation and cause changes in the dynamic-thermodynamic circulation. Figure 8 shows the effects of BC aerosols on the annual

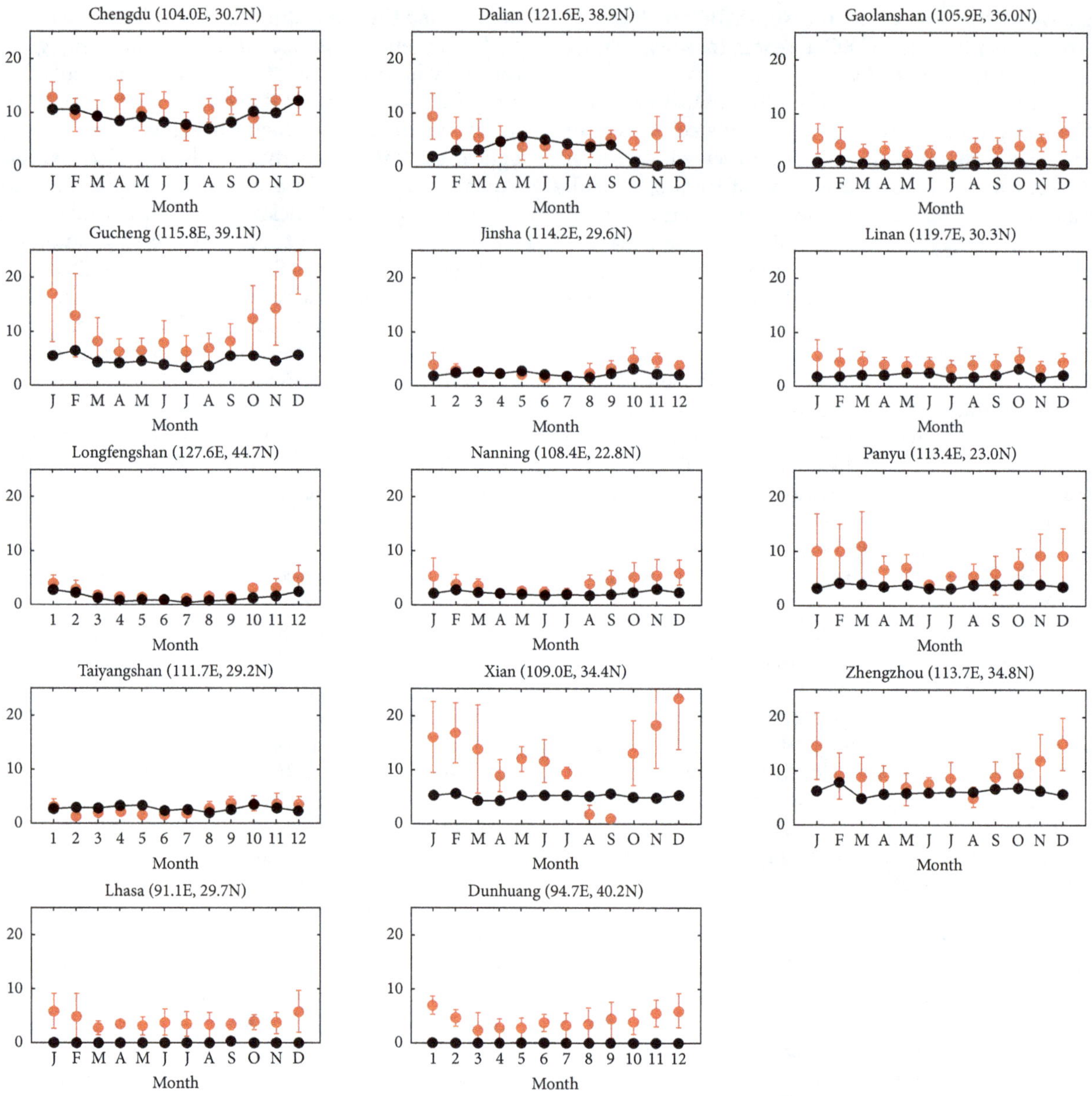

FIGURE 6: Comparison of the observed and simulated monthly changes in the BC aerosol concentrations ($\mu g/m^3$) in different regions of China. The black dotted lines represent the simulated BC aerosol ground surface concentrations, the red dots represent the observations, and the red vertical line segments represent the standard deviations.

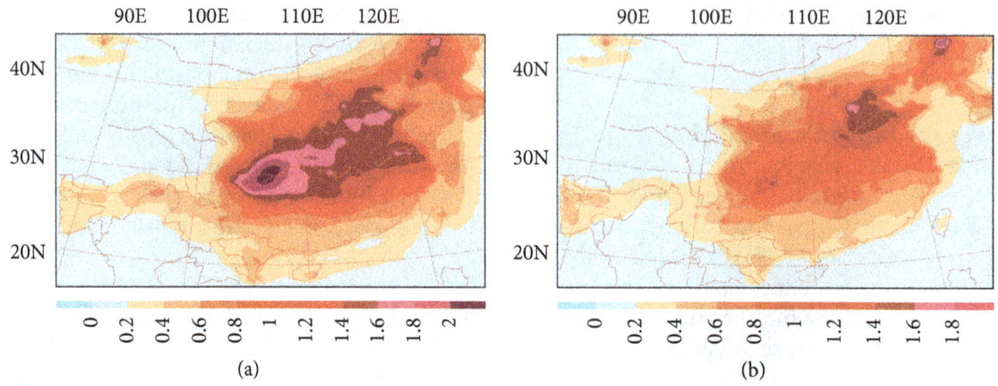

FIGURE 7: Top-of-atmosphere radiative forcing (TRF; W/m^2). (a) All-sky TRF. (b) Clear-sky TRF.

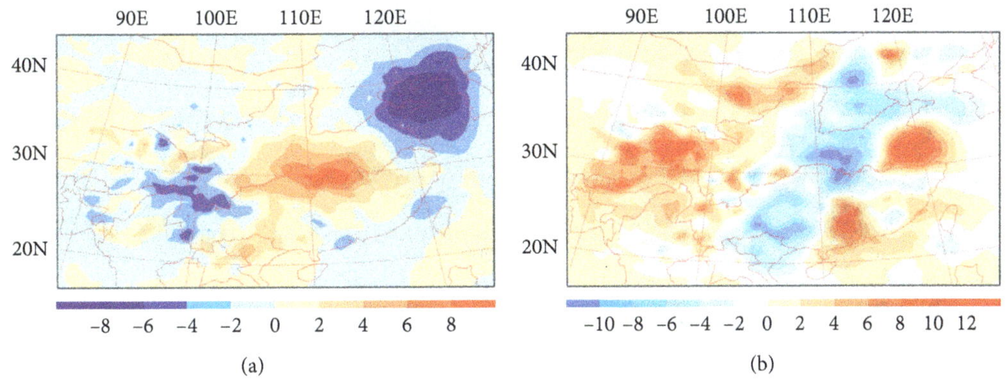

FIGURE 8: Sea surface pressure (Pa) changes caused by BC aerosols. (a) Summer. (b) Winter.

average ground surface temperature and annual average precipitation. In most regions of East China, North China, and central China, the BC aerosols cause increases in the ground surface temperature, and the changes range from 0.05 to 0.15 K. In particular, the warming is significant in some regions between the Yangtze River and Yellow River Basin and near the Sichuan Basin, and the maximum warming is 0.17 K. These regions also have high values of the BC aerosol radiative forcing. Therefore, there is the relatively good relationship between the influence of BC aerosols on the ground surface temperature and their radiative forcing. In parts of South China and Northwest China, BC aerosols cause cooling, varying from 0.05 to 0.15 K. At a regional scale, BC aerosols cause ground surface warming in North China, East China, and central China with average temperature changes of 0.034 K, 0.037 K, and 0.020 K, respectively, and has a cooling effect in South China. Because the emissions of BC aerosols are concentrated in central and eastern China, the radiative forcing of BC aerosols in the western area is relatively weak. However, Figure 9(a) shows that BC aerosols cause changes in the ground surface temperature in many regions of western China. This is because the large-scale transfer process of BC aerosols can change the properties of clouds in relatively distant regions and thus affect the climate, and it can also change the climate at short distances through its influences on the circulation, water vapor transportation, and cloud distribution [45]. It is worth noting that the distribution of strong BC aerosol radiative forcing is roughly consistent with the high value region of ground surface warming caused by BC aerosols. However, there are also several regions in which strong BC aerosol radiation causes decreases in the ground surface temperature. This phenomenon is possibly due to the influences of complicated regional cloud systems and precipitation, which also reflect the complicated BC aerosol-cloud feedback mechanism [40]. In several previous studies, the ranges of temperature changes caused by BC aerosols in East China and the eastern region of East Asia are −0.6 to 0.3 K and −0.4 to 0.1 K, respectively, which are greater than the climate effects of BC aerosols simulated in this study [46, 47]. Considering the particle size distribution of the optical parameters of BC aerosols causes the climate effect of BC aerosols to weaken.

The change in precipitation caused by BC aerosols is more complicated and has a greater regional uncertainty (Figure 9(b)). In parts to the south of the Yangtze River, the southeast coastal area, and north of the Yellow River, the BC aerosols increase the precipitation, varying from 0.5 to 2.5 mm/d. However, in some regions to the north of the Yangtze River and central China, the BC aerosols cause decreases in precipitation, and the range of the decreases is 0.5–3.0 mm/d. At a regional scale (Table 1), the BC aerosols cause decreases in precipitation in central China, North China, South China, and Northeast China, and the variations are −0.83, −0.05, −0.11, and 0.13 mm/d, respectively. In central China, the BC aerosols cause the precipitation to increase by 0.01 mm/d; for the entire simulation domain, the average decrease in precipitation caused by BC aerosols is 0.02 mm/d. BC aerosol emissions and precipitation in China are correlated over the past several years [7]. BC aerosols can cause increases in precipitation in southern China and decreases in precipitation in northern China. This study found that the precipitation in the northern regions decreases. However, the regional average precipitation in the southern regions decreases, which is likely associated with the differences between the models and different regional divisions. Due to the complexity of the influences of BC aerosols and precipitation in the climate system, there is still significant uncertainty in the relationship between BC aerosols and precipitation. Many studies have analyzed the influence of BC aerosols on precipitation in China. Several studies have indicated that BC aerosols cause decreases in precipitation in eastern China with a regional average reduction in precipitation of 0.09 mm/d [48], whereas other studies have indicated that BC aerosols can increase the precipitation and that the precipitation increased by 0.07 mm/d from 1950 to 2000 [44].

The influences of BC aerosols on the temperatures at different heights have been researched (figure omitted). At the 500 hPa height, BC aerosols cause warming in most regions of China. The range of warming is 0.01 to 0.05 K, which is significant in the region near the middle and lower reaches of the Yangtze River. On the other hand, in parts of Northeast China, BC aerosols cause cooling. The simulation at the height of 850 hPa indicates that BC aerosols

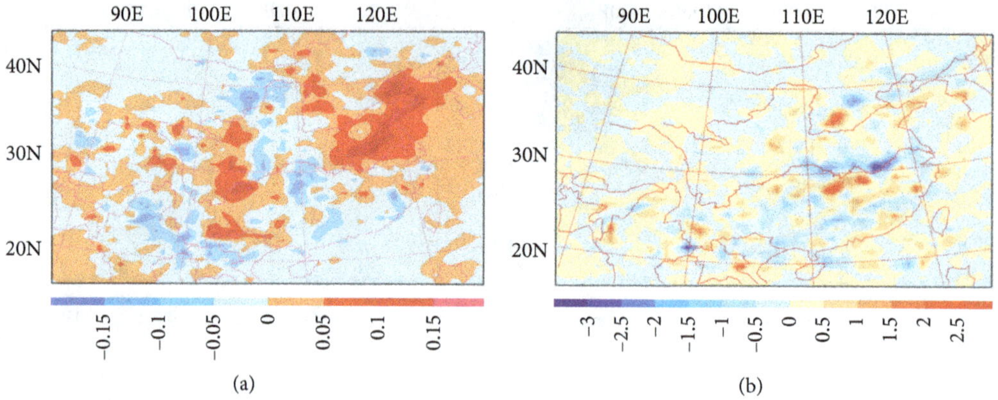

FIGURE 9: Annual average climate effects of BC aerosols. (a) Temperature (K). (b) Precipitation (mm/d).

TABLE 1: Regional average direct radiative forcing of BC aerosols and the climate effects.

	All-sky radiation at the top of atmosphere (W/m^2)	Clear-sky radiation at the top of atmosphere (W/m^2)	Temperature (K)	Precipitation (mm/d)
South China	0.70 (1.57)	0.47 (1.16)	−0.017 (−0.15)	−0.11 (−2.02)
Northeast China	0.90 (1.82)	0.79 (1.98)	0.002 (0.05)	−0.13 (−1.04)
North China	1.06 (1.73)	0.94 (1.85)	0.034 (0.12)	−0.05 (−2.19)
East China	1.36 (1.75)	0.94 (1.54)	0.037 (0.14)	−0.83 (−2.99)
Central China	1.01 (2.11)	0.75 (1.53)	0.020 (0.17)	0.01 (2.54)

Note: the values in the parentheses are the extreme values of the physical quantities.

cause increased temperatures near the Sichuan Basin and in the North China Plain as well as decreases in some regions of western China. A comparison of the influences of BC aerosols on the temperatures at these two heights shows that in the north of the Yellow River and near the Sichuan Basin, the warming caused by BC aerosols in the lower troposphere is more significant than that in the middle and upper troposphere. Therefore, the atmospheric instability in these regions increases, which may cause convective precipitation. Figure 8(b) shows that the precipitation in these regions also increases. A comparison of Figures 10(a) and 8(a) shows that at the 850 hPa height and at the ground surface, the distributions of the variations in the annual average temperatures caused by BC aerosols are similar.

4.2.5. Influences on the Circulation, Pressure, Temperature, and Precipitation in the Winter and Summer. The absorption of solar radiation by BC aerosols causes heating of the middle and lower atmosphere, and the airflow rises. The research indicates that BC aerosols can affect the wind changes in China [49]. Figure 11 shows the variations of the 850 hPa heights and the wind fields during the summer and winter. At the 850 hPa height, the decrease (increase) of the 850 hPa height field during the winter and summer caused by BC aerosols corresponds to a convergence (divergence) of the wind field. In the summer, the eastern region (near 110°E, 40°N) and central region (near 110°E, 30°N) of China correspond to low values and high values of height change, respectively. This region is located near the region of high BC aerosol emissions. In the winter, there are two dispersed regions of large height changes in central and eastern China. Correspondingly, BC aerosols cause the enhancement of the southwesterly airflow in the mid-latitude region during the summer and the intensification of the southerly airflow in the high-latitude region. The change of the wind field is the opposite during the winter, and the strength in the winter is stronger than that in the summer; it causes a clear divergence of the airflow in the region north of the Yangtze River and enhances the northerly wind during the winter. These changes in wind and height field caused by BC are similar to the results of other studies [37, 49].

Figures 12 and 8 show the influences of BC aerosols on the ground surface temperature and pressure field during the summer and winter. BC aerosols may cause the temperature increase or decrease between different areas [40]. Seen in the figure, in summer, except for some regions north of the Yellow River and a few areas along the eastern coast, the BC aerosols generally cause the ground surface temperature to decrease; at the 500 hPa height, however, the temperatures in most regions increase. It is assumed that the cooling effect caused by BC aerosols in South China and East China during the summer is induced by the complicated feedback and changes in the amount of clouds in the model, which merits further study [2, 50–52]. The region where the ground surface temperature decreases corresponds to an increase of the sea surface pressure. In most regions of South China and East China, the BC aerosols cause increases in the sea surface pressure, and the area of maximum pressure increase is located near the Sichuan Basin. This is because the cooling of the land over a large area causes the accumulation of the air mass, which results in increasing sea surface pressures

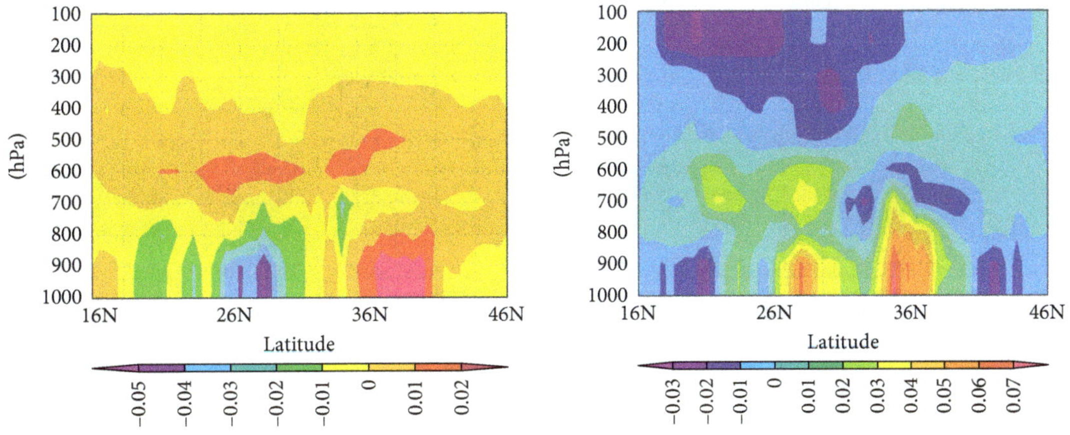

FIGURE 10: Zonal average temperature changes (K) due to BC aerosols. (a) Summer. (b) Winter.

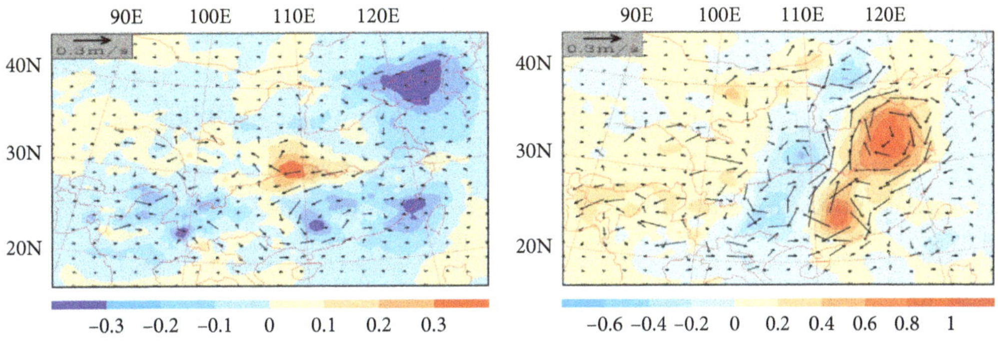

FIGURE 11: Changes of the average height field and wind field (m/s) at 850 hPa. (a) Summer. (b) Winter.

in most regions of central and southern China. The change of the ground surface temperature caused by BC aerosols is different in the winter than in the summer, and the temperatures increase in most regions of central and eastern China. In the region between the Yangtze River and the Yellow River Basin, the increases in ground surface temperature are significant, and the maximum warming is 0.42 K. In the Yangtze River Delta area, the ground surface temperatures decrease. The temperature decrease in winter can also be seen in Figure 10. In general, the changes of ground surface temperature corresponds relatively well to the changes of sea surface pressure in the winter (an increase in the ground surface temperature corresponds to a decrease in the sea surface pressure), which is coincident with other studies [37]. In the summer, the BC aerosols increase the temperature in some coastal regions in eastern China and intensify the difference in the thermodynamic properties of both the sea and mainland. In the winter, however, warming occurs in some coastal regions in eastern China, and the difference in the thermodynamic properties of the sea and mainland decreases. In general, the change in temperature at the 500 hPa height is smaller than that at the ground surface.

Figure 10 shows the vertical distribution of the zonal average temperature changes caused by BC aerosols during the summer and winter. During the summer, in the region from 21°N to 31°N, the BC aerosols cause cooling of the boundary layer, and the maximum temperature decrease is −0.05 K. Also as seen in Figure 12(a), the surface temperature decreases due to BC in summer. In the region from 33°N to 41°N, the warming caused by the BC aerosols reaches approximately the 300 hPa height, and the maximum warming is 0.02 K. At heights from 700 hPa to 300 hPa, the BC aerosols generally cause the temperatures to increase. At heights above 200 hPa, the BC aerosols cause weak cooling. In the winter, BC aerosols cause temperature increases in most regions at heights below 500 hPa, and the temperature changes in the middle and high atmosphere are relatively weak. This is coincident with the temperature changes shown in Figure 12. Warming centers are located near 26°N and 36°N, and the maximum warming exceeds 0.07 K.

Figure 13 shows the influence of BC aerosols on the precipitation in China during the summer and winter. Not the same as the temperature changes, the precipitation changes caused by BC show more uncertainty which make the distribution more complicated [21]. During the summer, the BC aerosols cause increases of precipitation in southern China and some regions north of the Yellow River. On the other hand, in the Yangtze River Basin and some regions of

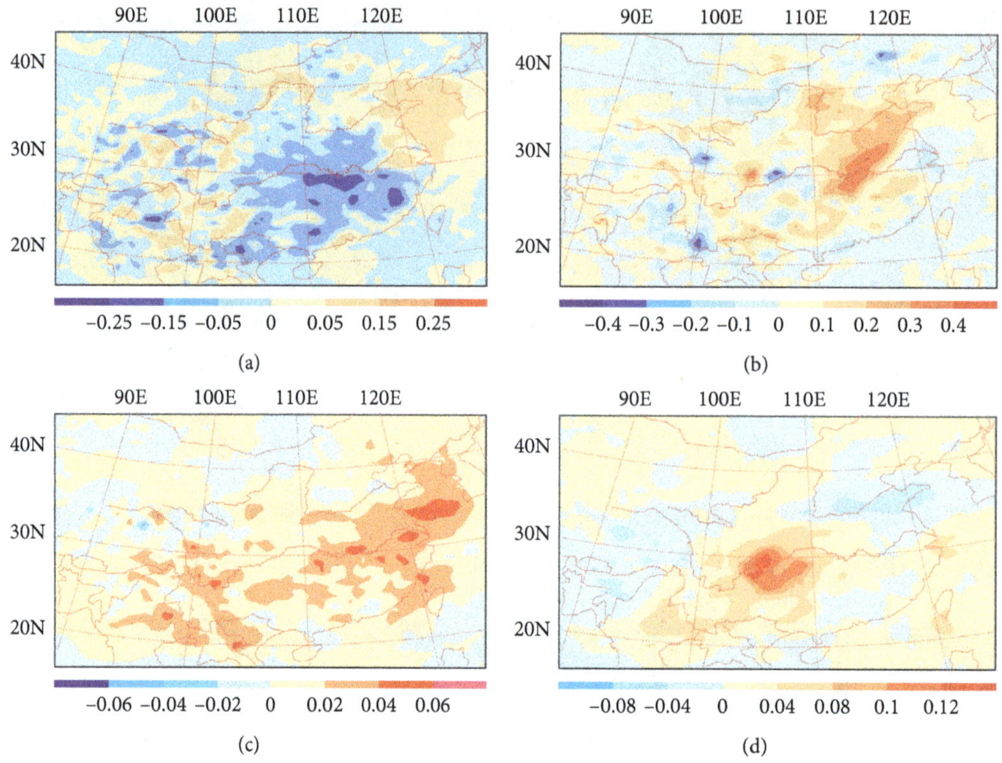

FIGURE 12: Changes of ground surface temperature (K) caused by BC aerosols. (a) Ground surface in the summer. (b) Ground surface in the winter. (c) 500 hPa height in the summer. (d) 500 hPa height in the winter.

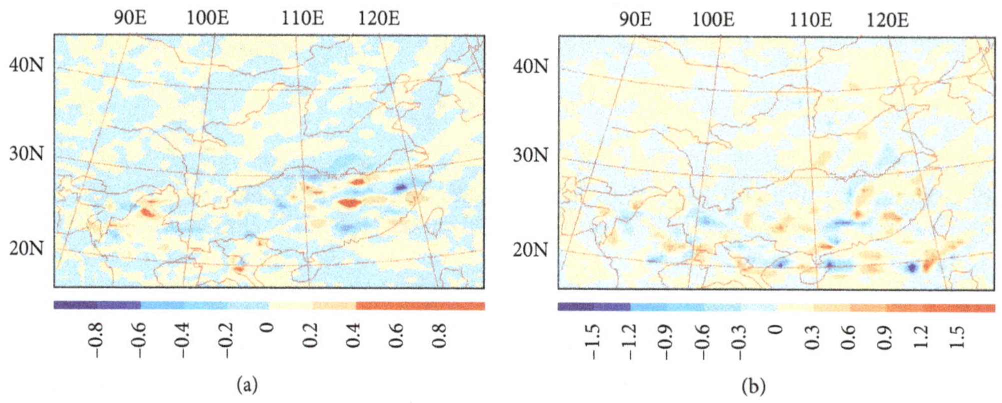

FIGURE 13: Distributions of the changes of precipitation (mm/d) caused by BC aerosols in China. (a) Summer. (b) Winter.

Northeast China, the precipitation decreases, which is similar to the simulation results of Zhuang et al. [37]. In the winter, the BC aerosols cause increases in precipitation in most areas of central and eastern China; however, in some regions near the Yangtze River Basin, the precipitation decreases.

4.2.6. Average Direct Radiation and the Climate Effect of Black Carbon in Different Regions of China. The climate effect of BC aerosols has significant regional uncertainty. Based on the regional division shown in Chapter 4, Table 1 lists the average direct radiative forcing and climate effects of BC aerosols in five typical regions. The statistical data indicate that the BC aerosols are relatively concentrated in East China and North China, and the all-sky TRF values of BC aerosols are high in East China and North China with regional averages of 1.36 and 1.06 W/m^2, respectively. The regional average clear-sky TRF values of BC aerosols are the same in East China and North China (0.94 W/m^2). The BC aerosols directly cause increases in the ground surface temperatures in East China, North China, and central China, and the average increases are 0.037 K, 0.034 K, and 0.02 K, respectively. In South China, however, the BC aerosols have a cooling effect, and the average temperature decrease is −0.017 K. At the regional scale, BC aerosols increase the ground surface temperatures over the entire domain region. The factors that control the precipitation are very complicated, such as circulation, radiation, and water vapor. Therefore, the distribution of the influence of BC aerosols on the precipitation is random and irregular. At the regional scale, BC aerosols decrease the precipitation in South China, Northeast China, North China, and East China by 0.11, 0.13, 0.05, and 0.83 mm/d, respectively. In central China, BC aerosols increase the precipitation by 0.01 mm/d. Over the entire domain, the regional average reduction of precipitation caused by BC aerosols is 0.02 mm/d.

5. Summary

In numerical simulations, the optical parameters of aerosols can directly affect the radiation properties. The optical parameters of BC aerosols are related to the particle size and the complex refractive index. Therefore, based on the size-resolved scheme used in previous studies, we use the principle of Mie scattering to calculate the optical parameters of BC aerosols of different scales and apply the new optical parameters to simulate the concentration distribution, radiative forcing, and climate effects of BC aerosols. We also compare the results with simulation results that use the conventional scheme and analyze the effects of the optical parameters on the simulation of the climate effects of BC aerosols.

In comparison with the conventional homogeneous parameterization scheme, the concentrations of the first mode of BC aerosols simulated with the optical parameters recalculated according to the particle size were higher, and the maximum increase was 3.30 $\mu g/m^3$. The concentrations of the other modes of BC aerosols were lower, the total BC aerosol concentration was also lower, and the range of the reductions was −0.2 to −2.8 $\mu g/m^3$. The regional average changes of the BC aerosol column concentrations for the different modes were 0.085, −0.095, −0.089, and −0.054 mg/m^2. The BC aerosol optical thicknesses varied by −0.008 to −0.0064.

In general, the changes in the clear-sky direct radiative forcing of BC aerosols are similar to the distribution of the change in its optical thickness and concentration changes. The regional average clear-sky radiative forcing of BC aerosols decreases by 0.034 W/m^2. The distributions of the influence of BC aerosols on the ground surface temperatures simulated with the two optical parameterization schemes are similar in many regions, but the warming region and the magnitude of warming simulated with the new scheme are smaller. However, the simulated cooling effect is enhanced in some regions, and the variation of the cooling is from −0.04 to −0.24 K.

The simulation of BC aerosols using Scheme 2 enhances the precipitation in South China even more, and the variation is from 0.4 to 2.4 mm/d. In some regions in the middle and lower reaches of the Yangtze River Basin, the BC aerosols cause greater reductions in precipitation (−0.4 to −2.4 mm/d). The influence of BC aerosols on precipitation is complicated and variable, and the difference in the optical parameterization scheme also causes various factors to influence the precipitation, which merits further study.

Comparisons with observations show that Scheme 2 can relatively accurately simulate the ground surface concentrations of BC aerosols. The simulation results indicate that the regional average ground surface concentrations for BC aerosols with the four particle sizes are 0.372, 0.264, 0.055, and 0.004 $\mu g/m^3$. The proportions of the first and second modes, which have relatively small radii, are 53% and 38%, respectively, and the proportions of the third and fourth modes are very small. The regional average total ground surface BC aerosol concentration is 0.69 $\mu g/m^3$. The distribution of the column concentrations is similar to that of the ground surface concentrations. The regional average total BC aerosol column concentration is 0.28 mg/m^2, and the maximum is 2.06 mg/m^2. Regions of high BC aerosol concentrations are located in the Sichuan Basin and the North China Plain and are related to the spatial distribution of their emissions and wind fields. The concentrations of BC aerosols in the spring and summer are higher than those in the other seasons. The regional average BC aerosol optical thickness is 0.0013, and the maximum is 0.0073. The maximum direct radiative forcing of BC aerosols under all-sky conditions is 2.11 W/m^2. The region of high values of the clear-sky radiative forcing is located to the north of the Yellow River, and the maximum value is 1.98 W/m^2. The regional average clear-sky and all-sky TRF values of BC aerosols are 0.49 and 0.36 W/m^2, respectively.

In most regions of East China, North China, and central China, BC aerosols cause increases in the annual average ground surface temperature, which ranges from 0.05 to 0.15 K. In South China and some regions of Northwest China, BC aerosols cause decreases in temperature, and the cooling ranges from 0.05 to 0.15 K. At the regional scale, BC aerosols cause decreases in the annual average precipitation in East China, North China, South China, and Northeast China of −0.83, −0.05, −0.11, and −0.13 mm/d, respectively. In central China, BC aerosols cause the annual average precipitation to increase by 0.01 mm/d. In the winter and summer, the region where the 850 hPa height field decreases corresponds to a convergence of the wind field. The change in airflow caused by BC aerosols in the winter is greater than that in the summer. In the winter, BC aerosols cause a greater increase in the ground surface temperature. In the region between the Yangtze River and the Yellow River Basin, the increase in the ground surface temperature is significant, and the maximum warming is 0.42 K. During the summer, BC aerosols increase the precipitation in southern China and some regions to the north of the Yellow River. However, in the Yangtze River Basin and some regions of Northeast China, BC aerosols decrease the precipitation. There are good correlations between the changes of circulation, pressure, and temperature induced by BC aerosols.

The climate effects of BC are related to its size distributions, and when using other size distributions, the results will change. This may be discussed in more detail in later studies. The results of this paper are related to the mixing sate of BC, and we just considered the simple situation. Taking the mixing state of BC and other aerosols into consideration may show different results.

Conflicts of Interest

The authors declare that they have no conflicts of interest.

Acknowledgments

This study was supported by the National Key Basic Research and Development Planning Program of China (Program 973) (2014CB441203), National Key Research and Development Program of China (2017YFC0209600), and National Natural Science Foundation of China (41575141). The numerical calculations in this paper have been done on the Blade cluster system in the High Performance Computing and Massive Data Center (HPC&MDC) of School of Atmospheric Science, Nanjing University. We thank the High-Performance Computing Center at Nanjing University. We acknowledge the BC emission inventory in the MEIC (Multi-resolution Emission Inventory for China) data provided by Professor Q. Zhang from Tsinghua University, China (http://www.meicmodel.org).

References

[1] V. Ramanathan and G. Carmichael, "Global and regional climate changes due to black carbon," *Nature Geoscience*, vol. 1, no. 1, pp. 221–227, 2008.

[2] T. Bond, S. Doherty, D. Fahey et al., "Bounding the role of black carbon in the climate system: a scientific assessment," *Journal of Geophysical Research: Atmospheres*, vol. 118, no. 11, pp. 5380–5552, 2013.

[3] IPCC, *Climate Change 2013: The Physical Science Basis. Contribution of Working Group I to the Fifth Assessment Report of the Intergovernmental Panel on Climate Change*, Cambridge University Press, Cambridge, UK, 2013.

[4] R. C. Levy, L. A. Remer, and O. Dubovik, "Global aerosol optical properties and application to moderate resolution imaging spectroradiometer aerosol retrieval over land," *Journal of Geophysical Research*, vol. 112, no. 112, pp. 3710-3711, 2007.

[5] F. S. Wei, "Some thinking on the formulation of SO_2 emission and its quality in China," *Chinese Journal of Environmental Monitoring*, vol. 11, no. 4, p. 22, 1995, in Chinese.

[6] J. R. Sun and Y. Liu, "Possible effects of aerosols from China on East Asia summer monsoon: the complex effects of black carbon aerosols and sulfate aerosols," *Journal of Climate Change Research*, vol. 4, no. 3, pp. 161–166, 2008.

[7] S. Menon, J. Hansen, L. Nazarenko, and Y. Luo, "Climate effects of black carbon aerosols in China and India," *Science*, vol. 297, no. 5590, pp. 2250–2253, 2002.

[8] M. Z. Jacobson, "Control of fossil-fuel particulate black carbon and organic matter, possibly the most effective method of slowing global warming," *Journal of Geophysical Research*, vol. 107, no. D19, p. 4410, 2002.

[9] J. E. Penner, H. Eddleman, and T. Novakov, "Towards the development of a global inventory for black carbon emissions," *Atmospheric Environment. Part A. General Topics*, vol. 27, no. 8, pp. 1277–1295, 1993.

[10] W. F. Cooke, C. Liousse, H. Cachier et al., "Construction of a 1×1 fossil fuel emission data set for carbonaceous aerosol and implementation and radiative impact in the ECHAM4 model," *Journal of Geophysical Research: Atmospheres*, vol. 104, no. D18, pp. 22137–22162, 1999.

[11] D. G. Streets, S. Gupta, S. T. Waldhoff, M. Q. Wang, T. C. Bond, and Y. Y. Bo, "Black carbon emissions in China," *Atmospheric Environment*, vol. 35, no. 25, pp. 4281–4296, 2001.

[12] J. M. Ge, Y. Z. Liu, and J. P. Huang, "Optical characteristics of black carbon and dust aerosols with HITRAN data," *Journal of Applied Optics*, vol. 30, no. 2, pp. 202–209, 2009.

[13] S. E. Bauer, S. Menon, D. Koch, T. C. Bond, and K. Tsigaridis, "A global modeling study on carbonaceous aerosol microphysical characteristics and radiative effects," *Atmospheric Chemistry and Physics*, vol. 10, no. 15, pp. 7439–7456, 2010.

[14] D. V. Spracklen, K. S. Carslaw, U. Pöschl, A. Rap, and P. M. Forster, "Global cloud condensation nuclei influenced by carbonaceous combustion aerosol," *Atmospheric Chemistry and Physics*, vol. 11, pp. 9067–9087, 2011.

[15] X. Ma, H. Liu, J. J. Liu et al., "Sensitivity of climate effects of black carbon in China to its size distributions," *Atmospheric Research*, vol. 185, pp. 118–130, 2016.

[16] E. P. Shettie and R. W. Fenn, *Models for the Aerosols of the Lower Annosphere and the Effect of Humidity Variations on their Optical Properties*, Air Force Geophysical Laboratory, AFGL-TR-79~0214, Bedford, MA, USA, 1979.

[17] J. H. Ma, Y. F. Zheng, and H. Zhang, "The optical depth global distribution of black carbon aerosol and its possible reason analysis," *Journal of the Meteorological Sciences*, vol. 12, no. 2, pp. 156–164, 2007, in Chinese.

[18] J. W. Winchester, W. X. Lu, L. X. Ren et al., "Fine and coarse aerosol composition from a rural area in northern China," *Atmospheric Environment*, vol. 15, no. 6, pp. 933–937, 1981.

[19] L. L. Li, J. Su, Q. Fu et al., "Parameterization of radiation properties of dust aerosols," *Atmospheric Physics and Atmospheric Environment*, vol. S18, 2012.

[20] K. N. Liou and R. Davies, "Radiation and cloud processes in the atmosphere," *Physics Today*, vol. 46, no. 46, pp. 66-67, 1993.

[21] J. J. Shang, H. Liao, Y. Fu et al., "The impact of sulfate and black carbon aerosols on summertime cloud properties in China," *Journal of Tropical Meteorology*, vol. 4, no. 16, pp. 451–466, 2017.

[22] P. B. Guan, H. D. Shi, Q. X. Gao et al., "Study on black carbon aerosol simulation of climate effect in China," *Journal of Environmental Engineering Technology*, vol. 7, no. 4, pp. 418–423, 2017, in Chinese.

[23] L. R. Chen, H. H. Zheng, H. D. Shi et al., "Simulation of impact of future black carbon aerosol emission on regional climate change," *Journal of Environmental Engineering Technology*, vol. 8, no. 1, pp. 1–11, 2018, in Chinese.

[24] T. M. Fu, J. J. Cao, X. Y. Zhang et al., "Carbonaceous aerosols in China: top-down constraints on primary sources and estimation of secondary contribution," *Atmospheric Chemistry and Physics*, vol. 12, no. 5, pp. 2725–2746, 2012.

[25] K. Li, H. Liao, Y. H. Ma, and A. R. David, "Source sector and region contributions to concentration and direct radiative forcing of black carbon in China," *Atmospheric Environment*, vol. 124, pp. 351–356, 2015.

[26] H. N. Liu, W. M. Jiang, J. P. Tang, and X. Li, "A numerical study on the troposphere photochemical oxidation process of sulfur dioxide over China," *Chinese Journal of Environmental Sciences*, vol. 21, no. 3, pp. 359–363, 2001.

[27] H. N. Liu, L. Zhang, and J. Wu, "A modeling study of the climate effects of sulfate and carbonaceous aerosols over China," *Advances in Atmospheric Sciences*, vol. 27, no. 6, pp. 1276–1288, 2010.

[28] L. Zhang, H. N. Liu, and N. Zhang, "Impacts of internally and externally mixed anthropogenic sulfate and carbonaceous aerosols on east Asian climate," *Acta Meteorol Sinica*, vol. 5, no. 5, pp. 639–658, 2011.
[29] H. Z. Chen, D. Wu, B. T. Liao, H. Y. Li, and F. Li, "Compare of black carbon concentration variation between Dongguan and Maofengshan," *China Environmental Science*, vol. 33, no. 4, pp. 605–612, 2013.
[30] Y. J. Li, L. Zhang, X. J. Cao, Y. Yue, and J. S. Shi, "Property of black carbon concentration over urban and suburban of Lanzhou," *China Environmental Science*, vol. 34, no. 6, pp. 1397–1403, 2014, in Chinese.
[31] H. N. Liu, Y. Zhu, H. J. Lin, and X. Y. Wang, "Observation and analysis of haze characteristics in Suzhou based on automatic station data," *China Environmental Science*, vol. 35, no. 3, pp. 668–675, 2015, in Chinese.
[32] G. L. Qiu, N. Liu, and X. B. Feng, "Pollution characteristics of atmospheric black carbon in Guiyang, Guizhouprovince, Southwest China," *Chinese Journal of Ecology*, vol. 30, no. 5, pp. 1018–1022, 2011, in Chinese.
[33] X. L. Sun and Y. F. Suo, "The observation study of black carbon concentration in Guilin during 2012," *Journal of Meteorological Research and Application*, vol. 34, pp. 130-131, 2013, in Chinese.
[34] Y. F. Wang, Y. J. Ma, Z. Y. Lu, D. P. Zhou, N. W. Liu, and Y. H. Zhang, "In situ continuous observation of atmospheric black carbon aerosol mass concentration in Liaoning region," *Research of Environmental Sciences*, vol. 24, no. 10, pp. 1088–1096, 2011, in Chinese.
[35] X. Zhang, J. Tang, Y. F. Wu, J. Wu, P. Yan, and R. J. Zhang, "Variations of black carbon aerosol observed in Beijing and surrounding area during 2006-2012," *China Powder Science and Technology*, vol. 21, no. 4, pp. 24–29, 2015, in Chinese.
[36] P. S. Zhao, F. Dong, Y. D. Yang et al., "Characteristics of carbonaceous aerosol in the region of Beijing, Tianjin, and Hebei, China," *Atmospheric Environment*, vol. 71, no. 3, pp. 389–398, 2013.
[37] B. L. Zhuang, F. Jiang, T. J. Wang, S. Li, and B. Zhu, "Investigation on the direct radiative effect of fossil fuel black-carbon aerosol over China," *Theoretical and Applied Climatology*, vol. 104, pp. 301–312, 2010.
[38] Y. Qian, L. R. Leung, S. J. Ghan, and F. Giorgi, "Regional climate effects of aerosols over China: modeling and observation," *Tellus B*, vol. 55, pp. 914–934, 2003.
[39] H. Zhang, Z. L. Wang, P. W. Guo, and Z. Z. Wang, "A modeling study of the effects of direct radiative forcing due to carbonaceous aerosol on the climate in east Asia," *Advances in Atmospheric Sciences*, vol. 26, no. 1, pp. 57–66, 2009.
[40] L. Liao, S. Lou, Y. Fu et al., "Radiative forcing of aerosols and its impact on surface air temperature on the synoptic scale in eastern China," *Chinese Journal of Atmospheric Sciences*, vol. 39, no. 1, pp. 68–82, 2015, in Chinese.
[41] H. Liao and J. H. Seinfeld, "Effect of clouds on direct aerosol radiative forcing of climate," *Journal of Aerosol Science*, vol. 28, no. D4, pp. 3781–3788, 1997.
[42] L. S. Chang and S. U. Park, "Direct radiative forcing due to anthropogenic aerosols in east Asia during April 2001," *Atmospheric Environment*, vol. 38, no. 27, pp. 4467–4482, 2004, in Chinese.
[43] J. Wu, W. M. Jiang, H. N. Liu, W. G. Wang, and Y. Luo, "Comparison of on-line and off-line simulation methods for direct radiative forcing of anthropogenic sulfate," *Acta Meteorologica Sinica*, vol. 62, no. 4, pp. 486–492, 2004, in Chinese.
[44] W. Y. Chang and H. Liao, "Anthropogenic direct radiative forcing of tropospheric ozone and aerosols from 1850 to 2000 estimated with IPCC AR5 emissions inventories," *Atmospheric and Oceanic Science Letters*, vol. 2, article 201e207, 2009.
[45] P. P. Wu and Z. W. Han, "Indirect Radiative and Climatic Effects of Sulfate and Organic Carbon Aerosols over East Asia Investigated by RIEMS," *Atmospheric and Oceanic Science Letters*, vol. 4, no. 1, pp. 7–11, 2011.
[46] L. Guo, E. J. Highwood, L. C. Shaffrey, and A. G. Turner, "The effect of regional changes in anthropogenic aerosols on rainfall of the east Asian summer monsoon," *Atmospheric Chemistry and Physics Discuss*, vol. 12, no. 9, pp. 23007–23038, 2012, in Chinese.
[47] Y. Q. Jiang, X. Q. Yang, and X. H. Liu, "Seasonality in anthropogenic aerosol effects on east Asian climate simulated with CAM5," *Journal of Geophysical Research*, vol. 120, no. 20, pp. 10837–10861, 2015.
[48] B. L. Zhuang, S. Li, T. J. Wang et al., "Direct radiative forcing and climate effects of anthropogenic aerosols with different mixing state over China," *Atmospheric Environment*, vol. 79, pp. 349–361, 2013.
[49] D. D. Wang, B. Zhu, Z. H. Jiang et al., "A modeling study of effects of anthropogenic aerosol on east Asian winter monsoon over eastern China," *Transactions of Atmospheric Sciences*, vol. 40, no. 4, pp. 541–552, 2017, in Chinese.
[50] A. S. Ackerman, O. B. Toon, D. E. Stevens et al., "Reduction of tropical cloudiness by soot," *Science*, vol. 288, no. 5468, pp. 1042–1047, 2000.
[51] J. E. Kristjánsson, "Studies of the aerosol indirect effect from sulfate and black carbon aerosols," *Journal of Geophysical Research*, vol. 107, no. D15, 2002.
[52] E. J. Highwood and R. P. Kinnersley, "When smoke gets in our eyes: the multiple impacts of atmospheric black carbon on climate, air quality and health," *Environment International*, vol. 32, no. 4, pp. 560–566, 2006.

10

Vertical Structure of Moisture Content over Europe

Agnieszka Wypych [1] and Bogdan Bochenek [2]

[1] *Jagiellonian University, 7 Gronostajowa St., 30-387 Krakow, Poland*
[2] *Institute of Meteorology and Water Management—National Research Institute, 14 Piotra Borowego St., 30-215 Krakow, Poland*

Correspondence should be addressed to Agnieszka Wypych; agnieszka.wypych@uj.edu.pl

Academic Editor: Stefania Bonafoni

The vertical structure of water vapor content in the atmosphere strongly affects the amount of solar radiation reaching the Earth's surface and processes associated with the formation of clouds and atmospheric precipitation. The purpose of this study was to assess the vertical differentiation of water vapor over Europe on a seasonal basis and also to evaluate the role of atmospheric circulation in changes therein. Daily values of specific humidity (SHUM) for the time period 1981–2015 were obtained from pressure levels available from ECMWF Era-Interim reanalysis data and used in the study. Eight grid points were analyzed in detail. Each point is representative of a region with different moisture conditions. SHUM profiles were then used to identify cases of moisture inversion. Horizontal flux of specific humidity (SHUMF) was analyzed for principal pressure levels that occur in both inversion-type and inversion-free situations. In addition, SHUM and SHUMF anomalies were identified for advection directions. The research results showed the existence of differences in the vertical structure of water vapor content in the troposphere over Europe, and the Northeastern Atlantic and the presence of moisture inversions not only in areas north of 60°N but also in temperate and subtropical zones. Inversions can occur in two different forms—surface-based and elevated. The occurrence of inversions varies with the seasons. The role of atmospheric circulation is observable in the winter and triggers both surpluses and shortages of moisture via the effect of specific pressure system types (significant role of seasonal pressure high) and via advection directions. In addition, there exists a clear difference between the structure of moisture in the atmospheric boundary layer and in the free atmosphere.

1. Introduction

The vertical structure of water vapor content in the Earth's atmosphere strongly impacts the amount of solar radiation reaching the surface of the planet and the process of formation of clouds and atmospheric precipitation. In light of this, water vapor is thought to be the most important trace gas in the atmosphere, and mutual feedbacks triggered by water vapor affect climate change the most by increasing its sensitivity. The total water vapor feedback factor is estimated to equal 0.47 [1, 2], although water vapor in the free atmosphere is believed to play a much more important role than that in the planet's boundary layers [2]. Increases in the quantity of water vapor increase the amount of longwave radiation trapped near the surface of the Earth and at the same time lead to an increase in air temperature and exacerbated warming. Given that it is air temperature that determines the pressure of saturated water vapor, the spatial distribution of water vapor and its vertical structure closely depend on temperature changes, which further implies an increased intensity of atmospheric processes.

The complexity of relationships between air temperature and water vapor content affects the radiation balance and water circulation patterns. For these very reasons, studies on the vertical structure of water vapor in the atmosphere, its variability over time, and differentiation across geographic space play a key role in the analysis of climate change including parametrization of climate models and weather forecasting [3, 4]. Studies are conducted first and foremost in areas where water vapor is especially vital due to its complex linkages with local conditions that also affect weather processes on a global scale in subtropical areas and polar areas. Many studies on the vertical structure of water vapor content focus on the key relevance of moisture inversions and their impact on selected processes related to water circulation and energy exchange. Kloesel and Albrecht [5] note the substantial role of the Earth's boundary layer in the regulation of water vapor transfers into the free atmosphere in the tropics, especially via convection, which is driven by low moisture inversions. The importance of studies in tropical areas is observable in the work of Holloway and Neelin [6] in the content of entrainment in convective boundary layers. In addition, Peters and Neelin [7] investigated factors that determine the formation of atmospheric precipitation. Finally, Wagner et al. [8] examined differences in the vertical structure of water vapor over the Atlantic Ocean. A large influx of solar energy along with strong air temperature inversions, cloud cover that is sensitive to multiple factors, and mesoscale circulation determining moisture transport through meridional inflows cause large differences in water vapor content across geographic space and in the vertical profile in polar areas. This subject is examined in multiple studies that emphasize the importance of the vertical structure of water vapor content in energy balances [9, 10]. Increases in the amount of water vapor present with altitude (moisture inversion) imply an increased share of the downward component of longwave radiation in the radiation balance, as shown by Devasthale et al. [11] for the winter season when water vapor from a moisture inversion constitutes more than 50% of the total column water vapor. This yields a significant effect on cloud cover [12]. The frequent occurrence of inversions in the Arctic and Antarctic was confirmed by Tomasi et al. [13], Vihma et al. [14], and Nygård et al. [15, 16] who showed their link with temperature inversions and the occurrence of low clouds.

Most research studies on this subject were conducted based on either regular weather balloon measurements or occasional field measurements. However, they also note the point nature of these measurements and the lack of opportunity to make measurements over areas of sea. One potential way to solve this problem is the use of reanalysis data, as noted by Serreze et al. [17] and Nygård et al. [15, 16]. Brunke et al. [18] compared data from five reanalysis attempts and were able to show that moisture inversions can also occur outside Arctic and Antarctic areas, although much less frequently and with less intensity. This is especially true in stratus regions and temperate regions—mostly in the cooler months of the year. On the contrary, frequencies exceed 80% in the winter, but only 10% in the summer, near the Arctic and Antarctic circles. The former are associated with lower (near the ground) temperature inversions over land areas and often are accompanied by the occurrence of fog [18, 19, 20]. These studies show that given the significance of water vapor content in key atmospheric processes, detailed analysis of the vertical structure of water vapor content in Arctic and Antarctic areas may provide additional information on factors modifying changes in moisture content in the atmosphere. At the same time, it may help in the parametrization of regional climate models (RCMs) and mesometeorological forecasting models.

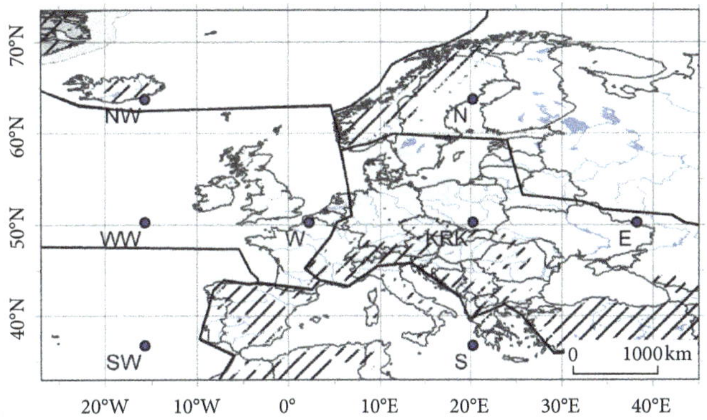

Figure 1: Study area with selected grid point locations representing different humidity regions in Europe (polygons); slanted lines are used to denote areas situated above the average 950 hPa level.

The purpose of this study was to evaluate the vertical differentiation of water vapor content in the atmosphere over Europe from a seasonal perspective and to examine the role of atmospheric circulation and its impact thereupon. Moisture content was identified at pressure levels up to 300 hPa with a special focus on cases of moisture inversion. Circulation patterns were described via advection directions for air masses including eastward intensity along with the northward moisture flux.

2. Data and Methods

Three spatial domains were examined: (1) the study area, (2) distinguished moisture regions, and (3) selected grid points. The study area consisted of Europe and the Northeastern Atlantic between 27°W and 45°E as well as 33°N and 73.5°N, with the exception of Greenland (Figure 1). Moisture regions (Figure 1; polygons) were identified by Wypych et al. [21] via the use of three variables describing moisture content, that is, total column water vapor of the entire atmosphere, specific humidity, and relative humidity at selected pressure levels (950 hPa, 850 hPa, 700 hPa, and 500 hPa). Cluster analysis helped identify six regions with different moisture conditions: Northern-Atlantic, Northern-Continental, Mid-Atlantic, Mid-Continental, Subtropical-Atlantic, and Mediterranean. Although the distinction was based on the grouping of the k-means method via largest possible differences and largest similarities within each type, internal cohesion tests for the identified moisture regions have shown them to be quite different. In addition, the studied moisture regions differ in the surface area (Figure 1). Hence, detailed analyses were carried out at key grid points (Figure 1). This included an analysis of the vertical structure of moisture conditions and the intensity of moisture flux. Moisture conditions at the studied grid points were similar to areal averages for the studied regions. At the same time, they also are representative of latitudinal (50.25°N) and longitudinal (15.75°W, 20.25°E) cross sections of the study area manifesting the effect of geographic

Table 1: Location of selected grid points.

Point ID	Latitude (°N)	Longitude (°)
NW	63.75	15.75 W
WW	50.25	15.75 W
SW	36.75	15.75 W
W	50.25	2.25 E
N	63.75	20.25 E
KRK	50.25	20.25 E
E	50.25	38.25 E
S	36.75	20.25 E

location on moisture conditions. Table 1 shows the location of all the grid points used in the study.

The study uses data from the period 1981–2015 obtained from ECMWF reanalysis data sets (ERA-Interim) at a base horizontal resolution of 0.75° [22]. Daily specific humidity values (q or SHUM) served as the basis for the study and were obtained for 18 pressure levels ranging from 950 hPa to 300 hPa.

The 1,000 and 975 hPa levels were not examined, as the average height of their location over the surface may be questionable with respect to their use in the analysis of large parts of the study area. In addition, the mean daily height of isobaric surfaces was checked in order to exclude cases where they were located underground at selected grid points.

The data were 6-hour averages. The use of daily data led to the omission of cases characterized as extremes. This is not an accidental omission, and it did make it possible to realize the principal goal of the study: the climatologic analysis of the vertical differentiation of water vapor content. The use of a 35-year-long daily data series (1981–2015) provides a sample large enough to produce a detailed climatologic analysis focused on extreme conditions. Wypych [23] was able to show differentiation in vertical profiles of specific humidity over Europe on both spatial and seasonal bases. Despite the smoothing of moisture fields in the course of reanalysis, this differentiation shows that moisture variables may be used to perform climatologic analyses while maintaining vital characteristics of mesoscale processes. Although climatologic reanalysis remains a somewhat questionable method due to its lack of full agreement with measured data, it represents a way to study moisture variables in the face of a lack of homogeneous data series. In addition, reanalysis remains insensitive to spatial changes and instrumental changes in the observation network. Numerous comparative studies [4, 17, 18] have shown the usefulness of reanalysis in the study of moisture fields.

The first step in the research study consisted of the preparation of a SHUM profile and the identification of cases of moisture inversion. The depth and strength of moisture inversions were determined using a formula provided by Brunke et al. [18] based on Vihma et al. [14] (1):

$$\mathrm{QIS} = \frac{q(p_{\min}) - q(p_{\max})}{|p_{\min} - p_{\max}|}, \tag{1}$$

where QIS is the inversion strength (g·kg^{-1}·(50 hPa)$^{-1}$), q is the specific humidity (g·kg^{-1}), $p_{\min}$ is the height of the inversion top (hPa), and $p_{\max}$ is the height of the inversion base (hPa).

The denominator of (1) describes the depth of an inversion layer. The strength of the inversion is calculated based on a layer with a depth of 50 hPa. Additionally, an analysis of inversion intensity based on normalized values was performed because of the general exponential decrease in humidity with height resulting in a decrease in inversion strength. Given the methods utilized in this study, only inversions of a certain intensity and those lasting over longer periods of time were examined (the positive gradient q present in averaged daily data). SHUM gradient (Δq) is calculated between consecutive 50 hPa layers. Inversion layers separated by a negative SHUM gradient were treated as separate moisture inversions. The study only considers cases where $\Delta q > 0.009$ g·kg^{-1} (10% positive lapse rate). Cases of inversion were classified as surface-based cases (inversion base ≥ 900 hPa) and elevated cases (inversion base < 900 hPa). The same method was used to identify inversions of the air temperature, although cases where $\Delta t < 0.6$ K were ignored (10% positive lapse rate). Cases where moisture inversions and temperature inversions occur simultaneously were also noted by calculating the frequency of occurrence and the coefficient of correlation for Δq and Δt. Water vapor transport was characterized via the specific humidity flux index (SHUMF, g·kg^{-1}m·s^{-1}) calculated using (2), specific humidity q, and zonal (u) and meridional (v) wind component values for every available vertical level up to 300 hPa:

$$\mathrm{SHUMF} = \sqrt{(q \cdot u)^2 + (q \cdot v)^2}. \tag{2}$$

The advective flux of specific humidity (g·kg^{-1}m·s^{-1}) was calculated for the zonal direction ($u \cdot \Delta q|_x$) and for the meridional direction ($u \cdot \Delta q|_y$), where u and v are the horizontal wind speeds in the respective directions and $\Delta q|_x$ and $\Delta q|_y$ are the horizontal humidity gradients in the respective directions.

The effect of atmospheric circulation on the vertical structure of water vapor content in the air was determined based on SHUM and SHUMF anomalies in advection directions. The study uses a simple division based on advection directions obtained directly from geostrophic wind directions. In addition, advection-free situations were identified with a wind speed not in excess of the 5th percentile, identified separately for each studied pressure level. Thresholds for the studied pressure levels were as follows: 2.0 m·s^{-1} for 950 hPa, 2.2 m·s^{-1} for 850 hPa, 2.7 m·s^{-1} for 700 hPa, 3.7 m·s^{-1} for 500 hPa, and 5.5 m·s^{-1} for the 300 hPa level. SHUM and SHUMF anomalies were calculated relative to mean monthly values and then standardized.

The use of climatologic reanalysis data in the study assures homogeneous spatial information for the study area in the context of water vapor data in the atmosphere. Multiple comparative studies [4, 17, 18] confirm the usefulness of reanalysis data in the examination of water vapor fields. A positive error was detected in temperature analysis results for the near-surface layer, which undoubtedly impacts moisture content values, and this is significant in cases of SHUM inversion occurrence [17, 24, 25]. In addition, research has shown a lack of adequate representativeness in moisture convection in research models [6]. However, despite a lack of agreement with respect to the occurrence of

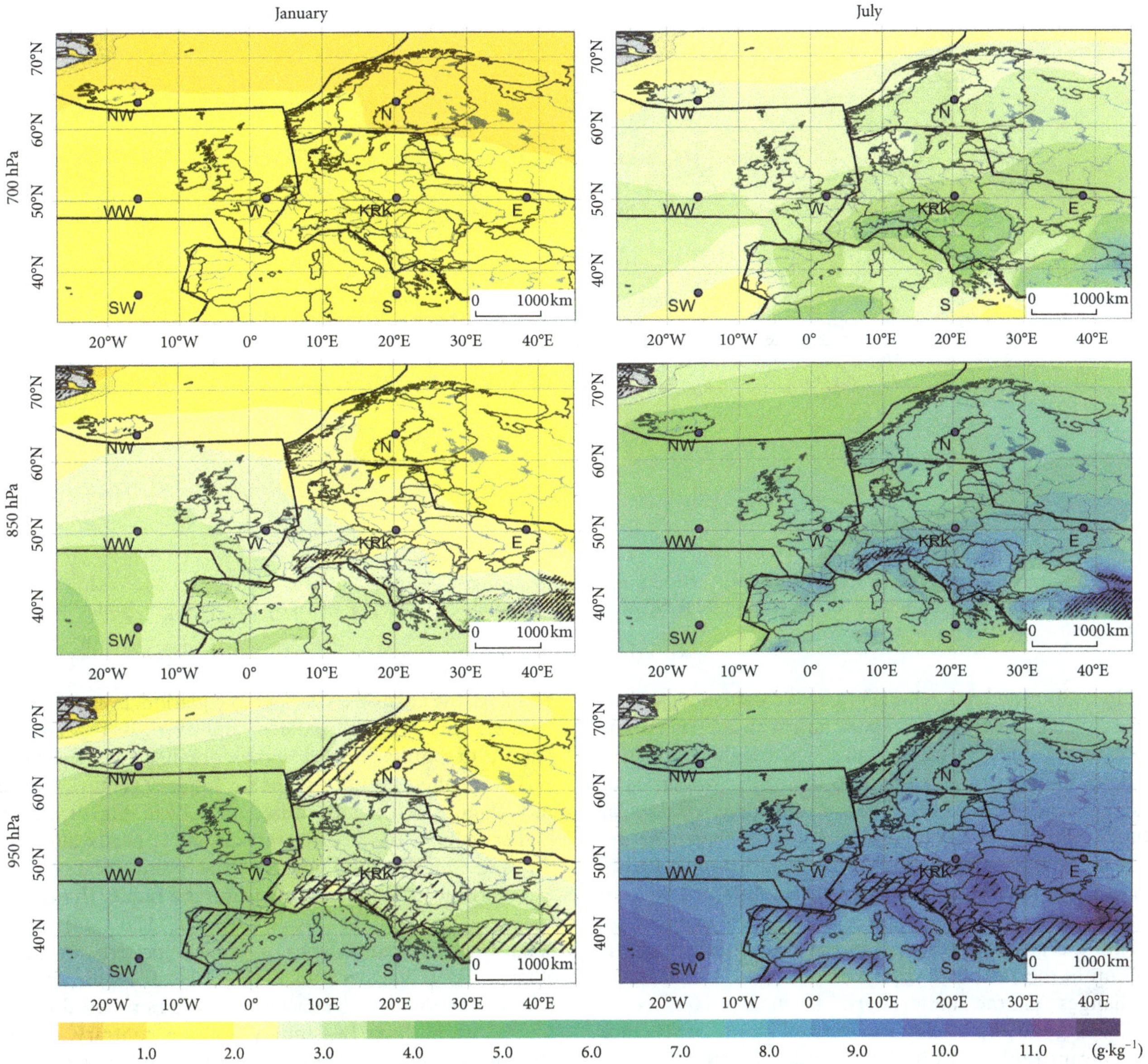

FIGURE 2: Monthly mean specific humidity at selected pressure levels (1981–2015); slanted lines are used to denote areas situated above the average of selected pressure levels.

SHUM inversions in reanalysis data and aerologic surveys, Brunke et al. [18] were able to show the usefulness of reanalysis data in climatologic studies or studies with a lower time resolution (daily or monthly averages), representing a larger geographic area. New knowledge on the moisture variable in newer reanalysis data including ERA-Interim has led to much improved parametrization of water circulation patterns. In turn, this results in a reduction in the size of errors in humidity-sensitive variables [4, 17, 18, 26].

3. Results and Discussion

3.1. Vertical Structure Climatology. Water vapor content in the air varies over time and across geographic space, which is closely linked with changes in temperature conditions. The relationship between air temperature and water vapor pressure (saturated) is described by the Clausius–Clapeyron equation. It assumes a decline of about 7% in water vapor content in the atmosphere due to a 1 K decrease in air temperature. This tendency, best observed over ocean areas, may be affected over land areas by rapid moisture transport, various processes affecting the near-surface zone of the air layer driven by local conditions, and convection.

The stated relationships indicate that the largest amount of water vapor in Europe is found in the area over the Atlantic Ocean at subtropical latitudes and in the south of the continent. Both areas are regions with the smallest fluctuations in water vapor content with high air temperatures throughout the year. The smallest amount of water vapor is detected in the atmosphere in the northern part of the study area, especially over the ocean, an area characterized by small fluctuations in air temperature on an annual

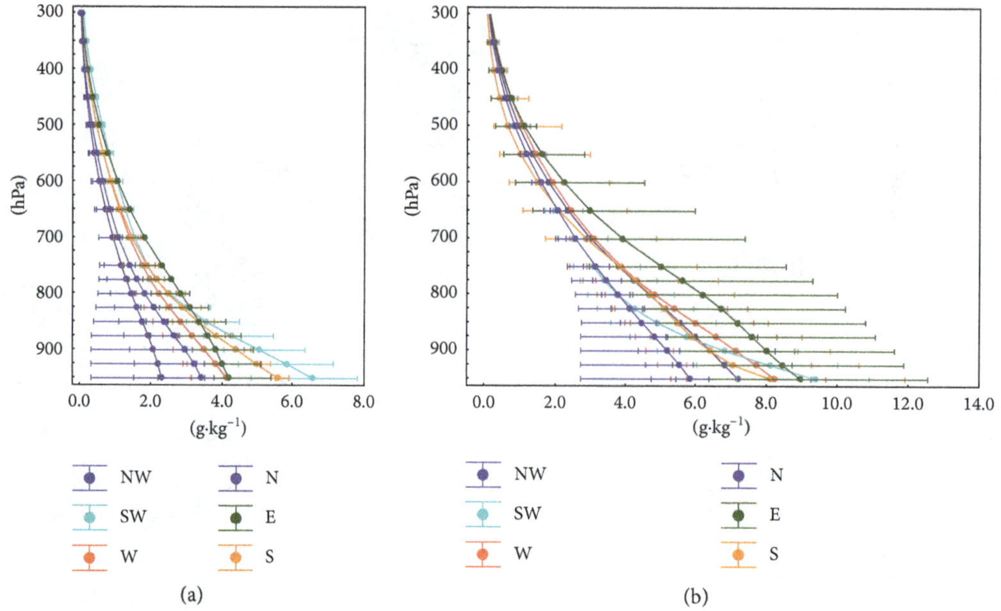

Figure 3: Monthly mean specific humidity (SHUM) vertical profiles at selected grid points (for details, see Figure 1 and Table 1); whiskers show the range defined by minimum and maximum SHUM values in a moisture region represented by the given grid point: (a) January; (b) July (1981–2015).

basis. In temperate zones, especially across continental Europe, annual fluctuations in water vapor content in the air are the highest in the study area due to the increasingly continental climate of the interior of Europe. In the winter, water vapor content over land areas corresponds to that detected at polar latitudes, while in the summer, it is greater than that determined for subtropical regions [21]. Given that the presence of water vapor in the Earth's atmosphere is substantial up to an altitude of about 2 km, determined by a negative temperature gradient, the greatest differentiation in the vertical profile is noted up to the 700 hPa level (Figure 2).

Changes in the atmospheric boundary layer yield 950 hPa and 850 hPa levels characterized by the occurrence of significant differences in SHUM values. This applies to differentiation in the vertical profile and also in the spatial sense, as expressed by an areal standard deviation (σ) calculated using all the domain grid points (Figures 2 and 3). In July, its values for mentioned pressure levels were as follows: $\sigma_{\text{areal}} = 1.6\,\text{g·kg}^{-1}$ and $\sigma_{\text{areal}} = 1.2\,\text{g·kg}^{-1}$, and in January, its values were $\sigma_{\text{areal}} = 1.4\,\text{g·kg}^{-1}$ and $\sigma_{\text{areal}} = 0.75\,\text{g·kg}^{-1}$. In January, the effect of warm North Atlantic and Norwegian currents may be observed along the western and northwestern coasts of the continent up to the 850 hPa level along with related increases in moisture levels across Western Europe (Figure 2). On the contrary, the interior of the European continent is affected by a seasonal high and characterized by much lower SHUM values. In July, the high SHUM content is readily observable in the troposphere over land areas, higher than that over the ocean at all three studied levels (Figure 2).

The stated characteristics of water vapor content are provided in detail in the vertical SHUM profiles produced for selected grid points in Figure 3.

From a climatologic perspective, the decline in the mean monthly water vapor content in the Earth's atmosphere, based on the Clausius–Clapeyron equation, is exponential in nature. This type of pattern, however, is representative mainly of points located across ocean areas. The vertical SHUM profile is quite similar for January (Figure 3(a)) and July (Figure 3(b)) and shifted in July in the direction of higher values. The Mediterranean Sea is partly affected by the surrounding land masses. The point located on the Mediterranean Sea (S) in July features a higher than oceanic moisture content at levels between 950 and 650 hPa (Figure 3(b)) and follows a similar pattern in January (Figure 3(a)). Northern areas (NW) have low air and water temperatures and are characterized by the smallest differences in SHUM in July. In January, on the contrary, northern areas have a higher water temperature than land temperature and experience evaporation off the ocean's surface, resulting in a moisture supply leading to a vertical SHUM gradient that is larger than that over continental areas (grids N and E). As stated previously, land areas are characterized by the largest differences in water vapor content. The SHUM gradient in layers from 950 hPa to 800 hPa at sites located deep inland equals about $2\,\text{g·kg}^{-1}\text{·}(150\,\text{hPa})^{-1}$, while its value in the troposphere over the ocean (grid SW) is about $6\,\text{g·kg}^{-1}\text{·}(150\,\text{hPa})^{-1}$ (Figure 3(b)).

There exist special cases in the troposphere characterized by increasing water vapor content in line with increasing altitude. Moisture inversions may assume variable depths and degrees of strength in relation to location and season of the year. In some cases, several inversion layers may be observed within one vertical SHUM profile [11, 15–18].

Given the procedural assumptions made in this study, only inversions lasting longer periods of time are considered (positive SHUM gradient in daily averages) as well as those of appropriate strength (Section 2). The frequency of inversions in the troposphere over Europe varies substantially

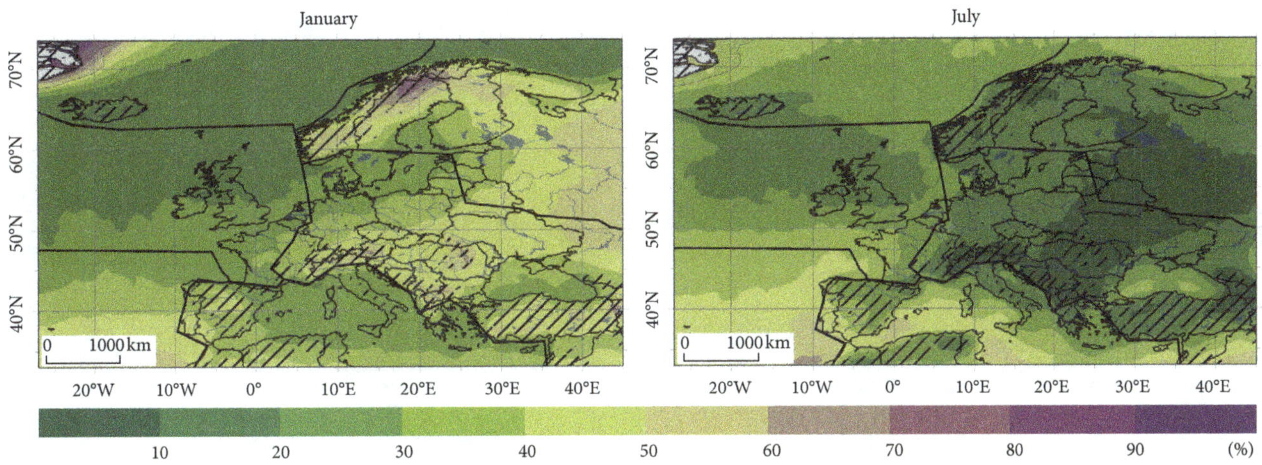

Figure 4: Frequency (%) of humidity inversion occurrence (1981–2015); slanted lines are used to denote areas situated above the average 950 hPa level.

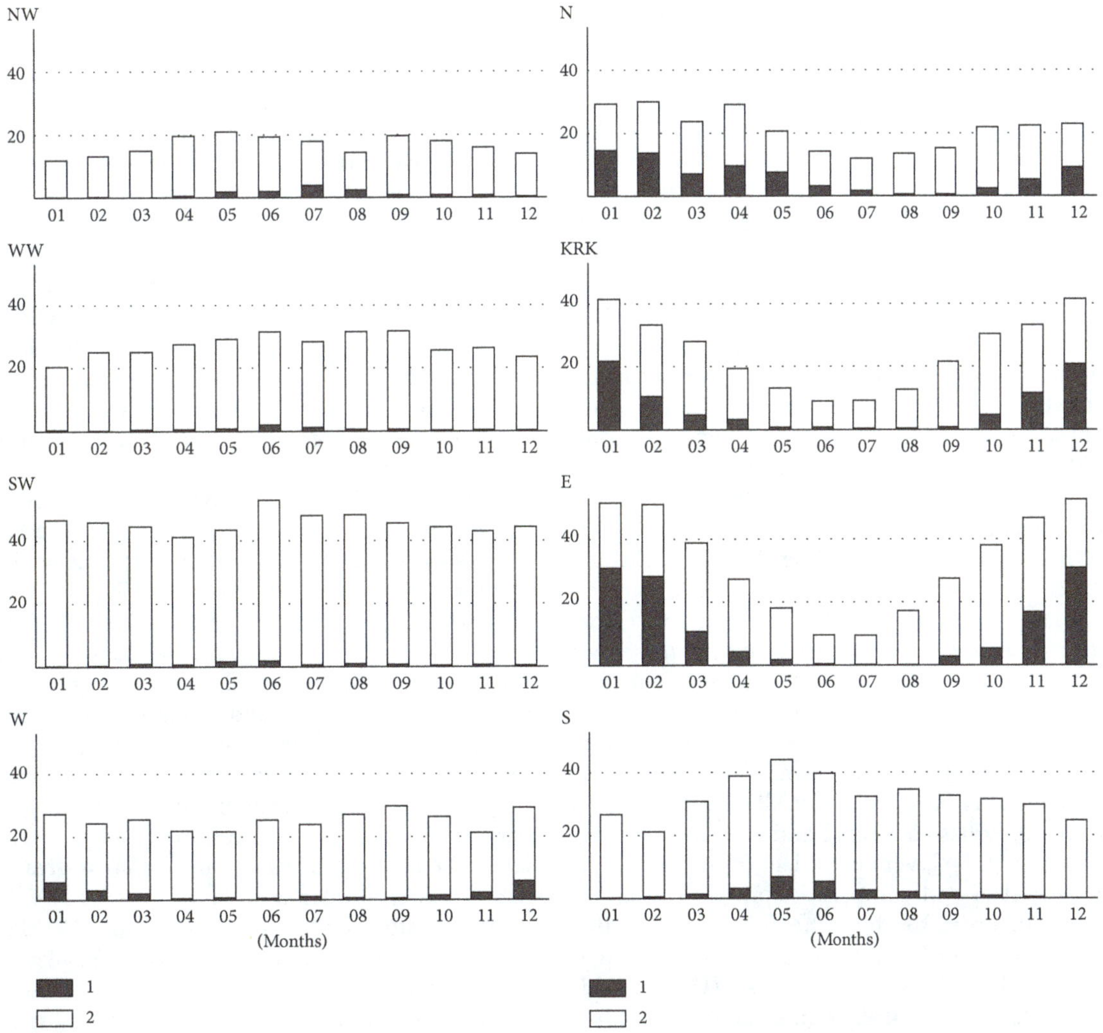

Figure 5: Frequency (%) of humidity inversion occurrence: (1) surface-based and (2) elevated at selected grid points (for details, see Figure 1 and Table 1) (1981–2015).

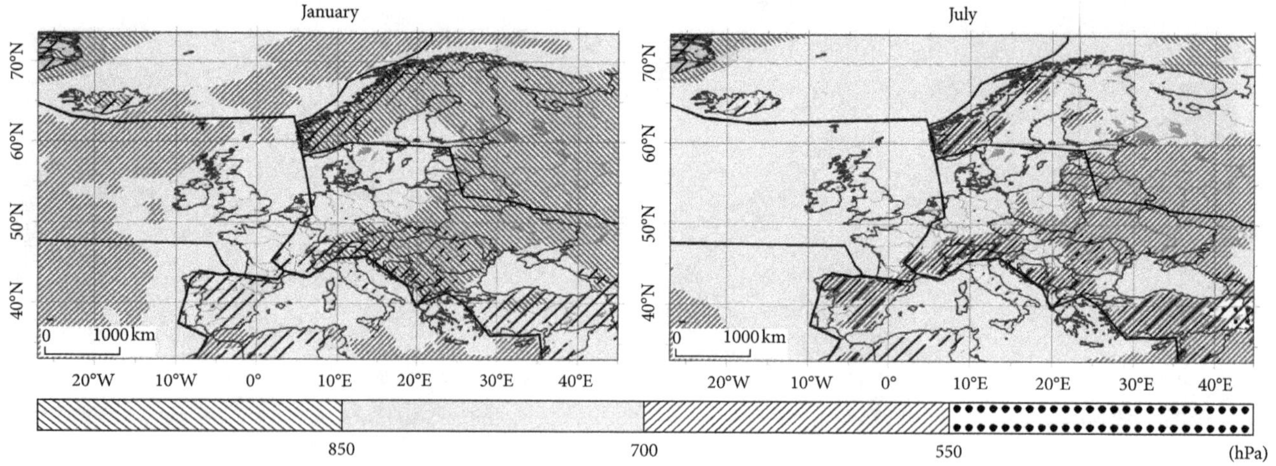

FIGURE 6: Mean height (hPa) of humidity inversion base (1981–2015); slanted lines are used to denote areas situated above the average 950 hPa level.

according to season and across geographic space. Inversions occur much more frequently in the winter (January) over land areas—sometimes in excess of 80% of days in the Scandinavian Peninsula and roughly 50% of days in Eastern and Southeastern Europe (Figures 4 and 5; KRK and E). In addition, inversions occur on 30% to 40% of days across the subtropical areas of the Atlantic (Figure 5; SW). However, in the summer, inversions are almost entirely limited to the subtropical zone across the Atlantic Ocean (40% to 50%) and the Mediterranean Sea (30% to 40%) (Figures 4 and 5; SW and S).

The temporal differentiation and spatial differentiation of the level of inversion occurrence are readily observable (Figure 6). In January, the mean base of the inversion layer over most land areas does not exceed 850 hPa, while in Western Europe, Southwestern Europe, and western coastal areas, it is 700 hPa. On the contrary, most inversions occurring over ocean areas are characterized by a lower level above 700 hPa (Figure 6). The mean base of the inversion layer in July in Eastern Europe and the Iberian Peninsula exceeds 700 hPa, while over the Caucasus, it is above 550 hPa (Figure 6). A large number of winter inversions take the form of surface-based inversions occurring over continental areas (Figure 5; W, N, KRK, and E). On the contrary, elevated-type inversions are the predominant type over sea and ocean areas regardless of season, while the few surface-based inversions that are detected may be designated marginal in significance (Figure 5; NW, SW, and S).

The current study has confirmed and expanded on the study by Brunke et al. [18], who used data from 5 different sets of reanalysis data and compared them with weather balloon surveys in order to show the occurrence of inversions in polar areas, mostly in the winter season, and the low base of these inversions. Nygård et al. [15, 16] and Vihma et al. [14] found statistically significant relationships between the occurrence of moisture inversions in relation to temperature inversions in near-polar areas in the northern and southern hemispheres. The similar conditions present in the winter in the atmospheric boundary layer over land areas (i.e., European interior areas) including negligible cloud cover, strong cooling of surfaces, and the associated water vapor condensation all suggest the simultaneous occurrence of moisture and temperature inversions also in polar regions.

Research has shown that surface-based SHUM inversions are accompanied by temperature inversions 70% of the time, which then increase to 90% in the direction of the continent's interior (grid E), while in Western Europe, they do not exceed 60%. This dependency declines to an average of 50% in the layer between 900 and 800 hPa. The correlation coefficient for the moisture gradient and air temperature on days with inversions reaches a statistically significant ($\alpha = 0.05$) average value of 0.5. For July, the absence of any type of relationship between the occurrence of moisture and temperature inversions confirms the dynamic origin of this particular linkage resulting from the advection of moisture or convection.

Although SHUM inversions occur more frequently in the winter, their depth is small, especially in privileged areas, and they range from <70 hPa in the Scandinavian Peninsula to about 80 hPa over other land areas (Figure 7). The deepest layer wherein water vapor content grows with altitude is most often detected over the North Atlantic (~100 hPa). Summer inversions are shallow and do not exceed 80 hPa over water and about 60 hPa over land (Figure 7). The strength of inversions varies negligibly over the study area. Somewhat higher values may be detected in July at an average of $0.2\,g{\cdot}kg^{-1}{\cdot}(50\,hPa)^{-1}$: from more than $0.3\,g{\cdot}kg^{-1}{\cdot}(50\,hPa)^{-1}$ over subtropical ocean areas to $0.1\,g{\cdot}kg^{-1}{\cdot}(50\,hPa)^{-1}$ over land. The strength of winter inversions is greatest over the Scandinavian Peninsula ($>0.3\,g{\cdot}kg^{-1}{\cdot}(50\,hPa)^{-1}$) and over parts of continental Europe (0.2-$0.3\,g{\cdot}kg^{-1}{\cdot}(50\,hPa)^{-1}$). Other parts of the study area do not exceed $0.15\,g{\cdot}kg^{-1}{\cdot}(50\,hPa)^{-1}$.

In order to reduce bias originating due to the exponential decrease in the amount of water vapor in the atmosphere, the analysis was repeated using standardized values. The procedure confirmed the most intensive inversions occurring in January in the northern (Scandinavian Peninsula) and northeastern parts of Europe (Northern Continental moisture region) but highlighted also the strength of

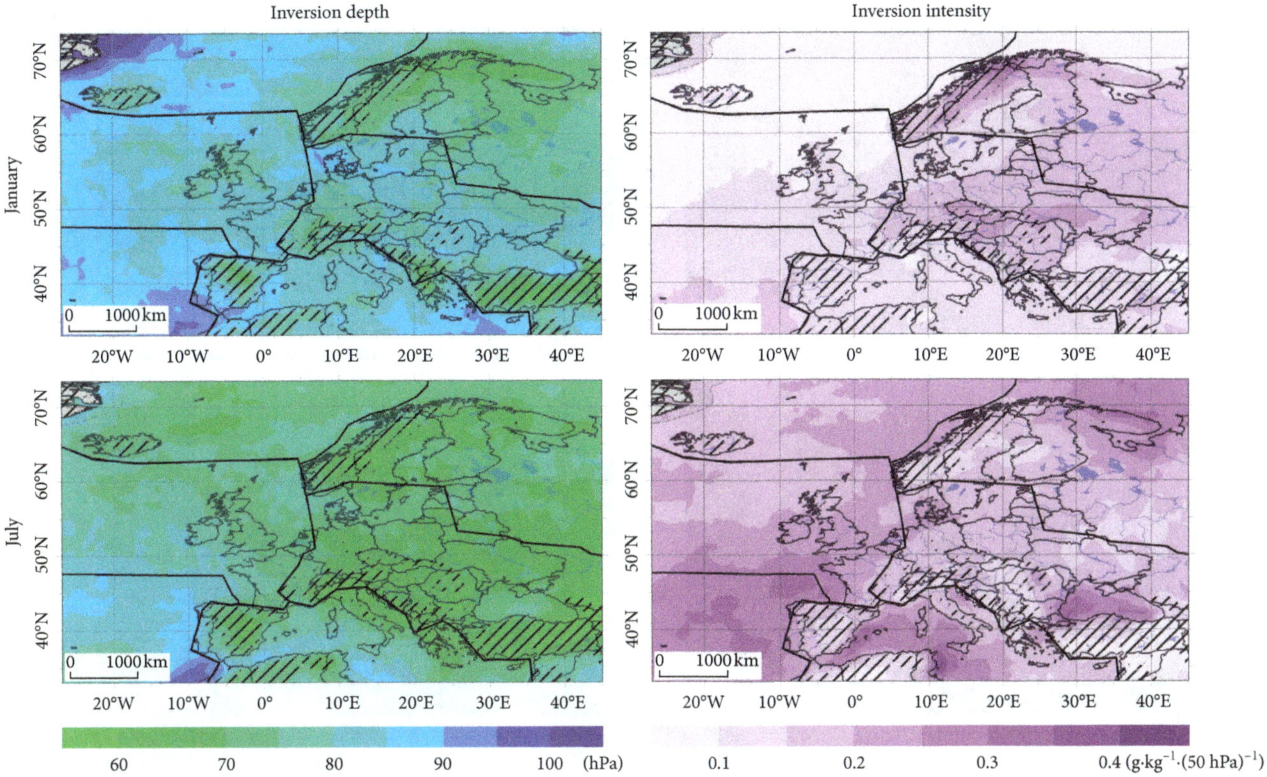

FIGURE 7: Mean humidity inversion depth (hPa) and intensity (g·kg^{-1}·(50 hPa)$^{-1}$) (1981–2015); slanted lines are used to denote areas situated above the average 950 hPa level.

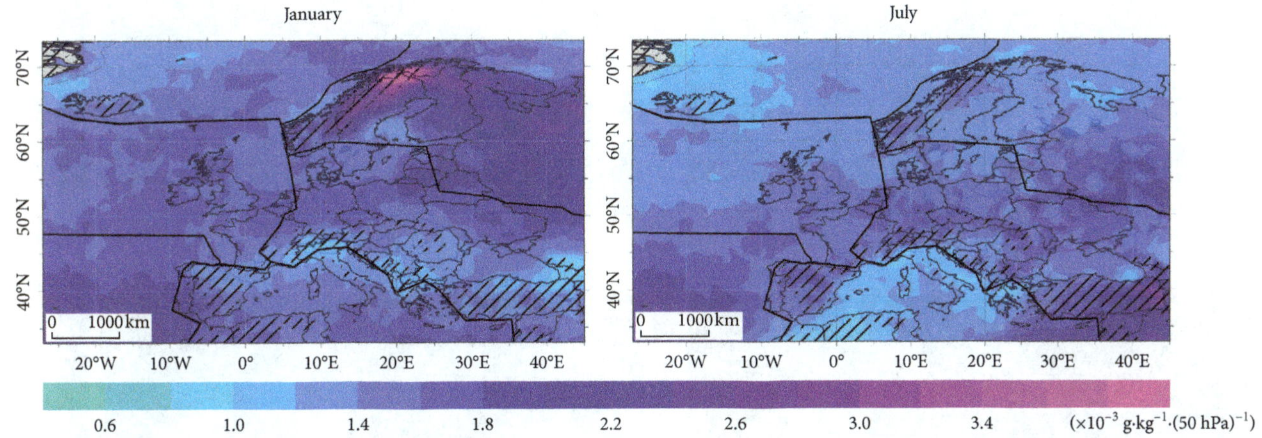

FIGURE 8: Mean standardized humidity inversion intensity (g·kg^{-1}·(50 hPa)$^{-1}$) (1981–2015); slanted lines are used to denote areas situated above the average 950 hPa level.

inversions over the Subtropical Atlantic region (Figure 8), which are mostly elevated, with the base between the levels of 700 and 550 hPa (Figure 6). The strength of summer inversions over the Subtropical Atlantic region was noted but was also determined over the continent where it was smoothed over by the moisture decrease bias (Figure 8).

3.2. Atmospheric Circulation Impact. Water vapor transport is one of the most important processes determining differences in its distribution across Europe. Existing research strongly suggests that atmospheric circulation, especially the North Atlantic Oscillation (NAO), shapes moisture content in the air in the winter [27–29]. Air saturated with water vapor is transported from over ocean areas in the direction of land, thanks to zonal circulation, and in the northerly direction, also known as meridional flux. Significantly weaker dependencies were noted in the warm half of the year when differences in water vapor content originate in convection, while horizontal moisture flux plays a secondary role.

In the present study, the analysis of water vapor transport concerns the advection of air masses or horizontal moisture flux at principal pressure levels: 950 hPa, 850 hPa, 700 hPa, and 500 hPa. Moisture transport is strongly aided

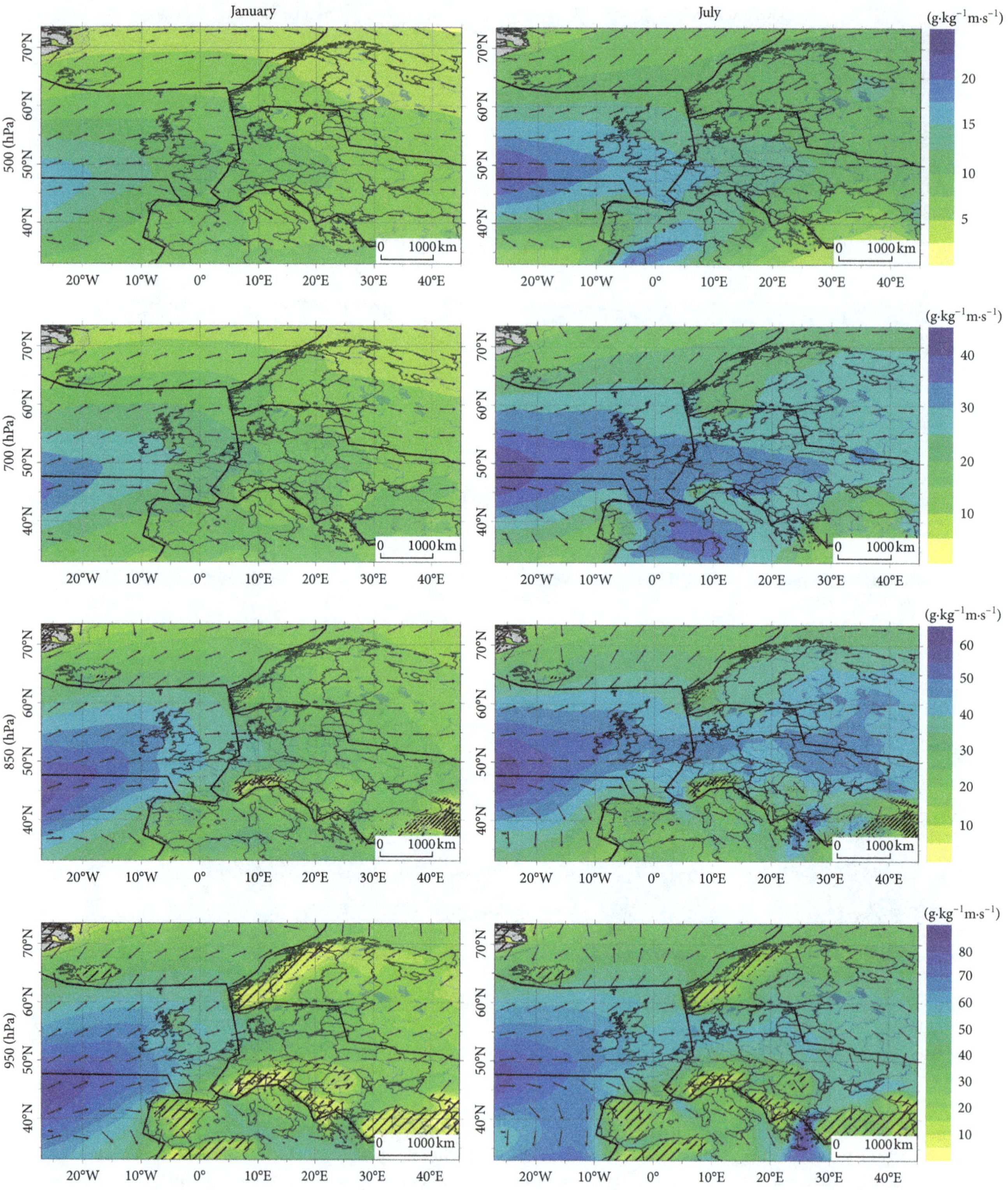

FIGURE 9: Monthly mean specific humidity flux and wind direction (arrows) at selected pressure levels (1981–2015); slanted lines are used to denote areas situated above the average of selected pressure levels.

by advection at the 850 hPa level. The limited impact of the atmospheric boundary layer enables the flux of air masses in the direction of the continent, and this includes not only Western Europe, as in the case of altitudes closer to the surface of the Earth (Figure 9), but also Central Europe.

Moisture transport strongly varies by season due to the amount of water vapor available per given air temperature and due to circulation in each given season. The substantial role of seasonal pressure highs is observable in January, in particular at the 950 hPa level, limiting the horizontal

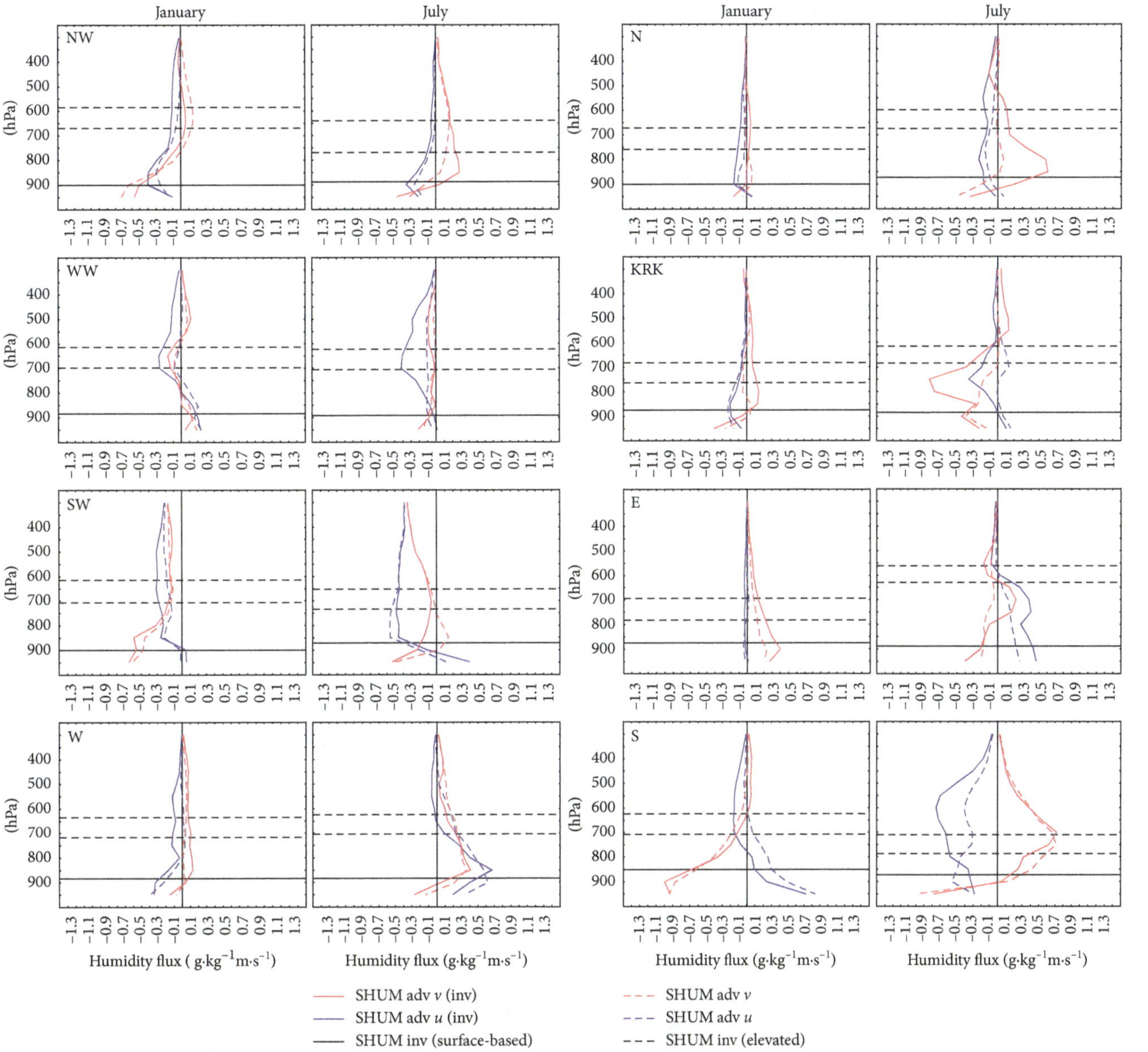

FIGURE 10: Climatologic mean profiles of specific humidity advective flux: eastward (u) and northward (v) during inversion (solid curve) and noninversion (dashed curve) events (detailed description in the text) at selected grid points (for details, see Figure 1 and Table 1).

motion of air masses (advection-free in many cases) and substantially affecting temperature conditions. The role of orographic barriers is also quite significant. Despite a high SHUM in July, SHUM transport is limited. In the south of Europe, this is associated with a shift in the intertropical convergence zone in the northerly direction and a predominance of longitudinal flux with a northerly component in the Mediterranean region. The influx of moisture becomes observable only in the free atmosphere (Figure 9).

Figure 10 uses dashed curves to show the significance of zonal and meridional transport of the moisture content in the air. Vertical profiles of specific humidity advective flux show a virtual lack of advection in January in continental Europe (grids W, N, KRK, and E). Small amounts of SHUM are carried by meridional flux along with relatively small differences in moisture transport over the ocean in the temperate and near-polar zones (grids WW and NW, resp.). On the contrary, large differences do occur in the subtropical zone, in particular over the Mediterranean Sea (grid S). Meridional flux assumes significant negative deviations, while zonal flux assumes equally large positive deviations (Figure 10). July vertical profiles show significant variability in specific humidity advective flux in continental Europe, especially in cases of days with SHUM inversions (solid lines).

Figure 11 shows in detail the role of selected advection directions in moisture transport based on selected grid points. Negative SHUM anomalies in January occur with advection from the north and east at all pressure levels, although in the interior of the continent (grids KRK and E), the deviations are largely closer to the surface of the Earth (levels 950 hPa and 850 hPa).

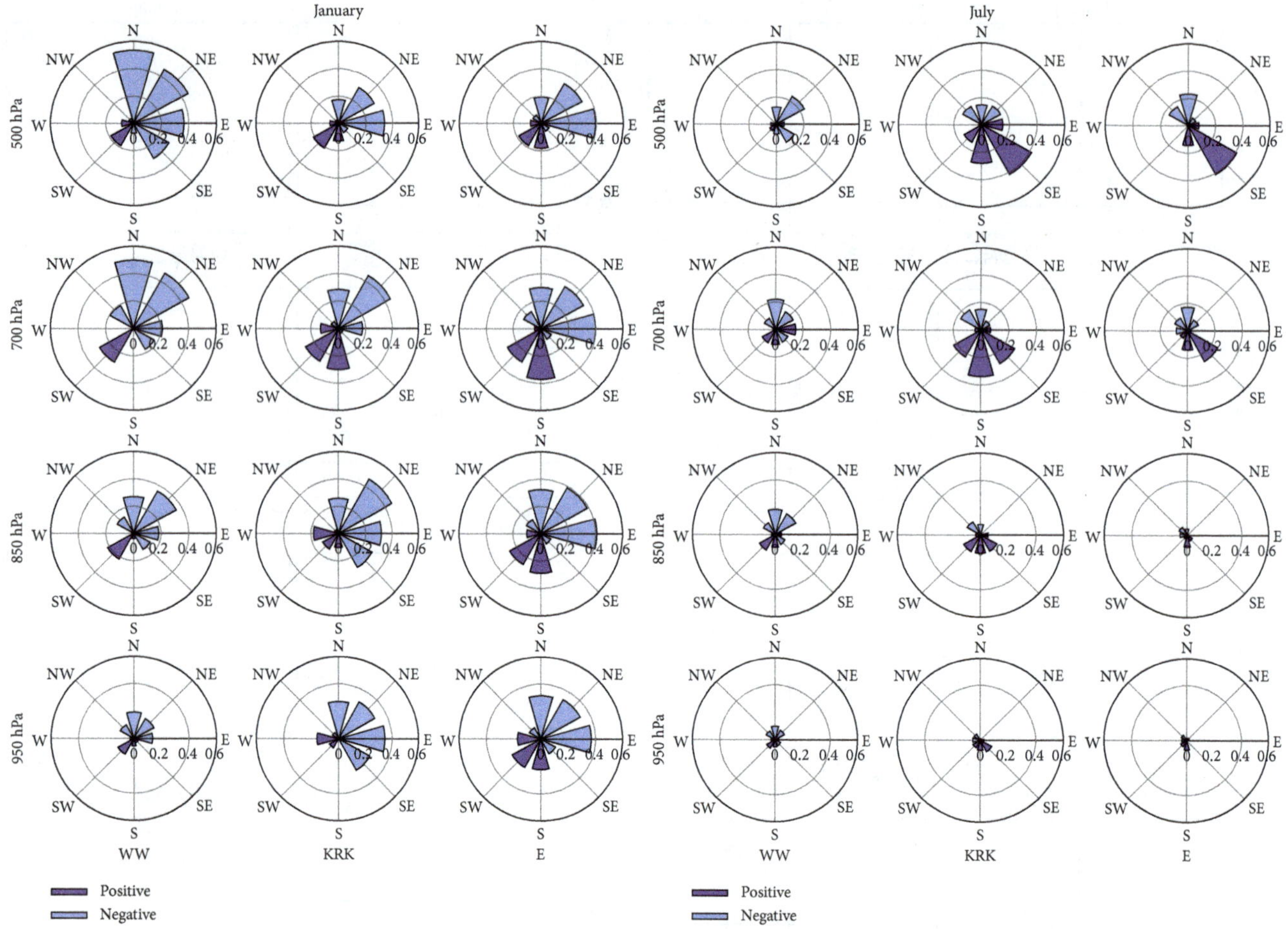

FIGURE 11: Specific humidity anomalies ($g{\cdot}kg^{-1}$) in advection directions (with respect to monthly mean, standardized) at selected pressure levels (1981–2015) and selected grid points (for details, see Figure 1 and Table 1), based on Wypych [23].

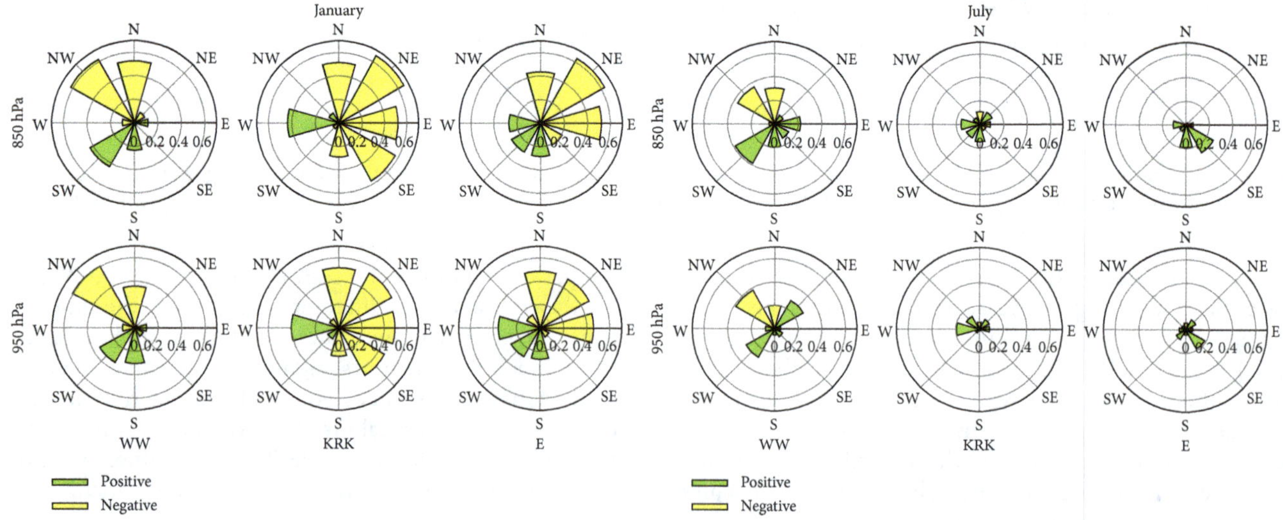

FIGURE 12: Specific humidity flux anomalies ($g{\cdot}kg^{-1}\ m{\cdot}s^{-1}$) in advection directions (with respect to monthly mean, standardized) at selected pressure levels (1981–2015) and selected grid points (for details, see Figure 1 and Table 1), based on Wypych [23].

The same holds true for water vapor transport (Figure 12), which indicates a significant role of the atmospheric boundary layer and mesoscale processes such as a seasonal high. Effects deemed continental may be observed in moisture transport also over the ocean (see grid point WW), although changes in the air mass associated with an increasing distance from a land mass yield a situation where moisture influx from the east produces a small positive anomaly (Figure 12).

Positive moisture anomalies and moisture transport in the winter are associated with advection from the west—or in the case of points located farthest to the east—and also from the south.

The insignificance of air circulation in the summer is confirmed by almost nonexistent SHUM and SHUMF anomalies in July at selected grid points (Figures 11 and 12).

One special case of the role of moisture advection is the occurrence of SHUM inversions (Figure 12). The significance of moisture inversions in the formation of total column water vapor (TCWV) has often been noted in the context of near-polar regions [14–16]. Given generally low water vapor content in the air, an inversion column may recharge TCWV up to 40%, whereas given a larger water vapor content, the effect of inversion on the TCWV value declines to about 15%. Therefore, the occurrence of inversions and share in the radiation balance are most substantial in the winter and spring. As noted previously, most cool season moisture inversions are associated with the occurrence of temperature inversions, often accompanying high pressure systems. The lowering of air temperature near the surface of the ground leads to the condensation of water vapor present in the air or the drying of the air mass resulting in an inversion of air moisture levels. Another factor that favors the occurrence of an inversion is the advection of moist air at higher levels or dry air near the surface of the Earth. Figure 10 shows mean profiles of specific humidity advective flux and mean inversion levels: top of surface-based inversions (solid lines) and bottom and top of elevated inversions (dashed lines). It may be readily observed that, in the case of surface-based inversions in January, the moisture flux effect is not significant. There is no meaningful difference between days with an inversion (solid curve) and days without an inversion (dashed curve). The grid point E is an exception where moisture advection is noted at the top of the surface-based inversion. On the contrary, in July, a month characterized by far fewer moisture inversions and especially by elevated inversions over regions of land, the significant role of meridional moisture flux may be observed on the top of surface-based inversions at points located in the northern part of the study area (grids NW and N) along with the advection of dry air (regardless of moisture flux direction) at the bottom of elevated inversions at the point KRK. Dependencies determined at selected grid points are consistent with results produced by studies in near-polar areas [14–16], as well as on the macroscale [21], which underscore the significance of meridional moisture flux and the convergence and convection of water vapor in the summer season.

4. Conclusions

Studies on the vertical structure of SHUM yield better parametrization of both climate models and mesometeorological models. Water vapor condensation in the atmospheric boundary layer is determined first and foremost by its vertical flux, as it is the structure of moisture both in the atmospheric boundary layer and in the free atmosphere that determines the strength and direction of water vapor transport [30]. For example, the presence of moisture inversions plays a decisive role in shifts in convection at subtropical latitudes [5] and also helps support the cloud layer by making evaporation off its upper part impossible [11] and assists in fog formation [12].

The present study constitutes a climatologic analysis based on averaged daily data and shows the existence of differences in the vertical structure of water vapor content in the troposphere over Europe and the North Atlantic, including the presence of moisture inversions beyond 60°N and at temperate and subtropical latitudes. Inversions may be surface-based or elevated, and their occurrence follows a seasonal pattern. Land areas are characterized by the presence of surface-based inversions in the winter and elevated inversions in the summer, while ocean areas are characterized by lower inversions in the summer versus winter, although these are not true surface-based inversions, which tend to occur sporadically over the ocean.

Atmospheric circulation plays an important role in changes in water vapor content, especially in the winter season, and determines both surpluses and shortages of moisture through effects generated by different pressure system types (especially important is the seasonal high) and advection directions. Anomalies in the atmospheric boundary layer occur primarily over land areas. Vertical profiles of water vapor content are characterized by large standard deviations due to the large variety of physical characteristics of incoming air masses. Over ocean areas, the vertical structure of moisture is quite stable at lower levels, with some deviations related to advection direction; however, at higher pressure levels, it becomes differentiated. In summer months, the vertical structure of moisture remains largely unaffected by changes in circulation conditions, which given a variable SHUM content at subsequent pressure levels indicating the significance of processes occurring in the atmospheric boundary layer in the shaping of moisture content in the troposphere.

The results of the present study suggest a need for further research on the subject of variances in the vertical structure of water vapor content over Europe, with a special focus on subdaily data that would make it possible to examine differences in water vapor content in a dynamic sense. The use of reanalysis data bears the burden of near-surface biases that also do impact the structure of moisture; nevertheless, given a large array of variables available, it allows for a comprehensive analysis of processes determining moisture levels and an evaluation of their role in relation to ongoing weather changes.

Conflicts of Interest

The authors declare that they have no conflicts of interest.

Acknowledgments

The authors would like to thank Mr. Michał Różycki for calculation assistance.

References

[1] S. Manabe and R. T. Wetherald, "Thermal equilibrium of the atmosphere with a given distribution of relative humidity,"

Journal of Atmospheric Sciences, vol. 24, no. 3, pp. 241–259, 1967.

[2] E. K. Schneider, B. P. Kirtman, and R. S. Lindzen, "Tropospheric water vapor and climate sensitivity," *Journal of Atmospheric Sciences*, vol. 56, no. 11, pp. 1649–1658, 1999.

[3] S. C. Sherwood, R. Roca, T. M. Weckwerth, and N. G. Andronova, "Tropospheric water vapor, convection, and climate," *Reviews of Geophysics*, vol. 48, no. 2, p. RG2001, 2010.

[4] K. E. Trenberth, J. T. Fasullo, and J. Mackaro, "Atmospheric moisture transports from ocean to land and global energy flows in reanalyses," *Journal of Climate*, vol. 24, no. 18, pp. 4907–4924, 2011.

[5] K. A. Kloesel and B. A. Albrecht, "Low-level inversions over the tropical Pacific-thermodynamic structure of the boundary layer and the above-inversion moisture structure," *Monthly Weather Review*, vol. 117, no. 1, pp. 87–101, 1989.

[6] C. Holloway and J. D. Neelin, "Moisture vertical structure, column water vapor, and tropical deep convection," *Journal of Atmospheric Sciences*, vol. 66, no. 6, pp. 1665–1683, 2009.

[7] O. Peters and J. D. Neelin, "Critical phenomena in atmospheric precipitation," *Nature Physics*, vol. 2, no. 6, pp. 393–396, 2006.

[8] D. Wagner, E. Ruprecht, and C. Simmer, "A combination of microwave observations from satellites and an EOF analysis to retrieve vertical humidity profiles over the ocean," *Journal of Applied Meteorology*, vol. 29, no. 11, pp. 1142–1157, 1990.

[9] M. C. Serreze, R. G. Barry, and J. E. Walsh, "Atmospheric water vapor characteristics at 70°N," *Journal of Climate*, vol. 8, no. 4, pp. 719–731, 1995.

[10] M. Gerding, C. Ritter, M. Müller, and R. Neuber, "Tropospheric water vapour soundings by lidar at high Arctic latitudes," *Atmospheric Research*, vol. 71, no. 4, pp. 289–302, 2004.

[11] A. Devasthale, J. Sedlar, and M. Tjernström, "Characteristics of water-vapour inversions observed over the Arctic by Atmospheric Infrared Sounder (AIRS) and radiosondes," *Atmospheric Chemistry and Physics*, vol. 11, no. 18, pp. 9813–9823, 2011.

[12] A. Solomon, M. D. Shupe, P. O. G. Persson, and H. Morrison, "Moisture and dynamical interactions maintaining decoupled Arctic mixed-phase stratocumulus in the presence of a humidity inversion," *Atmospheric Chemistry and Physics*, vol. 11, no. 19, pp. 10127–10148, 2011.

[13] C. Tomasi, B. Petkov, E. Benedetti et al., "Characterization of the atmospheric temperature and moisture conditions above Dome C (Antarctica) during austral summer and fall months," *Journal of Geophysical Research*, vol. 111, article D20305, 2006.

[14] T. Vihma, T. Kilpeläinen, M. Manninen et al., "Characteristics of temperature and humidity inversions and low-level jets over Svalbard Fjords in spring," *Advances in Meteorology*, vol. 2011, Article ID 486807, 14 pages, 2011.

[15] T. Nygård, T. Valkonen, and T. Vihma, "Antarctic low-tropospheric humidity inversions: 10-yr climatology," *Journal of Climate*, vol. 26, no. 14, pp. 5205–5219, 2013.

[16] T. Nygård, T. Valkonen, and T. Vihma, "Characteristics of Arctic low-tropospheric humidity inversions based on radio soundings," *Atmospheric Chemistry and Physics*, vol. 14, no. 4, pp. 1959–1971, 2014.

[17] M. C. Serreze, A. P. Barrett, and J. Stroeve, "Recent changes in tropospheric water vapor over the Arctic as assessed from radiosondes and atmospheric reanalyses," *Journal of Geophysical Research*, vol. 117, article D10104, 2012.

[18] M. A. Brunke, S. T. Stegall, and X. Zeng, "A climatology of tropospheric humidity inversions in five reanalyses," *Atmospheric Research*, vol. 153, pp. 165–187, 2015.

[19] H. Liu, H. Zhang, L. Bian et al., "Characteristics of micrometeorology in the surface layer in the Tibetan Plateau," *Advances in Atmospheric Sciences*, vol. 19, no. 1, pp. 74–87, 2002.

[20] D. Liu, J. Yang, S. Niu, and Z. Li, "On the evolution and structure of a radiation fog event in Nanjing," *Advances in Atmospheric Sciences*, vol. 28, no. 1, pp. 223–237, 2010.

[21] A. Wypych, B. Bochenek, and M. Różycki, "Atmospheric moisture content over Europe and the Northern Atlantic," *Atmosphere*, vol. 9, no. 1, p. 18, 2018.

[22] D. P. Dee, S. M. Uppala, A. J. Simmons et al., "The ERA-Interim reanalysis: configuration and performance of the data assimilation system," *Quarterly Journal of Royal Meteorological Society*, vol. 137, no. 656, pp. 553–597, 2011.

[23] A. Wypych, *Tropospheric Moisture Content over Europe*, Institute of Geography and Spatial Management Jagiellonian University, Krakow, Poland, 2018, in Polish.

[24] M. Tjernström and R.G. Graversen, "The vertical structure of the lower Arctic troposphere analysed from observations and the ERA-40 reanalysis," *Quarterly Journal of Royal Meteorological Society*, vol. 135, no. 639, pp. 431–443, 2009.

[25] E. Jakobson, T. Vihma, T. Palo, L. Jakobson, H. Keernik, and J. Jaagus, "Validation of atmospheric reanalyses over the central Arctic Ocean," *Geophysical Research Letters*, vol. 39, no. 10, p. L10802, 2012.

[26] H. Flentje, A. Dörnbrack, A. Fix, G. Ehret, and E. Hólm, "Evaluation of ECMWF water vapour analyses by airborne differential absorption lidar measurements: a case study between Brazil and Europe," *Atmospheric Chemistry and Physics*, vol. 7, no. 19, pp. 5033–5042, 2007.

[27] J. W. Hurrel, "Decadal trends in the North Atlantic Oscillation: regional temperatures and precipitation," *Science*, vol. 269, no. 5224, pp. 676–679, 1995.

[28] E. Ruprecht, S.S. Schröder, and S. Ubl, "On the relation between NAO and water vapour transport towards Europe," *Meteorologische Zeitschrift*, vol. 11, no. 6, pp. 395–401, 2002.

[29] A. Stohl, C. Forster, and H. Sodemann, "Remote sources of water vapour forming precipitation on the Norwegian west coast at 60°N—a tale of hurricanes and an atmospheric river," *Journal of Geophysical Research*, vol. 113, article D05102, 2008.

[30] H. Linné, B. Hennemuth, J. Bösenberg, and K. Ertel, "Water vapour flux profiles in the convective boundary layer," *Theoretical and Applied Climatology*, vol. 87, no. 1–4, pp. 201–211, 2007.

11

A Persistent Fog Event Involving Heavy Pollutants in Yancheng Area of Jiangsu Province

Yuying Zhu, [1,2] Chengying Zhu, [1,2] Fan Zu, [1,2] Hongbin Wang, [1,2] Chengsong Yuan, [1,2] Shengming Jiao, [1,2] and Linyi Zhou [1,2]

[1] *Key Laboratory of Transportation Meteorology, China Meteorological Administration, Nanjing 210009, China*
[2] *Jiangsu Institute of Meteorological Sciences, Nanjing 210009, China*

Correspondence should be addressed to Hongbin Wang; kaihren@163.com and Chengsong Yuan; 41235772@qq.com

Academic Editor: Pedro Salvador

In the early December 2013, dense fog involving heavy pollutants lasted for 9 days in the Yancheng area. The characteristics, formation, and lasting mechanisms of this persistent fog were analyzed based on observational data at the Sheyang site, reanalysis data, and final analysis data from NCEP/NCAR, combining with the weather background and meteorological and physical variable fields. Results include that (1) the fog process was characterized by long duration, low visibility, and high pollutants concentration, (2) the atmospheric general circulation contributed to the sustainability and development of the heavily polluted fog, (3) deep inversion was the key thermal factor causing the heavily polluted fog, (4) the fog exhibited obvious outbreaks with good visibility weather turned to severe fog several times, and (5) the weak cold air invasion and radiative cooling were the triggering factors to the sudden enhancement of the fog.

1. Introduction

Fog is a phenomenon due to lots of water droplets or ice crystals suspended in the air of the ground layer, which makes horizontal visibility decline to 1 km or less [1]. Haze is a phenomenon due to the aggregation of lots of tiny dust particles, soot, or salt particles suspended in the air, which makes horizontal visibility decline to 10 km or less [2]. Fog limits the visibility and thus affects human activities that rely on good visibility conditions. These activities are part of the core activities of modern societies, most notably aircraft operations, shipping [3], and road traffic [4]. Moreover, when the fog contains heavy pollutants, it is also harmful to human health [5]. This is why accurate forecasts of fog and haze have become an important issue.

The observation [6, 7] and modelling [8, 9] of dense fog have been studied in recent years. Several dynamic, thermal, and water vapor conditions including the layered structure of thermal inversion, the increase of temperature in the ground layer after sunrise, and the vertical transmission of heat, momentum, and water vapor caused by turbulent mixing have been found that trigger to dense fog [1, 10, 11]. Severe fog in China was first reported in a study on the Shanghai-Nanjing expressway [12]. From then on, more field observations were launched in different seasons and regions in China [13–17]. Results from these studies showed that the longwave radiation enhancement at night, evaporation from the wet surface after sunrise, and turbulent mixing are three main physical factors on fog enhancement [18, 19]. Liu et al. [20] analyzed the boundary layer features of a persistent advection fog process in the Yangtze River delta region and found that the double-inversion structure provided good thermal conditions for the thick fog, and the southeast vapor transport was not only conducive to maintaining the thickness of the fog but also sustained its long duration. Ma et al. [21] studied a fog case in central and southern Hebei Province and reported that the maintenance of the northeasterly winds was the main reason promoting the heavy fog. A rarely seen heavy fog occurred in Jiangsu Province during 24–27 December 2006, and it lasted for 64 h in Nanjing, with severe fog taking place over 41 h [22].

Recently, many studies revealed that when severe fog formed, it usually showed obvious explosive features, including sudden surge of fog drop number density, obvious increase in the fog drop scale and water content, sharp plunge of visibility from several hundred meters to below 50 meters, and fast changing from heavy fog to severe fog within a very short time period (about 30 min) [23]. With the improvement of temporal resolution of droplet spectrometer, Li et al. [24] analyzed microphysics processes and macroscopic conditions of fog droplet spectrum widening. He pointed out that the former stage of fog droplet spectrum widening mainly involved nucleation and condensation processes, while the latter stage mainly involved coagulation and condensation processes. Turbulence not only plays a significant role in heat, momentum, and moisture vertical transfers but also is the necessary condition for coagulation and increase of fog droplet. Researchers analyzed chemical characteristics of fog through conducting field observations [25–29] and improved forecasting capability of fog models by parameterizing microphysics features of fog over the sea and land. They also explained fog droplet excitation and fog formation through analysis on turbulence mixing and observations. A number of studies revealed the phenomenon of fog droplet increasing [30–35], mainly due to the factors of supersaturation and radiative cooling. Choularton et al. [36] found that radiative cooling alone cannot produce large fog droplets, while greater supersaturation and turbulence can promote the formation of large droplets.

Although fog and haze are different weather phenomena, their formations have similarity. They all form under breeze, high humidity, and static stability condition. Many studies indicated that the two can convert from one to the other [37–39]. Köhler [40] suggested that the haze aerosols can be transformed to fog droplets under certain conditions. Hygroscopic aerosols can become haze droplet before the saturation reaches the critical supersaturation S_c, and when the environmental saturation is equal or larger than S_c, the aerosol particles can be activated spontaneously. Before fog occurs, it is frequently accompanied with haze or heavy haze for a certain time; and after the fog dissipates, the haze will appear [41].

Despite there were efforts focusing on thick fog and haze, basic research is still needed with respect to the explosive development of fog and interactions between dense fog and heavy pollutants. In the early December 2013, a continuous heavy fog which was rare in record in terms of long duration, low visibility, and heavy pollution appeared in Yancheng, Jiangsu Province, China. In this study, the formation and maintaining mechanisms of this dense fog with strong pollutants are comprehensively analyzed by using the observational data at the Sheyang site (120.15°E, 33.46°N), reanalysis data, and final analysis data from NCEP/NCAR. The aim of this study was to discuss the causes of heavy pollution and analyze reasons and mechanism of the fog's explosiveness, which would be helpful to our better understanding of fog-related processes and fog-forecasting issues.

The article is structured as follows: Section 2 describes the data used in this study and the classification of grade of fog and air quality, respectively. Section 3 shows the features of this fog event, including long duration, explosive enhancement, and heavy pollution. Section 4 is concerned with the weather conditions contributing to the maintenance of strongly polluted fog, while Section 5 presents the thermal and dynamic conditions. Section 6 focuses on the causes for the explosive development of severe fog, and in Section 7, results are summarized, and conclusions are drawn.

2. Site and Data

2.1. Observational Site and Data. The visibility of less than 0.2 km due to fog is defined as severe fog, and that of less than 0.05 km, extremely dense fog [42]. The air quality is classified as four categories: excellent, good, slightly polluted, and severe polluted. According to the standards by China Meteorological Administration (CMA), if the primary pollutant is $PM_{2.5}$ and the daily-averaged concentration exceeds 250 $\mu g \cdot m^{-3}$, it is considered as severely polluted [43].

The thick fog with heavy pollutants was observed from 1 to 9 December 2013 in Yancheng, Jiangsu Province, China. This weather phenomenon lasted for 9 days, with severe fog which occurred 6 times. The duration was about 35 min, 2 h, 25 min, 9 h, 5.5 h, and 7.3 h, respectively. Moreover, five of them developed into extremely dense fog, with explosive enhancement feature each time. Furthermore, pollution was particularly heavy, with the maximum of the $PM_{2.5}$ concentration reaching 764 $\mu g \cdot m^{-3}$. It was also recorded by a meteorological observatory. This kind of weather phenomenon had not been observed in the recent decades. The rarity justifies our discussion.

The observational datasets are from the Sheyang site (elevation 1.8 m, 120.15°E, 33.46°N) in Yancheng, which is located in the eastern part of the Jiangsu Province (Figure 1). The meteorological variables being observed include 10 m wind speed and direction, 2 m air temperature, and 2 m relative humidity, which are sampled every 10 min. The visibility is obtained every 1 min at 2 m height. The boundary layer is observed via the radiosonde. The balloon takes the sensor up at speed of 400 meters per minute and records temperature, pressure, humidity, wind direction, and wind speed information at different altitudes. The observations are taken twice a day, 07:00 Beijing Time (BT) and 19:00 BT. The concentrations of three major pollutants, including $PM_{2.5}$, SO_2, and NO_2 are also observed.

2.2. Reanalysis Data and Final Analysis Data. To represent the atmospheric circulation fields, the daily data of the National Centers for Environmental Prediction/National Center for Atmospheric Research (NCEP/NCAR) global reanalysis from 1 December 1979 to 31 December 2013 are utilized. Variables used here include geopotential height, temperature, and U-component and V-component of wind, with 17 levels. The horizontal resolution for the NCEP/NCAR reanalysis is 2.5° × 2.5°.

The final analysis data of NCEP/NCAR are used during the fog event from 1 to 9 December 2013, including variables

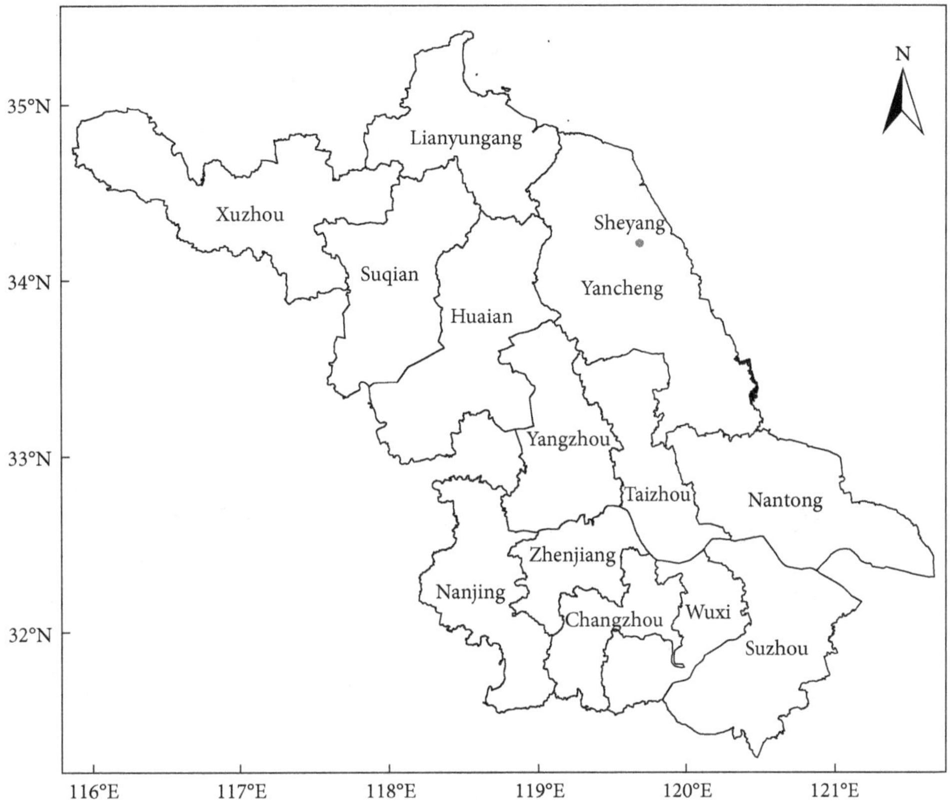

FIGURE 1: Map showing the location of the Sheyang site in Jiangsu Province.

of sea-level pressure, wind field at 10 m height, and U-component and V-component of wind with 17 levels. The horizontal resolution is $1^\circ \times 1^\circ$. The analysis data are taken four times a day: 02:00 BT, 08:00 BT, 14:00 BT, and 20:00 BT.

3. Fog Observation and Features

On the surface weather chart of 29-30 November 2013, Jiangsu Province was in front of a dry trough located over north China plain, with wind speed approximately 4–6 $m \cdot s^{-1}$, and depression of the dew point higher than 5°C. High-wind speed and dry air resulted in good visibility. During 1–9 December 2013, two high-pressure systems occurred from northern Xinjiang to west of Baikal Lake and from Northeast China to Southeast Russia, respectively. A low-pressure center appeared from Southwest China to Southeast Asian region, while Central and East China were under uniform pressure, with breeze or calm wind. Heavy fog and severe fog appeared successively in Jiangsu Province. Until 9 December, Mongolia high moved southward and eastward, and the cold air invaded, causing the fog to dissipate. The fog process was characterized by long duration, strong intensity, and heavy air pollution.

From 1 to 9 December, the Yancheng area was mostly foggy. During 1 December night, the sky was partly cloudy and the wind speed was 1 $m \cdot s^{-1}$. Due to radiative cooling, the fog began to form at about 23:00 BT. Extremely dense fog appeared at some sites in the southern Yancheng and expanded to northern Yancheng gradually. Until the early morning of 5 December, the extremely dense fog almost occupied the entire Yancheng area. What is worse was that as the fog turned into dense or even extremely dense fog, the air pollution also became severe. These dense fogs showed explosive development characteristics.

3.1. Long Duration. Long duration of severe fog was an important characteristic of this fog event. Figure 2 shows the temporal variation curves of visibility, relative humidity, temperature, wind direction, and speed at the Sheyang site. These curves reveal that since the morning of 2 December, the visibility at this station dropped sharply, and extremely dense fog appeared, which continued to 09:00 BT on 2 December. From 20:00 BT on 3 December to 02:00 BT on 4 December, some area in Yancheng had weak precipitation and thus supplied moisture for the fog development. When the sky cleared up, extremely dense fog appeared again. In the afternoon of 4 December, the visibility had improved; but at night, the fog broke out again and turned into an extremely dense one. In the afternoon of 5 December, the fog intensity weakened, and the visibility on 6 December remained above 1 km. From 00:00 BT to 08:00 BT on 7 December and from 18:00 BT on 8 December to 06:00 BT on 9 December, severe fog events occurred again and again. After 07:00 BT on 9 December, as cold air moved in, the visibility returned to 1 km or better. The whole fog event lasted for 9 days, and extremely dense fogs occurred frequently during the period, with the longest extremely dense fog lasting for as long as 9 hours.

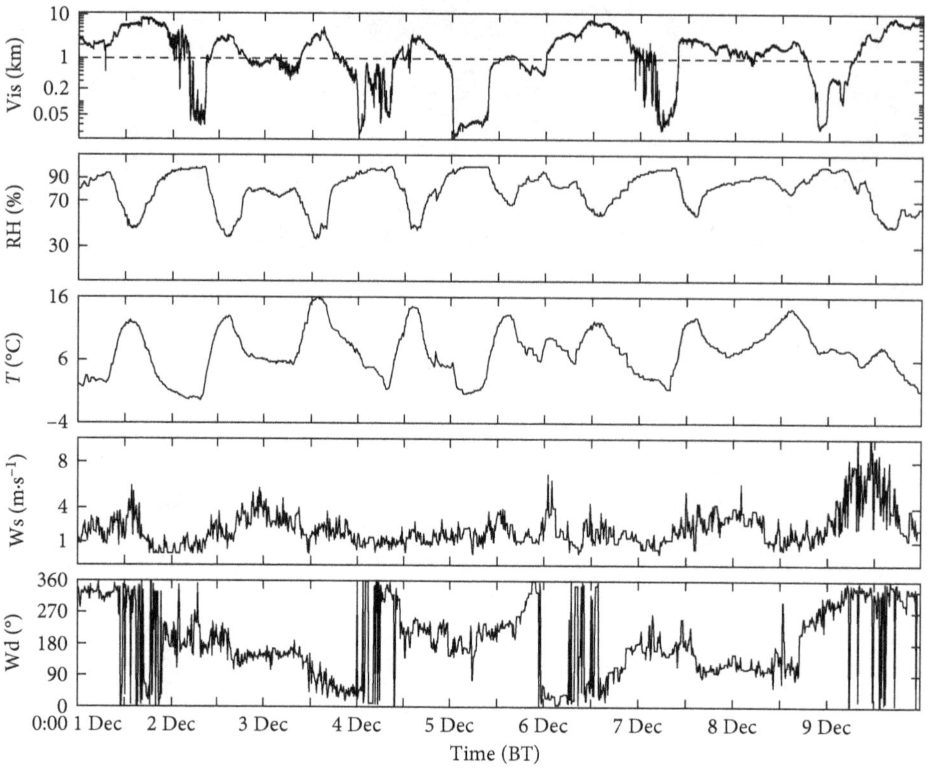

FIGURE 2: Temporal variation of visibility, relative humidity, temperature, wind speed, and wind direction at the Sheyang site from 1 to 9 December 2013.

3.2. Explosive Enhancement Characteristics. The characteristic of the formation of the dense fog in the Yancheng area was explosiveness, since the fog enhanced into severe fog or extremely dense fog within a very short time (about 30 min). This feature can be seen in the visibility variation curves of five fog cases recorded at the Sheyang site, which is shown in Figure 3. Table 1 lists the starting time, ending time, and the visibility of these five cases, indicating that explosive enhancement happened almost within 30 min and the shortest one was only 8 min. Another noticeable feature seen from Cases 1 and 4 is that light fog events with the visibility over 1 km turned into severe fog or extremely dense fog, which was rarely seen in previous observations. Figure 3 also shows that some fog processes had several explosive enhancements, such as Case 4 with four explosive enhancements, in which the second and third explosive enhancements burst into severe fogs again after the fog dissipated.

3.3. Heavy Pollution. This fog event in Yancheng had another important characteristic: heavy pollution. As the atmospheric stratification was stable and the inversion layer existed for a long time, a large amount of contaminated particles and polluted gases accumulated in the ground layer, causing heavy pollution. Figure 4 presents the hourly curves of $PM_{2.5}$ concentration and contaminated gases at the Yancheng site during 1–9 December. We can see that, on 2, 3, 5, and 9 December, the concentration of $PM_{2.5}$ was particularly high with an average value higher than 250 $\mu g \cdot m^{-3}$ that belongs to serious pollution category according to the CMA standard. In particular, at 21:00 BT on 4 December, the hourly concentration of $PM_{2.5}$ was up to 764 $\mu g \cdot m^{-3}$, about 1-2 times higher than the other fog or haze events reported [44, 45]. The situation was not improved until the midnight of 4 December, due to precipitation before the next fog occurrence. Thanks to the strong northwesterly winds, the fog began to dissipate on 9 December, and the $PM_{2.5}$ concentration dropped rapidly. Moreover, we may conclude that the dense fog and heavy pollution happened at the same time as a whole. Seriously polluted fog is not only harmful to human health but also has a strong extinction effect because of its hygroscopic property, which greatly reduces the visibility.

As is shown in Figure 4, the polluted gases generally began to rise after sunrise and reached their maximum concentrations during noon to dusk. The content of SO_2 reached a peak at 11:00 BT on 3 December, with the value of 142 $\mu g \cdot m^{-3}$, and the maximum content of NO_2 was 126 $\mu g \cdot m^{-3}$ at 18:00 BT on 2 December. Figure 4 also indicates that when the dense fog occurred, the polluted gases decreased obviously, which was significantly different from that of $PM_{2.5}$. Due to the large number of polluted gases dissolved in the fog, the polluted gases concentration decreased, which resulted in the increase of fog ion concentration and acidity [38, 41]. Previous research suggested that nitrate and sulfate are mostly contained in the noncoarse modes, with the conversion of NO_2 and SO_2 occurring mostly via gas-phase oxidation followed by condensation or through droplet mode sulfate produced from the fog process [46–48]. Shi et al. [41] found that the concentration of NO_3^- and SO_4^{2-} in foggy days are about 10 times and 2 times higher

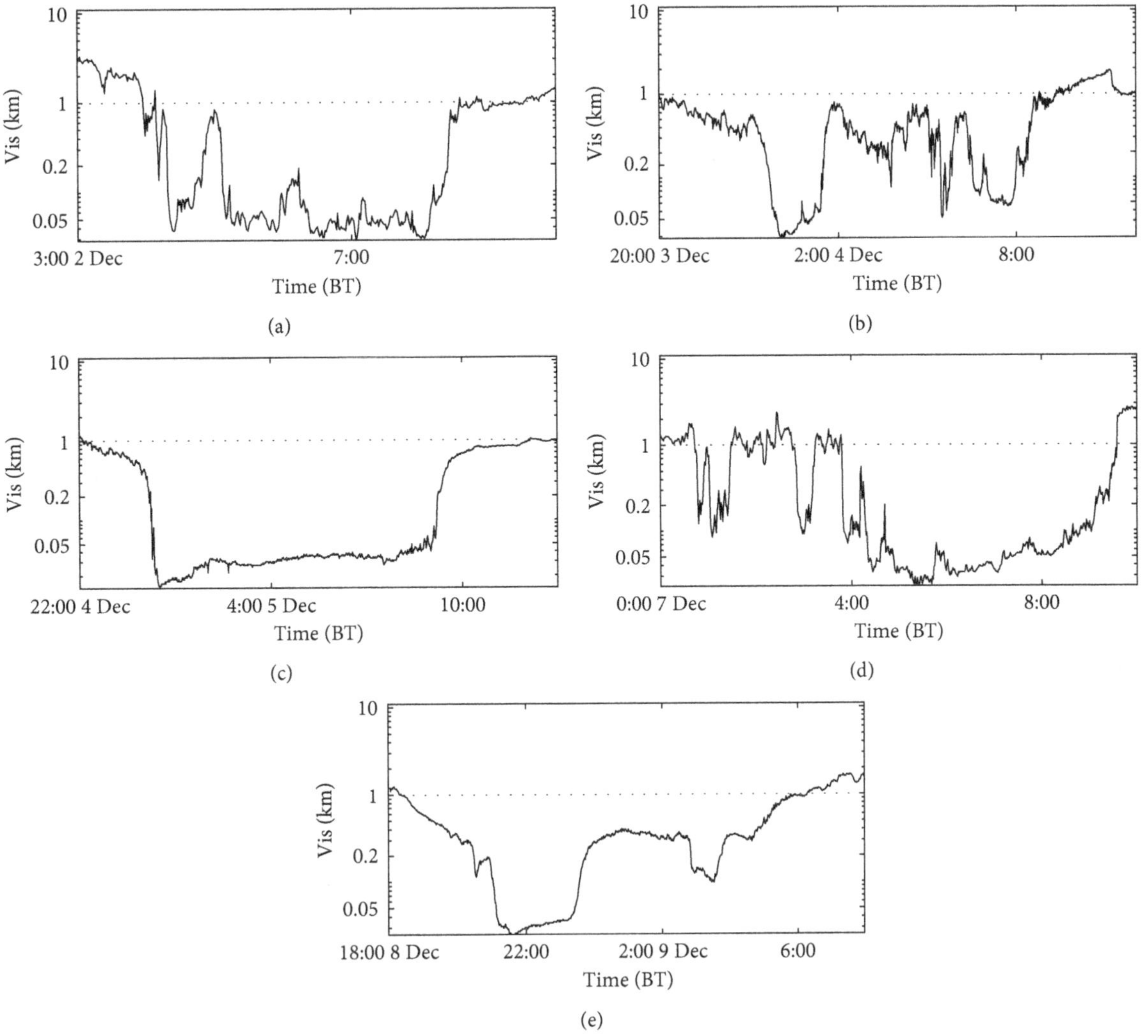

FIGURE 3: Temporal variation of visibility during five explosive processes at the Sheyang site. (a) Case 1: from 03:00 BT on 2 December to 10:00 BT on 2 December; (b) Case 2: from 20:00 BT on 3 December to 12:00 BT on 4 December; (c) Case 3: from 22:00 BT on 4 December to 13:00 BT on 5 December; (d) Case 4: from 00:00 BT on 7 December to 10:00 BT on 7 December; (e) Case 5: from 18:00 BT on 8 December to 08:00 BT on 9 December.

than those in clear days. This is the result of gas to particle phase conversion under fog conditions. In our study, the increase in $PM_{2.5}$ concentrations could be related to the formation of secondary aerosol in the atmosphere [49–51]. In addition, the strong inversion layer in the highly stable atmospheric condition is also a very important factor causing the high $PM_{2.5}$ concentrations.

4. Weather Condition Beneficial to the Maintenance of Strongly Polluted Fog

The severely polluted fog event lasted for 9 days, and the weather condition contributed to this event. Figure 5 shows the distributions of anomaly fields for geopotential height, temperature, wind speed, and wind field in the middle and lower troposphere in the early December 2013. In the lower troposphere (925 hPa), the air pressure over mainland China was lower, the anomalous westerly or southwesterly wind were located along the east coast in the South China Sea, the Malay Peninsula, and Indonesia, which indicated that the East Asian winter monsoon weakened at that time. Therefore, the southwesterly anomalies appeared in the lower troposphere in Jiangsu, with lower wind speed and higher temperature. In the middle troposphere (500 hPa), a significant positive height anomaly center was formed south of Lake Baikal (110°E, 50°N); meanwhile, a negative anomaly covered the area over (140°E, 27.5°N). Thus, the high-pressure center weakened, the corresponding surface circulation weakened, and the surface wind speed decreased. The East Asia trough enhanced, the cold air invasion strengthened, and the temperature in the middle troposphere became lower over Jiangsu Province, which was related to the frequent passages of the westerly trough. In addition, the area north of 30°N in East China was under easterly anomalies at 700 hPa, which was detrimental to the cold air invasion from the north and made the temperature higher at 700 hPa over Jiangsu Province. To reveal the vertical distribution characteristics of anomalous atmospheric background condition, Figure 6 shows the

Table 1: The beginning and ending times of five explosive processes and the visibility at each time.

	Time	Visibility (km)
Case 1		
Beginning	03:57 BT on 2 December	1.142
Ending	04:24 BT on 2 December	0.037
Case 2		
Beginning	23:11 BT on 3 December	0.517
Ending	23:47 BT on 3 December	0.042
Case 3		
Beginning	23:52 BT on 4 December	0.537
Ending	00:20 BT on 5 December	0.027
Case 4		
Beginning	00:36 BT on 7 December	1.772
Ending	01:04 BT on 7 December	0.085
Beginning	02:41 BT on 7 December	1.587
Ending	02:58 BT on 7 December	0.092
Beginning	03:47 BT on 7 December	1.29
Ending	03:55 BT on 7 December	0.071
Beginning	04:13 BT on 7 December	0.553
Ending	04:27 BT on 7 December	0.033
Case 5		
Beginning	20:18 BT on 8 December	0.298
Ending	21:08 BT on 8 December	0.04

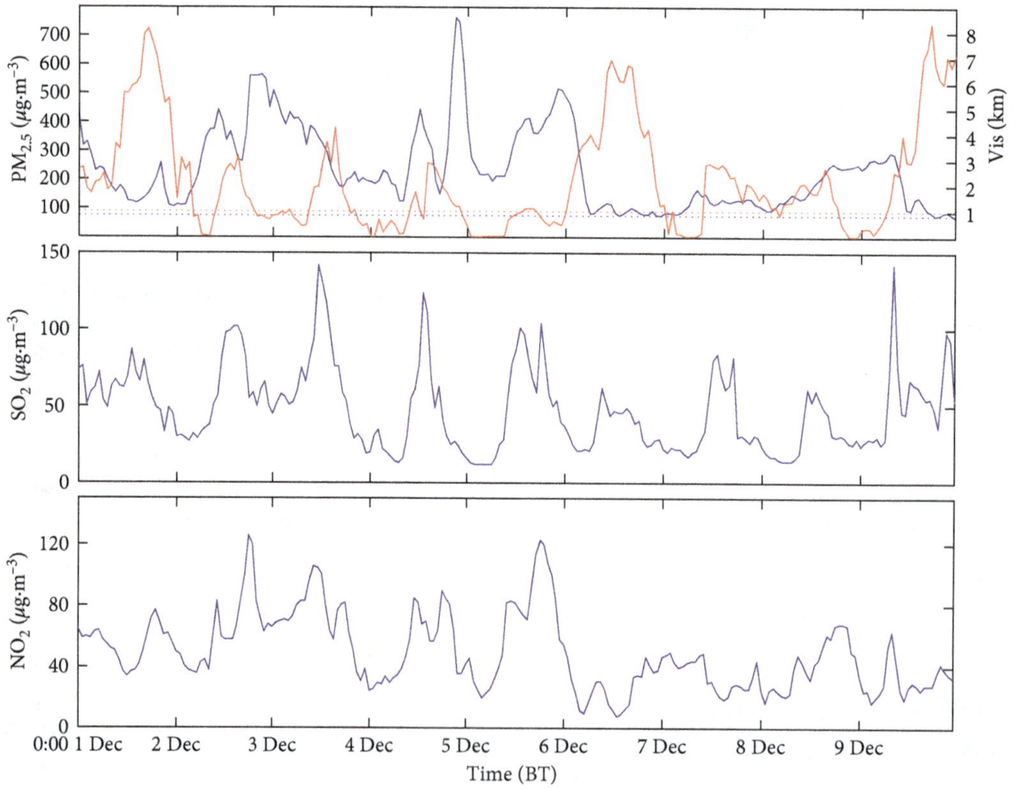

Figure 4: Time series of air pollutants concentration at the Yancheng site from 1 to 9 December 2013. The dotted red line represents 1 km visibility, and the dotted blue line represents 75 $\mu g \cdot m^{-3}$ $PM_{2.5}$ concentration.

anomalous variation of regionally averaged horizontal wind speed and regionally averaged temperature of Jiangsu Province (115–122.5°E, 30–35°N) with height. The horizontal wind speed weakened in the whole troposphere. In the middle and lower troposphere (850 hPa) and above the upper troposphere (200 hPa), the weakening of wind speed decreased with height. In the troposphere from 850 to 300 hPa, it was the opposite; that is, the weakening of wind speed increased with

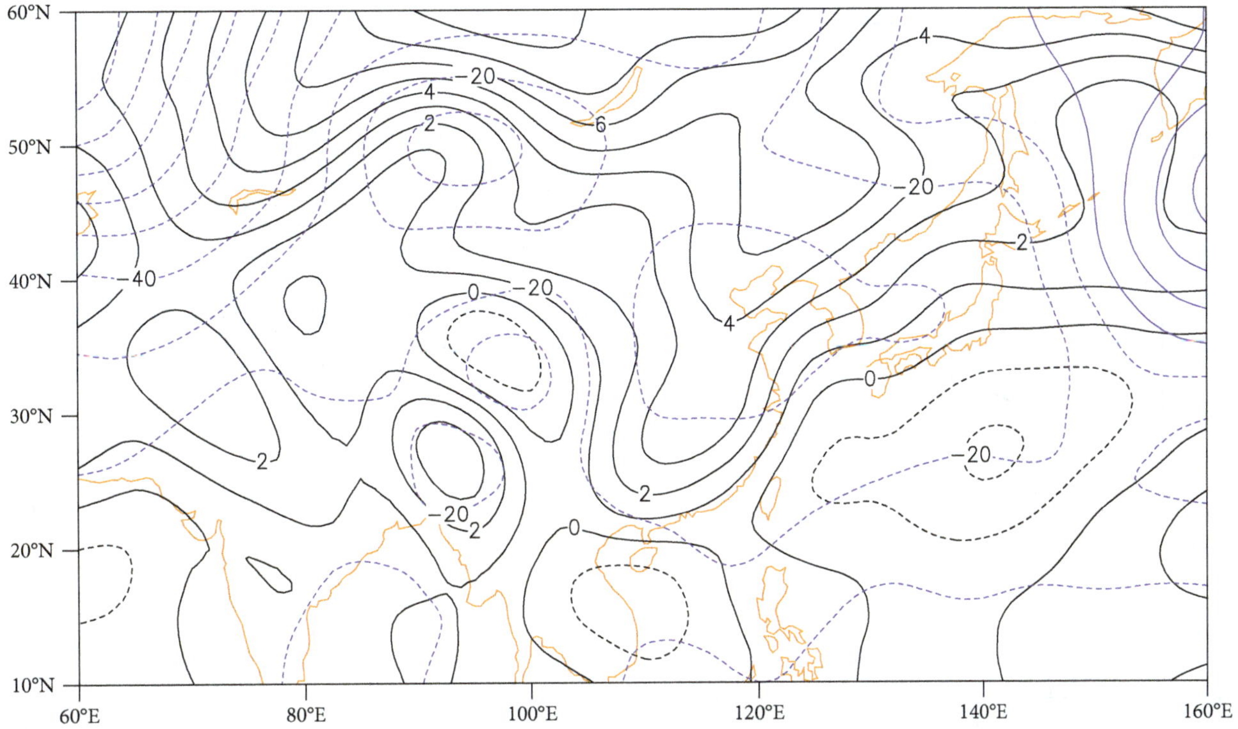

(a)

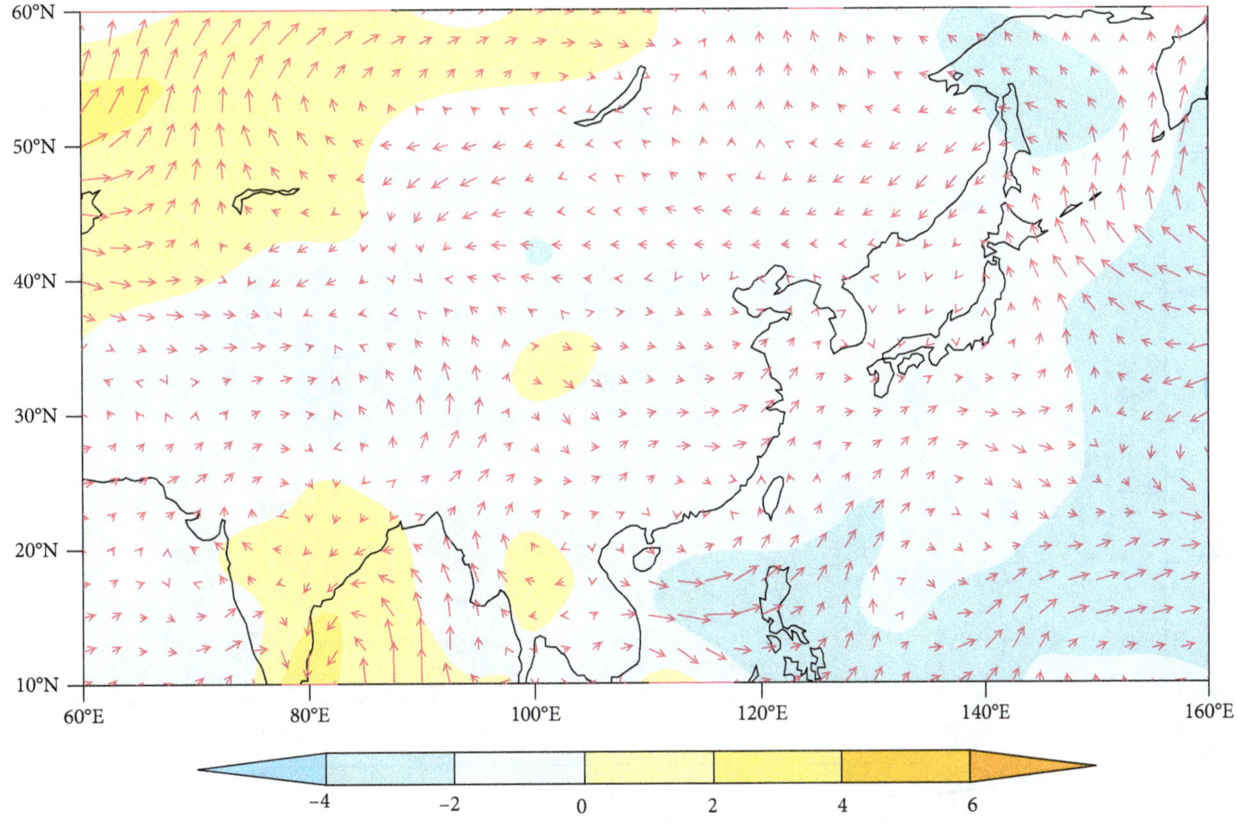

(b)

FIGURE 5: Continued.

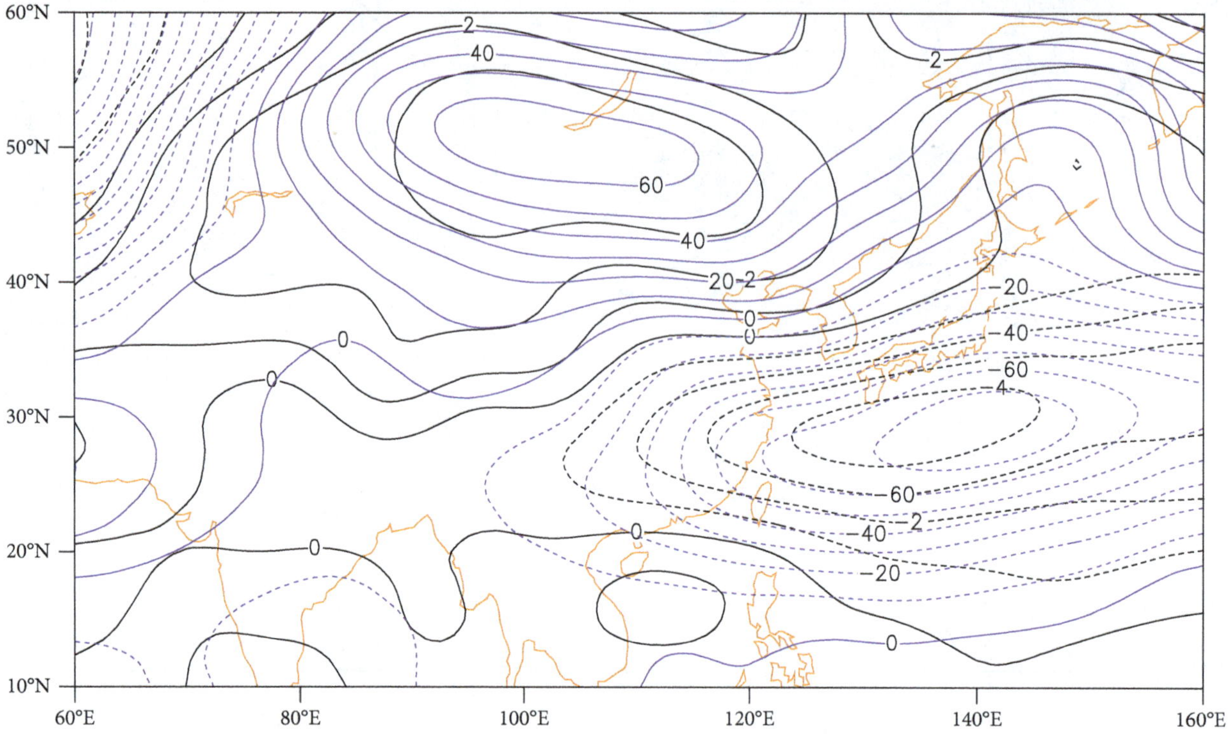

(c)

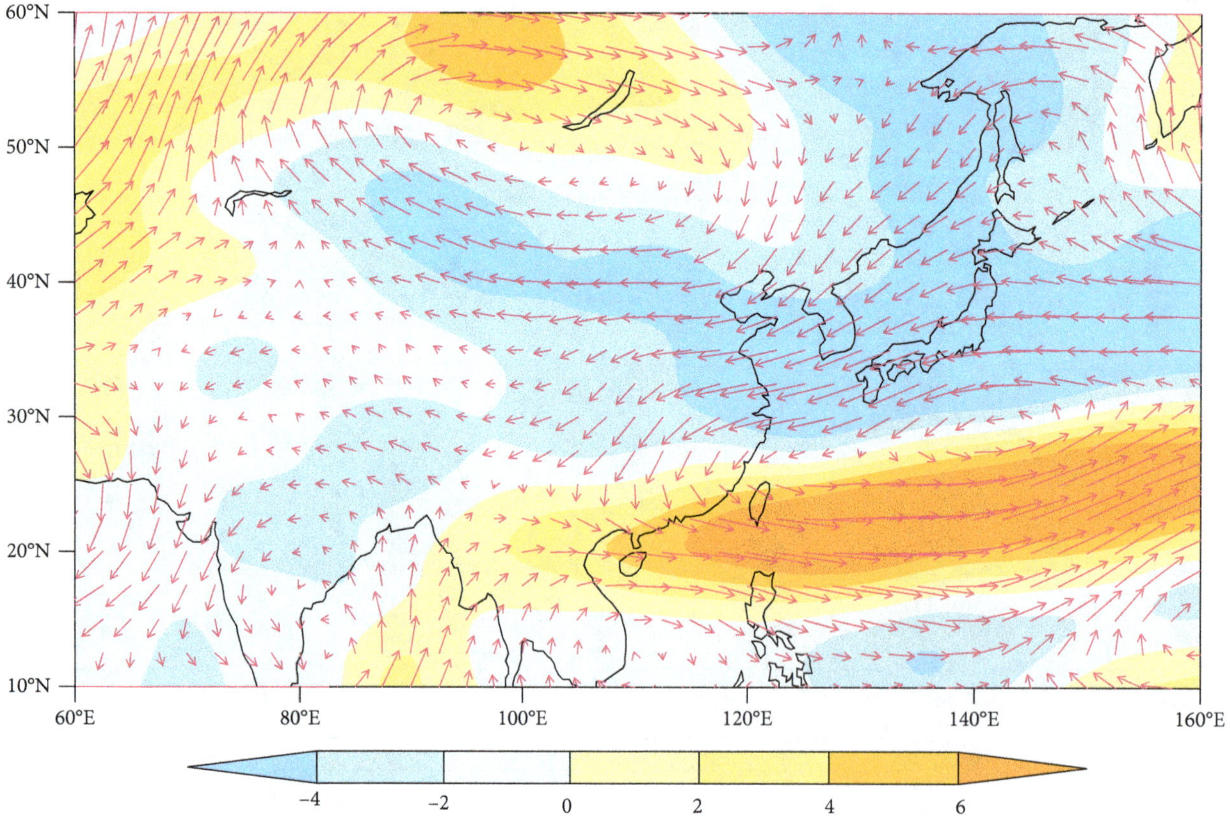

(d)

Figure 5: Spatial patterns of anomalous fields at 925 hPa from 1 to 9 December 2013. (a) Geopotential height and temperature. The blue contours are geopotential height (gpm), and the black contours are temperature (°C). (b) Wind speed and wind direction. The red arrows represent the wind direction (°), and the shading represents the wind speed ($m \cdot s^{-1}$). Subplots (c) and (d) are the same as subplots (a) and (b), respectively, but for anomalous fields at 500 hPa.

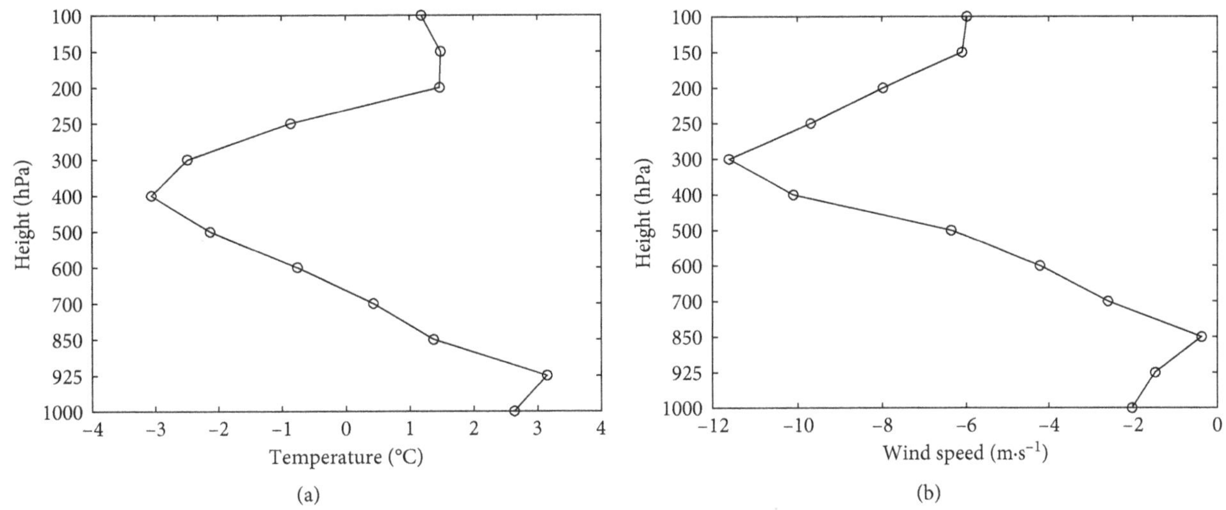

FIGURE 6: Vertical profiles of temperature anomaly (a) and wind speed anomaly (b) of Jiangsu area from 1 to 9 December 2013.

height. The distribution of regionally averaged temperature anomaly with height shows that the whole middle and lower troposphere (700 hPa) had uniform warming, while below 925 hPa, positive temperature anomaly increased with height, which indicated the presence of a thermal inversion layer.

Thus, the high-pressure control for a long time was the cause for the long-lasting polluted fog. Meanwhile, the southwesterly wind anomaly at the middle and lower troposphere was conductive to the transportation of moisture from South China to Jiangsu Province, which provided the moisture for the polluted fog. In addition, the thermal inversion layer in the lower troposphere made the atmosphere more stable and provided thermodynamic condition for the fog development. These atmospheric background conditions were all beneficial to the development and maintenance of the polluted fog.

5. Thermal and Dynamic Structures Beneficial to Maintaining the Strongly Polluted Fog

In the course of this fog event, why did the extremely dense fog with visibility below 50 m last for several days in succession, and why was the pollution so heavy? These are key issues that need to be researched.

Figure 7 shows the temperature and relative humidity profiles based on the radiosonde data at the Sheyang site. The thermal inversion structure was maintained both in the morning and in the evening during the entire fog event. The thickness of the inversion layer was between 100 and 700 m, and the average intensity was 3.5°C/100 m. During 2–5 December of the strongly polluted fog, the top of the inversion layer was under 200 m, and the intensity of inversion, especially at 19:00 BT on 4 December reached maximum of 7°C/100 m. Moreover, the temperature did not decrease above the inversion layer; there was a 100 m thick layer of high temperature, so it was difficult for the thermal inversion layer to disappear. At 07:00 BT and 19:00 BT on 8 December, multilayer inversion appeared, which made the fog vertical structure more stable. The ground-layer inversion was beneficial to the accumulation of moisture and pollutants near the surface, and the upper-layer inversion was not conductive to the upward diffusion of moisture and pollutants, which prevented the dissipation of polluted fog. In the morning of 9 December, with the invasion of cold air from the north, the inversion layer disappeared in the boundary layer, the fog disappeared, and the concentration of $PM_{2.5}$ decreased obviously. Thus, the long-term maintenance of the inversion layer was one of the important conditions for the formation and maintenance of the strongly polluted fog.

To further analyze the inversion structure in the boundary layer, Figure 8 shows the spatial and temporal profiles of wind barb, temperature, and humidity at Shenyang site from 07:00 BT on 1 December to 19:00 BT on 9 December. On 3 and 4 December, the temperature of the whole layer rose due to the effects of southerly and easterly airflows, and a high temperature center formed at 200–500 m, whose center value was over 10°C, about 8°C higher than the ground temperature. The warm advection not only transported warm air but also brought in plenty of moisture.

Figure 9 shows time-height cross section of divergence. When the fog occurred, the divergence was positive below 925 hPa, and the divergence center was less than $2 \times 10^{-5} \cdot s^{-1}$. Figure 9 indicates that there was a downdraft in the middle and lower layers, which not only helped to form the inversion, but also transported moisture from the upper layer to the near-surface layer, and the air pollutants could not diffuse upward.

The above results show that strong thermal inversion layer was an important cause for the formation of the heavily polluted fog. The formation of the ground-layer inversion resulted from radiative cooling, while the warm advection above the inversion layer and downdraft were beneficial to the enhancement of the inversion layer.

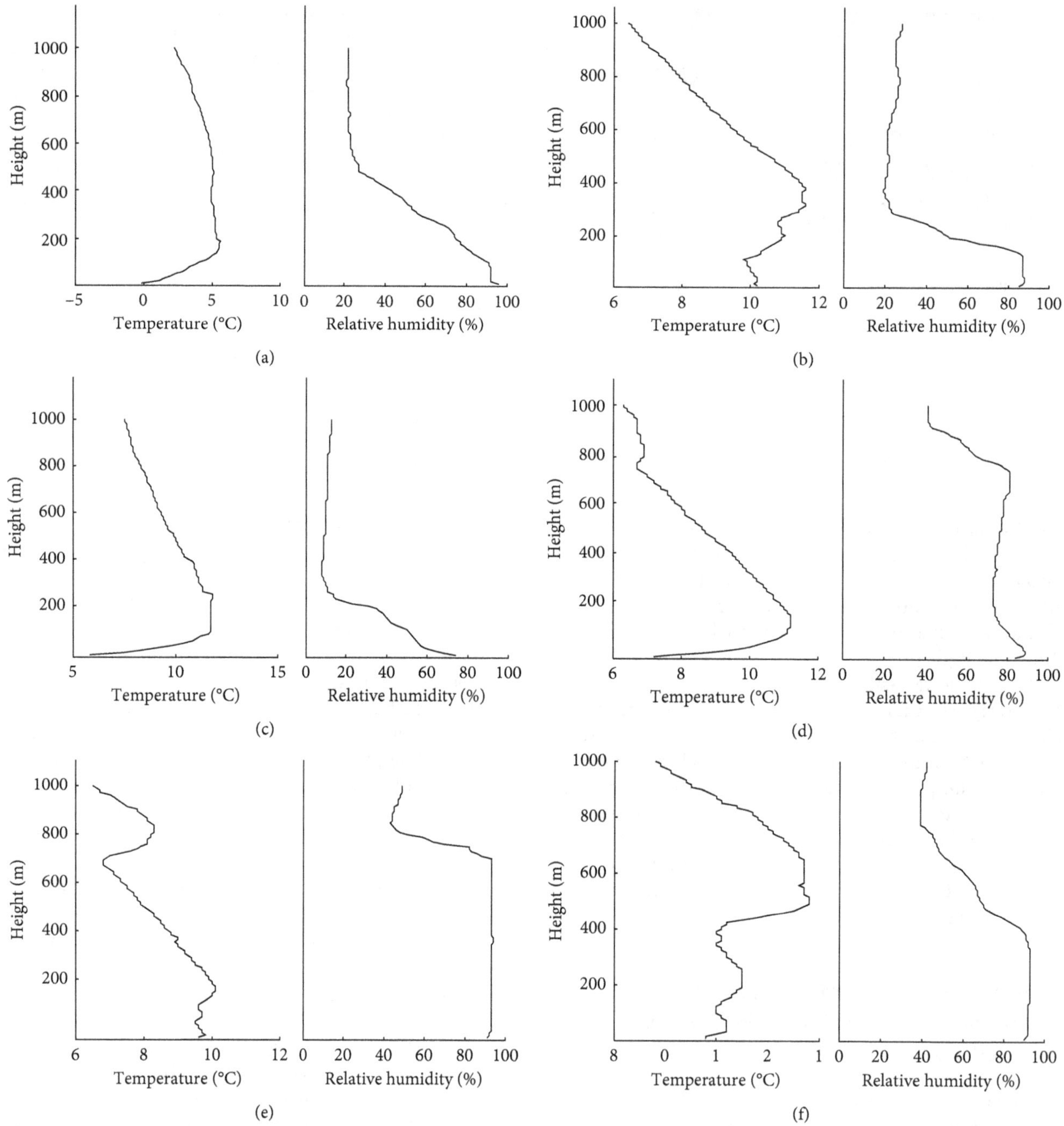

FIGURE 7: Vertical profiles of temperature and relative humidity at the Sheyang site from 1 to 9 December 2013: (a) 07:00 BT on 2 December, (b) 19:00 BT on 3 December, (c) 19:00 BT on 4 December, (d) 19:00 BT on 5 December, (e) 07:00 BT on 8 December, and (f) 19:00 BT on 8 December.

6. Causes for the Explosive Development of Severe Fog

Earlier, we pointed out that the five extremely dense fog processes all displayed explosive enhancement feature. So, what caused such explosiveness? Pu et al. [18, 19] noted that the longwave radiation enhancement at night and the sharp drop of temperature can lead to explosive enhancement; after sunrise, evaporation from the wet surface and cold advection near the surface layer make fog suddenly concentrated; the upper-layer warm advection, cold surface advection, and turbulent mixing can also lead to fog's explosive development. Four explosive fog processes happened during clear night, except for the last one, which was a frontal fog (Figure 10). The radiative cooling was one of the important factors, and the explosive development of fog resulted from accelerated cooling. At the same time, the wind direction suddenly turned to northwesterly, which indicated weak cold air invasion. As shown in Figure 2, in the evening of 4 December, it was clear sky at the Sheyang site with breeze and ground radiative cooling, and the temperature continued to fall by about 7°C from 17:00 BT to

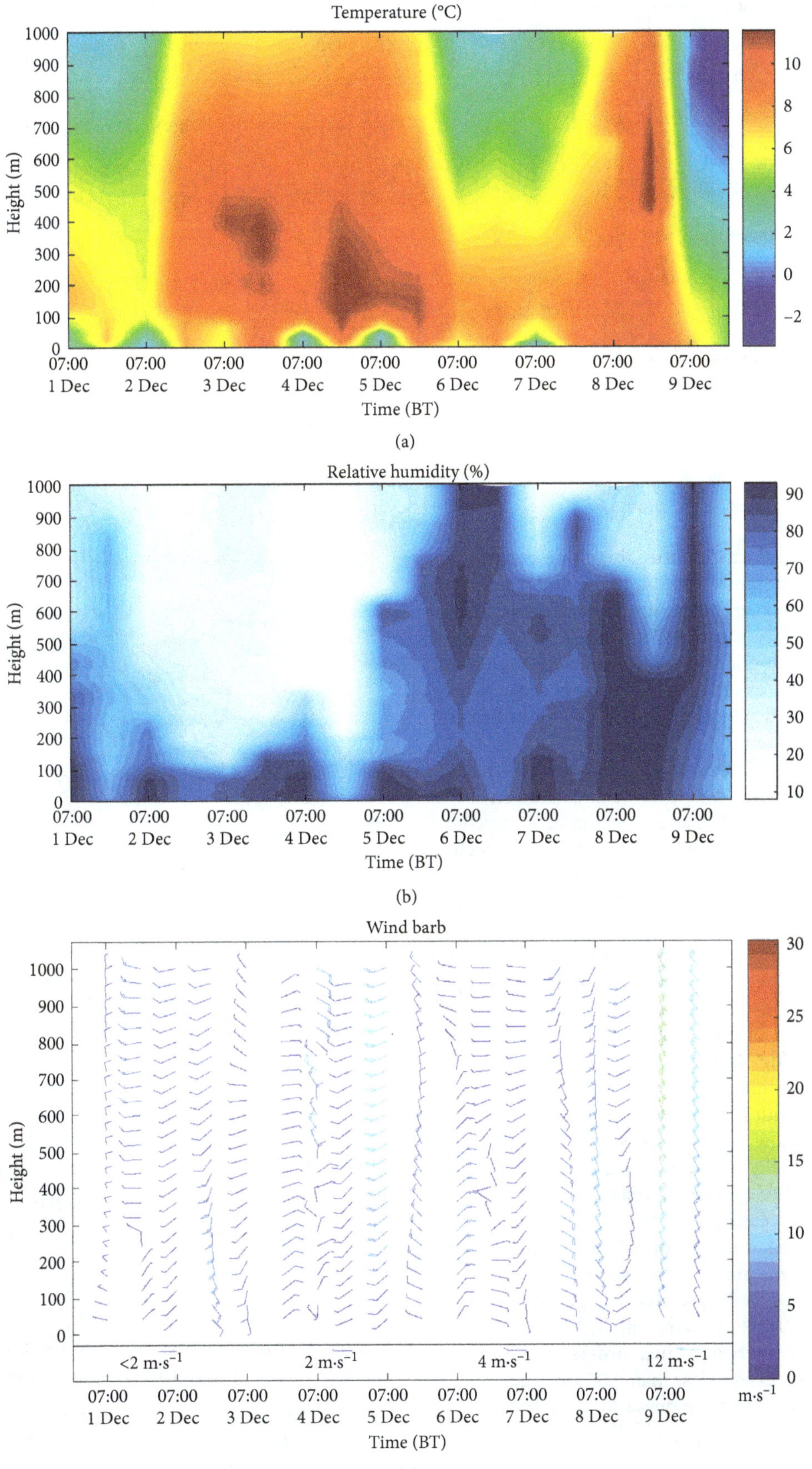

FIGURE 8: Time-height profile of (a) temperature, (b) relative humidity, and (c) wind barb at the Sheyang site from 07:00 BT on 1 December to 19:00 BT on 9 December 2013.

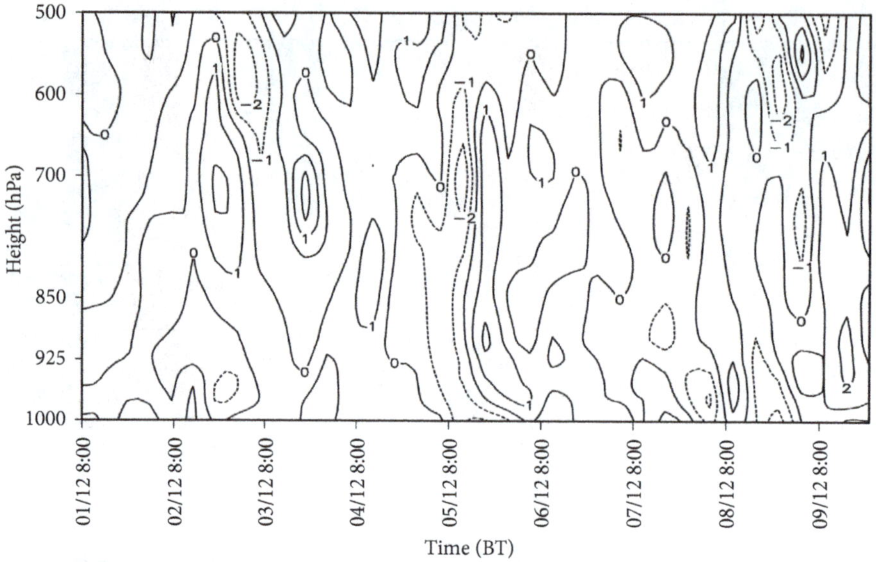

Figure 9: Time-height profile of divergence at the Sheyang site from 1 to 9 December 2013. Unit: $10^{-5} \cdot s^{-1}$.

20:00 BT; then, the fog formed. Since then, due to the latent heat release, the temperature rebounded slightly, and then both temperature and visibility dropped sharply. On 4 December, the visibility was 0.537 km at 23:52 BT, but it dropped to 0.027 km suddenly at 00:20 BT on 5 December, and the fog enhanced explosively. The sharp drop in temperature was due to radiative cooling and the wind direction changed to the northwesterly at the midnight of 00:00 BT, leading to the explosive enhancement. We can see that the explosive enhancement was related to weak cold air invasion. The severe fog that occurred in the morning of 7 December also explosively enhanced after the wind direction turned to westerly. The visibility at 00:36 BT was 1.772 km, but it dropped to 0.085 km quickly at 01:04 BT. It developed into severe fog in less than 30 min. Similarly, the explosive enhancement which occurred at 23:11 BT on 3 December was also related to the change of wind direction and sharp drop of temperature (Figure 2). The difference was that the weak precipitation started at 20:00 BT on 3 December, the cloud dissipated when the fog started, and the early precipitation increased surface humidity, which contributed to the fog enhancement. The continuous variation of fog droplet spectrum observations show that the burst broadening of fog droplet spectrum actually occurs under the condition of enhanced saturation which proceeds microphysical processes (e.g., condensation, collision, deposition, and nucleation) quickly. Therefore, the droplet concentration increases by about one order of magnitude. The spectral width of fog droplet spectrum exceeds 20 μm and can generally reaches 30–40 μm or even 50 μm. As a result, the LWC increases significantly (generally by two to three orders of magnitude). Thus, the visibility decreases to less than 0.05 km; that is, dense fog changes rapidly to extremely dense fog [18, 19].

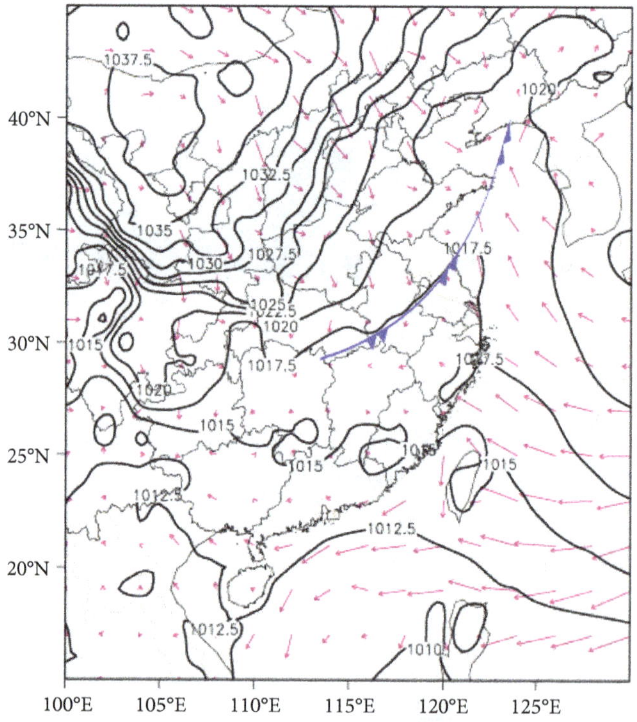

Figure 10: Surface fields: wind, sea-level pressure, and cold front at 20:00 BT on 8 December 2013.

7. Conclusions

In the early December 2013, a dense fog with heavily polluted air lasted for nine days in Yancheng, Jiangsu Province, China. Comprehensive analyses on the event were conducted in this study. The conclusions are as follows:

(1) The fog process was characterized by long duration and strong intensity. The $PM_{2.5}$ concentration increased obviously when severe fog occurred (except for the fog after the rain); that is, strong dense fog and heavy pollution appeared together. The fog was not transformed from haze. Only the polluted gases

(SO_2 and NO_2) exhibited obvious decline when absorbed by fog water.

(2) Atmospheric circulation contributed to the development and maintenance of heavily polluted fog. The existing southerly anomaly in the lower troposphere brought more moisture to the Jiangsu area, which provided moisture condition for fog formation. Meanwhile, the weakening of East Asian winter monsoon led to the weakening of surface wind, which favored fog formation and pollutants' accumulation, but not for transporting fog and pollutants outward via advection, resulting in the maintenance of the polluted fog process. The higher pressure in the middle troposphere controlled the development of convection and helped the polluted fog aggregated in the lower layer of the atmosphere. The decreasing vertical gradient of horizontal wind in the middle and lower troposphere not only suppressed the development of synoptic-scale disturbances by weakening the atmospheric baroclinic instability but also made the atmospheric stratification more stable by weakening vertical mixing.

(3) Under surface radiative cooling, sinking airflow in the lower level and warm, moist airflow from the south, a deep strong inversion of temperature layer was formed, which was the most important thermodynamic condition for the formation of these strong dense fogs, as well an important cause for the heavily polluted air.

(4) The fog process was accompanied by many times of explosive enhancements. Those enhancements happened almost within 30 min, and the shortest one was only 8 min with visibility dropping quickly from 1.29 km to 0.071 km, which was quite rare in previous observations. Weak cold air invasion and radiative cooling were the triggering factors for the explosiveness.

Conflicts of Interest

The authors declare that they have no conflicts of interest.

Acknowledgments

This work was jointly supported by the National Natural Science Foundation of China (Grant no. 41575135), the Natural Science Foundation of Jiangsu Province (Grant nos. BK20161073 and BE2016810), and the Beijige Foundation (Grant nos. BJG201503 and BJG201505).

References

[1] I. Gultepe, R. Tardif, S. C. Michaelides et al., "Fog research: a review of past achievements and future perspectives," *Pure and Applied Geophysics*, vol. 164, no. 6-7, pp. 1121–1159, 2007.

[2] J. Li, J. Sun, M. Zhou et al., "Observational analyses of dramatic developments of a severe air pollution event in the Beijing area," *Atmospheric Chemistry and Physics*, vol. 18, no. 6, pp. 3919–3935, 2018.

[3] J. Bartok, A. Bott, and M. Gera, "Fog prediction for road traffic safety in a coastal desert region," *Boundary-Layer Meteorology*, vol. 145, no. 3, pp. 485–506, 2012.

[4] G. Fu, P. Li, J. G. Crompton et al., "An observational and modeling study of a sea fog event over the Yellow Sea on 1 August 2003," *Meteorology and Atmospheric Physics*, vol. 107, no. 3-4, pp. 149–159, 2010.

[5] H. Tanaka, S. Honma, M. Nishi et al., "Acid fog and hospital visits for asthma: an epidemiological study," *European Respiratory Journal*, vol. 11, no. 6, pp. 1301–1306, 1998.

[6] C. Román-Cascón, C. Yague, M. Sastre et al., "Estimating fog-top height through near-surface micrometeorological measurements," *Atmospheric Research*, vol. 170, pp. 76–86, 2016.

[7] J. Cuxart, C. Yagüe, G. Morales et al., "Stable atmospheric boundary-layer experiment in Spain (SABLES 98): a report," *Boundary-Layer Meteorology*, vol. 96, no. 3, pp. 337–370, 2000.

[8] R. Tardif, "The impact of vertical resolution in the explicit numerical forecasting of radiation fog: a case study," *Pure and Applied Geophysics*, vol. 164, no. 6-7, pp. 1221–1240, 2007.

[9] S. Rémy and T. Bergot, "Assessing the impact of observations on a local numerical fog prediction system," *Quarterly Journal of the Royal Meteorological Society*, vol. 135, no. 642, pp. 1248–1265, 2009.

[10] D. Liu, J. Yang, S. Niu, and Z. Li, "On the evolution and structure of a radiation fog event in Nanjing," *Advances in Atmospheric Sciences*, vol. 28, no. 1, pp. 223–237, 2011.

[11] Z. Li, J. Huang, B. Sun et al., "Burst characteristics during the development of radiation fog," *Chinese Journal of Atmospheric Science*, vol. 23, no. 5, pp. 623–631, 1999, in Chinese.

[12] Z. Li, J. Huang, Y. Zhou et al., "Physical structures of the five-day sustained fog around Nanjing in 1996," *Acta Meteorologica Sinica*, vol. 57, no. 5, pp. 622–631, 1999, in Chinese.

[13] S. Niu, C. Lu, H. Yu, L. Zhao, and J. Lü, "Fog research in China: an overview," *Advances in Atmospheric Sciences*, vol. 27, no. 3, pp. 639–663, 2010.

[14] Q. Fan, A. Wang, S. Fan et al., "Numerical prediction experiment of an advection fog in Nanling mountain area," *Journal of Meteorological Research*, vol. 17, no. 3, pp. 337–349, 2003.

[15] S. Gao, H. Lin, B. Shen, and G. Fu, "A heavy sea fog event over the Yellow Sea in March 2005: analysis and numerical modeling," *Advances in Atmospheric Sciences*, vol. 24, no. 1, pp. 65–81, 2007.

[16] Z. Li, J. Huang, Y. Huang et al., "Study on the physical process of winter valley fog in Xishuangbanna Region," *Journal of Meteorological Research*, vol. 13, no. 4, pp. 494–508, 1999.

[17] L. Wang, S. Chen, and A. Dong, "The distribution and seasonal variations of fog in China," *Acta Geographica Sinica*, vol. 60, no. 4, pp. 134–139, 2005, in Chinese.

[18] M. Pu, G. Zhang, W. Yan et al., "Features of a rare advection-radiation fog event," *Science in China Series D-Earth Sciences*, vol. 38, no. 6, pp. 776–783, 2008, in Chinese.

[19] M. Pu, W. Yan, Z. Shang et al., "Study on the physical characteristics of burst reinforcement during the winter fog of Nanjing," *Plateau Meteorology*, vol. 27, no. 5, pp. 1111–1118, 2008, in Chinese.

[20] D. Liu, W. Yan, J. Yang, M. Pu, S. Niu, and Z. Li, "A study of the physical processes of an advection fog boundary layer," *Boundary-Layer Meteorology*, vol. 158, no. 1, pp. 125–138, 2016.

[21] C. Ma, B. Wu, Y. Li et al., "Mechanisms of formation and maintenance of 12-day long-drawn fog in central and southern Hebei Province," *Plateau Meteorology*, vol. 31, no. 6, pp. 1663–1674, 2012, in Chinese.

[22] D. Liu, M. Pu, J. Yang et al., "Microphysical structure and evolution of a four-day persistent fog event in Nanjing in December 2006," *Aerosol and Air Quality Research*, vol. 24, no. 1, pp. 104–115, 2010.

[23] D. Liu, Z. Li, W. Yan, and Y. Li, "Advances in fog microphysics research in China," *Asia-Pacific Journal of Atmospheric Sciences*, vol. 53, no. 1, pp. 1–18, 2017.

[24] Z. Li, J. Yang, C. Shi, and M. Pu, "Urbanization effects on fog in china: field research and modeling," *Pure and Applied Geophysics*, vol. 169, no. 5-6, pp. 927–939, 2012.

[25] J. L. Collett, D. E. Sherman, K. F. Moore, M. P. Hannigan, and T. Lee, "Aerosol particle processing and removal by fogs: observations in chemically heterogeneous central California radiation fogs," *Water Air and Soil Pollution Focus*, vol. 1, no. 5-6, pp. 303–312, 2001.

[26] I. Gultepe, M. D. Muller, and Z. Boybeyi, "A new visibility parameterization for warm-fog applications in numerical weather prediction models," *Journal of Applied Meteorology and Climatology*, vol. 45, no. 11, pp. 1469–1480, 2006.

[27] I. Gultepe and J. A. Milbrandt, "Microphysical observations and mesoscale model simulation of a warm fog case during FRAM project," *Pure and Applied Geophysics*, vol. 164, no. 6-7, pp. 1161–1178, 2007.

[28] I. Gultepe, G. Pearson, J. A. Milbrandt et al., "The fog remote sensing and modeling field project," *Bulletin of the American Meteorological Society*, vol. 90, no. 3, pp. 341–359, 2010.

[29] M. Haeffelin, T. Bergot, T. Elias et al., "Parisfog: shedding new light on fog physical processes," *Bulletin of the American Meteorological Society*, vol. 91, no. 6, pp. 767–783, 2010.

[30] R. G. Eldridge, "A few fog drop-size distributions," *Journal of Meteorology*, vol. 18, no. 5, pp. 671–676, 1961.

[31] R. G. Eldridge, "Haze and fog aerosol distributions," *Journal of the Atmospheric Sciences*, vol. 23, no. 5, pp. 605–613, 1966.

[32] R. G. Eldridge, "The relationship between visibility and liquid water content in fog," *Journal of the Atmospheric Sciences*, vol. 28, no. 7, pp. 1183–1186, 1971.

[33] J. Goodman, "The microstructure of California coastal fog and stratus," *Journal of Applied Meteorology*, vol. 16, no. 10, pp. 1056–1067, 1977.

[34] K. E. Pickering and J. E. Jiusto, "Observations of the relationship between dew and radiation fog," *Journal of Geophysical Research Oceans*, vol. 83, no. C5, pp. 2430–2436, 1978.

[35] H. Gerber, "Supersaturation and droplet spectral evolution in fog," *Journal of the Atmospheric Sciences*, vol. 48, no. 24, pp. 2569–2588, 1991.

[36] T. W. Choularton, G. Fullarton, J. Latham, C. S. Mill, M. H. Smith, and I. M. Stromberg, "A field study of radiation fog in Meppen, West Germany," *Quarterly Journal of the Royal Meteorological Society*, vol. 107, no. 452, pp. 381–394, 1981.

[37] M. Mohan and S. Payra, "Influence of aerosol spectrum and air pollutants on fog formation in urban environment of megacity Delhi, India," *Environmental Monitoring and Assessment*, vol. 151, no. 1-4, pp. 265–277, 2009.

[38] S. N. Pandis, C. Pilinis, and J. H. Seinfeld, "The smog-fog-smog cycle and acid deposition," *Journal of Geophysical Research Atmospheres*, vol. 95, no. 11, pp. 18489–18500, 1990.

[39] D. J. Jacob, J. M. Waldman, J. W. Munger, and M. R. Hoffmann, "A field investigation of physical and chemical mechanisms affecting pollutant concentrations in fog droplets," *Tellus Series B-Chemical and Physical Meteorology*, vol. 36, no. 4, pp. 272–285, 1984.

[40] H. Köhler, "The nucleus in and the growth of hygroscopic droplets," *Transactions of the Faraday Society*, vol. 32, pp. 1152–1161, 1936.

[41] C. Shi, X. Deng, B. Zhu et al., "Physical and chemical characteristics of atmospheric aerosol under the different weather conditions in Hefei," *Acta Meteorologica Sinica*, vol. 74, no. 1, pp. 149–163, 2016, in Chinese.

[42] China Meteorological Administration, *Grade of Fog Forecast*, The State Standard of the People's Republic of China, GB/T 27964 -2011, Beijing, China, 2011, in Chinese.

[43] China Meteorological Administration, *Observation and Forecasting Levels of Haze*, China Meteorological Press, Beijing, China, 2010, in Chinese.

[44] D. Liu, M. Pu, W. Yan et al., "Study on the formation and the cause of the fog-haze transformation in the lover reaches of Huaihe River," *China Environmental Science*, vol. 34, no. 7, pp. 1673–1683, 2014, in Chinese.

[45] D. Liu, X. Liu, H. Wang et al., "A new type of haze? The December 2015 purple (magenta) haze event in Nanjing, China," *Atmosphere*, vol. 8, no. 12, p. 76, 2017.

[46] Q. Jiang, Y. Sun, Z. Wang, and Y. Yin, "Aerosol composition and sources during the Chinese Spring Festival: fireworks, secondary aerosol, and holiday effects," *Atmospheric Chemistry and Physics*, vol. 15, no. 11, pp. 6023–6034, 2015.

[47] M. C. Barth, D. A. Hegg, and P. V. Hobbs, "Numerical modeling of cloud and precipitation chemistry associated with two rainbands and some comparisons with observations," *Journal of Geophysical Research Atmospheres*, vol. 97, no. D5, pp. 5825–5845, 1992.

[48] Q. Zhang, X. Tie, W. Lin et al., "Variability of SO_2 in an intensive fog in North China Plain: evidence of high solubility SO_2," *Particuology*, vol. 11, no. 1, pp. 41–47, 2013.

[49] W. Birmili and A. Wiedensohler, "New particles formation in the continental boundary layer: meteorological and gas phase parameter influence," *Geophysical Research Letters*, vol. 27, no. 20, pp. 3325–3328, 2000.

[50] J. Du, T. Cheng, M. Zhang et al., "Aerosol size spectra and particle formation events at urban Shanghai in eastern China," *Aerosol and Air Quality Research*, vol. 12, no. 6, pp. 1362–1372, 2012.

[51] H. Guo, D. Wang, K. Cheung, Z. H. Ling, C. K. Chan, and X. H. Yao, "Observation of aerosol size distribution and new particle formation at a mountain site in subtropical Hong Kong," *Atmospheric Chemistry and Physics*, vol. 12, no. 20, pp. 9923–9939, 2012.

12

Combining of the *H*/*A*/Alpha and Freeman–Durden Polarization Decomposition Methods for Soil Moisture Retrieval from Full-Polarization Radarsat-2 Data

Qiuxia Xie,[1,2,3] Qingyan Meng,[1,3] Linlin Zhang,[1,2,3] Chunmei Wang,[1,3] Qiao Wang,[4] and Shaohua Zhao[4]

[1]*Institute of Remote Sensing and Digital Earth, Chinese Academy of Sciences, Beijing 100101, China*
[2]*University of Chinese Academy of Sciences, Beijing 100101, China*
[3]*Sanya Institute of Remote Sensing, Sanya 572029, China*
[4]*Satellite Environment Center, Ministry of Environmental Protection, Beijing 100094, China*

Correspondence should be addressed to Qingyan Meng; mengqy@radi.ac.cn

Academic Editor: Stefania Bonafoni

Soil moisture (SM) plays important roles in surface energy conversion, crop growth, environmental protection, and drought monitoring. As crops grow, the associated vegetation seriously affects the ability of satellites to retrieve SM data. Here, we collected such data at different growth stages of maize using Bragg and X-Bragg scattering models based on the Freeman–Durden polarization decomposition method. We used the *H*/*A*/Alpha polarization decomposition approach to extract accurate threshold values of decomposed scattering components. The results showed that the H and Alpha values of bare soil areas were lower and those of vegetated areas were higher. The threshold values of the three scattering components were 0.2–0.4 *H* and 7–24° Alpha for the surface scattering component, 0.6–0.9 *H* and 22–50° Alpha for the volume scattering component, and other values for the dihedral scattering component. The SM data retrieved (using the X-Bragg model) on June 27, 2014, were better than those retrieved at other maize growth stages and were thus associated with the minimum root-mean-square error value (0.028). The satellite-evaluated SM contents were in broad agreement with data measured in situ. Our algorithm thus improves the accuracy of SM data retrieval from synthetic-aperture radar (SAR) images.

1. Introduction

Soil moisture content (SMC) is a key in study of agricultural production, environmental protection, and surface energy conversion, such as drought monitoring and dust storm monitoring [1–3]. SMC can be retrieved by evaluating the backscattering coefficients associated with microwave remote sensing data. A variety of theoretical, empirical, and semiempirical SM retrieval models have been developed, including the classical IEM, AIEM, Oh, Dubois, and Shi models [4]. However, vegetation adversely affects the reliability of SMC retrieval [5]. It is difficult to distinguish the various scattering components of vegetation-covered surfaces [6]. The radar was frequently used in detecting SMC; however, the signals are attenuated by vegetation. Therefore, the sensitivity of the radar signal detecting SMC is reduced by vegetation. This is a major problem for soil moisture retrieval from radar data [7]. Water cloud model is widely used in evaluating low vegetation area. It is the most common retrieval method used [5]. This model features only two scattering mechanisms: surface scattering from bare soil and volume scattering from vegetation [6]. As the crop grows, the scattering mechanisms change. Surface scattering changes with the growth of the crop. Therefore, we need to study the scattering at different growth stages. It is difficult to retrieve the SMC using only water cloud model at different growth stage of maize [8].

C-band Radarsat-2 has the characteristics of full polarization with higher resolution (8 m). C-band Radarsat-2 could provide more information in SMC retrieval by full-polarization characteristics (HH, VV, HV, and VH) [9]. The data can be decomposed into various scattering mechanisms (two-, three-, and four-component mechanisms) [10]. Currently, the principal decomposition methods used are the Freeman–Durden, Yamaguchi, Vanzly, *H*/*A*/Alpha, and Arii algorithms [11], particularly the Freeman–Durden method [12]. The SAR data are divided into a surface scattering component associated with bare soil, a volume scattering component associated with vegetation, and a dihedral scattering component associated with both bare soil and vegetation [13]. The advantages of the method include simple direct separation of the components and elimination of vegetation scattering signals. The surface scattering component pertains to bare soil only. Theoretically, SMC retrieval is more accurate when the surface scattering component is used. Therefore, it is essential to maximize the surface scattering component when building an SMC retrieval model. However, the decomposition of the data is a challenge. The *H*/*A*/Alpha polarization decomposition method divides the full-polarization SAR data into a scattering entropy (*H*), an antientropy (*A*), and an alpha angle (Alpha) [11]. Using this full-polarization model, Cloude and Pottier divided the feature spaces of *H* and Alpha into eight effective areas [14], each of which corresponded to a specific scattering mechanism. Usually, the cutoff value of *H* is 0.5 and that of Alpha is 45°, between surface and vegetation scattering [15]. Based on the feature spaces of *H* and Alpha, surface, volume, and dihedral scattering components can be accurately extracted [16].

In this study, we used the *H*/*A*/Alpha and Freeman–Durden polarization decomposition methods to retrieve SMC at different growth stages of maize (seeding, jointing, heading, and flowering). We incorporated the *H*/*A*/Alpha method to increase the accuracies of the three scattering components (surface, volume, and dihedral). Backscattering and threshold analyses of typical objects were evaluated by calculating backscattering coefficients and the feature spaces of *H* and Alpha. Using both the Bragg and X-Bragg scattering models, SMC was retrieved from the surface scattering data obtained at different maize growth stages. The advantages of this approach are that the *H*/*A*/Alpha method yields quantitative SMC measures using three highly accurate scattering components to retrieve SMC data at different maize growth stages. Thus, SMC retrieval was possible employing a polarization decomposition technique.

2. Study Site and Dataset

2.1. Study Site. Our study site was near Hengshui City, Hebei Province, in the northern plain of China (38°3′00″N, 115°27′54″E) (Figure 1). The plain is also termed the Huang Huai Hai plain and is one of the three major plains of China. Our study site was a typical agricultural region. The major summer crops are maize (ca. 80%), cotton, and peanuts. This study site was approximately 25×25 km in area, and it is dominated by a loamy soil (37.9% sand, 44.6% silt, and 17.5% clay (all w/w)) [17]. The terrain is almost flat, the soil is homogeneous, and maize is the preferred crop. The climate is continental, characterized by high temperature and severe drought in summer. Most crop growth occurs in summer, at which time the SMC is critical [17].

The maize growing period runs from June to September. Therefore, we performed four field experiments on June 24, July 24, August 14, and September 4, when maize was at the sowing, jointing, heading, and flowering stages, respectively. At the sowing stage, vegetation is lacking (the soil is bare). However, small areas were covered with residual wheat stalks, grass, fruit trees, or early-planted maize. Therefore, use of the bare soil moisture content retrieval method alone was inappropriate; it was necessary to distinguish between various land cover types and adopt our methods accordingly to ensure data accuracy. At the jointing stage, the leaf area index (LAI) of maize is about 1.5, the mean maize height (MH) about 50 cm, and the mean leaf age index (LAGI) about 30%; the wheat stalks disappear gradually. At the heading stage, the LAI is about 3.0, the MH about 150 cm, and the LAGI about 70%. At the flowering stage, the LAI is about 3.5, the MH about 200 cm, and the LAGI about 80% [17].

2.2. Dataset. In this study, data from Radarsat-2 and GF-1 (the first high-resolution satellite) were used to retrieve SMC based on a combination of the *H*/*A*/Alpha and Freeman–Durden polarization decomposition methods. The C-band full-polarization synthetic-aperture radar (SAR) data of the Radarsat-2 high-resolution radar satellite were acquired by the Canadian Space Agency and the MDA Corporation in 2007 [18]. Of the four Radarsat-2 polarization products, the single-look complex (SLC) product is a slanted-range dataset affording optimal resolution, amplitude, and phase information for all beam patterns. Four SLC product scenarios were selected with 8-meter spatial resolution, time resolution of 24 days, central incidence angle of 39°, and swath width of 25 km in the quad-polarization beam mode (Q19) [19]. The GF-1 satellite is a multispectral satellite launched by China in 2013 and affords 8-meter spatial resolution and a time resolution of 4 days [20, 21]. Various ground objects can be distinguished using the GF-1 multispectral data [22].

It was necessary to preprocess both datasets to allow SMC retrieval via polarization decomposition methods [23]. We used ENVI 5.3 software and the NEXT ESA SAR Toolbox to preprocess Radarsat-2 SLC data in terms of multilooking, radiometric correction, geocoding, filtering, and processing. Filtering was used to eliminate speckle noise. During processing of multilooking, the spatial resolutions of SLC data are reduced, but the radiation resolution (which imparts intensity information) improves. GF-1 preprocessing includes radiometric calibration, atmospheric and geometric corrections, and image cropping [23]. After such preprocessing, GF-1 distinguished different ground-cover types well. We used the GF-1 data as auxiliary in terms of SMC retrieval. The data reveal the spatial characteristics and differences among ground objects such as bare soil, urban/rural areas, orchards, grassland, and water bodies.

The four filed experimental days were chosen to coincide with the passage of Radarsat-2 (Table 1). We measured soil and vegetation parameters at 23 study sites (Figure 1). All of 23 study sites were the maize fields. For the convenience of

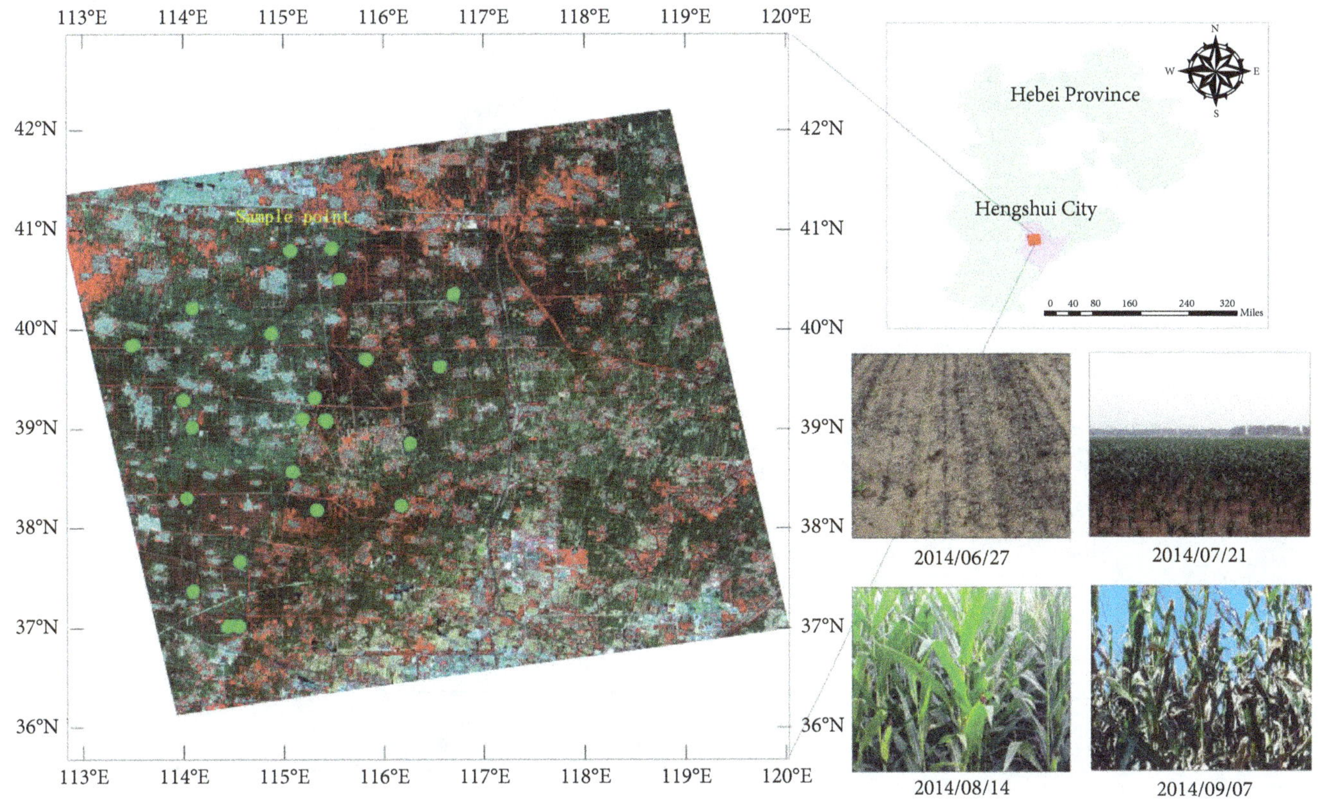

Figure 1: Location of the study and the four experiments.

Table 1: Radarsat-2 and in situ measurement.

Date (yyyy/mm/dd)	Time UTM	Dominant objects	SPN	SMC (g/cm^3)	EWT (kg/m^2)	ε_r	H (m)	ρ_{soil}	RMS_h (cm)	LAI
2014/06/27	10:17:30	Bare soil/grasses/orchard	64	5.62–48.1	0	2.3–20.1	0	1.578	0.4–3.4	0
2014/07/21	10:17:30	Bare soil/maize area/orchard	60	6.4–58.4	0.05–2.93	1.3–23.7	0.57	1.27	0.4–5.0	1.37
2014/08/14	10:17:30	Maize area/orchard	60	9.1–40.9	0.48–5.13	3.7–25.0	2.19	1.33	0.2–3.0	2.89
2014/09/07	10:17:30	Maize area/orchard	60	8.6–39.3	1.53–3.90	3.4–17.8	2.58	1.33	0.4–2.2	3.35

SPN: sample point number; ε_r: soil dielectric constant; ρ_{soil}: mean soil bulk density; EWT: vegetation canopy water content; H: mean maize height.

the experiment, the route of 23 study sites was set up along the road. We assessed SMC, the soil dielectric constant, surface roughness, vegetation water content (EWT), the LAI, and maize height. All parameters were measured three times, and the means were calculated. Half of the data were used to build SMC retrieval model, and the other half were used to verify the models.

The SMC is the most important field parameter. The C-band penetrates the surface to only about 5 cm. Therefore, we measured SMC at that depth using an aluminum box and a 200 cm diameter cutting ring. Then, the soil samples were weighed, brought to the laboratory, and dried at 105°C for about 24 h. The SMC was calculated as the difference between the dry and wet weights. The soil dielectric constant was measured in the same laboratory samples with the aid of an E5071C vector network analyzer (Keysight Technologies, Inc., Santa Rosa, CA, USA). Roughness was estimated using a plate of 1 m in length bearing 101 pins; we measured the root-mean-square heights (RMS_h) and the correlation lengths (L) in two directions (N to S and E to W) and calculated the means. We also measured vegetation parameters at different growth stages. LAI was measured five times with the aid of an LAI-2200C plant canopy analyzer (LI-COR Biosciences, Lincoln, NE, USA), and the data were averaged. Vegetation water content was calculated as the difference between the wet and dry weights. Fresh maize (including the green portions) was dried at 105°C for 40 min and then at 85°C for 48 h.

3. Methodology

Vegetation greatly affects SMC retrieval from SAR data. As ground scattering signals are attenuated by vegetation, ground data are difficult to obtain, compromising SMC retrieval. To eliminate the effects of vegetation, we combined two polarization decomposition techniques ($H/A/$Alpha and Freeman–Durden) [24]. The specific step is shown in Figure 2 for soil moisture retrieval.

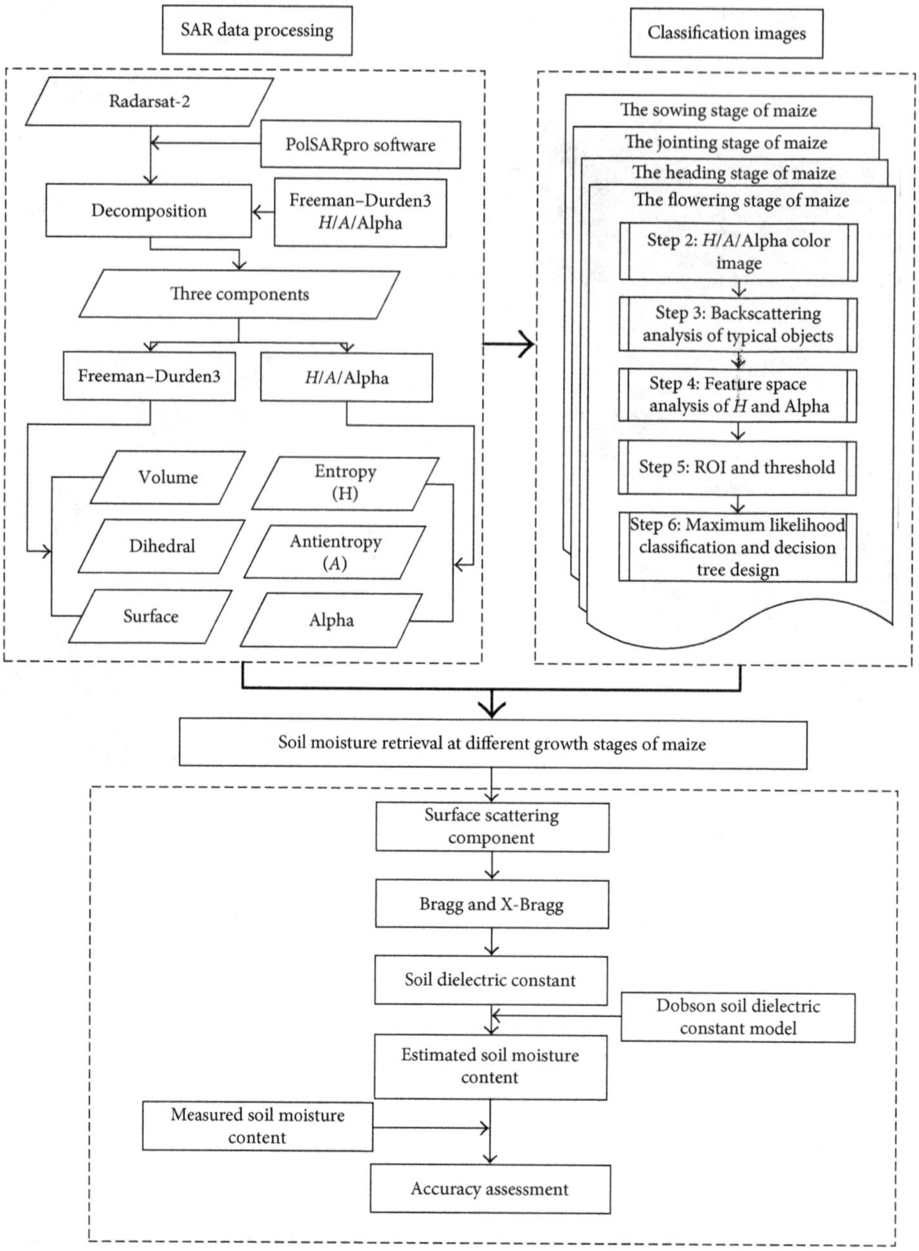

FIGURE 2: Flowchart of soil moisture content retrieval.

3.1. The H/A/Alpha Method. Based on the electromagnetic theory, the different polarization types of SAR data can be obtained, such as HH, VV, and HV. Different polarization SAR data reflected the different characteristics between the object and electromagnetic wave. When ground objects and radar signals interact, backscattering of objects is affected by their characteristics and the type of polarization used. Therefore, ground objects appear different depending on the chosen polarization, allowing the various objects to be distinguished and classified.

H/*A*/Alpha is a polarization decomposition method based on eigen decomposition of polarized SAR data [18]. The biggest advantage of eigen decomposition is that it is not limited by the specific scattering mechanism employed; the eigenvalues do not change with transformation of the antenna coordinate system [25]. Therefore, the method is associated with invariance in the basis, including rotation invariance, rendering the data more reliable.

Backscattering information obtained by radar can be expressed as a scattering matrix, as follows:

$$\mathbf{S} = \begin{bmatrix} S_{hh} & S_{hv} \\ S_{vh} & S_{vv} \end{bmatrix}. \quad (1)$$

After Pauli decomposition of this matrix, the **K** vector is given by

$$\mathbf{K} = \frac{1}{\sqrt{2}} \left[S_{hh} + S_{vv} \quad S_{hh} - S_{vv} \quad 2S_{hv} \right]. \quad (2)$$

As the Pauli basis is orthogonal, the three components of the **K** vector are interrelated. Therefore, SAR data are usually decomposed based on the amplitudes or powers of these three items. The two-order matrix of the **K** vector is regarded as a $[\mathbf{T}_3]$ matrix and also as a **U** matrix [26]. The **U** matrix is given by

$$\mathbf{T}_3 = \mathbf{U}_3 \sum \mathbf{U}_3^*, \tag{3}$$

where $\sum$ is a diagonal matrix consisting of the eigenvalues (λ_i ($i = 1, 2, 3$)) of $\mathbf{T}_3$; $\mathbf{U}_3^*$ is a complex conjugate transposition matrix; and $\mathbf{U}_3$ is an **U** matrix including the eigenvector of $[\mathbf{T}_3]$ [27].

Two important parameters (the scattering entropy and the antientropy) were defined by Cloude [28] in 1986 using the eigenvalues of the coherence matrix based on $\mathbf{T}_3$. The scattering entropy (H) was obtained from the three eigenvalues as

$$\begin{aligned} P_i &= \frac{\lambda_i}{\sum_j \lambda_j}, \\ H &= \sum_{i=1}^{3} -P_i \log_3^{P_i}. \end{aligned} \tag{4}$$

From (4), the range of H is 0-1. The H parameter can be regarded as the output of a stochastic calculation within a resolution unit, based on the backscattering mechanism. In other words, H represents the probability of effective scattering. When H approaches 0, the main types of associated land surfaces are isotropically pure and mediate effective scattering. When H is equal to 0, there is one only nonzero eigenvalue in $[\mathbf{T}_3]$, with rank 1, $\lambda_2 = \lambda_3 = 0$. When H approaches 1, random scattering is in play. When H is equal to 1, the three nonzero eigenvalues are equal [26]. When H lies between 0 and 1, the system changes from isotropic simple scattering to completely random scattering [28]. Therefore, H also reflects coherent nondepolarization backscattering based on the object matrix. In fact, the H values of most natural objects lie between 0 and 1 [26]. Empirically, when the H value is <0.3, the scattering object is weakly depolarized, and the dominant scattering mechanism can be recovered based on the features of a specifically recognizable, equivalent point object. Here, the eigenvector corresponding to the maximum eigenvalue is selected and the others ignored. However, when H is higher, the set of scattering objects is depolarized, and no single equivalent point object can be identified. Thus, the number of recognizable classes falls as H increases [27].

The second important parameter is the antientropy (A) value, defined as the value associated with normalization of the second and third object components [27]. A is given by

$$A = \frac{P_2 - P_3}{P_2 + P_3} = \frac{\lambda_2 - \lambda_3}{\lambda_2 + \lambda_3}. \tag{5}$$

A also ranges from 0 to 1 and can be taken to complement H. A reflects the relative importance of the second and third eigenvalues and can also be viewed as a source of difference when H is >0.7, as the second and third eigenvalues are greatly affected by noise when H is less than this value [29]. Therefore, A also represents the noise level. As H increases, the types of ground objects recognized become fewer in number. Therefore, the use of H alone will be inadequate. Then, A can assist in recognition of the object type [28]. A is particularly key in PolSAR applications. When A reaches a maximum, A can be used to distinguish scattering objects. Usually, when A is higher, only the second scattering process is in play. When A is lower, both scattering processes are equally strong. Therefore, a combination of H and A parameters greatly aids in solving polarization scattering problems because the eigenvalues of the **U** matrix are invariant. Using the A parameter, the object scattering information imparted by the three eigenvalues is evident on polarization spectra created using different combinations of H and A [26]. The four such spectra are given by

$$S \Rightarrow \begin{cases} (1-H) * (1-A) \\ H * (1-A) \\ A * (1-H) \\ H * A. \end{cases} \tag{6}$$

These spectra, termed polarization characteristic decomposition spectra [26], contain all "random" information about the scattering object. The satellite polarization electromagnetic power can be divided into four parts by reference to the spectra, and each is a component of the scattering mechanism.

The scattering angle is the third relevant parameter; this represents the type of scattering mechanism in play. The angle is defined as

$$\alpha = p_1 * \alpha_1 + p_2 * \alpha_2 + p_3 * \alpha_3. \tag{7}$$

When α approaches $\pi/4$, this represents the volume scattering component; when α is $\pi/2$, this represents the dihedral scattering component; and when α is 0, this represents the odd scattering component [29, 30].

3.2. The Freeman–Durden Method. The three-component Freeman–Durden polarization decomposition method was developed in 1998 [31, 32]. The method differs from the $H/A/$Alpha method. The surface, dihedral, and volume scattering components are obtained directly from a scattering coherent matrix [33]. The surface scattering component contains only ground soil information (thus, no vegetation information) [34]. Therefore, the scattering information from vegetation and soil is completely separated. In other words, the method eliminates the vegetation effects on the SMC. The coherent matrices differ for different surface scattering models [34]. Here, we used the Bragg and X-Bragg models. The Bragg model can be viewed as a special case of the X-Bragg model. The X-Bragg model is an improved Bragg model, resolving the problems of decomposition and nonzero cross-polarization [35]. When $\delta = 0$, the Bragg model is equivalent to the X-Bragg model [33]. The coherent matrices are as follows:

$$[\mathbf{T}_{\text{Bragg}}] = f_S \begin{bmatrix} 1 & \beta^* & 0 \\ \beta & |\beta|^2 & 0 \\ 0 & 0 & 0 \end{bmatrix} + f_D \begin{bmatrix} |\alpha|^2 & \alpha & 0 \\ \alpha^* & 1 & 0 \\ 0 & 0 & 0 \end{bmatrix} + \frac{f_V}{4} \begin{bmatrix} 2 & 0 & 0 \\ 0 & 1 & 0 \\ 0 & 0 & 1 \end{bmatrix},$$

$$[\mathbf{T}_{\text{X-Bragg}}] = f_S \begin{bmatrix} 1 & \beta^* \sin c(2\delta) & 0 \\ \beta \sin c(2\delta) & \frac{1}{2}|\beta^2|(1+\sin c(4\delta)) & 0 \\ 0 & 0 & \frac{1}{2}|\beta^2|(1-\sin c(4\delta)) \end{bmatrix} + f_D \begin{bmatrix} |\alpha^2| & \alpha & 0 \\ \alpha^* & 1 & 0 \\ 0 & 0 & 0 \end{bmatrix} + f_V \begin{bmatrix} 2 & 0 & 0 \\ 0 & 1 & 0 \\ 0 & 0 & 1 \end{bmatrix}, \quad (8)$$

where $[\mathbf{T}_{\text{Bragg}}]$ and $[\mathbf{T}_{\text{X-Bragg}}]$ are the Bragg and X-Bragg coherent matrices; f_S, f_D, and f_V are the scattering amplitudes of the odd, dihedral, and volume components, respectively; β and α are the odd and dihedral scattering parameters; and δ is the extent of depolarization [36].

4. Results and Discussion

4.1. Backscattering of Typical Objects. The backscattering characteristics of different objects differ; this is of fundamental importance in terms of image classification using SAR data. It is important to analyze the backscattering characteristics of different objects when simulating backscattering coefficients.

We chose four typical objects and analyzed the spatiotemporal changes in backscattering coefficients; the objects were bare soil fields, maize fields, orchards, and cotton fields. Figure 1 shows the 27 random sample points (10 in maize fields, 9 in orchards, and 8 in cotton fields) and 22 points along tree-lined roads. GF-1 data were used as the auxiliary data to distinguish land cover types. We evaluated changes in backscattering coefficients of Radarsat-2 data for different objects, such as maize field, bare field, and orchard (Figures 3–6, 20140627: the sowing period of maize, 20140721: the jointing period of maize, 20140814: the heading period of maize, and 20140907: the flowering period of maize).

Figure 3 shows that the C-band backscattering coefficients of maize changed over time. The HH coefficient fell as maize grew, peaking at about −5 dB at the jointing stage on July 21, 2014. During flowering, the rate of change was reduced and then ceased. Thus, the C-band coefficient was greatly affected by maize growth. After the heading stage, it is almost impossible to obtain soil surface backscattering coefficients using C-band Radarsat-2 data. Thus, it is essential to reduce the effects of vegetation after heading. The HH and VV coefficients ranged from −4 to −16 dB, and those of HV and VH from −10 to 35 dB. The HH or VV changes (about 16 dB) were less than those of HV or VH (about 20 dB). Therefore, when we compared the changes in the four polarization backscattering coefficients, those of HV and VH were more obvious and coherent than those of HH and VV at all sampling times. Changes in HH or VV backscattering were reflected by a simple cubic regression equation with a correlation coefficient of approximately unity (HV in Figure 3). This equation showed the relationship between the backscattering coefficient of SAR data and maize growth periods. This regression equation indicated how the backscattering coefficients of the C-band Radarsat-2 sensors changed, approximately simulating backscattering at various maize growth stages. With the growth of maize, the HV backscattering coefficient of SAR data is descending in Figure 3. This afforded the theoretical support required for image classification using the *H*/*A*/Alpha polarization decomposition method of C-band Radarsat-2 data.

Figure 4 shows that there is no change rule of the C-band HH and VV backscattering coefficients of orchards. As maize grew, HV and VH backscattering decreased. In particular, HV backscattering ranged from about −25 to −5 dB, a range of 20 dB. Figure 5 shows that the cotton backscattering changes were similar to those of maize. Therefore, maize and cotton cannot be distinguished using only backscattering coefficients; furthermore, those of both crops changed with growth. Thus, maize and cotton fields can be divided into only bare soil, full-cover vegetation field (vegetation field), and mixed fields. However, Figure 6 shows that the backscattering coefficients of tree-lined roads obviously changed with tree growth. On June 27, 2014, the HH and VV backscattering coefficients ranged from 0 to −20 dB. Thereafter, the HH or VV ranged from −4 to −12 dB; the maximal variation fell from 20 to 8 dB, indicating that roadside trees did not change much from July to September. The changes in HV or VH backscattering coefficients were similar to those of HH or VV.

The backscattering characteristics of typical objects over time and space are shown in Figures 3–6. The backscattering characteristics of maize and cotton clearly changed as the crops grew, and this change could be expressed by regression equations. Maize and cotton cannot be distinguished by backscattering coefficients alone, but we show below that maize and cotton fields can be distinguished in the three-field state (bare soil, full-cover (maize and cotton), and mixed). The advantage of this approach is that the vegetation type is unimportant, allowing retrieval of SMC data via polarization decomposition of C-band Radarsat-2 data.

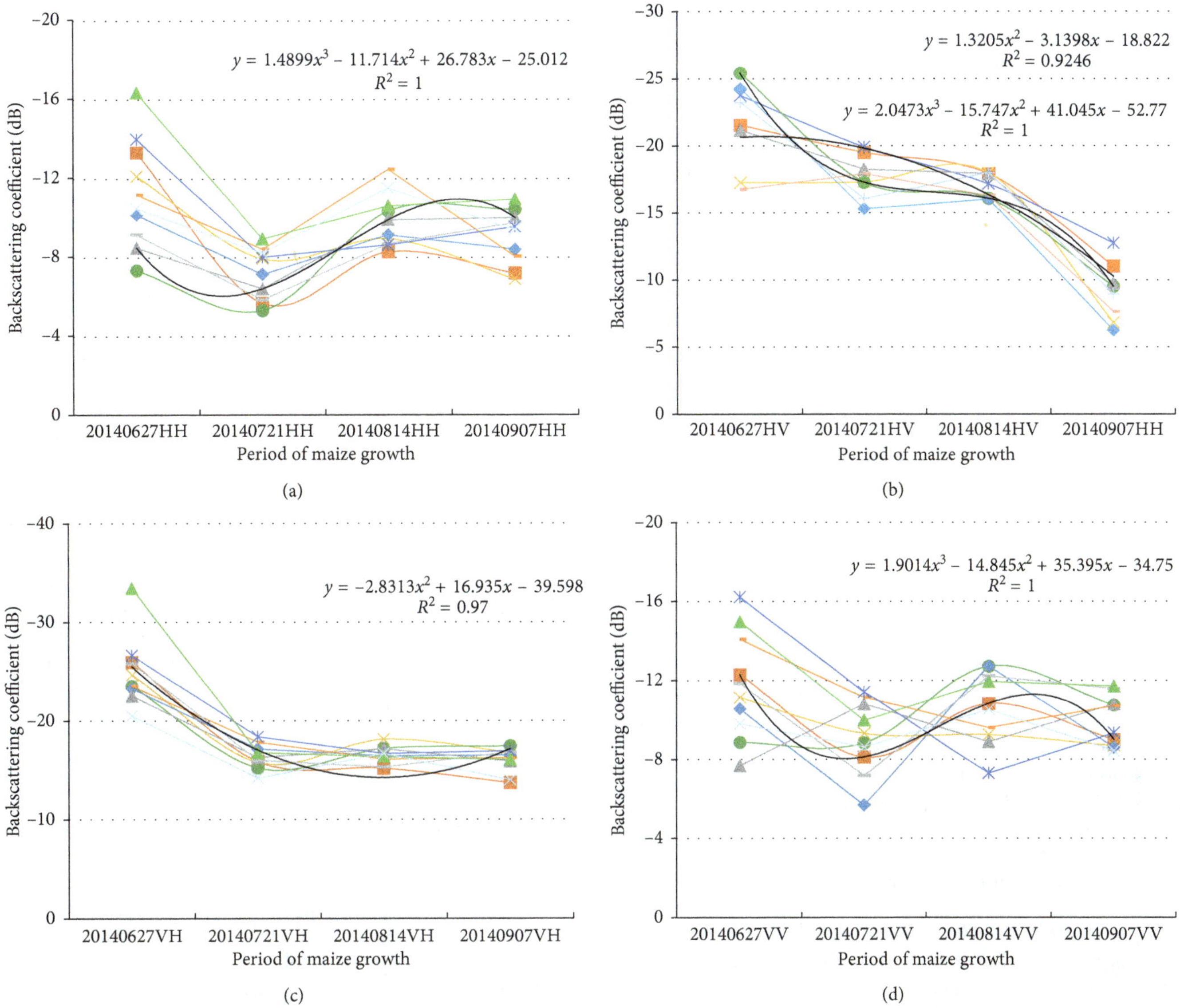

FIGURE 3: Backscattering coefficients used to evaluate maize fields.

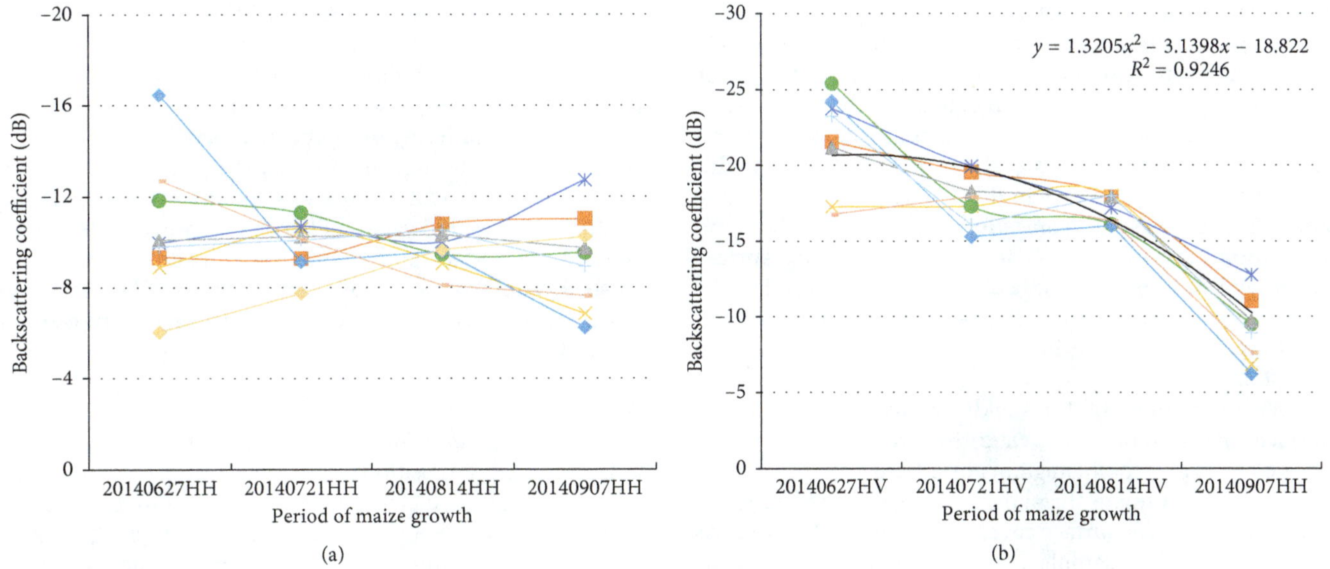

FIGURE 4: Continued.

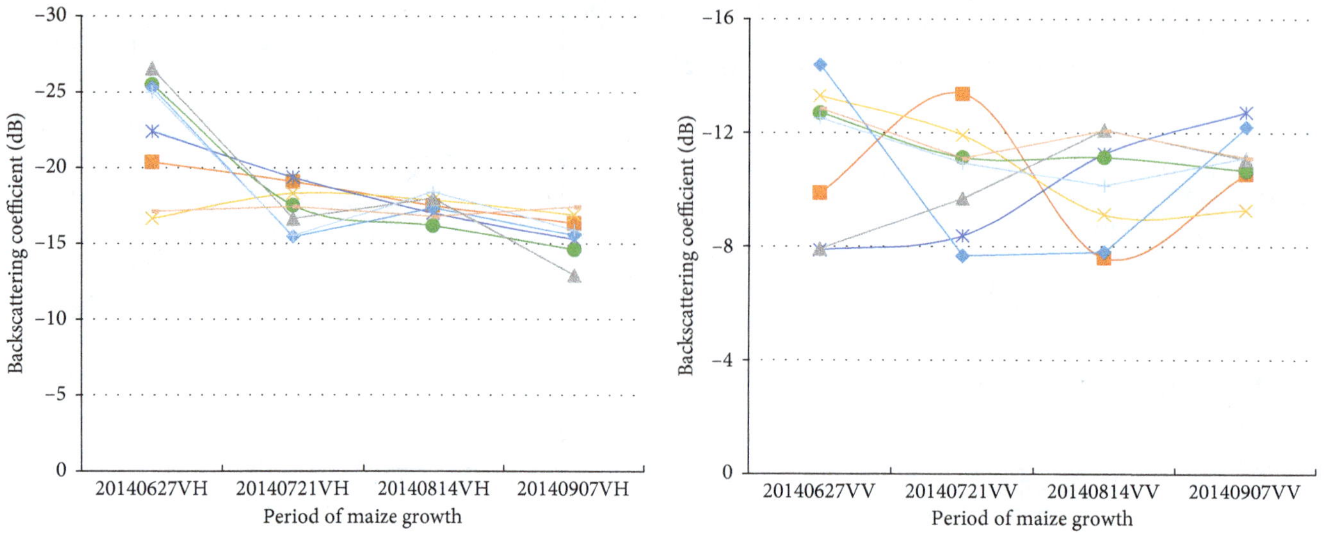

(c) (d)

FIGURE 4: Backscattering coefficients used to evaluate orchards.

4.2. Thresholds of Typical Objects. To optimize SMC estimates, the retrieval methods differ by surface types. By reference to the backscattering characteristics of different objects, namely, bare soil fields, vegetation, mixed fields, and urban/city areas, we used color composites of the June 2014 Radarsat-2 data to this end (Figure 7). We derived the two-dimensional feature spaces of *H* and the Alpha decompositional components of these four surface cover types (Figure 8).

Figure 8 shows that the feature space is occupied by the *H* and Alpha decompositional components of the various surface cover types. Differences in surface cover can generally be distinguished by their ranges of *H* and Alpha values. The bare soil sample points (red) cluster at the lower left of the feature space. Vegetation sample points (gray) cluster at the upper right. Urban/city sample points (yellow) cluster at the upper left. The bare soil and vegetation sample points are closer to each other than the urban/city points. In particular, a distributional rule controlling sample point mixing (orange points) is evident. As vegetation increases, the distribution also increases in a manner that can be expressed by a simple regression equation with a correlation coefficient of about 0.9. This equation showed the relationship between *H* and alpha parameters. By this equation, there is a change in the characteristic that *H* and alpha are increasing from the bare soil to mixing to vegetation. The distributional differences among surface cover types support the image classification afforded by the *H*/*A*/Alpha polarization decomposition method and lay the foundation for the improved SMC retrieval method that follows.

We used different thresholds to distinguish different surface covers. In order to obtain the thresholds of different land cover types, the histograms of different surface covers were used to estimate thresholds. Histograms showing the features of different surface covers are shown in Figure 9. As not all pixels of the sample area are pure, we also show the normal distributions of the *H* and Alpha decomposition components. The histogram distributions are closer (more concentrated) in the sampled areas, and the numbers of pure pixels are higher. Here, threshold values were determined by reference to *H* or Alpha decomposition component frequencies >100, except in urban/city areas. Therefore, the *H* and alpha thresholds of different surface covers can be obtained.

Figure 9 shows that the *H* and Alpha threshold values of bare soil fields were 0.18–0.42 and 7–24°C, respectively. The peak *H* and Alpha values were 0.3 and 15°C. The *H* and Alpha values of vegetated fields were 0.61–88 and 22–50°C, respectively. The peak *H* and Alpha values were 0.65 and 40°C. Bare and vegetated fields could be distinguished by their *H* and Alpha threshold values. However, the *H* and Alpha threshold values of urban/city areas were 0-1 and 18–90°C, respectively. The *H* and Alpha threshold values of mixed fields were 0.2–0.9 and 9–25°C, respectively. The peak *H* and Alpha values were 0.55 and 22°C. These ranges were rather wide. The thresholds of mixed fields and urban/city areas overlapped with those of bare soil and vegetation fields (Table 2).

The main scattering mechanism in play in bare soil areas was surface scattering. In contrast, the principal scattering mechanism in vegetated areas was volume scattering. The *H* and Alpha threshold values of bare soils and vegetated fields allowed the surface and volume mechanisms to be distinguished, affording a useful foundation for SMC retrieval based on the Freeman–Durden polarization decomposition method.

4.3. Image Classification. To retrieve SMC, we used a polarization decomposition technique to decompose the backscattering coefficients of Radarsat-2 into three components: surface, volume, and dihedral scattering. In this study, the two classification methods (maximum-likelihood

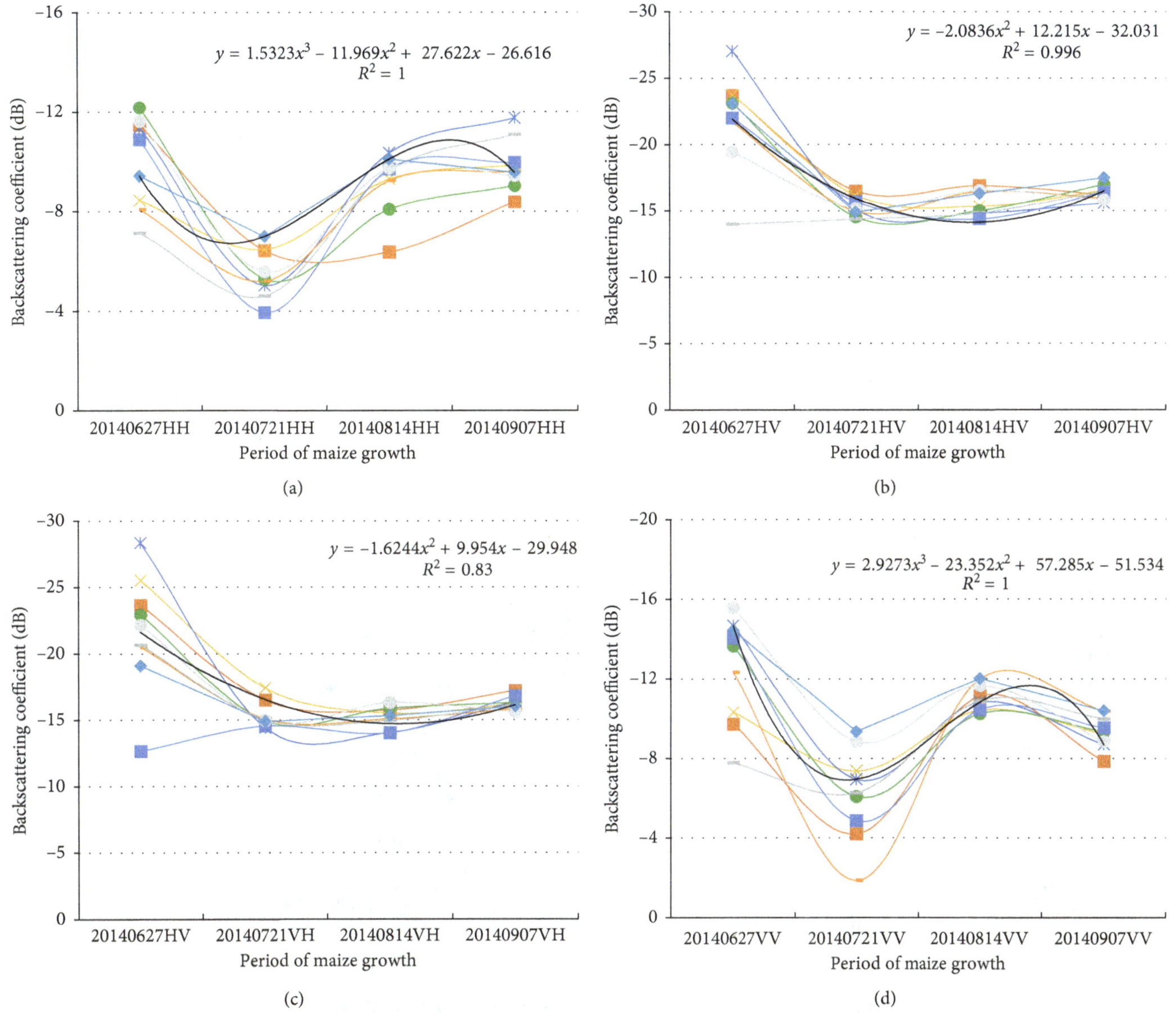

Figure 5: Backscattering coefficients used to evaluate cotton fields.

and threshold classification) were used to acquire the three components (Figure 10). The maximum-likelihood classification is a statistical method employed to calculate correlation probability density functions. This method is commonly used in remote sensing. However, the regions of interest are manually selected. The threshold classification method is more objective; thresholds are determined by the feature spaces of H and Alpha, thus reflecting information characteristics of different objects (Figure 8).

Figure 10 shows that the three components differed as maize grew. The classifications differed in terms of their spatial components. Generally, the proportion of the surface scattering component in the maximum-likelihood method was higher than that of the threshold classification. As maize grew, the surface scattering proportion decreased gradually. On August 14 and September 7, the surface scattering proportions were close to 0. Apart from the dihedral scattering of urban/city areas, volume scattering was almost exclusively in play. On July 21, the dihedral scattering proportion of the maximum-likelihood classification was lower than that of the threshold classification. In reality, some mixed areas were evident between low-growing maize and bare soil surfaces. In such areas, maize stalks and bare soil readily formed dihedral corner reflectors. The threshold classification is then more appropriate.

4.4. Soil Moisture Retrieval. To retrieve SMCs as maize grew, we used the Freeman–Durden method to obtain three backscattering components: surface, dihedral, and volume. These components were then refined using the data of Figure 10. Finally, we used the Bragg and X-Bragg models to calculate soil dielectric constants and the Dobson model to

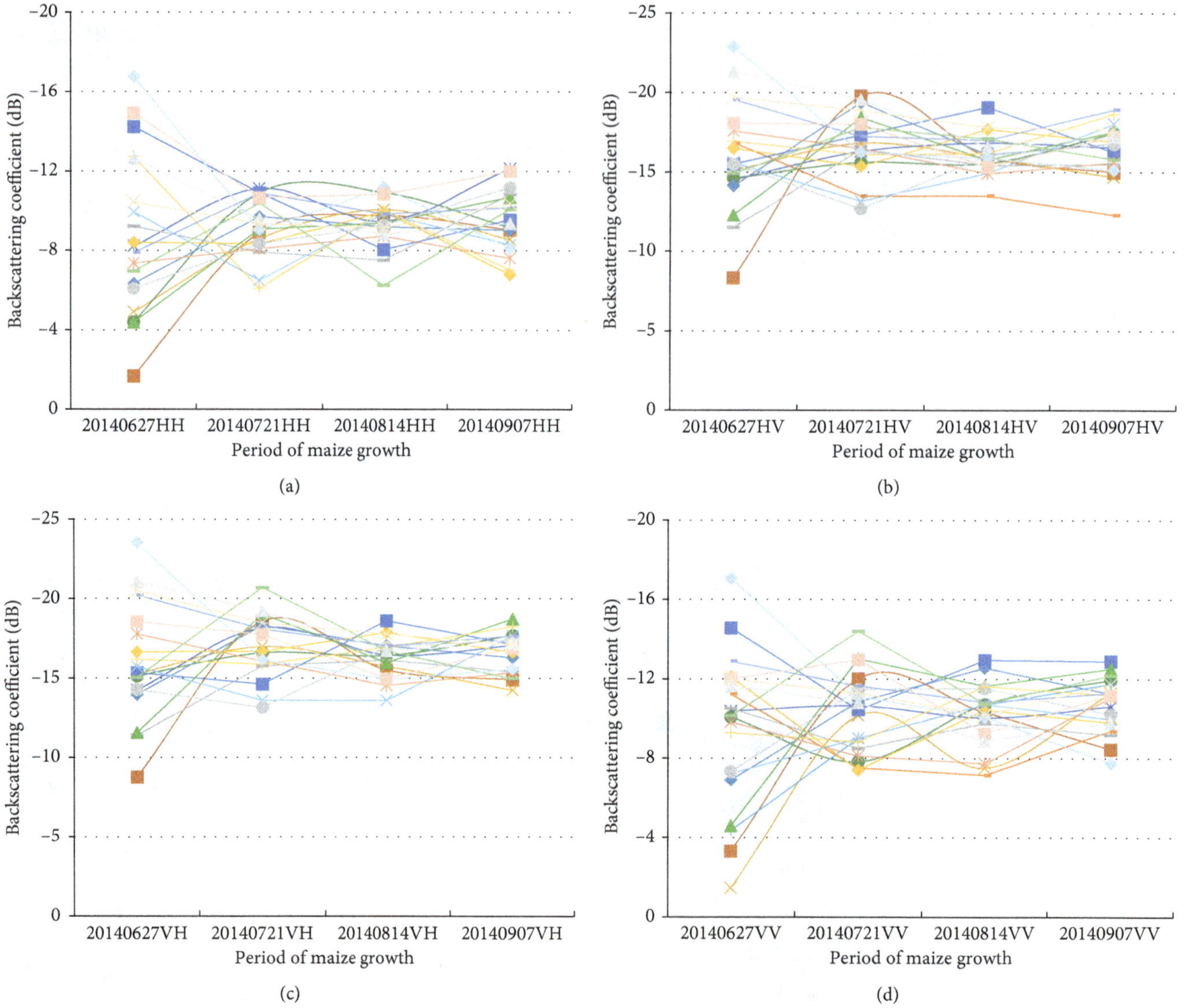

FIGURE 6: Backscattering coefficients used to evaluate roads bordered by trees.

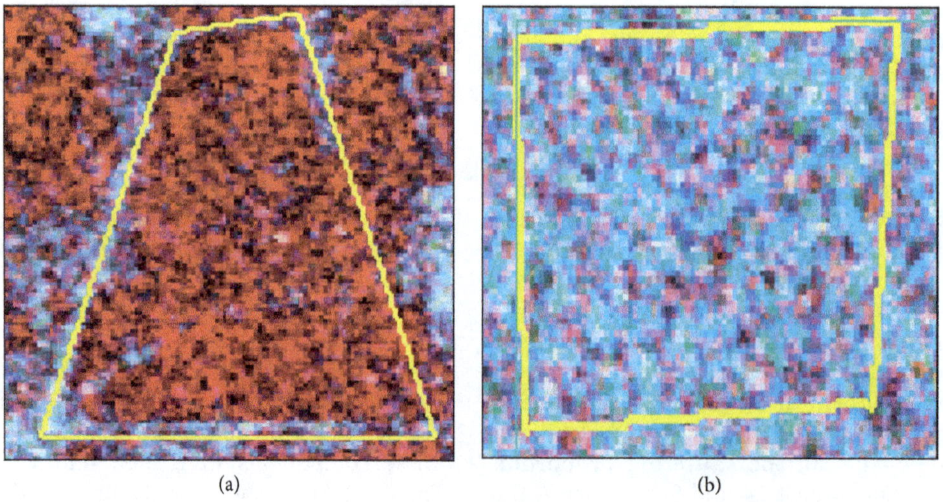

FIGURE 7: Continued.

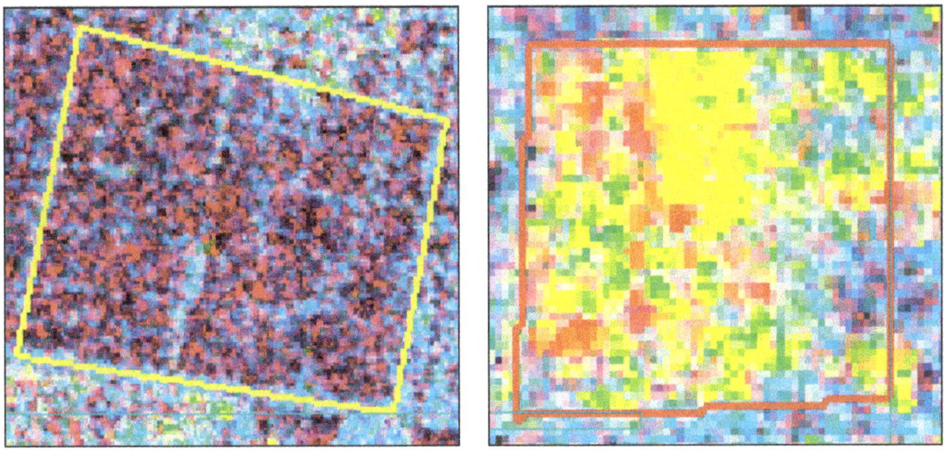

FIGURE 7: Typical examples of the four types of area examined using *H*/*A*/Alpha composite color images based on June 2014 Radarsat-2 data (*H*: red layer; *A*: green layer; Alpha: blue layer). (a) Bare soil, (b) vegetation, (c) mixing, and (d) urban/city.

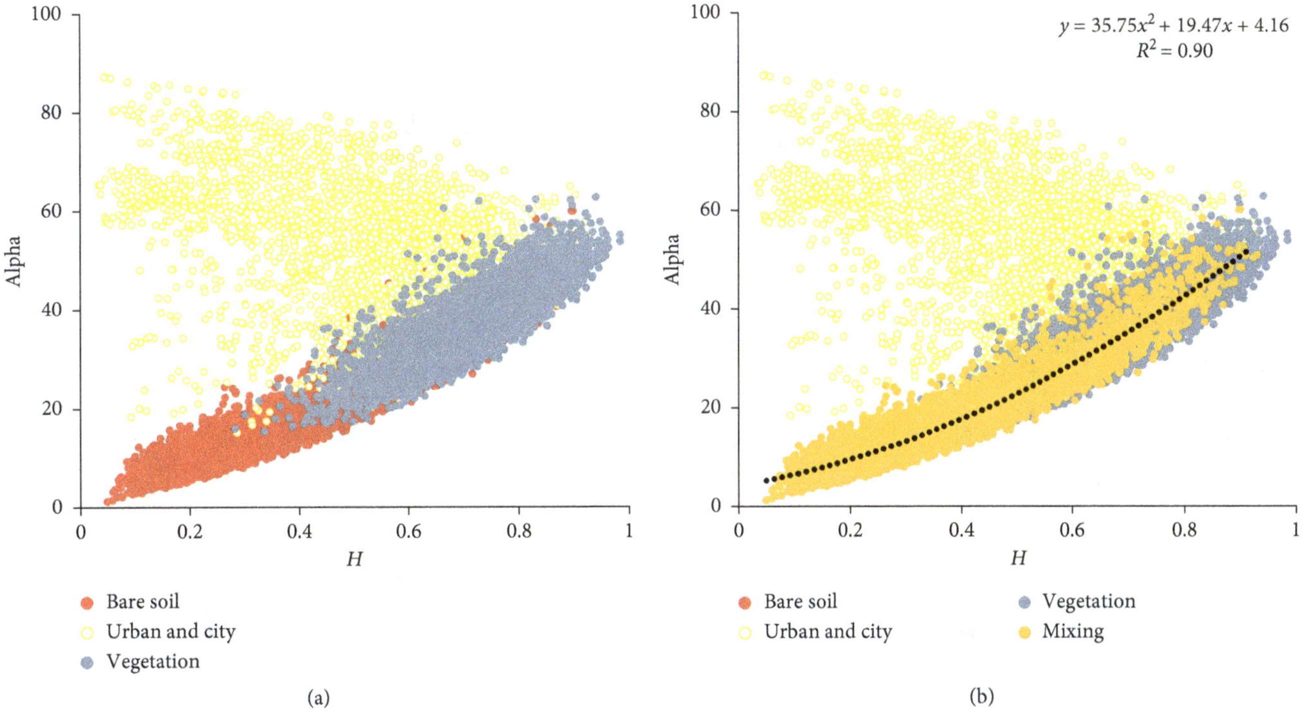

FIGURE 8: Two-dimensional feature space plots of *H* and Alpha (red points: bare soil field; yellow points: vegetated field; orange points: a field with mixed bare soil and vegetation).

estimate soil moisture. We then calculated RMSEs to evaluate data accuracy (Figure 11).

Figure 11 shows that the X-Bragg soil moisture data retrieved on June 27, 2014, were the most accurate of all data obtained during the maize growth period (thus associated with the lowest RMSE value, 0.028). Overall, the X-Bragg model was better than the Bragg model at all maize growth stages. From July to September 2014, the data were less reliable because, as maize grew, moisture signals from the surface soil become increasingly difficult to obtain. Also, the C-band penetration of Radarsat-2 is limited; soil moisture data cannot be reliably retrieved using C-band microwave data, especially in the late stages of maize growth. Although the results are thus variable, our concept is valuable. The L-band penetrated the soil better than did the C-band. Our method can thus be used to retrieve soil moisture data delivered by the L-band.

5. Conclusions

When retrieving soil moisture data employing decomposition technology, it is essential that the three

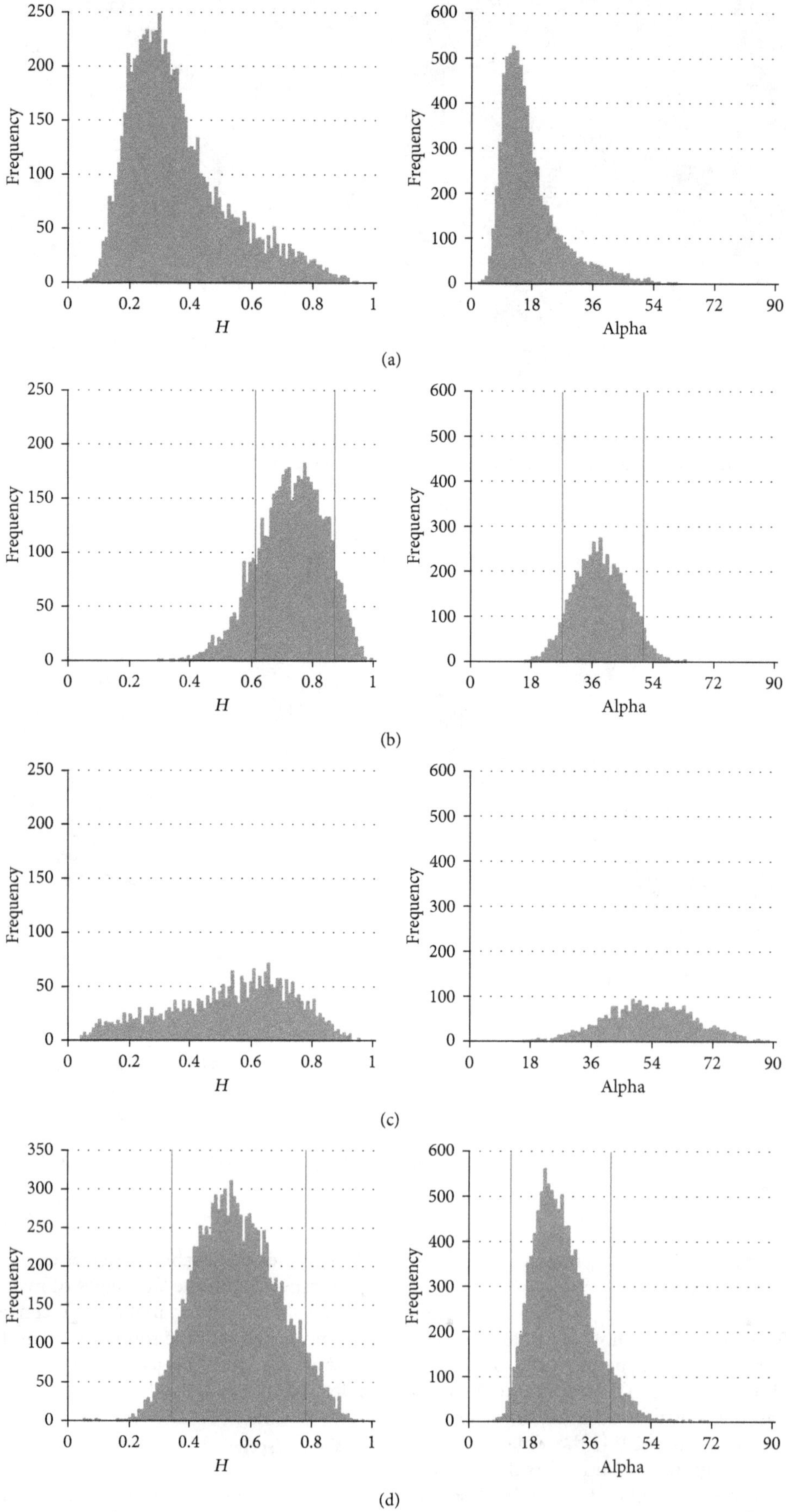

FIGURE 9: Histograms of H and Alpha for four typical types of surface cover. (a) A bare soil field; (b) a vegetated field; (c) an urban/city area; (d) a field with mixed bare soil and vegetation.

Table 2: Values obtained when evaluating surfaces differing in terms of cover.

Types	*H*_min	*H*_max	*H*_mean	al_min (°C)	al_max (°C)	al_mean (°C)	*H*_th	al_th (°C)	*H*_p	al_p (°C)
The bare soil field	0.049	0.930	0.350	1.27	59.96	16.32	0.2–0.4	7–24	0.3	15
The vegetation field	0.282	0.984	0.720	15.40	62.70	37.45	0.6–0.9	22–50	0.7	40
The urban/city field	0.034	0.957	0.526	15.10	97.34	52.47	0-1	18–90	0.65	50
The mixing field	0.041	0.973	0.544	6.65	69.09	26.48	0.2–1	9–50	0.5	25

*H*_min: *H* minima of sample areas; *H*_max: *H* maxima; *H*_mean: *H* mean values; al_min: Alpha minima; al_max: Alpha maxima; al_mean: Alpha mean values; *H*_th: *H* threshold ranges; al_th: Alpha threshold ranges; *H*_p: *H* peak values; al_p: Alpha peak values.

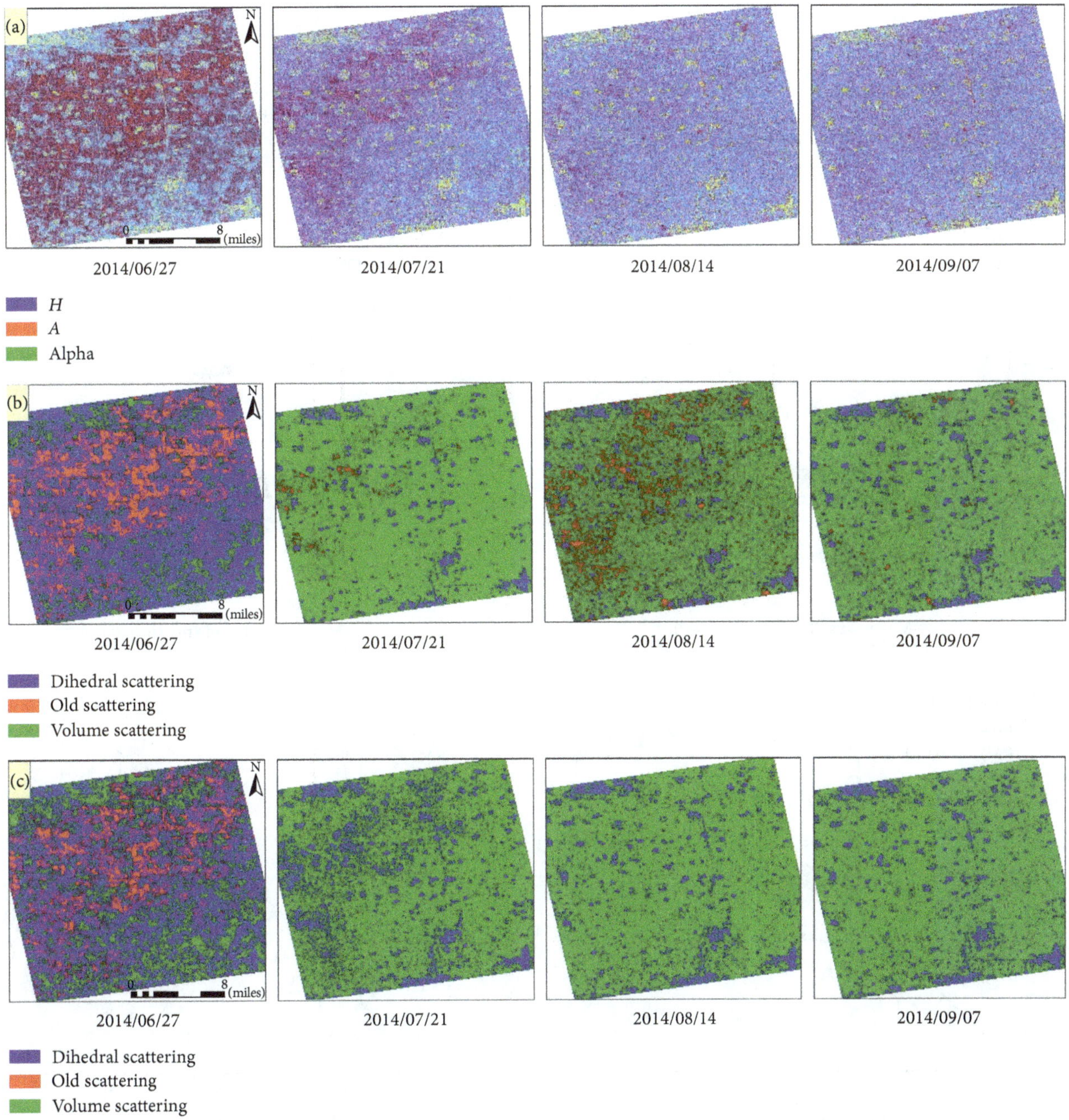

Figure 10: Classification of images based on the *H*/*A*/Alpha decomposition method. (a) Synthetic color images of *H*, *A*, and Alpha. (b) Images obtained using the maximum-likelihood method. (c) Images obtained using the threshold method.

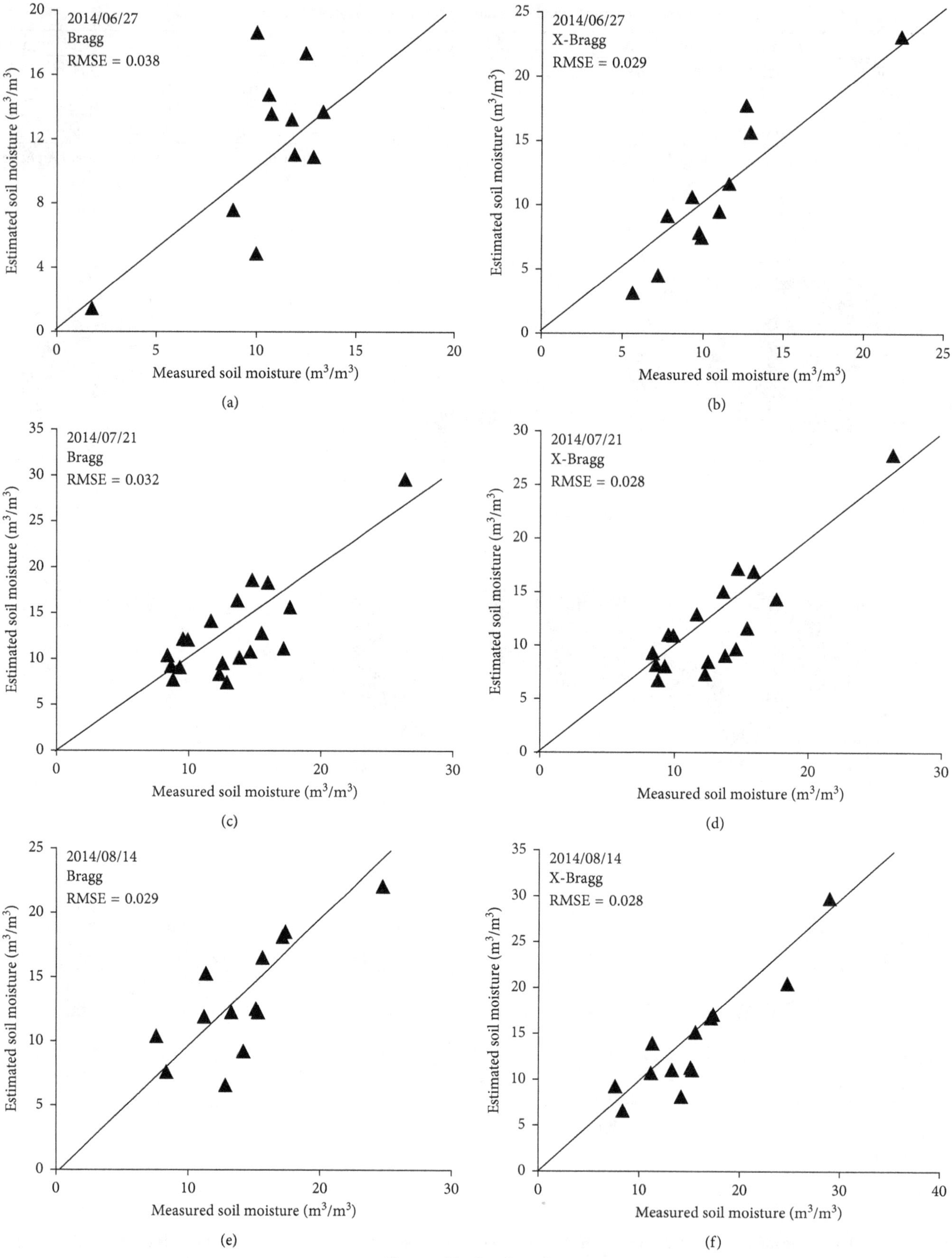

Figure 11: Continued.

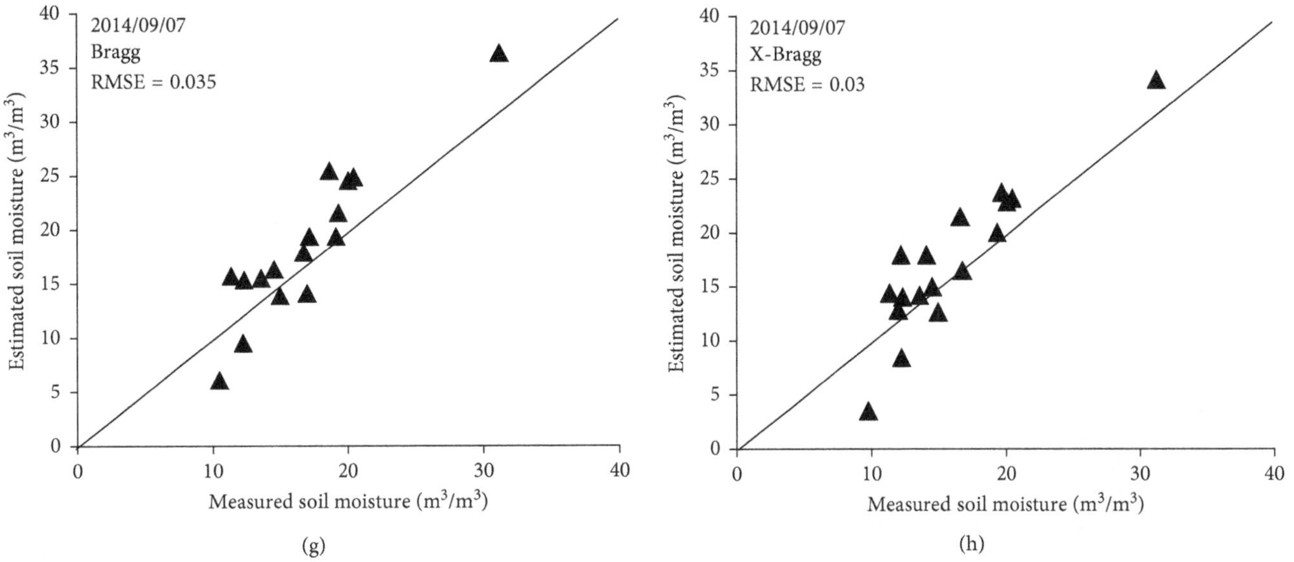

FIGURE 11: Accuracy of estimated soil moisture levels at different growth stages of maize.

decomposition components used in calculation (surface, dihedral, and volume scattering) are accurate. Here, we used the *H*/*A*/Alpha decomposition method to extract the surface, dihedral, and volume scattering components. The backscattering characteristics and the feature spaces of typical objects were analyzed to optimize the scattering components. We then retrieved soil moisture data using a combination of the Freeman–Durden decomposition method and the results of *H*/*A*/Alpha image classification.

Our analysis shows that the backscattering characteristics of maize and cotton changed with growth in a manner expressed by our regression equations (Figures 3 and 5). Based on the threshold values of typical objects as revealed by the *H*/*A*/Alpha polarization decomposition method, the nature of the surface and the volume of vegetation (if any) were apparent in the space plots of both *H* and Alpha. Bare soil data clustered in the lower left corner of the feature space and vegetation data in the upper right corner. Thus, the *H* and Alpha values of bare soil are lower than those of vegetation. The threshold values are approximately 0.5 *H* and 25° Alpha. Therefore, the thresholds of the three scattering components are 0.2–0.4 *H* and 7–24° Alpha for the surface scattering component; 0.6–0.9 *H* and 22–50° Alpha for the volume scattering component, and other values for the dihedral scattering component. We compared the soil moisture retrieval data obtained using the Bragg and X-Bragg models, employing the Freeman–Durden polarization decomposition method, in fields in which maize was at different stages of growth. The X-Bragg data of June 27, 2014, were better (having the lowest RMSE value, 0.028) than those derived at times when maize was at other stages of growth. The data obtained from July to September 2014 were less reliable because the C-band penetrated poorly over this time interval. However, our algorithm shows great potential for retrieval of soil moisture data using L-band polarization.

Conflicts of Interest

The authors declare that they have no conflicts of interest.

Authors' Contributions

This study was performed with collaboration among all the authors. Qiuxia Xie and Qingyan Meng determined the research theme and developed the methodology of soil moisture retrieval. Linlin Zhang, Chuimei Wang, Qiao Wang, and Shaohua Zhao were involved in the field experiments and paper modification. All authors agreed to the manuscript being submitted.

Acknowledgments

The authors would like to thank the National Key Research and Development Program (China's 13th Five Year Plan) "Spatial information service and application demonstration of comprehensive monitoring of urban and rural ecological environment" (2017YFB0503900); the National Key Research and Development Program (China's 13th Five Year Plan) "Integration and application demonstration of comprehensive monitoring technology of urban and rural ecological environment" (2017YFB0503905); Hainan Province Natural Science Foundation: Urban Impervious Surface Remote Sensing Extraction and Study on the Characteristics of Multi Temporal and Spatial Evolution (417218); Hainan Province Natural Science Foundation: Urban Heat Island Remote Sensing Retrieval Technology in Urban Main Construction Area based on Night Light Index (417219); and National Natural Science Foundation Project (41501400).

References

[1] H. Xu, F. Zheng, and W. Zhang, "Variability in dust observed over China using A-train CALIOP instrument," *Advances in Meteorology*, vol. 2016, Article ID 1246590, 11 pages, 2016.

[2] H. Xu, T. Cheng, X. Gu, T. Yu, D. Xie, and F. Zheng, "Spatiotemporal variability in dust observed over the Sinkiang and Inner Mongolia regions of Northern China," *Atmospheric Pollution Research*, vol. 6, no. 4, pp. 562–571, 2015.

[3] H. Xu, T. Cheng, X. Gu, T. Yu, Y. Wu, and H. Chen, "New Asia dust storm detection method based on the thermal infrared spectral signature," *Remote Sensing*, vol. 7, no. 1, pp. 51–71, 2014.

[4] S. Tomer, A. Al Bitar, M. Sekhar et al., "Retrieval and multi-scale validation of soil moisture from multi-temporal SAR data in a semi-arid tropical region," *Remote Sensing*, vol. 7, no. 6, pp. 8128–8153, 2015.

[5] E. P. W. Attema and F. T. Ulaby, "Vegetation modeled as a water cloud," *Radio Science*, vol. 13, no. 2, pp. 357–364, 1978.

[6] N. N. Baghdadi, M. El Hajj, M. Zribi, and I. Fayad, "Coupling SAR C-band and optical data for soil moisture and leaf area index retrieval over irrigated grasslands," *IEEE Journal of Selected Topics in Applied Earth Observations and Remote Sensing*, vol. 9, no. 3, pp. 1229–1243, 2016.

[7] G. Bertoldi, S. Della Chiesa, C. Notarnicola, L. Pasolli, G. Niedrist, and U. Tappeiner, "Estimation of soil moisture patterns in mountain grasslands by means of SAR RADARSAT2 images and hydrological modeling," *Journal of Hydrology*, vol. 516, pp. 245–257, 2014.

[8] I. Gherboudj, R. Magagi, A. A. Berg, and B. Toth, "Soil moisture retrieval over agricultural fields from multi-polarized and multi-angular RADARSAT-2 SAR data," *Remote Sensing of Environment*, vol. 115, no. 1, pp. 33–43, 2011.

[9] T. Jagdhuber, I. Hajnsek, and K. P. Papathanassiou, "An iterative generalized hybrid decomposition for soil moisture retrieval under vegetation cover using fully polarimetric SAR," *IEEE Journal of Selected Topics in Applied Earth Observations and Remote Sensing*, vol. 8, no. 8, pp. 3911–3922, 2015.

[10] Y. Kumar, S. Singh, R. S. Chatterjee, and M. Trivedi, "A comparative analysis of extended water cloud model and backscatter modelling for above-ground biomass assessment in Corbett Tiger Reserve," in *Proceedings of the SPIE Asia-Pacific Remote Sensing*, vol. 9880, New Delhi, India, April 2016.

[11] J. X. Zhang, G. M. Huang, J. J. Wei, and Z. Zhao, "Alternative to four-component decomposition for polarimetric SAR," *ISPRS Annals of Photogrammetry, Remote Sensing and Spatial Information Sciences*, vol. III-7, pp. 207–211, 2016.

[12] A. Freeman and S. L. Durden, "A three-component scattering model for polarimetric SAR data," *IEEE Transactions on Geoscience and Remote Sensing*, vol. 36, no. 3, pp. 963–973, 1996.

[13] A. Freeman and S. L. Durden, "Three-component scattering model to describe polarimetric SAR data," in *Proceedings of the SPIE, Radar Polarimetry*, vol. 1748, pp. 213–224, San Diego, CA, USA, February 1993.

[14] C. Fang and H. Wen, "A new classification method based on Cloude Pottier eigenvalue/eigenvector decomposition," in *Proceedings of the IEEE International Geoscience and Remote Sensing Symposium, IGARSS'05*, p. 4, Seoul, Korea, July 2005.

[15] R. Cloude and K. P. Papathanassiou, "Surface roughness and polarimetric entropy," in *Proceedings of the International Geoscience and Remote Sensing Symposium, IGARSS'99*, vol. 5, pp. 2443–2445, Hamburg, Germany, June-July 1999.

[16] F. Cao, W. Hong, and Y. R. Wu, "Unsupervised classification for fully polarimetric SAR data using Cloude-Pottier decomposition and agglomerative hierarchical clustering algorithm," *Acta Electronica Sinica*, vol. 36, no. 3, pp. 543–546, 2008.

[17] Q. Xie, Q. Meng, L. Zhang, C. Wang, Y. Sun, and Z. Sun, "A soil moisture retrieval method based on typical polarization decomposition techniques for a maize field from full-polarization Radarsat-2 data," *Remote Sensing*, vol. 9, no. 2, p. 168, 2017.

[18] L. Pasolli, C. Notarnicola, L. Bruzzone et al., "Polarimetric RADARSAT-2 imagery for soil moisture retrieval in alpine areas," *Canadian Journal of Remote Sensing*, vol. 37, no. 5, pp. 535–547, 2011.

[19] H. Lievens and N. E. C. Verhoest, "Spatial and temporal soil moisture estimation from RADARSAT-2 imagery over Flevoland, The Netherlands," *Journal of Hydrology*, vol. 456-457, pp. 44–56, 2012.

[20] Q. B. Zhou, Y. U. Qiang-Yi, L. Jia, W. U. Wen-Bin, and H. J. Tang, "Perspective of Chinese GF-1 high-resolution satellite data in agricultural remote sensing monitoring," *Journal of Integrative Agriculture*, vol. 16, no. 2, pp. 242–251, 2017.

[21] X. Zhang and N. Chen, "Reconstruction of GF-1 soil moisture observation based on satellite and in situ sensor collaboration under full cloud contamination," *IEEE Transactions on Geoscience and Remote Sensing*, vol. 54, no. 9, pp. 5185–5202, 2016.

[22] N. Chen, J. Li, and X. Zhang, "Quantitative evaluation of observation capability of GF-1 wide field of view sensors for soil moisture inversion," *Journal of Applied Remote Sensing*, vol. 9, no. 1, article 097097, 2015.

[23] Y. Guo, W. U. Xi-Hong, Y. Z. Cheng, L. G. Wang, and T. Liu, "Maize recognition and accuracy evaluation based on high resolution remote sensing (GF-1) data," *Remote Sensing Information*, vol. 30, no. 6, pp. 31–36, 2015.

[24] P. Wang, Z. F. Zhou, J. Liao, and G. N. University, "Study on soil moisture retrieval of tobacco field in Karst plateau mountainous area based on freeman decomposition," *Geography and Geo-Information Science*, vol. 32, pp. 72–76, 2016.

[25] I. Brown, S. Mwansasu, and L. Westerberg, "L-band polarimetric target decomposition of mangroves of the Rufiji Delta, Tanzania," *Remote Sensing*, vol. 8, no. 2, p. 140, 2016.

[26] H. Kankwamba, J. Mangisoni, F. Simtowe, and K. Mausch, "Quantitative estimation of surface soil moisture in agricultural landscapes using spaceborne synthetic aperture radar imaging at different frequencies and polarizations," Ph.D. thesis, Natural Sciences and Mathematics, Universität zu Köln, Cologne, Germany, 2011.

[27] N. Baghdadi, R. Cresson, E. Pottier et al., "A potential use for the c-band polarimetric SAR parameters to characterize the soil surface over bare agriculture fields," *IEEE Transactions On Geoscience and Remote Sensing*, vol. 50, no. 10, pp. 3844–3858, 2012.

[28] S. R. Cloude, "Polarimetry: the characterisation of polarisation effects in EM scattering," Ph.D. thesis, University of Birmingham, Birmingham, UK, 1986.

[29] N. Baghdadi, P. Dubois-Fernandez, X. Dupuis, and M. Zribi, "Sensitivity of main polarimetric parameters of multifrequency polarimetric SAR data to soil moisture and surface roughness over bare agricultural soils," *IEEE Geoscience and Remote Sensing Letters*, vol. 10, no. 4, pp. 731–735, 2013.
[30] T. Jagdhuber, I. Hajnsek, A. Bronstert, and K. P. Papathanassiou, "Soil moisture estimation under low vegetation cover using a multi-angular polarimetric decomposition," *IEEE Transactions On Geoscience and Remote Sensing*, vol. 51, no. 4, pp. 2201–2215, 2013.
[31] A. Freeman, S. Durden, and R. Zimmerman, "Mapping subtropical vegetation using multi-frequency, multi-polarization SAR data," in *Proceedings of the International Geoscience and Remote Sensing Symposium, IGARSS'92*, pp. 1686–1689, Houston, TX, USA, May 1992.
[32] H. Wang, R. Magagi, and K. Goita, "Comparison of different polarimetric decompositions for soil moisture retrieval over vegetation covered agricultural area," *Remote Sensing of Environment*, vol. 199, pp. 120–136, 2017.
[33] H. Wang, R. Magagi, K. Goita, T. Jagdhuber, and I. Hajnsek, "Evaluation of simplified polarimetric decomposition for soil moisture retrieval over vegetated agricultural fields," *Remote Sensing*, vol. 8, no. 2, p. 142, 2016.
[34] H. Wang, R. Magagi, K. Goita, T. Jagdhuber, and N. Djamai, "Evaluation of polarimetric decomposition for soil moisture retrieval over vegetation agriculture fields," in *Proceedings of the IEEE International Geoscience and Remote Sensing Symposium, IGARSS'15*, pp. 689–692, Milan, Italy, July 2015.
[35] X. U. Xing-Ou and N. Shu, "Water content information extraction from quad-polarization SAR images via Freeman–Durden decomposition," *Computer Engineering and Applications*, vol. 46, no. 27, pp. 17–20, 2010.
[36] J. C. Shi, J. S. Lee, K. Chen, and Q. Q. Sun, "Evaluate usage of decomposition technique in estimation of soil moisture with vegetated surface by multi-temporal measurements," in *Proceedings of the IEEE 2000 International Geoscience and Remote Sensing Symposium, IGARSS 2000*, pp. 1098–1100, Honolulu, HI, USA, July 2000.

13

Intercomparison of Downscaling Techniques for Satellite Soil Moisture Products

Daeun Kim,[1] Heewon Moon,[2] Hyunglok Kim,[3] Jungho Im,[4] and Minha Choi[5]

[1]*Center for Built Environment, Sungkyunkwan University, Suwon, Gyeonggi-do, Republic of Korea*
[2]*Institute for Atmospheric and Climate Science, ETH Zürich, Zürich, Switzerland*
[3]*School of Earth, Ocean and the Environment, University of South Carolina, Columbia, SC, USA*
[4]*School of Urban and Environmental Engineering, UNIST, Ulsan, Republic of Korea*
[5]*Graduate School of Water Resources, Sungkyunkwan University, Suwon, Gyeonggi-do, Republic of Korea*

Correspondence should be addressed to Minha Choi; mhchoi@skku.edu

Academic Editor: Jifu Yin

During recent decades, various downscaling methods of satellite soil moisture (SM) products, which incorporate geophysical variables such as land surface temperature and vegetation, have been studied for improving their spatial resolution. Most of these studies have used least squares regression models built from those variables and have demonstrated partial improvement in the downscaled SM. This study introduces a new downscaling method based on support vector regression (SVR) that includes the geophysical variables with locational weighting. Regarding the in situ SM, the SVR downscaling method exhibited a smaller root mean square error, from 0.09 to 0.07 $m^3{\cdot}m^{-3}$, and a larger average correlation coefficient increased, from 0.62 to 0.68, compared to the conventional method. In addition, the SM downscaled using the SVR method had a greater statistical resemblance to that of the original advanced scatterometer SM. A residual magnitude analysis for each model with two independent variables was performed, which indicated that only the residuals from the SVR model were not well correlated, suggesting a more effective performance than regression models with a significant contribution of independent variables to residual magnitude. The spatial variations of the downscaled SM products were affected by the seasonal patterns in temperature-vegetation relationships, and the SVR downscaling method showed more consistent performance in terms of seasonal effect. Based on these results, the suggested SVR downscaling method is an effective approach to improve the spatial resolution of satellite SM measurements.

1. Introduction

Remotely sensed soil moisture (SM) offers increased spatial coverage and improved temporal continuity and has thus resulted in substantial changes in our understanding of the global water cycle [1, 2]. Nevertheless, the relatively large spatial resolution of approximately 10 km for passive/active microwave satellite remote sensing datasets is the main reason they cannot be effectively applied to hydrological studies at a regional scale [3]. The issue of scale mismatch between remotely sensed and in situ SM has also been considered unavoidable and has been critically evaluated using coarse satellite measurements, particularly in areas with nonhomogeneous land cover [4]. Thus, downscaling techniques that focus on the spatial resolution of remotely sensed SM are important to match with an in situ dataset and enable practical applications.

To resolve this problem, synergistic approaches to disaggregate microwave remote sensing SM measurements using visible/infrared (VIS/IR) sensors with enhanced spatial resolution have been performed in previous studies [5–9]. This approach is based on the relationship of SM between the land surface temperature (T_s) and the normalized difference vegetation index (NDVI) that theoretically forms a triangular shape because of the evaporative cooling effect [10, 11]. However, the downscaling methods based on this relationship are considered semiempirical. Previous SM downscaling researches have consisted largely of variations in the regression formula based on these three related variables. Chauhan et al. [6] introduced surface albedo into this method to strengthen the relationship between SM and land parameters and applied it to 25 km SM data from a special sensor microwave imager and 1 km land parameters from the advanced very high resolution radiometer. A comparison of the 1 km SM and in situ SM revealed fairly similar trends, with a root mean square error (RMSE) that ranged from 0.005 to 0.037 $m^3 \cdot m^{-3}$. In addition, the introduction of surface albedo was later adopted in Yu et al. [7] and Choi and Hur [5]. Piles et al. [8] introduced brightness temperature (TB) instead of surface albedo to downscale the SM with other variables and the Ocean Salinity (SMOS) mission.

The polynomial regression formula applied in previous downscaling studies has been shown to have good performance. However, the method features innate errors resulting from the regression of a highly complex and nonlinear relationship of T_s in nonhomogeneous vegetation conditions and SM into a polynomial model [12, 13]. Thus, there is a need to find and employ a different regression model to better capture the inherent complexity.

A support vector machine (SVM), an alternative method to downscale the SM, is a machine-learning algorithm that provides a nonlinear generalization solution to datasets through structural risk minimization and is based on the solid theoretical foundation of Vapnik–Chervonenkis theory [14–17]. The initial applications of SVM have targeted optical characteristic recognition and object recognition tasks using support vector (SV) classifiers [18–20]; its application in regression and time series prediction was subsequently adopted [18]. Support vector regression (SVR) in remote sensing research has often been applied to predict variables that appear as responses to other input variables [21, 22]. Kaheil et al. [23] suggested using downscaling algorithms for the Southern Great Plains 1997 (SGP 97) with SVM and assimilation with ground SM measurements. The SVM method was specifically used to tune the downscaled image based on the relationship between the original and approximated coarse scale image. Keramitsoglou et al. [24] applied an SVR to downscale the meteosat second generation T_s using moderate resolution imaging spectroradiometer (MODIS) NDVI, emissivity, and other regression methods to find the preferred methodology. These studies using SVR have evaluated SM downscaling methods by comparing them within identically structured calculation methods in which only the input variables varied [5, 8]. However, the application of SVR in East Asia is insufficient using a remote sensing dataset. Thus, the comparison of downscaling in this area is necessary because the various methods for downscaling SM are inadequate.

In this study, a methodology to downscale active microwave SM based on T_s and NDVI using SVR is suggested to build an optimized regression model that considers the spatial pattern of the original dataset to obtain finer, more accurate SM distribution relative to the conventional VIS/IR downscaling methods. This research is unique because it offers a cross comparison between the newly suggested SVR downscaling method and conventional methods. The downscaled SM was evaluated by taking in situ measurements from nine measurement sites within a 150 km × 125 km study area of the Korean Peninsula from March to November 2012. The polynomial regression downscaling method was also applied in the same study area for comparative evaluation.

2. Study Area and Dataset Descriptions

2.1. Study Area. The study area in southwestern South Korea encompasses the area from 35.0 to 36.3°N and 126.6 to 128.4°E for a total of 18,750 km^2 (Figure 1). Cropland and mixed forest are the dominant land covers. The area was selected for its representative land cover characteristics and the availability of in situ measurements, the locations and characteristics of which are described in Table 1. The land cover types were considered because surface properties such as vegetation types, soil types, land uses, and topography could affect the SM retrieval algorithm that is based on microwave sensor observations [25, 26].

The annual precipitation at the measurement sites ranged from 1300 to 1800 mm, with the heaviest rainfall occurring during the summer, and the annual mean temperature ranges from 10.6 to 13.2°C [27]. The western part of the study area generally consists of plains that are used as cropland, while the eastern part is of higher altitude and mostly forested (Figure 1). The land cover is classified using the MODIS yearly land cover type data with the international Geosphere-Biosphere Programme (IGBP) global vegetation classification scheme [28].

2.2. In Situ SM Measurements. The nine in situ SM measurements stations that were used in this study were installed by the Rural Development Administration (RDA), Korea. The measurement sites were approximately distributed to cover the study area, and SM was measured within 0 to 10 cm depth with time-domain reflectometry (TDR) at hourly time-step from March to November 2012. TDR and frequency-domain reflectometry (FDR) sensors are the most commonly used techniques to measure soil water content [29]. The TDR measures the propagation time of an electromagnetic wave along the transmission line to determine the dielectric permittivity, while FDR measures the capacitance. Previous studies have demonstrated good agreement in SM measurements between the two approaches [30–32]. Note that there is an unavoidable limitation in the difference in the measurement depth of microwave satellite SM data and in situ data [4]. However, because the geophysical variables adopted in this study represent the surface properties observed using an optical satellite sensor, the measuring depth difference was disregarded.

2.3. Advanced Scatterometer SM. The advanced scatterometer (ASCAT) is an active microwave sensor aboard the European Space Agency's (ESA) meteorological operation (MetOp-A) satellite. It began operation in 2006 and

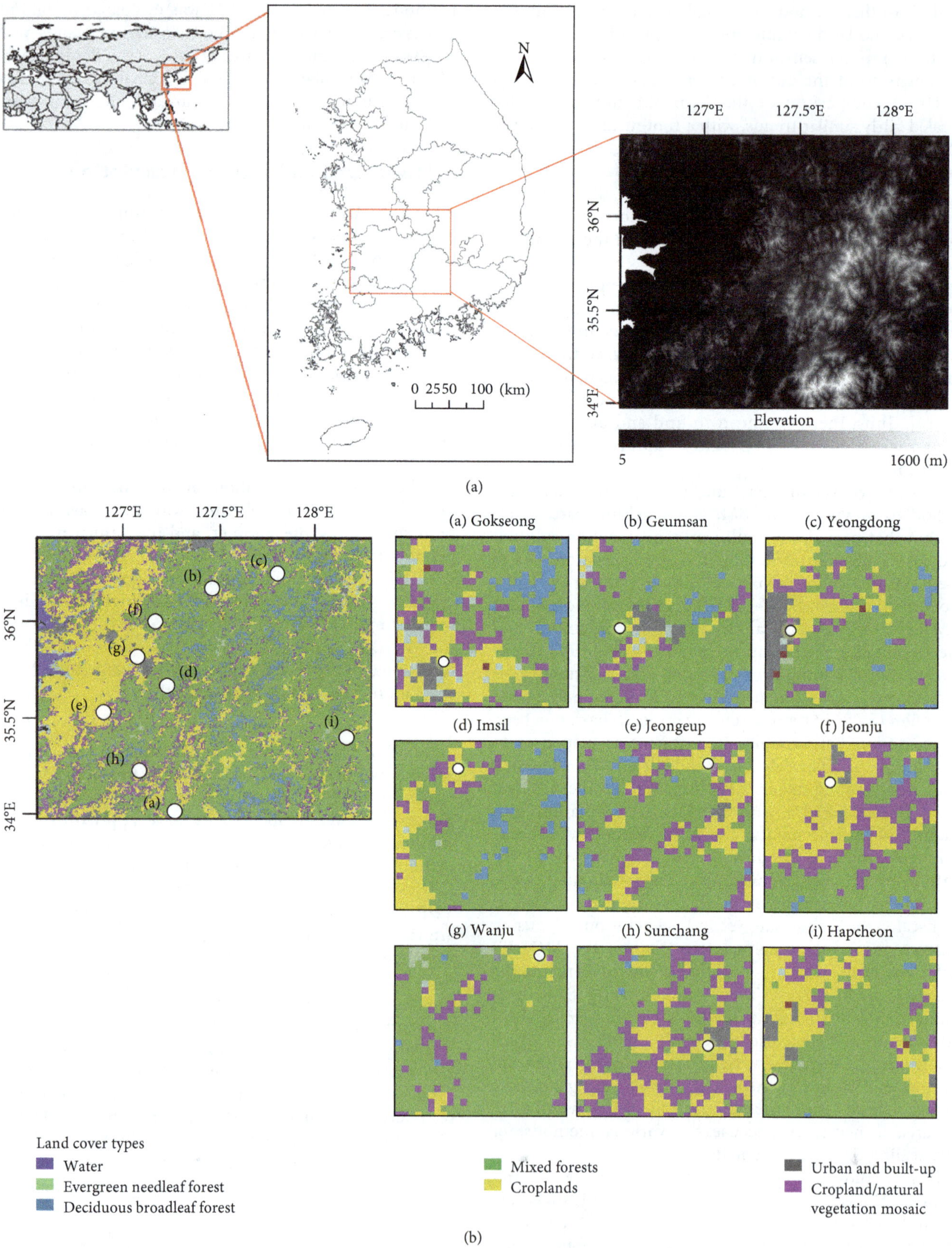

Figure 1: (a) Study area on the Korean Peninsula and (b) the in situ sites (white circles).

Table 1: Descriptions of the nine in situ sites.

Site	Latitude (°)	Longitude (°)	Soil texture	Land use	Annual rainfall (mm)	Mean air temperature (°C)
Geumsan	36.130	127.492	Loam	Mixed forest	1434.5	11.1
Yeongdong	36.173	127.752	Silt loam	Cropland	1427.4	10.8
Jeonju	35.824	127.158	Loam	Urban	1359.7	13.2
Wanju	35.985	127.220	Sandy loam	Cropland	1308.5	12.0
Imsil	35.654	127.272	Loam	Cropland	1503.9	10.6
Jeongeup	35.622	126.896	Loam	Cropland	1473.4	13.1
Hapcheon	35.547	128.110	Sandy loam	Cropland	1795.4	12.9
Sunchang	35.374	127.137	Loam	Mixed forest	1819.5	10.7
Gokseong	35.267	127.304	Loam	Cropland	1464.0	11.3

measures the radar backscatter at C-band (5.255 GHz) with vertical transmit–vertical receive (VV) polarization. Since the measurement is performed using two satellite tracks, dual 550 km-wide swaths are produced, covering 82% of the Earth daily. While the SM retrieval from passive microwave observations with 12.5 km resolution is mainly based on the linkage between TB and geophysical variables, the retrieval of the SM from ASCAT uses a time series-based approach to scale the backscattering coefficient between the lowest and the highest values which are presented as the degree of saturation [33, 34]. The SM values are estimated as the relative variation between the wettest (100%) and driest (0%) values. The dataset used in this study was daily ASCAT-relative SM processed by the Integrated Climate Data Centre in Hamburg. Relative SM was converted to volumetric SM ($m^3 \cdot m^{-3}$) by applying the porosity of each soil texture to enable comparison with the in situ measurements (Table 2).

2.4. Moderate Resolution Imaging Spectroradiometer. The MODIS on board the Earth observation system (EOS) Terra (10:30/22:30) and Aqua (01:30/13:30) satellites uses 36 spectral bands to observe characteristics of the atmosphere, land, and ocean. The MODIS products used in this study were 1 km resolution daily daytime T_s (MOD11A1) and 1 km resolution 16-day NDVI (MOD13A2) from the Terra satellite. The T_s is retrieved from TB using the generalized split-window algorithm [35]. Cloudy pixels are excluded from the T_s retrieval process since thermal infrared signals do not penetrate clouds and are thus confounded with cloud-top temperature. The NDVI can be calculated as the normalized ratio of the near IR and red bands, reflecting the chlorophyll and mesophyll in the vegetation canopy [36, 37]. The level 2 daily surface reflectance product, from which the 16-day period of the MOD13A2 NDVI product is generated, is the adjusted data for ozone absorption, molecular scattering, and aerosols [38]. To establish statistically significant regression models, only days with more than 90% cloud-free pixels were used.

3. Methods

3.1. Preprocessing of Remote Sensing Images. The T_s and NDVI from MODIS which originally have 1 km spatial resolution were uniformly disaggregated to a spatial resolution of 500 m and were then aggregated to have a 12.5 km resolution by applying arithmetic means as follows:

Table 2: Green-Ampt infiltration parameters for various soil textures.

	Porosity ($m^3 \cdot m^{-3}$)
Loam	0.463
Sandy loam	0.453
Silt loam	0.501

$$\mathrm{NDVI}_{12.5} = \frac{\sum_{i=1}^{n}\sum_{j=1}^{m}\mathrm{NDVI}_{ij}}{mn}, \qquad T_{s,12.5} = \frac{\sum_{i=1}^{n}\sum_{j=1}^{m}T_{s,ij}}{mn}, \tag{1}$$

where $\mathrm{NDVI}_{12.5}$ is the 12.5 km averaged NDVI, $T_{s,12.5}$ is the 12.5 km averaged T_s, and m and n are the number of 500 m pixels in ith rows and jth columns in 12.5 km ASCAT, respectively. For downscaling the 12.5 km ASCAT SM, the difference between the 500 m and 12.5 km spatial resolution of the LST and NDVI dataset was required; thus, the 500 m LST and NDVI products were upscaled to 12.5 km resolution for calculating the difference between the products that had different resolutions.

3.2. SM Downscaling Using Polynomial Regression. The performance of the suggested downscaling method using SVR was evaluated by calculating the downscaled SM from the conventional polynomial method using the same input variables. Carlson et al. [10] suggested a relationship among SM, NDVI, and T_s, a polynomial regression formula, under the different climatic conditions and land cover types as follows:

$$\mathrm{SM} = \sum_{i=0}^{i=n}\sum_{j=0}^{j=n} a_{ij}\mathrm{NDVI}^{*(i)}T_s^{*(j)}, \tag{2}$$

where n is the number of a reasonable dataset and a_{ij} is the regression coefficient at a specific day and scene for analysis.

$$\mathrm{NDVI}^* = \frac{\mathrm{NDVI}_{max} - \mathrm{NDVI}}{\mathrm{NDVI}_{max} - \mathrm{NDVI}_{min}}, \qquad T_s^* = \frac{T_{s,max} - T_s}{T_{s,max} - T_{s,min}}. \tag{3}$$

In this study, the equation is applied with $n = 2$ and $i + j \leq 2$ to yield second-order polynomial equation as follows:

$$\begin{aligned} SM = {} & a_{00}NDVI^{*0}T_s^{*0} + a_{10}NDVI^{*1}T_s^{*0} + a_{01}NDVI^{0}T_s^{1} \\ & + a_{11}NDVI^{*1}T_s^{1} + a_{20}NDVI^{*2}T_s^{*0} + a_{02}NDVI^{*0}T_s^{*2}. \end{aligned} \tag{4}$$

3.3. SM Downscaling Using Support Vector Regression. The SM downscaling procedure using SVR consisted of two parts. The remote sensing images (ASCAT SM, MODIS T_s, and NDVI) were preprocessed for application during the SVR process, and high-resolution SM data were produced using the training and prediction procedure in the SVR. The downscaling methodology suggested in this study combines the conventional VIS/IR synergistic downscaling method with the image approximation concept by introducing the locational information of latitude and longitude as an additional input variable. Figure 2 shows the entire procedure for the suggested downscaling.

The SVM is among the machine learning based on covariates' nonlinear transformations developed by Vapnik in the early 1990s [39]. The SVM for regression was also updated by Vapnik [14]. This model included a training phase to train the associated input and target output dataset based on statistical learning theory [40].

Of the various versions of SVM tools, the LibSVM that was built by Chang and Lin [41] was used in this study. The radial basis function (RBF) was selected for the kernel as follow:

$$K(x, x_i) = \exp\left(-\gamma \|x_i - x\|^2\right), \tag{5}$$

where γ is the bandwidth that determines the under- or overfitting loss [42]. The x consists of $\{u, v, T_s$, and NDVI$\}$, where u is the x position of the pixel and v is the y position of a pixel. The selection of the RBF kernel was based on previous studies that showed its superiority over other kernel functions for both classification and regression tasks [43–45]. Two RBF parameters—gamma and penalty—were optimized using a grid search algorithm and n-fold cross validation, both of which have been widely used in the literature [46–48]. The original sample in the n-fold cross validation is randomly divided into n subsamples of equal size. A single sample among the n subsamples is maintained with validation data for assessment of the model, and the $n - 1$ subsamples are used for training data. Then, the process of cross validation is repeated n times, and the n subsamples are used at once for validation. This approach has been widely used in SVM research [49–51], and it is regarded as a basic application in the LibSVM tool as previously mentioned. A three-fold cross validation was then used, and the selected parameters are showed in Table 3. Since the selection of geophysical variables (T_s and NDVI) was theoretically conducted [3, 6], variable selection was omitted. All of the variables were scaled to [0, 1] to even out quantitative differences among them. To obtain valid regression models with a sufficient number of samples, only satellite images from cloud-free days were used; thus, a total of 55 days from March to November 2012 were available.

3.4. Statistical Analysis Methods. The four following indices were used for the statistical evaluation as follows:

$$R = \sqrt{1 - \frac{\sum (SM_{\text{satellite/model}} - SM_{\text{in situ}})^2}{\sum (SM_{\text{satellite/model}} - \overline{SM}_{\text{in situ}})^2}},$$

$$\text{Bias} = \overline{\sum \text{SM}_{\text{satellite/model}} - \text{SM}_{\text{in situ}}},$$

$$\text{RMSE} = \sqrt{\overline{\sum (\text{SM}_{\text{satellite/model}} - \text{SM}_{\text{in situ}})^2}},$$

$$\text{IOA} = 1 - \frac{\sum (\text{SM}_{\text{satellite/model}} - \text{SM}_{\text{in situ}})^2}{\sum \left(\left|\text{SM}_{\text{satellite/model}} - \overline{\text{SM}_{\text{in situ}}}\right| + \left|\text{SM}_{\text{in situ}} - \overline{\text{SM}_{\text{in situ}}}\right|\right)^2}, \tag{6}$$

where $SM_{\text{satellite/model}}$ is the satellite-observed or modeled SM and SM in situ is the ground-measured SM. In this study, the averaged R value instead of each R value from each site was employed in accordance with previous studies [5, 52–54] because of the limitations of a lack of SM samples from both the ground and satellites. In case of in situ measurements, it is difficult to obtain simultaneous data with the over pass time of the satellite, and vice versa, as cloud cover causes an absence of the visible band-based MODIS land data (LST and NDVI) making it impossible to get a downscaled SM. The index of agreement (IOA) ranges from 0 to 1, with higher index values indicating a smaller mean square error and better agreement between the modeled values and observations [54].

4. Results and Discussion

The polynomial regression and SVR models were established to perform daily-scale evaluations of SM variability. The averaged linear correlation coefficient value between the original and downscaled products was 0.55 for both models. This model performed relatively well in disaggregating the coarse-scale original SM product; thus, both models were considered suitable to downscale the original SM dataset.

4.1. Evaluation of Downscaled SM Compared with In Situ SM. The original ASCAT SM and each downscaling algorithm were compared against nine in situ SM measurements in the study region. Figure 3 shows the temporal variation in the SM measurements, 12.5 km SM from ASCAT, and downscaled 1 km SM using SVR and polynomial regression and their response to daily rainfall events. Although the characteristics of the temporal patterns are site-specific, all three remotely sensed measurements approximately followed the patterns of the in situ SM. The 1 km SM downscaled data using polynomial regression also showed a similar temporal pattern to that of the 12.5 km and in situ SM measurements, but crucially underestimated some values on occasion. This was most visibly demonstrated in comparison to the downscaled SM from SVR. In Geumsan, Yeongdong, Wanju, and Jeonju, the underestimation of the polynomial

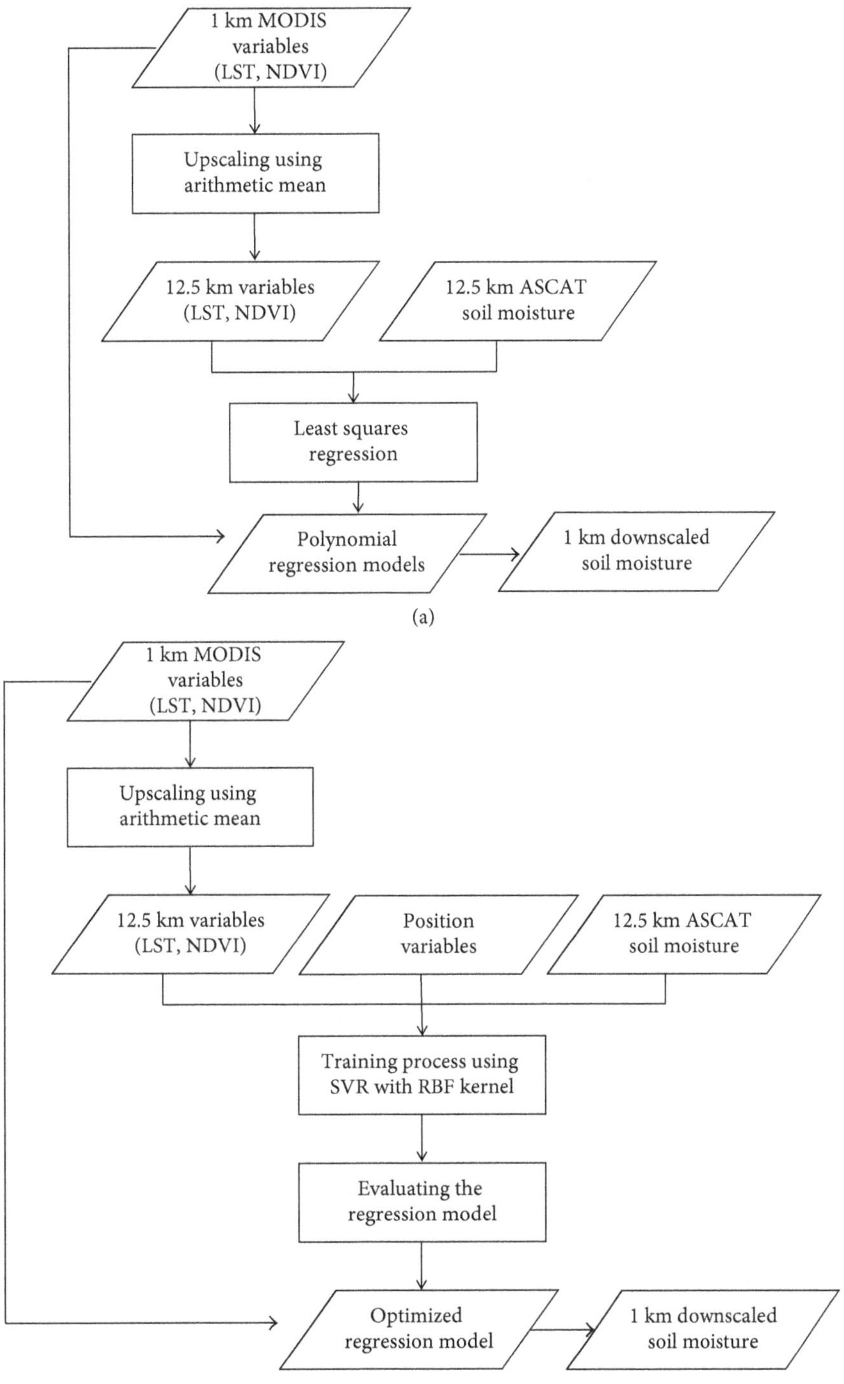

FIGURE 2: Flow charts of (a) the conventional and (b) the support vector machine (SVR) downscaling algorithms.

TABLE 3: Parameter characteristics of the SVR regression model.

Parameter	Value
C	$n \cdot \max(\lvert\overline{y}+3\sigma_y\rvert, \lvert\overline{y}-3\sigma_y\rvert)$
Gamma	2^k (for integer k, $k = [-5, -2]$)
Epsilon	(0, 0.1]

downscaling results are more apparent than elsewhere. Overall, the 1 km SM downscaled using SVR had more realistic trend values than that using polynomial regression compared with the in situ SM, and its pattern was very similar to the original 12.5 km SM.

Since the results of the comparison of the downscaled SM methods appeared to have clear differences, particularly at some sites, the correlation between the two independent variables (NDVI and T_s) and the 12.5 km SM residual magnitudes of each regression model prediction were analyzed to evaluate the manner in which the variables affect the regression models using the p-test. Figure 4 shows the time series for R between the residual magnitude and the

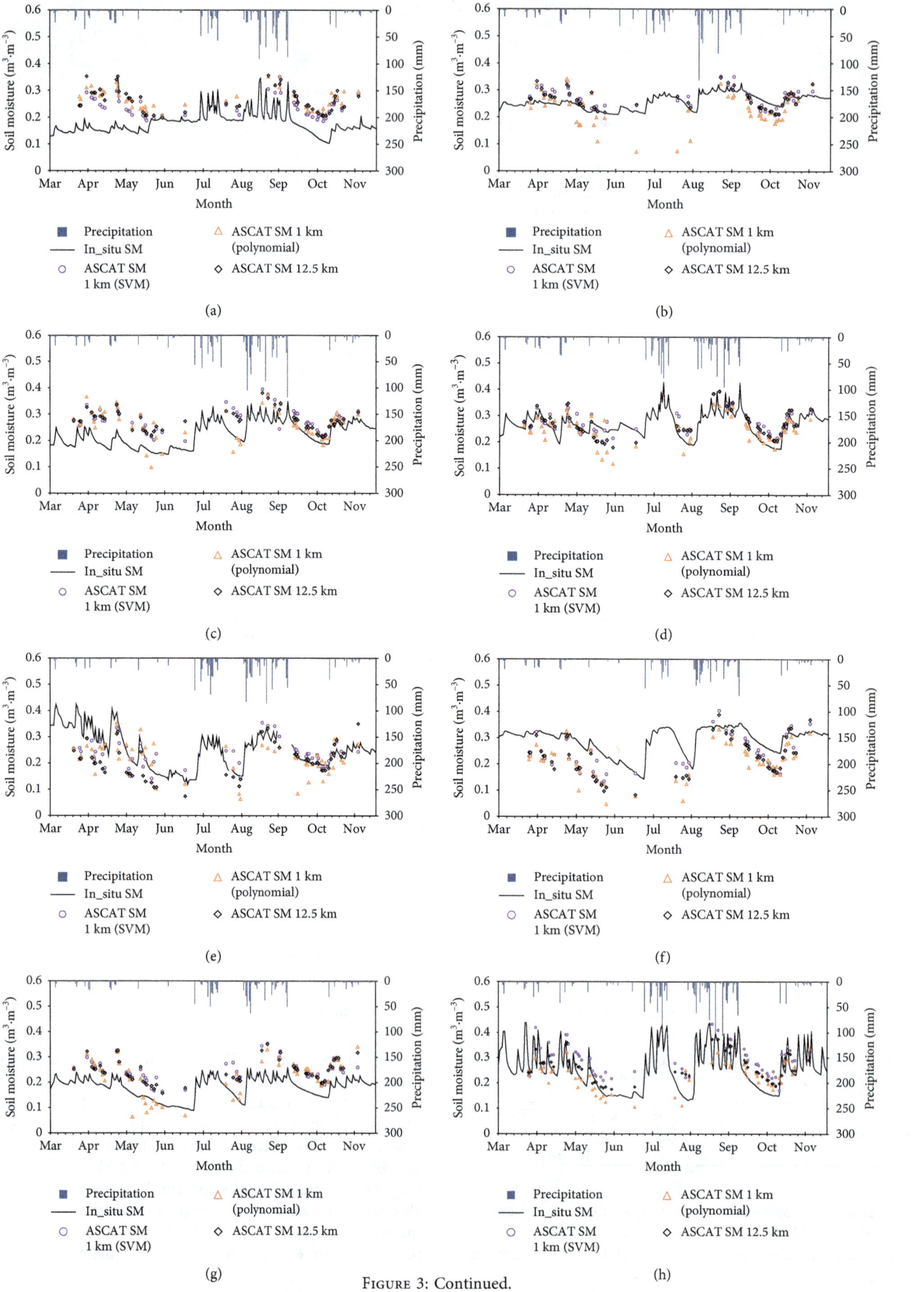

Figure 3: Continued.

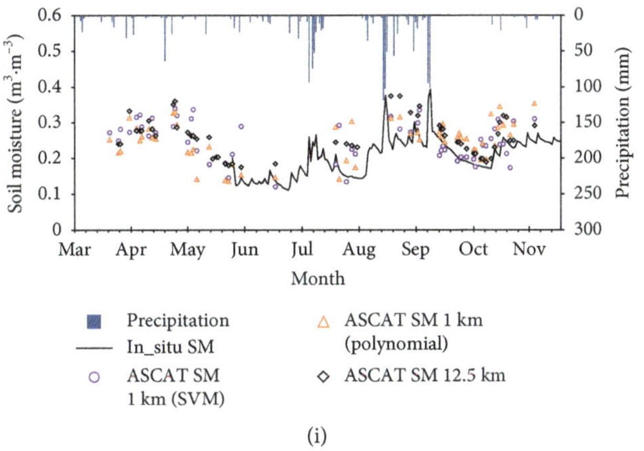

FIGURE 3: Time series of remotely sensed and in situ soil moisture (SM) and precipitation at the (a) Gokseong, (b) Geumsan, (c) Yeongdong, (d) Imsil, (e) Jeonju, (f) Jeongeup, (g) Wanju, (h) Sunchang, and (i) Hapcheon sites.

variables with the corresponding significance. The statistical results are summarized in Table 4. The p value, a statistical significance, is the marginal significance level under the assumption that true for the null hypothesis stands for occurrence probability of the given event and the slope means whether there is linear relationship between the independent variable x and the dependent variable y. For the polynomial regression model, the residual magnitude showed a significant and strong correlation on average with both independent variables with average R values of 0.32 and 0.41. Comparatively, the SVR model showed a weaker and insignificant correlation with averaged R values of 0.10 and 0.17. The difference between the degrees of correlation of the two models used in this study was likely caused by the methodological difference of the SVR model that additionally considered locational weighting.

Meanwhile, the signs of the slope coefficients for each variable were found to be opposing for each regression model. For the polynomial regression model, NDVI showed a positive relationship and T_s showed a negative relationship with the SM residual magnitude, while the signs were opposing for the SVR model. Considering that for most dates, the relationship between the two variables and the residual magnitudes of the SVR models was insignificant, with large p values (average values of 0.41 and 0.24, resp.), and the signs were only meaningful for the polynomial regression models. The positive slope coefficients between the NDVI and the residual magnitude can be explained as a result of the increased uncertainty of the microwave SM retrieval for areas with denser vegetation [55]. The oppositely negative signs of the slope coefficients between T_s and the residual magnitude were partially due to their relationship with vegetation. While the relationship between the two variables was assumed to be mutually independent in the regression models, similar water stress conditions produced two variables that were negatively correlated, partly due to evaporative cooling [12, 56]. In addition, the highest R values were found during the growing season, from mid-May to mid-September, demonstrating that seasonal patterns occurred in the residual of the polynomial regression models. This result was also partly explained by the crucial underestimating tendency of the polynomial regression model found at some sites. In the case of Jeonju, this pattern could be attributed to the highest annual mean air temperature based on the observed significant positive correlation between the regression model's residual magnitude and T_s (Table 1). In addition, in a pixel-by-pixel inspection for days with extremely underestimated performance at each site (not shown here), the underestimations were found to have occurred in pixels in which T_s was substantially higher than the average for that particular day.

As shown in Tables 5–7, each of the remotely sensed SM measurements were quantitatively evaluated by comparing them with the in situ SM measurements. The average value of nine in situ SM measurement sites was 0.23 m^3·m^{-3}, and among the remotely sensed products, the 1 km SM downscaled using polynomial regression had the nearest value (0.24 m^3·m^{-3}) but with the highest RMSE. The average downscaled SM using SVR was 0.26 m^3·m^{-3} with a standard deviation (SD) of 0.05 m^3·m^{-3}, similar to that of the 12.5 km ASCAT SM measurement (0.05 m^3·m^{-3}) (Table 7). The R values between in situ SM and 12.5 km ASCAT SM, downscaled SM using polynomial regression, and downscaled SM using SVR were 0.66, 0.62, and 0.68, respectively.

Figure 5 presents the overall error distribution for each remotely sensed SM measurement. Since a difference in SM indicates an error in the remotely sensed SM relative to the in situ SM measurement, an ideal histogram would have a steep and narrow form centered on zero, thus indicating a normal distribution with a zero mean [57]. While the original coarse scale SM had a positive bias on average with an RMSE of 0.072 m^3·m^{-3}, for the downscaled SM using polynomial regression, the RMSE was the same as that of the original ASCAT SM but with a higher SD (0.072 m^3·m^{-3}). In the case of the SVR, it also had a positive bias with a decreased RMSE (0.065 m^3·m^{-3}) and SD (0.056 m^3·m^{-3}) (Figure 5). Thus, these results indicate that SVR offers better performance in reducing the error of the downscaled satellite SM. The R values between the satellite SM and the corresponding in situ measurements showed better results for the SVR downscaling method, with an increase from 0.62 to 0.68 as previously mentioned (Tables 6 and 7). The IOA

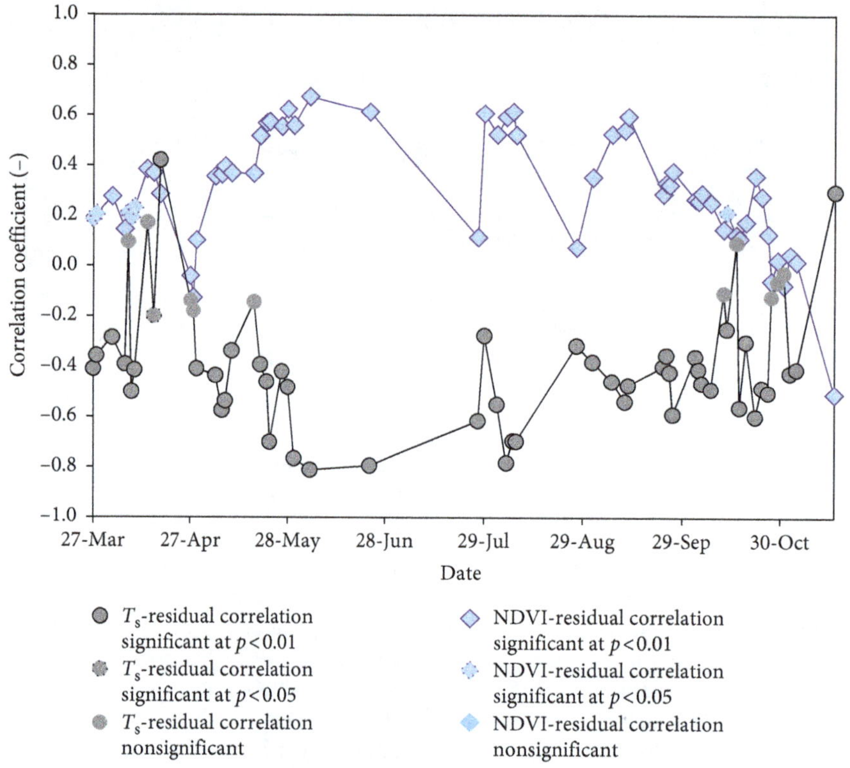

(a)

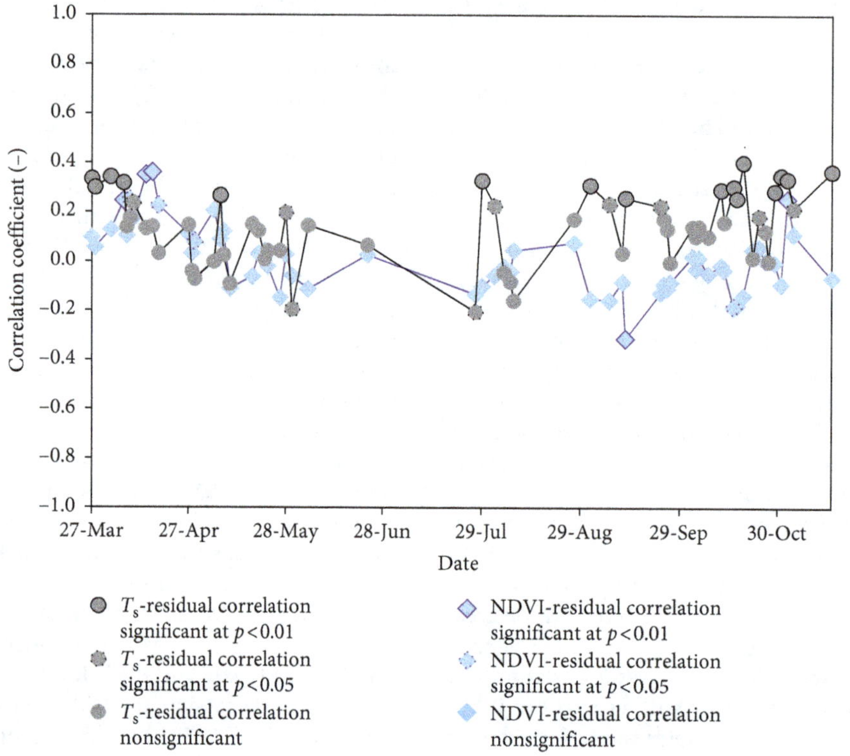

(b)

FIGURE 4: Correlation time series of the (a) polynomial regression and (b) SVR models between residual magnitudes and land surface temperature (T_s) and normalized vegetation difference index (NDVI).

Table 4: Correlation analysis between the independent variables and residual magnitude.

Residual magnitude	Polynomial						SVR					
	NDVI			T_s			NDVI			T_s		
	Slope	R	p value	Slope	R	p value	Slope	R	p value	Slope	R	p value
Average	17.35	0.32	0.10	−19.21	0.41	0.05	0.84	0.10	0.41	4.47	0.17	0.24
SD	16.52	0.19	0.21	17.68	0.19	0.14	6.75	0.08	0.30	7.93	0.11	0.31

Table 5: Comparison between in situ and original scale (12.5 km) ASCAT SM.

	In situ SM		ASCAT SM (12.5 km)		$y=ax+b$ (x: in situ SM, y: downscaled SM)	R	Bias ($m^3 \cdot m^{-3}$)	RMSE ($m^3 \cdot m^{-3}$)	IOA
	Average ($m^3 \cdot m^{-3}$)	SD ($m^3 \cdot m^{-3}$)	Average ($m^3 \cdot m^{-3}$)	SD ($m^3 \cdot m^{-3}$)					
Gokseong	0.16	0.03	0.27	0.05	$y=0.67x+0.16$	0.40**	−0.11	0.12	0.29
Geumsan	0.26	0.02	0.24	0.03	$y=0.82x+0.03$	0.59**	0.01	0.03	0.72
Yeongdong	0.22	0.04	0.28	0.04	$y=0.61x+0.14$	0.66**	−0.06	0.07	0.58
Imsil	0.26	0.04	0.25	0.05	$y=0.77x+0.04$	0.73**	0.01	0.03	0.83
Jeongeup	0.29	0.04	0.20	0.07	$y=1.40x-0.20$	0.85**	0.09	0.10	0.53
Jeonju	0.24	0.07	0.20	0.07	$y=0.65x+0.04$	0.67**	0.04	0.08	0.70
Wanju	0.17	0.03	0.24	0.04	$y=1.21x+0.03$	0.89**	−0.07	0.08	0.54
Sunchang	0.23	0.06	0.24	0.05	$y=0.61x+0.09$	0.78**	−0.01	0.04	0.86
Hapcheon	0.21	0.09	0.30	0.04	$y=0.40x+0.20$	0.40*	−0.15	0.20	0.10
Average	*0.23*	*0.04*	*0.25*	*0.05*		*0.66*	*−2.92*	*0.08*	*0.57*

*The significance level of the p value > 0.05; **the significance level of the p value > 0.01.

Table 6: Comparison between in situ and downscaled (1 km) ASCAT SM_Polynomial.

	In situ SM		ASCAT SM (1 km) _Polynomial		$y=ax+b$ (x: in situ SM, y: downscaled SM)	R	Bias ($m^3 \cdot m^{-3}$)	RMSE ($m^3 \cdot m^{-3}$)	IOA
	Average ($m^3 \cdot m^{-3}$)	SD ($m^3 \cdot m^{-3}$)	Average ($m^3 \cdot m^{-3}$)	SD ($m^3 \cdot m^{-3}$)					
Gokseong	0.16	0.03	0.26	0.04	$y=0.57x+16.90$	0.40**	−0.10	0.11	0.31
Geumsan	0.26	0.02	0.23	0.06	$y=1.25x-8.837$	0.47**	0.02	0.07	0.45
Yeongdong	0.22	0.04	0.26	0.06	$y=0.85x+7.60$	0.60**	−0.04	0.06	0.65
Imsil	0.26	0.04	0.25	0.05	$y=0.71x+6.29$	0.63**	0.00	0.06	0.63
Jeongeup	0.29	0.04	0.21	0.07	$y=1.32x-17.23$	0.82**	0.08	0.09	0.56
Jeonju	0.24	0.07	0.22	0.06	$y=0.50x+9.38$	0.55**	0.02	0.09	0.60
Wanju	0.17	0.03	0.22	0.07	$y=1.85x-9.31$	0.85**	−0.05	0.07	0.64
Sunchang	0.23	0.06	0.23	0.06	$y=0.61x+8.93$	0.68**	0.00	0.06	0.77
Hapcheon	0.21	0.09	0.25	0.06	$y=0.84x+8.10$	0.60*	−0.12	0.16	0.23
Average	*0.23*	*0.04*	0.24	0.06		*0.62*	−0.02	0.08	*0.54*

*The significance level of the p value > 0.05; **the significance level of the p value > 0.01.

Table 7: Comparison between in situ and downscaled (1 km) ASCAT SM_SVR.

	In situ SM		ASCAT SM (1 km)_SVR		$y=ax+b$ (x: in situ SM, y: downscaled SM)	R	Bias ($m^3 \cdot m^{-3}$)	RMSE ($m^3 \cdot m^{-3}$)	IOA
	Average ($m^3 \cdot m^{-3}$)	SD ($m^3 \cdot m^{-3}$)	Average ($m^3 \cdot m^{-3}$)	SD ($m^3 \cdot m^{-3}$)					
Gokseong	0.16	0.03	0.25	0.04	$y=0.70x+13.32$	0.50**	−0.08	0.09	0.37
Geumsan	0.26	0.02	0.25	0.03	$y=0.85x+2.74$	0.61**	0.01	0.03	0.74
Yeongdong	0.22	0.04	0.28	0.04	$y=0.73x+12.07$	0.71**	−0.06	0.07	0.58
Imsil	0.26	0.04	0.25	0.04	$y=0.69x+7.74$	0.73**	0.00	0.03	0.85
Jeongeup	0.29	0.04	0.22	0.06	$y=1.16x-11.75$	0.77**	0.07	0.08	0.59
Jeonju	0.24	0.07	0.23	0.06	$y=0.57x+9.46$	0.70**	0.00	0.06	0.72
Wanju	0.17	0.03	0.25	0.04	$y=1.16x+4.76$	0.87**	−0.08	0.08	0.52
Sunchang	0.23	0.06	0.37	0.06	$y=0.80x+10.50$	0.80**	−0.04	0.05	0.79
Hapcheon	0.21	0.09	0.25	0.06	$y=0.55x+12.35$	0.44*	−0.11	0.17	0.22
Average	*0.23*	*0.04*	*0.26*	*0.05*		*0.68*	−0.03	0.07	0.60

*The significance level of the p value > 0.05; **the significance level of the p value > 0.01.

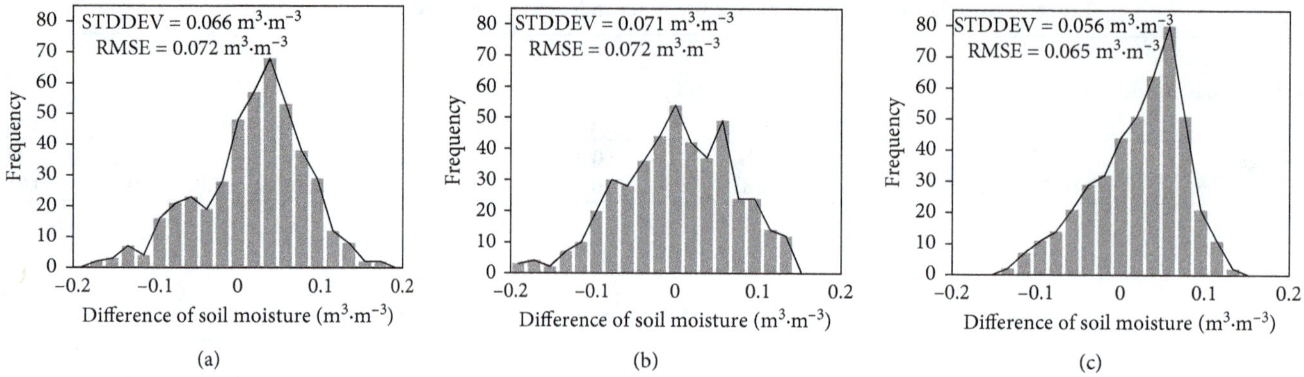

Figure 5: Histograms of the difference between remotely sensed SM such as (a) 12.5 km ASCAT SM, (b) 1 km ASCAT SM (polynomial), and (c) 1 km ASCAT SM (SVR) and in situ SM measurements.

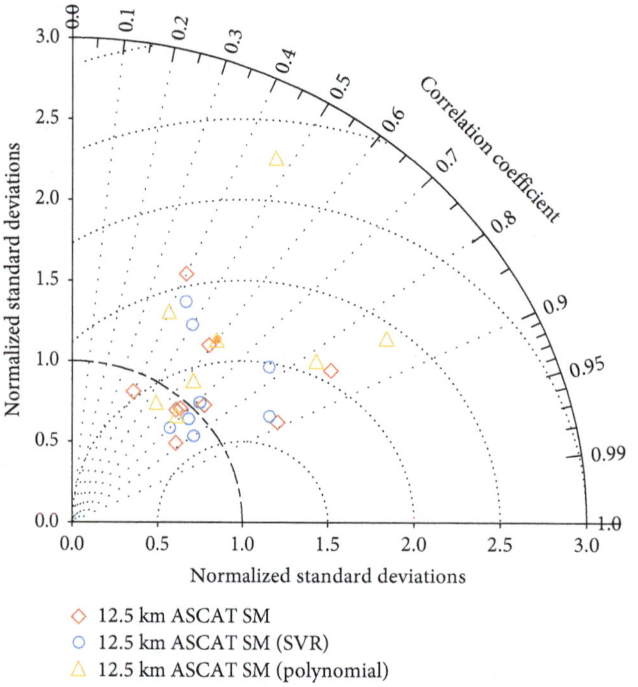

Figure 6: Taylor diagram (correlation coefficient, normalized standard deviation, and root mean square error) of each SM measurement from 12.5 km ASCAT, 1 km ASCAT with SVR, and 1 km ASCAT SM with polynomial.

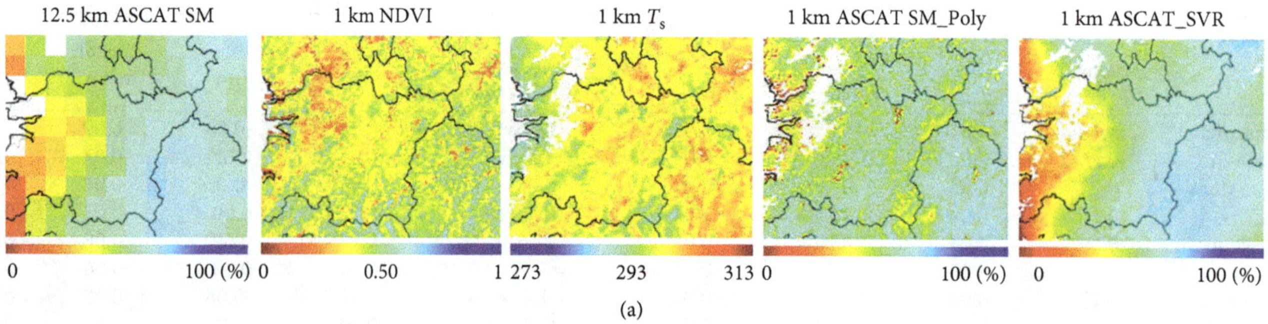

(a)

Figure 7: Continued.

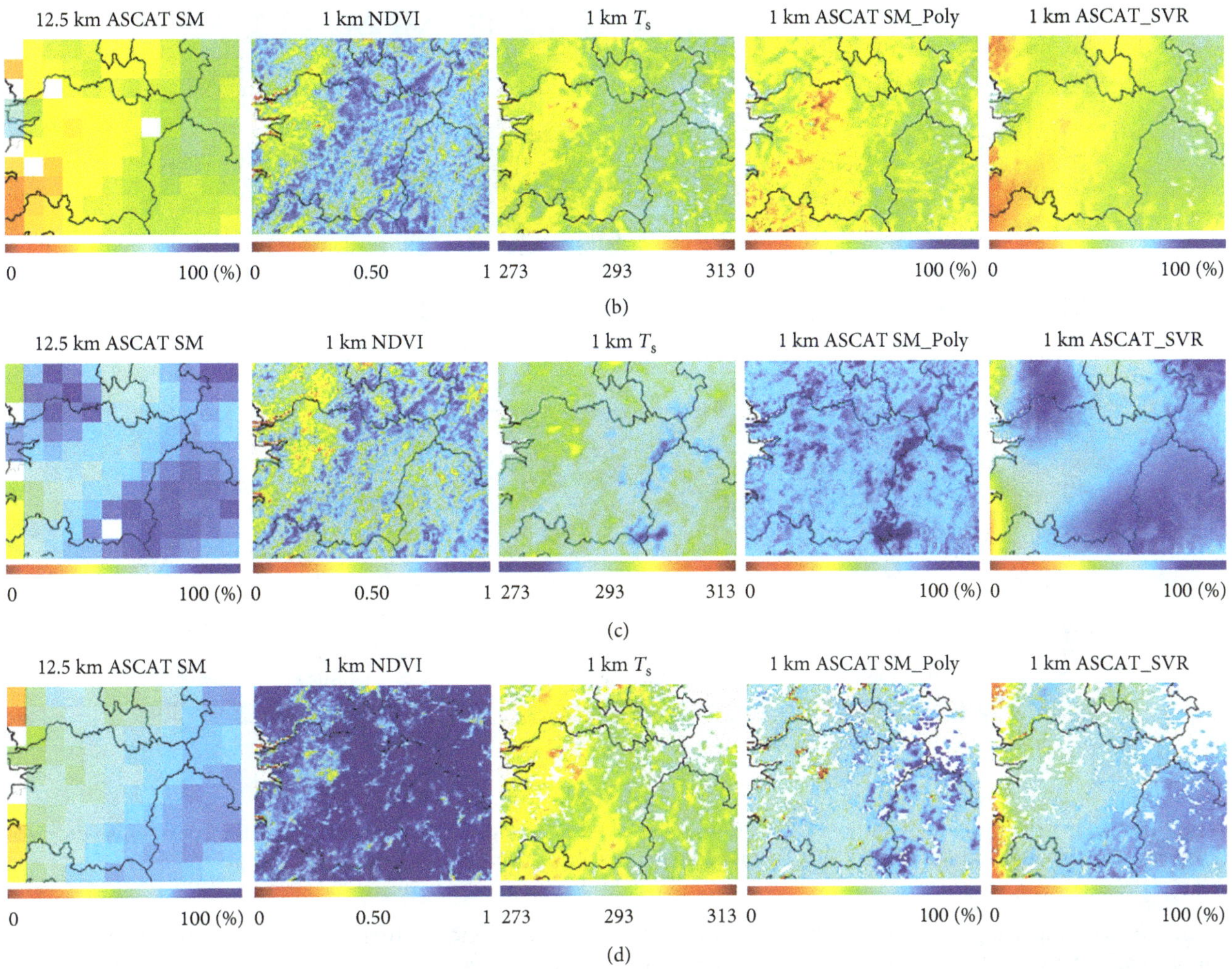

FIGURE 7: Spatial distributions of the original and downscaled SM, NDVI, and T_s on dry days on April 16 and October 16 and wet days on April 26 and September 10.

values, which are more sensitive to extreme values in estimating the model agreement, showed differences with the *R* value at some sites (Jeongeup and Wanju). However, on average, they also indicated the SVR results to be a better estimation for in situ SM.

Figure 6 shows the two-dimensional Taylor diagram [58] summarizing the statistics for the three ASCAT SM products compared with the in situ SM measurements from nine sites. This diagram shows the statistical values between the original SM and downscaled SM using SVR and the polynomial method and in situ data. While the ranges of the *R* values for the three SM products were similar, from approximately 0.4 to 0.9, there were apparent differences in the distributions of the ratio of the SDs and RMSE. Although statistical resemblances were found between the results of the 12.5 km ASCAT SM (diamond) and 1 km SVR SM (circle), the diagram indicates a clearly higher SD for the 1 km poly SM (triangle). In particular, the SVR SM points were found to be most closely around the ideal arc drawn with a dashed line. The results of the polynomial downscaling were more sparsely distributed on the diagram with an isolated point representing the result at the Hapcheon site, and the larger RMSE was probably a result of the geophysical characteristics at Hapcheon site since the corresponding ASCAT pixel contained a mixed land cover of forest and cropland. Generally, they showed some weak agreement between SM retrievals with in situ measurements such as at the Hapcheon site; however, the *R* values of the downscaled SM were largely improved even if the range of that improvement was small.

4.2. Spatial Distribution of Downscaled SM. The 12.5 km and 1 km ASCAT SM measurements obtained using two different downscaling methods were spatially compared with daily mappings of each type of data on dry and wet days (Figure 7). The overall spatial variations of the 1 km ASCAT SM measurement were approximately similar to those of the 12.5 km data, but with more finely distributed characteristics. While the eastern part of the study area with forested land cover had a higher average SM of approximately 0.5-0.6 $m^3 \cdot m^{-3}$, the western part with primarily cropland land cover had more temporal variation according to meteorological events. A comparison of the spatial distributions in

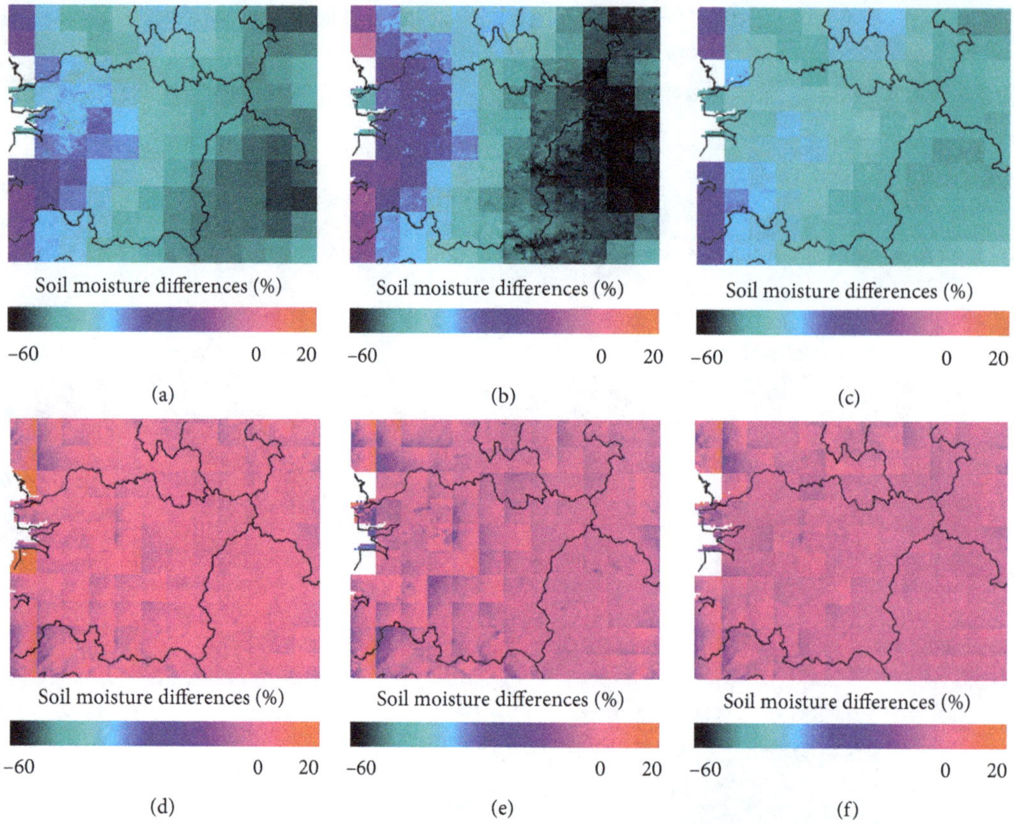

Figure 8: Seasonal spatial distributions during 2012 of the residual between the original and downscaled SM using polynomial downscaling for (a) spring, (b) summer, and (c) fall and SVR downscaling for (d) spring, (e) summer, and (f) fall.

the 1 km SM mapping using polynomial regression revealed a clear similarity between the mappings of the 12.5 km SM and 1 km SM for SVR caused by the downscaling algorithm that uses each pixel's position as a predictive variable. Under wet conditions (Figures 7(c) and 7(d)), the spatial patterns of the downscaled SM from the polynomial regression are evenly distributed and rely on the distribution of T_s compared with that under dry conditions (Figures 7(a) and 7(b)). Under wet conditions, the original ASCAT SM shows relatively dry patterns in the western part of the study area, while the TS and downscaled SM using polynomial regression show no higher temperature or drier patterns in the same region, respectively. Piles et al. [8] also reported a more consistent and similar spatial variability of the downscaled SM product relative to the original SMOS SM under dry soil conditions. Thus, a consideration of positional weighting would allow substantial performance improvement of the SM downscaling based on T_s-NDVI.

Figure 8 shows the distribution of the seasonal mean differences between the 1 km ASCAT, uniformly disaggregated, and downscaled SM measurements generated using each methodology. On average, the difference between the original data and the downscaled SM using the SVR was clearly less than that using the polynomial downscaling. Although there were clear per-pixel discrepancies in the polynomial-downscaled SM, the differences in the SVR-downscaled SM were more evenly distributed, regardless of location. This characteristic was a result of the methodological difference between those that the SVR downscaling considered as the positional weight while the polynomial downscaling did not. The difference in the polynomial-downscaled SM was negatively biased and was found to be larger in the southeastern region of the study area where the elevation was higher with a land cover dominated by mixed forests. It is probable that the higher uncertainty in the NDVI for densely vegetated areas erroneously affected the regression model. The largest differences in SM for both products were found during the summer (June to August) when the vegetation growth reached its peak, and this might have affected the relationship between the SM and T_s. Similar seasonal differences in the error pattern for the downscaled SM were also found in Merlin et al. [3], and that study adopted a separate downscaling algorithm to reduce the seasonal discrepancy in downscaling performance by considering the controlling variable of SM for each pixel. In addition, in a future study, the depth discrepancy between satellite- and ground-based SM measurements should be corrected when comparing downscaled SM product with in situ data by estimating the profile satellite SM values [59, 60].

5. Conclusions

The downscaling methods for remotely sensed SM dataset are among the most important topics in related research fields since they provide a solution to low spatial resolution.

This s tudy p roposed a nd e valuated a n ew downscaling method using SVR by comparing with in situ SM measurements and results of a conventional downscaling method. The RMSE decreased after downscaling using SVR from 0.08 to 0.07 $m^3 \cdot m^{-3}$, and the R increased from 0.66 to 0.68; the bias remained the same at 0.03 $m^3 \cdot m^{-3}$. Considering that the improvements and deterioration of the downscaled SM evened out on average, valid improvements in accuracy should be noticed at the nine sites selected for validation. The s tatistics w ere b etter t han t hose o f the polynomial downscaling method, which had an RMSE of 0.09 $m^3 \cdot m^{-3}$, a nR of 0.62, and a bias of −0.02 $m^3 \cdot m^{-3}$. In the correlation analysis between the independent variables (NDVI and T_s) and the residual magnitude between the 12.5 km ASCAT SM and predicted SM from each regression method, only the polynomial regression residual magnitudes showed significant r esults t hat w ere p ositively correlated with NDVI and negatively correlated with T_s. In a spatial comparison among the SM mappings at two scales, the 1 km SM using SVR better followed the spatial distribution of the original scale (12.5 km) than the 1 km SM using a polynomial regression. In the spatial distribution of the seasonally averaged differences b etween t he o riginal a nd t he downscaled SM contents, the SVR downscaling method showed a more consistent performance, given the seasonal effect. Based on these results, the suggested SVR downscaling method can be used to improve the spatial resolution of satellite SM while offering b etter p erformance t han the conventional downscaling method. However, this study did have several limitations; first, t he r emote s ensing data were difficult to obtain due to missing products; second, it took considerable time to preprocess the dataset and execute the model to obtain downscaled SM; and lastly, the algorithm's complexity needed considerable memory requirements for a wide range of tasks [61].

In a future study, the limitations of this study will be improved by applying various remote sensing and assimilation datasets. This method can be extended to apply to various fields that require fine-resolution SM datasets such as large-scale water-related natural disasters. This is because anteced-ent SM information can be effectively used to predict landslides, droughts, dust outbreaks, and agricultural water deficiencies [62, 63].

Conflicts of Interest

The authors declare that they have no conflicts of interest.

Acknowledgments

This work was supported by the National Research Foundation of Korea (NRF) grant funded by the Korean government (MSIT) (NRF-2016R1A2B4008312); Space Core Technology Development Program through the National Research Foundation of Korea (NRF) funded by the Ministry of Science and ICT (NRF-2014M1A3A3A02034789); and Basic Science Research Program through the National Research Foundation of Korea (NRF) funded by the Ministry of Education (2017R1A6A3A11034250; NRF-2017R1D1A1B03028129).

References

[1] T. J. Jackson, "III measuring surface soil moisture using passive microwave remote sensing," *Hydrological Processes*, vol. 7, no. 2, pp. 139–152, 1993.

[2] E. G. Njoku, T. J. Jackson, V. Lakshmi, T. K. Chan, and S. V. Nghiem, "Soil moisture retrieval from AMSR-E," *IEEE Transactions on Geoscience and Remote Sensing*, vol. 41, no. 2, pp. 215–229, 2003.

[3] O. Merlin, A. Chehbouni, Y. H. Kerr, and D. C. Goodrich, "A downscaling method for distributing surface soil moisture within a microwave pixel: application to the Monsoon'90 data," *Remote Sensing of Environment*, vol. 101, no. 3, pp. 379–389, 2006.

[4] T. J. Jackson, M. H. Cosh, R. Bindlish et al., "Validation of advanced microwave scanning radiometer soil moisture products," *IEEE Transactions on Geoscience and Remote Sensing*, vol. 48, no. 12, pp. 4256–4272, 2010.

[5] M. Choi and Y. Hur, "A microwave-optical/infrared disaggregation for improving spatial representation of soil moisture using AMSR-E and MODIS products," *Remote Sensing of Environment*, vol. 124, pp. 259–269, 2012.

[6] N. S. Chauhan, S. Miller, and P. Ardanuy, "Spaceborne soil moisture estimation at high resolution: a microwave-optical/IR synergistic approach," *International Journal of Remote Sensing*, vol. 24, no. 22, pp. 4599–4622, 2003.

[7] G. Yu, L. Di, and W. Yang, "Downscaling of global soil moisture using auxiliary data," in *Proceedings of the IEEE International Geoscience and Remote Sensing Symposium, IGARSS 2008*, Boston, MA, USA, July 2008.

[8] M. Piles, A. Camps, M. Vall-llossera et al., "Downscaling SMOS-derived soil moisture using MODIS visible/infrared data," *IEEE Transactions on Geoscience and Remote Sensing*, vol. 49, no. 9, pp. 3156–3166, 2011.

[9] O. Merlin, C. Rudiger, A. Al Bitar, P. Richaume, J. P. Walker, and Y. H. Kerr, "Disaggregation of SMOS soil moisture in Southeastern Australia," *IEEE Transactions on Geoscience and Remote Sensing*, vol. 50, no. 5, pp. 1556–1571, 2012.

[10] T. N. Carlson, R. R. Gillies, and E. M. Perry, "A method to make use of thermal infrared temperature and NDVI measurements to infer surface soil water content and fractional vegetation cover," *Remote Sensing Reviews*, vol. 9, no. 1-2, pp. 161–173, 1994.

[11] R. R. Gillies, W. P. Kustas, and K. S. Humes, "A verification of the 'triangle' method for obtaining surface soil water content and energy fluxes from remote measurements of the normalized difference vegetation index (NDVI) and surface," *International Journal of Remote Sensing*, vol. 18, no. 15, pp. 3145–3166, 1997.

[12] I. Sandholt, K. Rasmussen, and J. Andersen, "A simple interpretation of the surface temperature/vegetation index space for assessment of surface moisture status," *Remote Sensing of Environment*, vol. 79, no. 2, pp. 213–224, 2002.

[13] Q. Weng, D. Lu, and J. Schubring, "Estimation of land surface temperature–vegetation abundance relationship for urban

heat island studies," *Remote Sensing of Environment*, vol. 89, no. 4, pp. 467–483, 2004.

[14] V. N. Vapnik, *The Nature of Statistical Learning Theory*, Springer-Verlag New York Inc., New York, NY, USA, 1995.

[15] V. Cherkassky and Y. Ma, "Practical selection of SVM parameters and noise estimation for SVM regression," *Neural Networks*, vol. 17, no. 1, pp. 113–126, 2004.

[16] A. J. Smola and B. A. Schölkopf, "A tutorial on support vector regression," *Statistics and Computing*, vol. 14, no. 3, pp. 199–222, 2004.

[17] S. Ghosh, "SVM-PGSL coupled approach for statistical downscaling to predict rainfall from GCM output," *Journal of Geophysical Research: Atmospheres*, vol. 115, p. D22, 2010.

[18] L. Xun and L. Wang, "An object-based SVM method incorporating optimal segmentation scale estimation using Bhattacharyya Distance for mapping salt cedar (Tamarisk spp.) with QuickBird imagery," *GIScience and Remote Sensing*, vol. 52, no. 3, pp. 257–273, 2015.

[19] Z. Lin and L. Yan, "A support vector machine classifier based on a new kernel function model for hyper spectral data," *GIScience and Remote Sensing*, vol. 53, no. 1, pp. 85–101, 2016.

[20] C. Wang, R. Pavlowsky, Q. Huang, and C. Chang, "Channel bar feature extraction for a mining-contaminated river using high-spatial multispectral remote sensing imagery," *GIScience and Remote Sensing*, vol. 53, no. 3, pp. 283–302, 2016.

[21] C. J. Gleason and J. Im, "Forest biomass estimation from airborne LiDAR data using machine learning approaches," *Remote Sensing of Environment*, vol. 125, pp. 80–91, 2012.

[22] Y. H. Kim, J. Im, H. K. Ha, J. K. Choi, and S. Ha, "Machine learning approaches to coastal water quality monitoring using GOCI satellite data," *GIScience and Remote Sensing*, vol. 51, no. 2, pp. 158–174, 2014.

[23] Y. H. Kaheil, M. K. Gill, M. McKee, L. A. Bastidas, and E. Rosero, "Downscaling and assimilation of surface soil moisture using ground truth measurements," *IEEE Transactions on Geoscience and Remote Sensing*, vol. 46, no. 5, pp. 375–384, 2008.

[24] I. Keramitsoglou, C. T. Kiranoudis, and Q. Weng, "Downscaling geostationary land surface temperature imagery for urban analysis," *IEEE Geoscience and Remote Sensing Letters*, vol. 10, no. 5, pp. 1253–1257, 2013.

[25] A. Al-Yaari, J. P. Wigneron, A. Ducharne et al., "Global-scale comparison of passive (SMOS) and active (ASCAT) satellite based microwave soil moisture retrievals with soil moisture simulations (MERRA-Land)," *Remote Sensing of Environment*, vol. 152, pp. 614–626, 2014.

[26] H. Kim, R. Parinussa, A. G. Konings et al., "Global-scale assessment and combination of SMAP with ASCAT (active) and AMSR2 (passive) soil moisture products," *Remote Sensing of Environment*, vol. 204, pp. 260–275, 2018.

[27] Korea Meteorological Administration, *Annual Climatological Report*, Korea Meteorological Administration, Seoul, Republic of Korea, 2012.

[28] M. A. Friedl, D. Sulla-Menashe, B. Tan et al., "MODIS Collection 5 global land cover: algorithm refinements and characterization of new datasets," *Remote Sensing of Environment*, vol. 114, no. 1, pp. 168–182, 2010.

[29] G. C. Topp, "State of the art of measuring soil water content," *Hydrological Processes*, vol. 17, no. 14, pp. 2993–2996, 2003.

[30] H. Eller and A. Denoth, "A capacitive soil moisture sensor," *Journal of Hydrology*, vol. 185, no. 1–4, pp. 137–146, 2016.

[31] C. M. K. Gardner, T. J. Dean, and J. D. Cooper, "Soil water content measurement with a high-frequency capacitance sensor," *Journal of Agricultural Engineering Research*, vol. 71, no. 4, pp. 395–403, 1998.

[32] H. Nguyen, H. Kim, and M. Choi, "Evaluation of the soil water content using cosmic-ray neutron probe in a heterogeneous monsoon climate-dominated region," *Advances in Water Resources*, vol. 108, pp. 125–138, 2017.

[33] W. Wagner, G. Lemoine, and H. Rott, "A method for estimating soil moisture from ERS scatterometer and soil data," *Remote Sensing of Environment*, vol. 70, no. 2, pp. 191–207, 1997.

[34] C. Albergel, C. Rüdiger, D. Carrer et al., "An evaluation of ASCAT surface soil moisture products with in-situ observations in Southwestern France," *Hydrology and Earth System Sciences*, vol. 13, no. 2, pp. 115–124, 2009.

[35] Z. Wan, Y. Zhang, Q. Zhang, and Z. L. Li, "Quality assessment and validation of the MODIS global land surface temperature," *International Journal of Remote Sensing*, vol. 25, no. 1, pp. 261–274, 2004.

[36] A. Huete, K. Didan, T. Miura, E. P. Rodriguez, X. Gao, and L. G. Ferreira, "Overview of the radiometric and biophysical performance of the MODIS vegetation indices," *Remote Sensing of Environment*, vol. 83, no. 1, pp. 195–213, 2002.

[37] Y. Gu, J. F. Brown, J. P. Verdin, and B. Wardlow, "A five-year analysis of MODIS NDVI and NDWI for grassland drought assessment over the central Great Plains of the United States," *Geophysical Research Letters*, vol. 34, no. 6, article L06407, 2007.

[38] D. Tanré, Y. J. Kaufman, M. Herman, and S. Mattoo, "Remote sensing of aerosol properties over oceans using the MODIS/EOS spectral radiances," *Journal of Geophysical Research: Atmospheres*, vol. 102, no. D14, pp. 16971–16988, 1997.

[39] M. K. Gill, T. Asefa, M. W. Kemblowski, and M. McKee, "Soil moisture prediction using support vector machines," *Journal of the American Water Resources Association*, vol. 42, pp. 1033–1046, 2006.

[40] I. Yilmaz, "Comparison of landslide susceptibility mapping mythologies for Koyulhisar, Turkey: conditional probability, logistic regression, artificial neural networks, and support vector machine," *Environmental Earth Sciences*, vol. 61, no. 4, pp. 821–836, 2010.

[41] C. C. Chang and C. J. Lin, "LIBSVM: a library for support vector machines," *ACM Transactions on Intelligent Systems and Technology*, vol. 2, no. 3, pp. 1–27, 2011.

[42] Q. Chang, Q. Chen, and X. Wang, "Scaling Gaussian RBF kernel width to improve SVM classification," in *Proceeding of the International Conference on IEEE Neural Networks and Brain, ICNN&B'05*, vol. 1, pp. 19–22, Beijing, China, 2005.

[43] X. Li, C. Zhang, and W. Li, "Building block level urban land-use information retrieval based on Google Street View images," *GIScience and Remote Sensing*, vol. 54, no. 6, pp. 819–835, 2017.

[44] H. Chu, C. Wang, S. Kong, and K. Chen, "Integration of full-waveform LiDAR and hyper spectral data to enhance tea

Areca classification," *GIScience and Remote Sensing*, vol. 53, no. 4, pp. 542–559, 2016.
[45] L. C. J. Moreira, A. D. S. Teixeira, and L. S. Galvão, "Potential of multispectral and hyperspectral data to detect saline-exposed soils in Brazil," *GIScience and Remote Sensing*, vol. 52, no. 4, pp. 416–436, 2015.
[46] G. Mountrakis, J. Im, and C. Ogole, "Support vector machines in remote sensing: a review," *ISPRS Journal of Photogrammetry and Remote Sensing*, vol. 66, no. 3, pp. 247–259, 2011.
[47] R. Sonobe, Y. Yamaya, H. Tani, X. Wang, N. Kobayashi, and K. Mochizuki, "Assessing the suitability of data from Sentinel-1A and 2A for crop classification," *GIScience and Remote Sensing*, vol. 54, no. 6, pp. 918–938, 2017.
[48] S. Georganos, T. Grippa, S. Vanhuysse et al., "Less is more: optimization classification performance through feature selection in a very-high resolution remote sensing object-based urban application," *GIScience and Remote Sensing*, vol. 55, no. 2, pp. 221–242, 2018.
[49] A. Moghaddamnia, M. Ghafari, J. Piri, and D. Han, "Evaporation estimation using support vector machines technique," *World Academy of Science, Engineering and Technology*, vol. 43, pp. 14–22, 2008.
[50] S. S. Eslamian, S. A. Gohari, M. Biabanaki, and R. Malekian, "Estimation of monthly pan evaporation using artificial neural networks and support vector machines," *Journal of Applied Sciences*, vol. 8, no. 19, pp. 3497–3502, 2008.
[51] P. C. Deka, "Support vector machine applications in the field of hydrology: a review," *Applied Soft Computing*, vol. 19, pp. 372–386, 2014.
[52] P. K. Srivastava, D. Han, M. R. Ramirez, and T. Islam, "Machine learning techniques for downscaling SMOS satellite soil moisture using MODIS land surface temperature for hydrological application," *Water Resources Management*, vol. 27, no. 8, pp. 3127–3144, 2013.
[53] R. Parinussa, M. Yilmaz, M. Anderson, C. Hain, and R. de Jeu, "An intercomparison of remotely sensed soil moisture products at various spatial scales over the Iberian Peninsula," *Hydrological Processes*, vol. 28, no. 18, pp. 4865–4876, 2014.
[54] C. Song, L. Jia, and M. Menenti, "Retrieving high-resolution surface soil moisture by downscaling AMSR-E brightness temperature using MODIS LST and NDVI data," *IEEE Journal of Selected Topics in Applied Earth Observations and Remote Sensing*, vol. 7, no. 3, pp. 935–942, 2014.
[55] R. Bindlish and A. P. Barros, "Parameterization of vegetation backscatter in radar-based, soil moisture estimation," *Remote Sensing of Environment*, vol. 76, no. 1, pp. 130–137, 2001.
[56] S. J. Goetz, "Multi-sensor analysis of NDVI, surface temperature and biophysical variables at a mixed grassland site," *International Journal of Remote Sensing*, vol. 18, no. 1, pp. 71–94, 1997.
[57] D. W. Scott, *Multivariate Density Estimation: Theory, Practice, and Visualization*, John Wiley & Sons Inc., New York, NY, USA, 2015.
[58] K. E. Taylor, "Summarizing multiple aspects of model performance in a single diagram," *Journal of Geophysical Research: Atmospheres*, vol. 106, no. D7, pp. 7183–7192, 2001.
[59] H. Kim and V. Lakshmi, "Use of Cyclone Global Navigation Satellite System (CYGNSS) observations for estimation of soil moisture," *Geophysical Research Letter*, 2018, In press.
[60] M. Zohaib, H. Kim, and M. Choi, "Evaluating the patterns of spatiotemporal trends of root zone soil moisture in major climate regions in East Asia," *Journal of Geophysical Research: Atmospheres*, vol. 122, no. 15, pp. 7705–7722, 2017.
[61] J. Valyon and G. Horváth, "A weighted generalized LS-SVM," *Periodica Polytechnica Electrical Engineering*, vol. 47, pp. 229–252, 2003.
[62] L. Brocca, L. Ciabatta, C. Massari, S. Camici, and A. Tarpanelli, "Soil moisture for hydrological applications: open questions and new opportunities," *Water*, vol. 9, no. 2, p. 140, 2017.
[63] H. Kim, M. Zohaib, E. Cho, Y. H Kerr, and M. Choi, "Development and assessment of the sand dust prediction model by utilizing microwave-based satellite soil moisture and reanalysis datasets in East Asian desert areas," *Advances in Meteorology*, vol. 2017, Article ID 1917372, 13 pages, 2017.

14

Changes in Different Classes of Precipitation and the Impacts on Sediment Yield in the Hekouzhen-Longmen Region of the Yellow River Basin, China

Suzhen Dang,[1,2] Xiaoyan Liu,[3] Xiaoyu Li,[4] Manfei Yao,[5] and Dan Zhang[6]

[1]*Yellow River Institute of Hydraulic Research, Yellow River Conservancy Commission, Zhengzhou 450003, China*
[2]*Key Laboratory of Soil and Water Loss Process and Control on the Loess Plateau, MWR, Yellow River Institute of Hydraulic Research, Zhengzhou 450003, China*
[3]*Yellow River Conservancy Commission, Zhengzhou 450003, China*
[4]*Hydrology Bureau of Yellow River Conservancy Commission, Zhengzhou 450004, China*
[5]*School of Water Conservancy, North China University of Water Resources and Electric Power, Zhengzhou 450045, China*
[6]*Key Laboratory of Watershed Geographic Sciences, Nanjing Institute of Geography and Limnology, Chinese Academy of Sciences, Nanjing 210008, China*

Correspondence should be addressed to Suzhen Dang; dangsz_hky@163.com

Academic Editor: James Cleverly

The sediment yield of the Yellow River Basin has obviously decreased since the 1980s, and the impacts of precipitation on sediment yield changes have become increasingly important with the global climate change. The spatial and temporal variations in annual precipitation and different classes of precipitation in the Hekouzhen-Longmen region (HLR) in the middle reaches of the Yellow River Basin were investigated using data collected from 301 rainfall stations from 1966 to 2016. The impacts of precipitation variation on sediment yield were evaluated, and the hydrological modeling method was used to quantitatively assess the attribution of precipitation and other factors to sediment yield changes in the HLR. The results show that the annual precipitation and P_{10} increased from the northwest to the southeast of the HLR, suggesting it was drier in the northwest region of the HLR. P_{25} and P_{50} were mainly concentrated in the northwestern and southwestern parts of the HLR, reflecting that heavy rain was more likely to occur in these regions of the HLR. All of the annual precipitation and different classes of precipitation had no significant changing trends from 1966 to 2016, and the relationship between rainfall and sediment yield obviously changed in 2006. Compared with the average annual mean values from 1966 to 2016, both the annual precipitation and the different classes of precipitation were higher in the HLR during 2007–2016. The sediment yield decrease during 1990–1999 was mainly influenced by precipitation, while other factors were the main driving factor for the sediment yield decrease in the periods of 1980–1989, 2000–2009, and 2010–2016, and other factors have become the dominant driving factors of the sediment yield change in the HLR since 2000.

1. Introduction

The Yellow River is known for its large sediment discharge and high sediment concentration, and 98% of its sediment originates from the area above the Shaanxian (Tongguan) station on the main channel of the Yellow River. During the natural period (i.e., 1919–1960), the average annual sediment discharge of the Shaanxian was 1.6 billion tons. However, the sediment discharge of the Yellow River has obviously decreased since the 1980s [1]. The average annual sediment discharge of the Tongguan station was only 539 million and 248 million tons for the periods of 1980–2016 and 2000–2016, representing decreases of 66.3% and 84.5%, respectively. Since precipitation is a direct influencing factor on sediment yield, determining changes in precipitation and identifying their impact on sediment discharge have become a topic of great concern in recent years.

In this paper, we selected the region from Hekouzhen to Longmen in the middle reaches of the Yellow River Basin (hereinafter referred to as the HLR) as the study area, which is located in the Loess Plateau and represents the most seriously affected area of water and soil loss and is the concentrated source area of sediment and coarse sediment [2]; additionally, this region is one of the three storm flood source areas in the middle reaches of the Yellow River [3]. The soil erosion area occupies 83.45% of the total area in the HLR, and the average annual sediment discharge is 908 million tons, accounting for 57% of the sediment discharge from the Tongguan station; furthermore, the average annual coarse sediment discharge is 223.6 million tons, accounting for 72% of the annual coarse sediment discharge of the Yellow River [2].

To date, many scholars have conducted relevant studies on the changes in precipitation in this region. Kang et al. analyzed the spatial distribution and temporal variation of precipitation in the HLR from 1955 to 1995 [4]. Hu et al. found out that compared with the baseline period before 1969, the precipitation in the HLR decreased by 7%, 11%, and 13% in the 1970s, 1980s, and 1990s, respectively [5]. Compared with 1956–1996, the precipitation in the upper and middle reaches of the Yellow River decreased from 1997 to 2006; additionally, the rainfall in July and August in the HLR decreased by 17% [6]. Based on the precipitation data collected from 74 rainfall stations located in the five tributaries of the HLR between 1980 and 2009, the precipitation increased slightly during the flood season, but the rainfall intensity was significantly smaller [7]. Sun et al. [8] analyzed changes in the mean and extreme temperature and precipitation values in the Loess Plateau between 1961 and 2011 using a gridded dataset and found out that the total amount of precipitation on wet days decreased over a large area of the Loess Plateau, and there were only minor changes in extreme precipitation over the Loess Plateau. Other studies have analyzed the changes in precipitation in the Yellow River Basin by using the observed precipitation data collected at dozens of weather stations by the National Meteorological Bureau [9–12], and most of the studies focused on the changes in annual precipitation and precipitation during the flood season [13–15]. In contrast, little research has been focused on the different classes of precipitation that have significant impacts on sediment yield. In view of the uneven spatial distribution of precipitation, we collected data from more additional rainfall stations, which provided more information than the previous studies.

Using the observed precipitation data collected at 301 rainfall stations in the HLR from 1966 to 2016, the objectives of this study were (i) to analyze the spatial distribution and temporal variation in the annual precipitation, different precipitation classes, and rainstorm frequency and (ii) to explore the impacts of precipitation on sediment yield and quantitatively analyze the contribution of precipitation and other factors to sediment yield change.

2. Study Area and Data

2.1. Study Region. The study area is the Hekouzhen-Longmen region, which is situated in the middle reaches of the Yellow River (Figure 1), with an area of 111,586 km^2 between 108°02′~112°44′E and 35°40′N~40°34′N, accounting for 14.8% of the total area of the Yellow River Basin [14]. The length of the main stream of the Yellow River in the study area is 723 km. There are 21 large tributaries with a catchment area larger than 1000 km^2. The high sediment yield area in the HLR is 71,600 km^2 and accounts for 60.1% of the total high sediment yield area in the entire Yellow River Basin; additionally, the high sediment yield area in the HLR accounts for 64.2% of the study area. The terrain is dominated by the gully region of the Loess Plateau and belongs to the temperate continental monsoon climate. The annual average temperature is 6–14°C, and the average annual precipitation is 290–620 mm and is mainly concentrated in July–September.

2.2. Data

2.2.1. Rainfall Stations. Considering the equal distribution of rainfall stations and the integrity of data, this study used daily precipitation data collected from 301 selected rain gauge stations during the period from 1966 to 2016. The rainfall station locations are shown in Figure 1. The precipitation data of 292 rainfall stations and the sediment discharge data are from the hydrological yearbook and the Hydrological Bureau of Yellow River Conservancy Commission; additionally, data from 9 meteorological stations were provided by China Meteorological Administration (CMA). Among the 301 rainfall stations, a total of 150 rainfall stations were established before 1966; the remaining rainfall stations were established between 1967 and 1976.

2.2.2. Rainfall Factor. In consideration of the impact of precipitation on sediment yield in the study area, the rainfall factors selected in this paper included annual precipitation (P) and different classes of precipitation. The different precipitation classes refer to the total annual rainfall values, and categories were divided based on daily precipitation amounts that were greater than 10 mm, 25 mm, 50 mm, and 100 mm, which were expressed as P_{10}, P_{25}, P_{50}, and P_{100}, respectively, and were measured in mm. The different classes of precipitation not only reflect the impact of total precipitation on sediment yield but also reflect the impact of precipitation intensity on sediment yield [16].

2.2.3. Regional Average Precipitation. First, the annual precipitation, P_{10}, P_{25}, P_{50}, and P_{100} of each rainfall station were calculated, and then the average precipitation in the entire study area was interpolated by the Thiessen polygon method.

2.2.4. Rainstorm Frequency. To objectively reflect the change in the frequency of heavy rain events with daily precipitation higher than 50 mm, the ratio of the number of rainfall stations

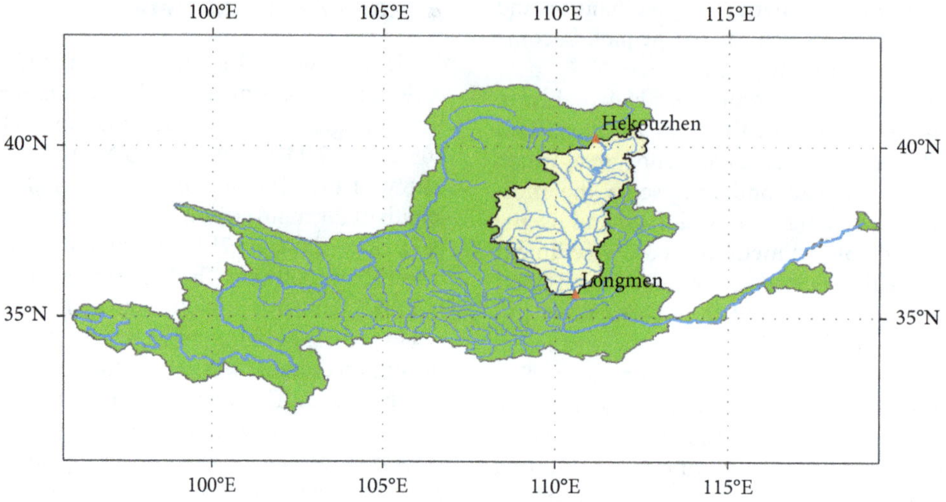

(a)

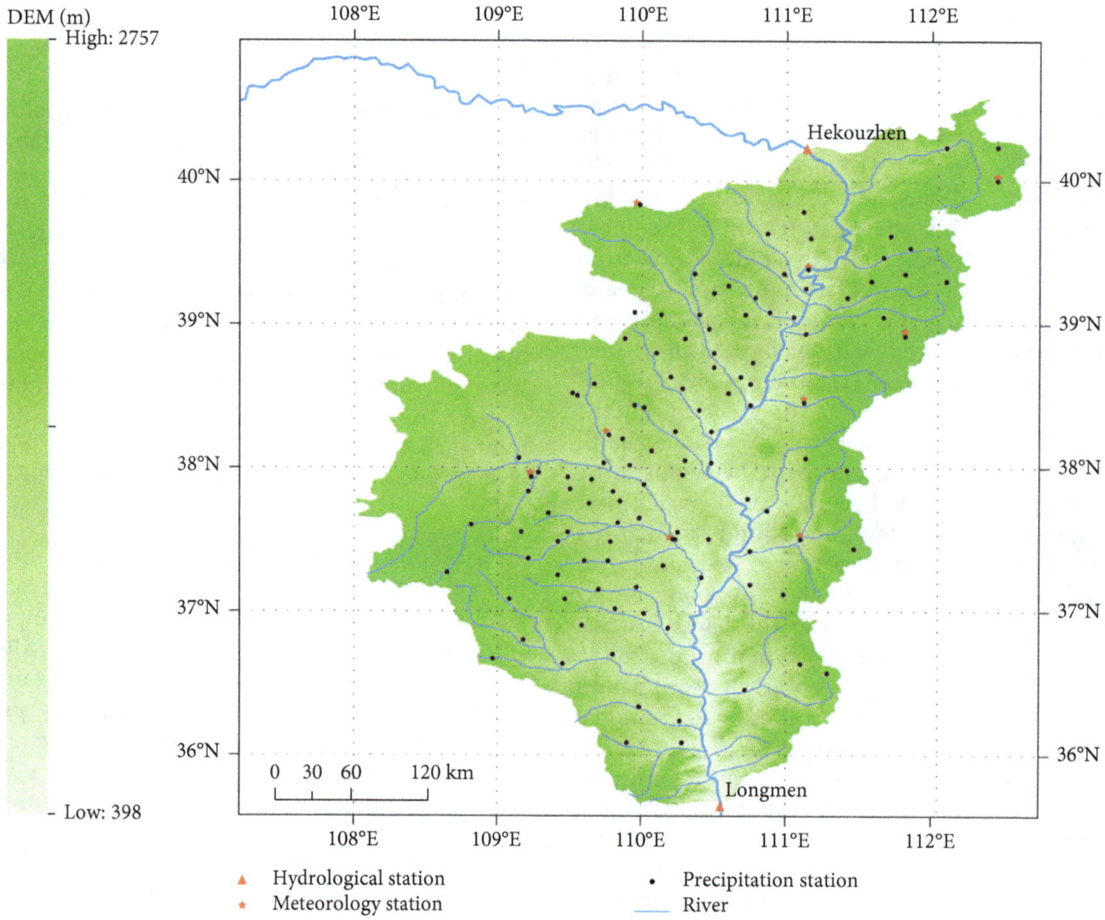

(b)

FIGURE 1: Sketch map of the study area.

with daily precipitation greater than 50 mm in a given year to the total number of rainfall stations participating in rainfall measurements in the same year was referred to as the rainstorm frequency. Similarly, the frequency of heavy rainstorms can be calculated by the number of rainfall stations with daily precipitation greater than 100 mm.

3. Methodology

3.1. Double Mass Curve. The double mass curve (DMC) method is the simplest, most intuitive, and most widely used method for consistency analysis of long-term evolutionary trends of hydrological and meteorological elements [17, 18].

By establishing a double cumulative curve that excludes the influence of the reference variable, whether another factor leads to significant trend changes in the tested variable is revealed [19]. The DMC can analyze changing trends in runoff and sediment discharge, and this method has been widely used in the study on water and sediment effects of water and soil conservation measures [20–22]. Changes in the slope of the double mass curve between precipitation and sediment discharge reflect changes in sediment discharge that are produced by unit of rainfall.

3.2. Mann–Kendall Test. The Mann–Kendall (MK) test is a nonparametric method and is recommended by the World Meteorological Organization [23–25]. The MK test does not require samples to follow a specific distribution, and the results are not affected by a few abnormal values. The MK test is suitable for nonnormally distributed data and has been widely used in assessing the changing trends of hydrological and meteorological time series data [26–28]. The MK test was applied to analyze the changes in precipitation in the study area. The MK test will not be discussed here, as its detailed description can be found in many studies [29, 30]. Mann–Kendall Z statistics greater than 1.96 indicated a significant increasing trend at the significance level of $P = 0.05$, while a Z statistic less than -1.96 indicated a significant decreasing trend.

3.3. Calculation Method of Precipitation Change. To eliminate the influence of changes affiliated with the rainfall stations on the calculated average watershed precipitation as much as possible, the establishment time and spatial distribution of the rainfall stations were considered, and the multiyear average annual precipitation from 1966 to 2016 was identified as the reference precipitation. Daily precipitation was collected from 150 rainfall stations established before 1966 for the period from 1966 to 2016, and the spatial distribution of the annual precipitation, P_{10}, P_{25}, P_{50}, and P_{100} was spatially interpolated by the Thiessen polygon method. The average annual precipitation for the period from 2007 to 2016 was calculated by using the precipitation data collected from all rainfall stations.

The measured data from 1966 to 2016 were used to calculate the average precipitation for all the rainfall stations established before 1966. For the rainfall stations without measured data from 1966 to 1976, the average precipitation from 1966 to 2016 was extracted from the precipitation spatial distribution maps. By comparing the current annual precipitation with the multiyear average precipitation at each station, the annual abundance and spatial distribution of precipitation could be determined for each station.

3.4. Quantitative Assessment of the Changes in Sediment Yield. For a given basin, the changes in observed sediment yield under the impacts of precipitation and other factors can be expressed as follows:

$$\Delta W_S = \Delta W_{SP} + \Delta W_{SH}, \qquad (1)$$

where ΔW_S is the observed sediment yield difference between the impacted period and the baseline period and ΔW_{SP} and ΔW_{SH} represent the changes in sediment yield due to precipitation and other factors, respectively. To quantitatively identify the impact of precipitation and other factors on the sediment yield changes, the hydrological modeling method can be used. Here, the empirical model was considered due to its good performance in modeling the sediment yield in the middle reaches of the Yellow River Basin [1, 31]. The relationship between precipitation and sediment yield of the HLR in the baseline period is given as follows [1]:

$$W_S = 481.1 \times P_{2.5}^{1.059}, \qquad (2)$$

where W_S is the sediment yield and $P_{2.5}$ is the rainfall factor.

The effects of precipitation change and other factors on sediment yield can be calculated as follows:

$$\begin{aligned} \Delta W_{SP} &= W_{S0} - W_{S2}, \\ \Delta W_{SH} &= W_{S2} - W_{S1}, \end{aligned} \qquad (3)$$

where W_{S0} is the sediment yield in the baseline period and W_{S1} and W_{S2} are the observed and calculated sediment yields in impacted periods, respectively. Taking into account the change in the underlying surface and other factors such as the change in the relationship between precipitation and sediment yield, the baseline period of the HLR is from 1956 to 1977; for more details, the readers can refer to Liu [1].

4. Results and Discussion

4.1. Spatial-Temporal Variations in Precipitation from 1966 to 2016

4.1.1. Spatial Distribution of Precipitation. Figure 2 shows the spatial distribution of the average annual precipitation and the different classes of precipitation in the study area based on the measured precipitation data from 1966 to 2016. The annual precipitation tended to increase gradually from the northwest to the southeast. P_{10} and the annual precipitation had basically the same spatial distribution. However, P_{25} and P_{50} were mainly concentrated in the northwestern and southwestern parts of the HLR, with annual precipitation amounts of 450–600 mm.

4.1.2. Change in Annual Precipitation. Figure 3 shows the change in annual precipitation in the HLR from 1966 to 2016. The average annual precipitation during the 51-year period was 443.6 mm, the maximum annual precipitation was 617.2 mm (in 2016), and the minimum annual precipitation was 291.2 mm (in 1999).

Since the beginning of the twenty-first century, precipitation in the HLR has been more abundant than the overall multiyear average annual precipitation. The average annual precipitation during 2000–2016 was 483.5 mm; however, the precipitation was higher in the years 2003, 2007, 2012, 2013, and 2016. In terms of the precipitation

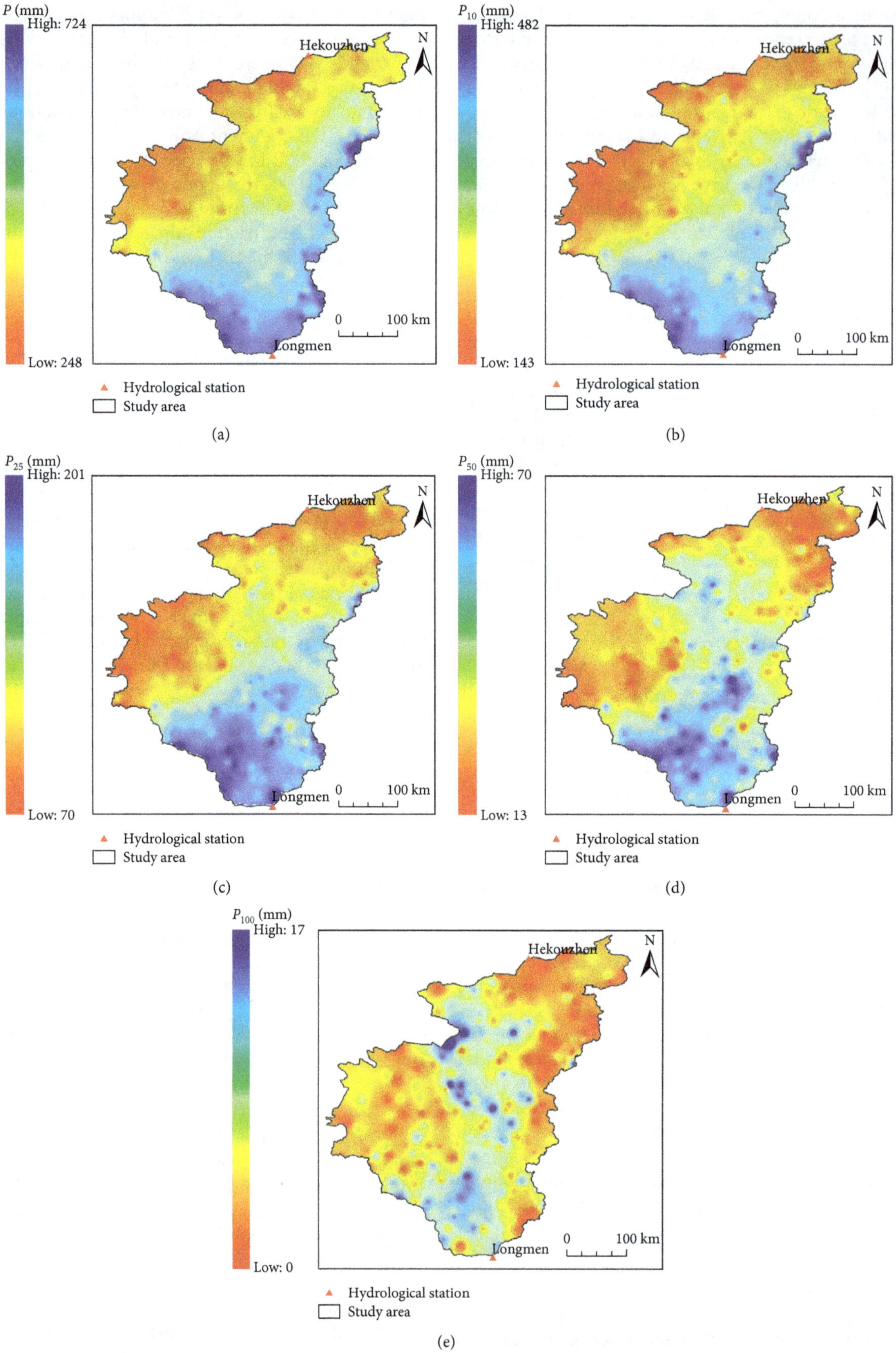

FIGURE 2: Spatial distribution of (a) P, (b) P_{10}, (c) P_{25}, (d) P_{50}, and (e) P_{100} in the HLR.

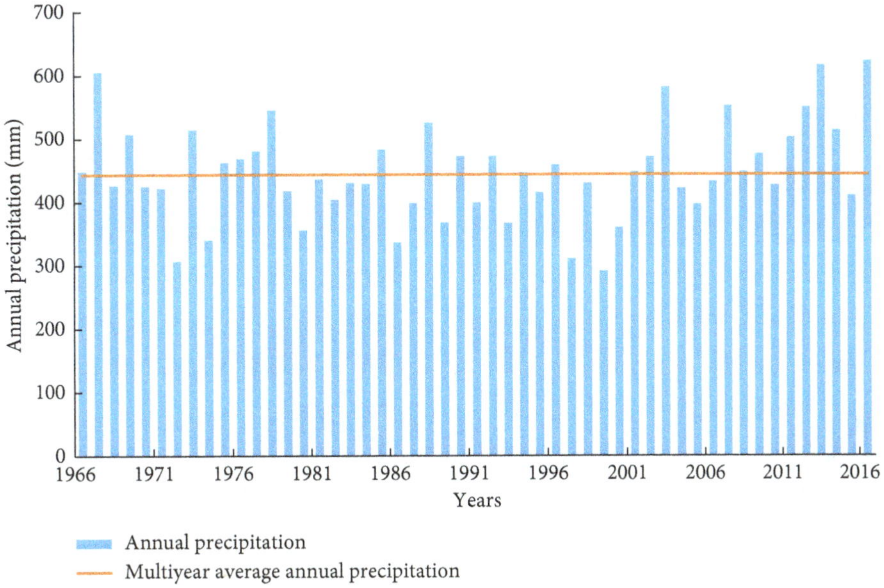

FIGURE 3: Time series of annual precipitation in the HLR from 1966 to 2016.

series from 1966 to 2016, four of the five years with the heaviest rainfall occurred after the year 2000.

4.1.3. Changes in Different Classes of Precipitation. From 1966 to 2016, the average annual P_{10}, P_{25}, P_{50}, and P_{100} values for the HLR were 251.1 mm, 128.2 mm, 39.1 mm, and 5.4 mm, respectively, accounting for 56.6%, 28.9%, 8.8%, and 1.2% of the annual precipitation, respectively (Table 1).

Approximately 80%–95% of P_{10} occurred from June to September, and few of these events occurred in May and October. Almost all of the P_{25}, P_{50}, and P_{100} events occurred between June and September. The precipitation from June to September in the HLR is mostly characterized by short duration and high rainfall intensity.

The variation in the different classes of precipitation from 1966 to 2016 is shown in Figure 4. It can be seen that, except for P_{25}, the different classes of precipitation in 2016 reached the highest observed values since 1966. P_{25} reaches its maximum level in 2013, followed by 2016.

According to the MK test results shown in Table 1, there was no significant trend in annual precipitation, P_{10}, P_{25}, P_{50}, and P_{100} in the HLR from 1966 to 2016.

4.1.4. Change in Rainstorm Frequency. Figure 5 shows the time series of the frequency of rainstorms and heavy rainstorms from 1966 to 2016 in the HLR. The occurrence of rainstorms and heavy rainstorms was high, and the past 51 years can be divided into three periods. The occurrence of heavy rainstorms in 1982–2000 was obviously lower than normal, and only in 1995 and 1996, a large area of heavy rainfall occurred. Rainstorms and heavy rainstorms occurred frequently in the two periods of 1966–1981 and 2001–2016. Especially in 2007–2016, both the average annual frequency of rainstorms and heavy rainstorms and the proportion of precipitation exceeded the multiyear average values (Table 2).

4.2. Precipitation Changes in the Focus Period

4.2.1. Selection of the Focus Period. Since 1998, the rate of change in the sediment-producing environment in the Loess Plateau has been unprecedented [1, 32]. The turning point in the relationship between rainfall and sediment discharge is an important basis for selecting a focus period.

The double mass curve between rainfall and sediment discharge in the HLR is shown in Figure 6. Based on the change in the slope of the double mass curve of rainfall and sediment discharge, the relationship between rainfall and sediment discharge in the study region obviously changed in 1979 and in 2006, and after 2006, the transition of the relationship between precipitation and sediment discharge was much more significant than the previous turning point.

Since the 1990s, the effects of large-scale soil and water conservation measures have significantly enhanced the reductions in sediment discharge. The relationship between rainfall and sediment discharge has changed significantly and is manifested as the double accumulation points of rainfall and sediment discharge that significantly deviate from the sediment transport axis (Figure 6).

The rainfall-sediment relationship in the major sediment tributaries in the HLR has an inflection point in the 1970s and 1980s [33], but the turning points in 2004–2008 were more prominent than previously observed [1]. Although the annual sediment discharge decreased in the 1970s and the 1980s, the change in sediment concentration was not significant during the same periods [1]. In fact, until 2004–2008, the annual sediment discharge, the average sediment concentration during the flood season, and the annual

Table 1: Average precipitation in the HLR from 1966 to 2016.

Rainfall factor	P	P_{10}	P_{25}	P_{50}	P_{100}
Precipitation (mm)	443.6	251.1	128.2	39.1	5.4
Proportion of different precipitation classes of annual precipitation (%)	—	56.6	28.9	8.8	1.2
MK Z value	1.129	0.382	0.755	0.381	−0.317

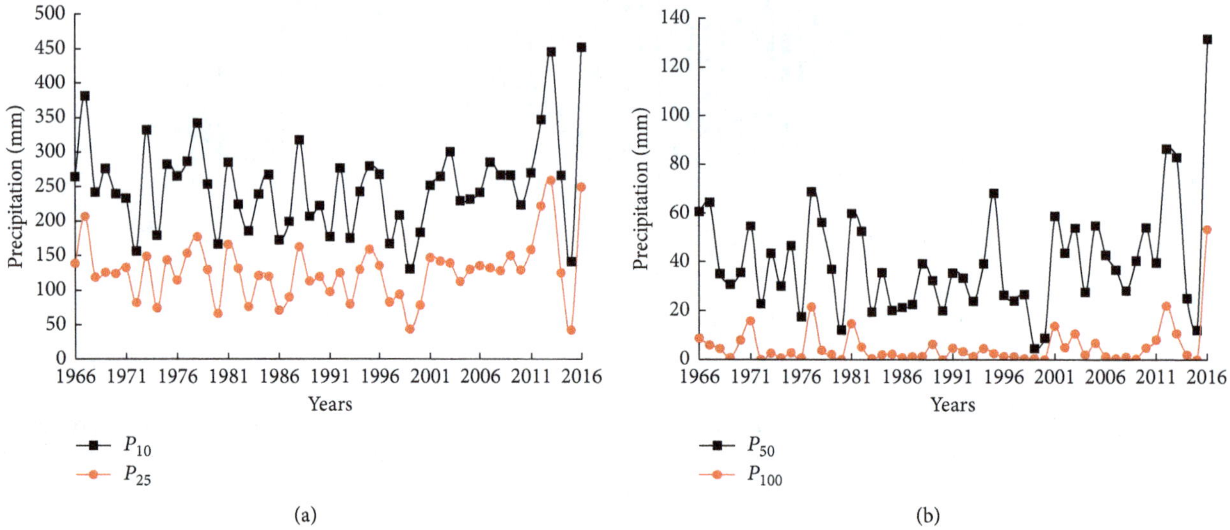

Figure 4: Time series of different classes of precipitation in the HLR from 1966 to 2016.

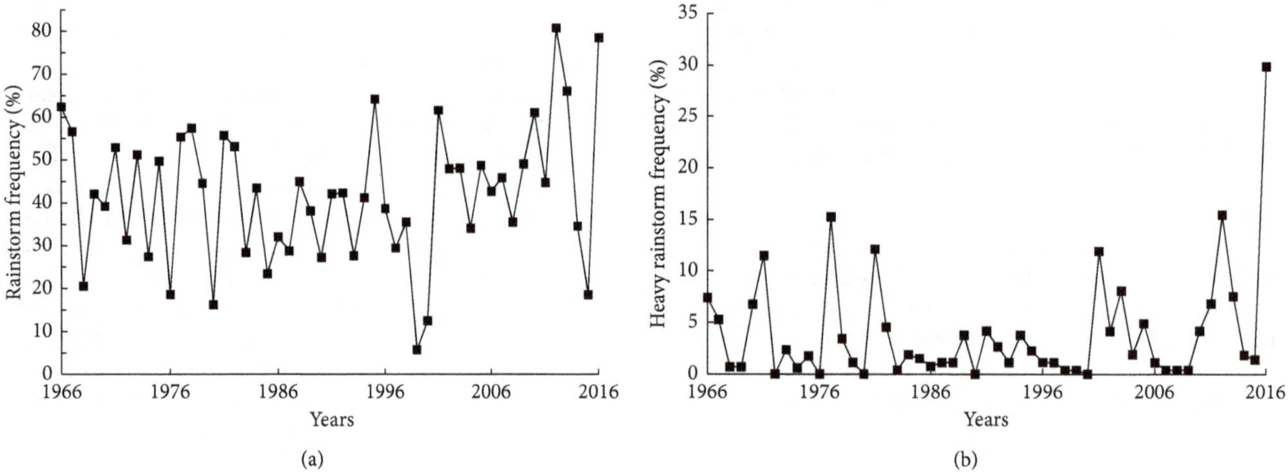

Figure 5: Time series of frequency of rainstorms (a) and heavy rainstorms (b) in the HLR from 1966 to 2016.

Table 2: The frequency and rainfall proportion of rainstorms and heavy rainstorms during different periods in the HLR (%).

Period	Rainstorm		Heavy rainstorm	
	Average annual frequency	Rainfall proportion	Average annual frequency	Rainfall proportion
1966–2016	41.9	8.8	3.41	1.2
2007–2016	51.5	10.3	6.83	2.0

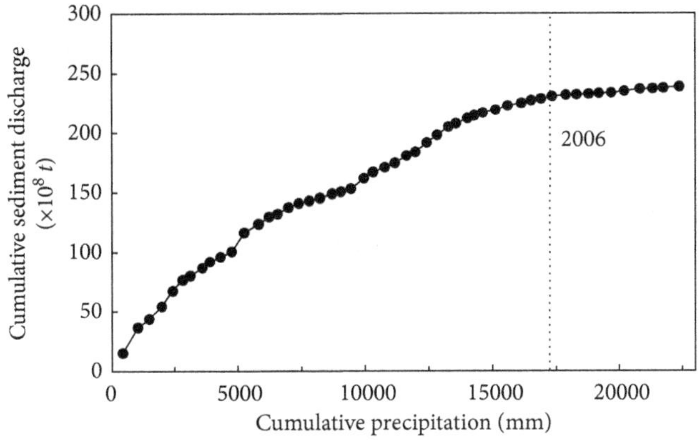

FIGURE 6: The double mass curve between rainfall and sediment discharge in the HLR.

maximum sediment concentration of the major tributaries of the study area decreased considerably [1, 34].

Therefore, we selected 2007–2016 as the focus period, emphasizing the analysis of changes in precipitation in the HLR during this period.

4.2.2. Changes in Precipitation from 2007 to 2016. Compared with 1966–2016, the annual precipitation and different classes of precipitation were all generally more abundant during the focus period. In 2007–2016, the average annual precipitation was 507.6 mm, which was 14.4% higher than the average annual precipitation from 1966 to 2016. Additionally, P_{10}, P_{25}, P_{50}, and P_{100} were 20%, 26.4%, 33.5%, and 90.7% higher, respectively (Table 3), and P_{100} was particularly abundant.

Except for the northern part of the HLR, the different classes of precipitation in the other areas were more abundant (Figure 7). The areas with the most abundant precipitation were mainly located in the central part of the HLR and in the southern part of the region. The area of increasing rainfall intensity was mainly concentrated in the central part of the HLR.

Table 4 shows the proportion of the area with different degrees of change in precipitation from 2007 to 2016 compared with 1966–2016 in the HLR. In most of the study area, the annual precipitation was abundant in 2007–2016, which accounted for 86.58% of the study area, while the area with the annual precipitation reduction of more than 5% accounted for 3.56% of the study area (Table 4). From 2007 to 2016, the area with P_{25}, P_{50}, and P_{100} reduction of more than 5% accounted for 7.52%, 11.4%, and 41.67% of the study area, respectively (Table 4). In contrast, the area where P_{25}, P_{50}, and P_{100} were more than 5% higher accounted for 84.40%, 80.11%, and 56.41% of the study area, respectively.

4.3. Impacts of Precipitation on Sediment Yield

4.3.1. Contribution of P_{50} to Sediment Yield. To learn more about the impact of precipitation on sediment yield, the sediment yield of typical tributaries with daily rainfall higher than 50 mm and the proportion of annual sediment yield in the corresponding year were calculated from 1966 to 1985 (Table 5). The average proportion of sediment yield produced by P_{50} was 50.5% in the HLR. Compared with Table 1, although P_{50} accounted for only 8.8% of the annual precipitation, the amount of sediment yield produced by P_{50} was 50.5% of the total annual sediment discharge, making P_{50} the key driving force for sediment yield.

TABLE 3: Changes in precipitation in the HLR from 2007 to 2016.

Average precipitation in 2007–2016 (mm)					Changes in precipitation (%)				
P	P_{10}	P_{25}	P_{50}	P_{100}	P	P_{10}	P_{25}	P_{50}	P_{100}
507.6	301.4	162.0	52.2	10.3	14.4	20.0	26.4	33.5	90.7

The annual sediment yield of P_{50} varied greatly, accounting for 10%–98% of the annual sediment yield in the HLR (Figure 8).

4.3.2. Changes in the Relationship between Rainfall and Sediment Discharge. It has been proposed that 10 mm of rainfall is the erosive rainfall standard [35], and this value is well correlated with sediment yield [31]. Therefore, P_{10} was selected to analyze the relationship between rainfall and sediment yield in the HLR (Figure 9). Compared with the period before the 1970s, the relationship between rainfall and sediment yield in 2007–2016 experienced great changes. The rainfall-sediment yield relationship in 2007–2016 obviously departed from that before the 1970s, and it is difficult to see the response of sediment yield to rainfall.

4.3.3. Attribution of Changes in Sediment Yield. The hydrological modeling method was employed to estimate the attribution of precipitation and other factors to the changes in sediment yield in the HLR since 1980. The sediment yield reached 936 million tons in the baseline period in the HLR. The hydrological modeling results showed that precipitation led to 225.23, 259.10, 106.30, and −167.79 million ton changes in sediment yield for 1980–1989, 1990–1999, 2000–2009, and 2010–2016, respectively (Table 6). Other factors led to 339.07, 202.47, 650.03, and 1017.14 million ton changes in sediment yield for the four periods, respectively (Table 6). Figure 10 shows the relative contributions of precipitation and other factors to the sediment yield changes since 1980. During the period of 1980–1989, the impact of precipitation was responsible for 39.9% of the sediment yield decrease, while the effects of other factors were responsible for 60.1% of the sediment yield decrease. Other factors were the main driving factor for the sediment yield changes in the HLR. During the period of 1990–1999, precipitation and other factors were responsible for 56.1% and 43.9% of the sediment yield reduction, respectively, which indicated that precipitation was the main driving factor for the sediment yield change. During the period of 2000–2009, precipitation and other factors accounted for 14.1% and 85.9% of the sediment yield reduction, respectively. Other factors were obviously the driving factor for the sediment yield reduction. During the period of 2010–2016, precipitation and other factors were responsible for −19.8% and 119.8% of the

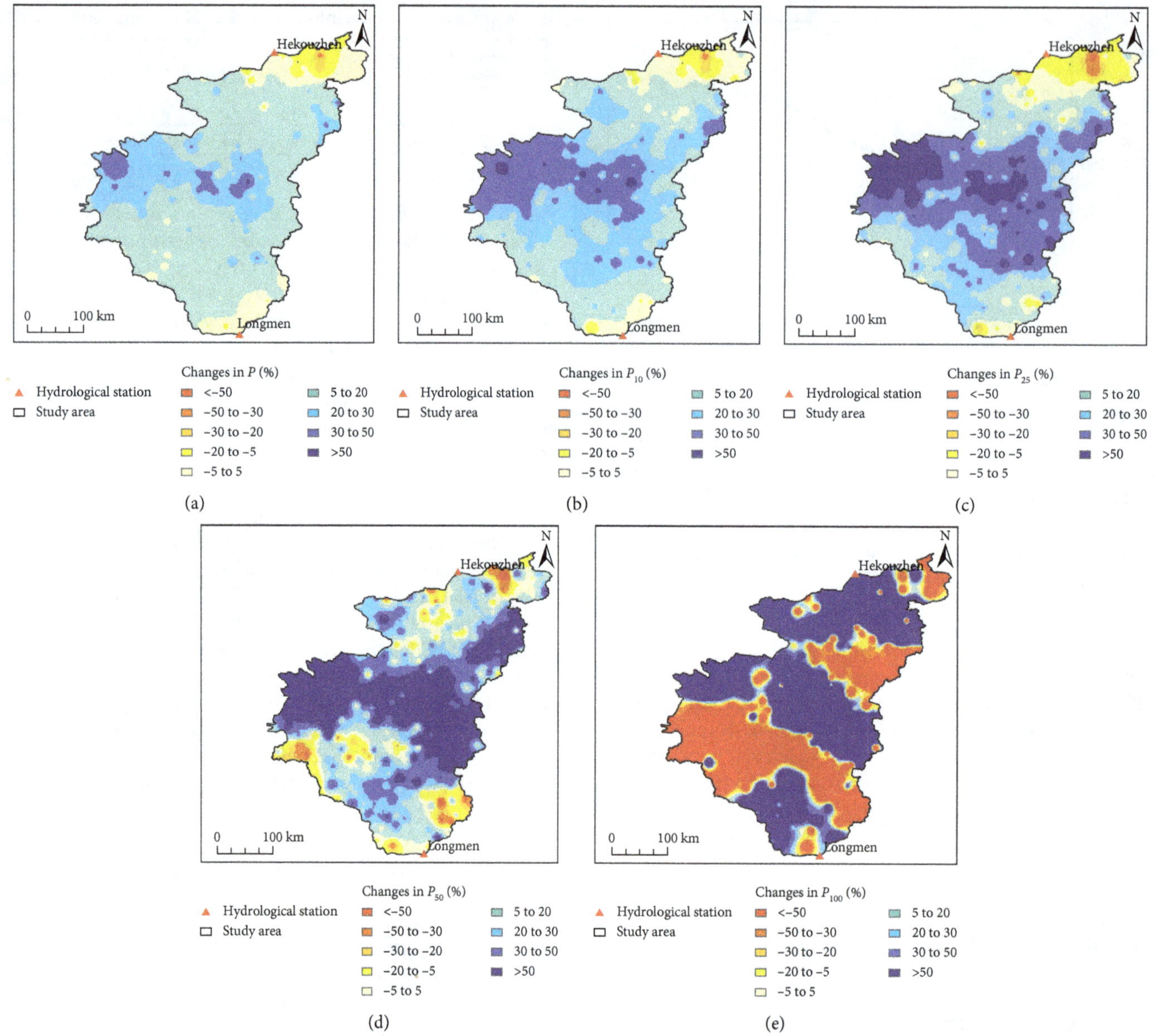

FIGURE 7: Spatial distribution of changes in precipitation: (a) P, (b) P_{10}, (c) P_{25}, (d) P_{50}, and (e) P_{100}, in the HLR from 2007 to 2016 compared with 1966–2016.

sediment yield reduction, respectively. Precipitation increased the sediment yield, while other factors reduced the sediment yield amount and were still the main driving factor for the sediment yield changes.

4.3.4. The Impacts of Precipitation on Sediment Yield in Typical Years. In early July and early August 1977, there were two large-scale and high-intensity rainfall events in the HLR, covering most of the major sediment-yielding areas in the middle reaches of the Yellow River, as shown in Figure 11. Under this rainfall situation, the annual sediment yield of the HLR reached 1.592 billion tons, and the value at the Tongguan station reached 2.21 billion tons in 1977; this was the largest amount of sediment measured since 1968. Therefore, 1977 became a year of great concern to all researchers in terms of changes in the water and sediment changes of the Yellow River.

Rainfall in 2012, 2013, and 2016 was relatively abundant in recent years, and their rainstorm covered most of the major sediment-yielding areas in the middle reaches of the Yellow River. Specifically, the heavy rainstorms and rainfall intensities measured in 2016 were the largest from 1966 to 2016 (Figure 11). In 2012 and 2016, high-intensity rainfall events were mainly distributed in the middle and northern parts of the HLR. In 2013, high-intensity rainfall was mainly distributed in the middle and southern parts of the HLR.

In the rainfall situations of 2012, 2013, and 2016, the sediment yield of the HLR reached 142 million tons, 178 million tons, and 108 million tons (Table 7), respectively; additionally, the sediment yield at the Tongguan station

TABLE 4: The proportion of the area with different degrees of change in precipitation from 2007 to 2016 compared with 1966–2016 in the HLR.

Change	Degree of change	Proportion of the area (%)				
		P	P_{10}	P_{25}	P_{50}	P_{100}
Less precipitation	−5 to −20%	2.99	2.64	6.00	6.87	3.18
	−20 to −30%	0.51	0.45	0.79	2.09	2.19
	−30 to −50%	0.06	0.09	0.62	2.01	5.36
	>−50%	0.00	0.00	0.11	0.43	30.95
	Subtotal	3.56	3.18	7.52	11.40	41.67
Abundant precipitation	5 to 20%	63.57	36.14	22.16	19.01	2.98
	20 to 30%	19.54	30.83	19.90	13.37	2.05
	30 to 50%	3.43	20.06	30.22	16.09	4.26
	>50%	0.04	0.68	12.12	31.64	47.13
	Subtotal	86.58	87.71	84.40	80.11	56.41
Slight change	−5 to 5%	9.87	9.12	8.07	8.49	1.92

TABLE 5: The proportion of sediment yield produced by P_{50} in typical tributaries (%).

Tributary	Huangpu-chuan	Gushan-chuan	Kuye River	Jialu River	Yanhe River	Qingjian River	Qiushui River	Xinshui River	Average weight of 11 tributaries
Proportion of sediment discharge	52.4	36.0	66.6	32.4	42.0	33.9	42.7	66.1	50.5

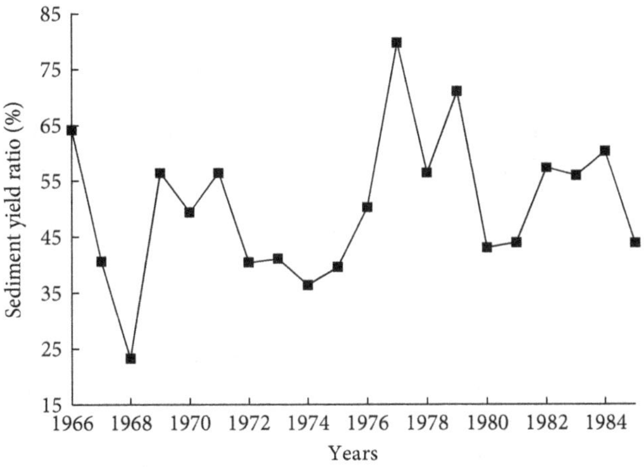

FIGURE 8: Proportion of sediment produced by P_{50} in the HLR from 1966 to 1985.

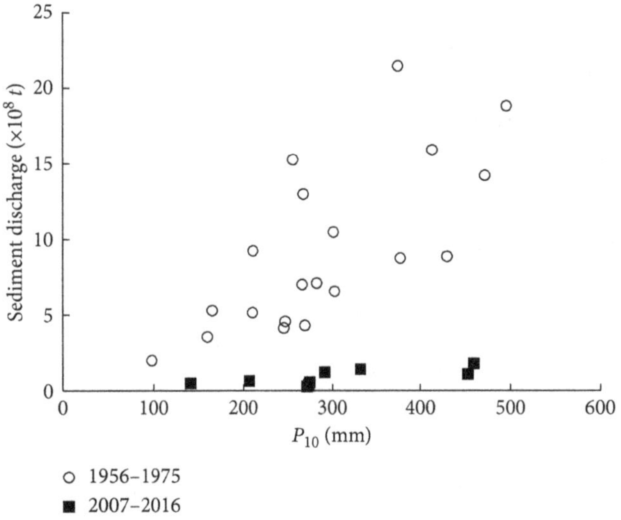

FIGURE 9: Relationship between rainfall and sediment yield in two periods in the HLR.

reached 206 million tons, 305 million tons, and 108 million tons, respectively.

To better understand the impact of precipitation on sediment yield, P_{25}, P_{50}, and P_{100} were calculated for all rainfall stations in 1977, 2012, 2013, and 2016 (Table 7). The results showed that P_{25} and P_{50} were greater in 2012 and 2013 than in 1977, and P_{100} was the same as or lower than the value recorded in 1977; however, the sediment yield of the HLR was 90% and 89% lower than the amount recorded in 1977, respectively. In 2016, P_{25}, P_{50}, and P_{100} were all greater than those in 1977, while the sediment yield of the HLR declined by 93% compared with that in 1977. Through comparative analysis of the precipitation changes in typical years in the HLR, it can be seen that precipitation is not the main cause of sediment yield reduction in recent years, the current underlying surface has changed a lot compared with the period before 1970s [36], and the sediment yield capacity of most tributaries has decreased significantly [1].

5. Discussion

5.1. Comparisons with Similar Studies. Several studies have investigated the spatiotemporal variation in annual precipitation, extreme precipitation, and erosive rainfall in the Loess Plateau or in the middle reaches of the Yellow River Basin in different periods. In this study, we investigated the variations in annual precipitation and different classes of

Table 6: Attribution of the change in sediment yield in the HLR since 1980.

Period	W_{S1} (10^6 t)	W_{S2} (10^6 t)	ΔW_S (10^6 t)	ΔW_{SP} (10^6 t)	ΔW_{SH} (10^6 t)	ΔW_{SP} (%)	ΔW_{SH} (%)
1980–1989	372.21	711.28	564.30	225.23	339.07	39.9	60.1
1990–1999	474.93	677.40	461.58	259.10	202.47	56.1	43.9
2000–2009	180.18	830.21	756.33	106.30	650.03	14.1	85.9
2010–2016	87.16	1104.30	849.35	−167.79	1017.14	−19.8	119.8

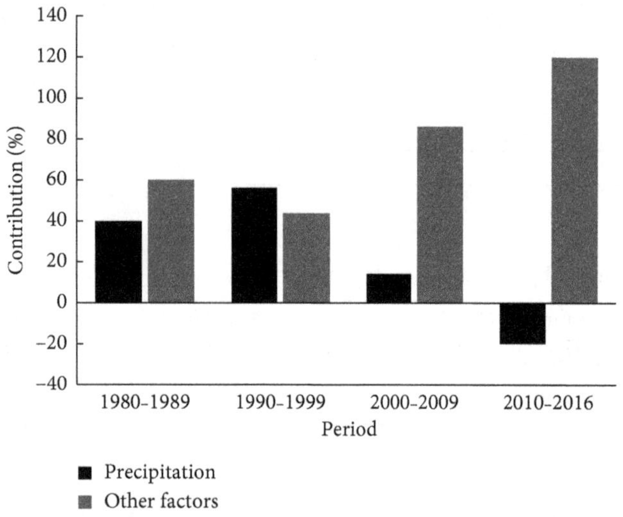

Figure 10: Contributions of precipitation and other factors to the changes in sediment yield in the HLR since 1980.

precipitation using high-resolution data, which provided more detailed information than other studies.

In general, annual precipitation exhibited no obvious trend in the Loess Plateau for the last decades. For example, Sun et al. and Zhang et al. found that the annual total precipitation showed no obvious trends in the Loess Plateau and in the middle reaches of the Yellow River Basin during 1960–2013 [37, 38], respectively; Wang et al. suggested that the region-averaged annual precipitation shows a non-significant negative trend in the Loess Plateau in the period of 1961–2010 [39], which all agree with the results of our study.

Besides, the results reported by Sun et al. and Xin et al. were not completely consistent with the conclusions of our study [8, 40]. Xin et al. suggested that the annual rainfall and erosive rainfall decreased in the Loess Plateau from 1956 to 2008 [40]. Sun et al. found that the total amount of precipitation on wet days decreased over a large area of the Loess Plateau during 1961–2011, particularly in the southeast region [8]. The differences between our findings and those reported by Xin et al. and Sun et al. are probably related to the different research periods and spatial domains adopted by these studies. The decreasing trends were mainly due to the relatively dry period in the 2000s in the Loess Plateau. In this study, we found different trends using updated time series, and the precipitation data extended into 2016. According to Figures 3 and 4, annual precipitation and different classes of precipitation had increased in the last several years (2012–2016). This was due to increased precipitation and more frequent storms occurring in recent years in the HLR. Zhao et al. has found the similar pattern for extreme precipitation indices [41].

5.2. Changes in the Underlying Surface. The sediment yield reduction caused by other factors is the total amount of sediment reduction produced by the changes in the underlying surface in the HLR. Vegetation changes, terraces, check dams, reservoirs, irrigation, and channel scouring and siltation are the main underlying surface factors in the study area. Since the late 1970s, numerous soil conservation practices have been implemented in the Loess Plateau (Tables 8 and 9), such as afforestation and construction of level terraces and check dams, to reduce soil erosion [42, 43]. Some studies indicated that sediment yield reduction in major tributaries in the middle reaches of the Yellow River Basin was mainly caused by the SCP [44, 45], which can effectively reduce the sediment yield by increasing intercepted precipitation and water infiltration, retarding surface runoff and trapping sediment [46, 47].

With the implementation of the nationwide ecological recovery program (i.e., the "Grain for Green Project" (GGP)) since 1999, the vegetation coverage in the HLR has greatly increased [48]. The major factors impacting the effects of vegetation on sediment reduction are the vegetation coverage and thickness of litter layer and plant roots [49, 50]. When the percentage of effective vegetation is less than 35~40%, vegetation improvement has an obvious impact on reducing sediment [16]. Regardless of precipitation, vegetation type, and other underlying surface factors, surface erosion is extremely weak when the vegetation coverage is greater than 70% [51]. At the end of the twentieth century, the vegetation coverage in the HLR was mostly 12~30% [16]. With the implementation of GGP, the vegetation coverage in many areas of the HLR in 2010 has reached 40~60% [16], which has exactly experienced the sensitive period of vegetation change and sediment yield response. The results of turning point detection further reveal that abrupt changes likely have associations with the implementation of SCP and GGP (Figure 6). Different types of underlying surface factors had different influences on the sediment yield changes. The proportional effect of different underlying surface factors on sediment yield changes should be further investigated.

With the continuous increase of vegetation coverage, the check dam would be damaged during heavy rains, and the reservoirs and check dams will be gradually filled up and lose the function of sediment retention; the future change trend of sediment yield in the Yellow River Basin still should be further investigated.

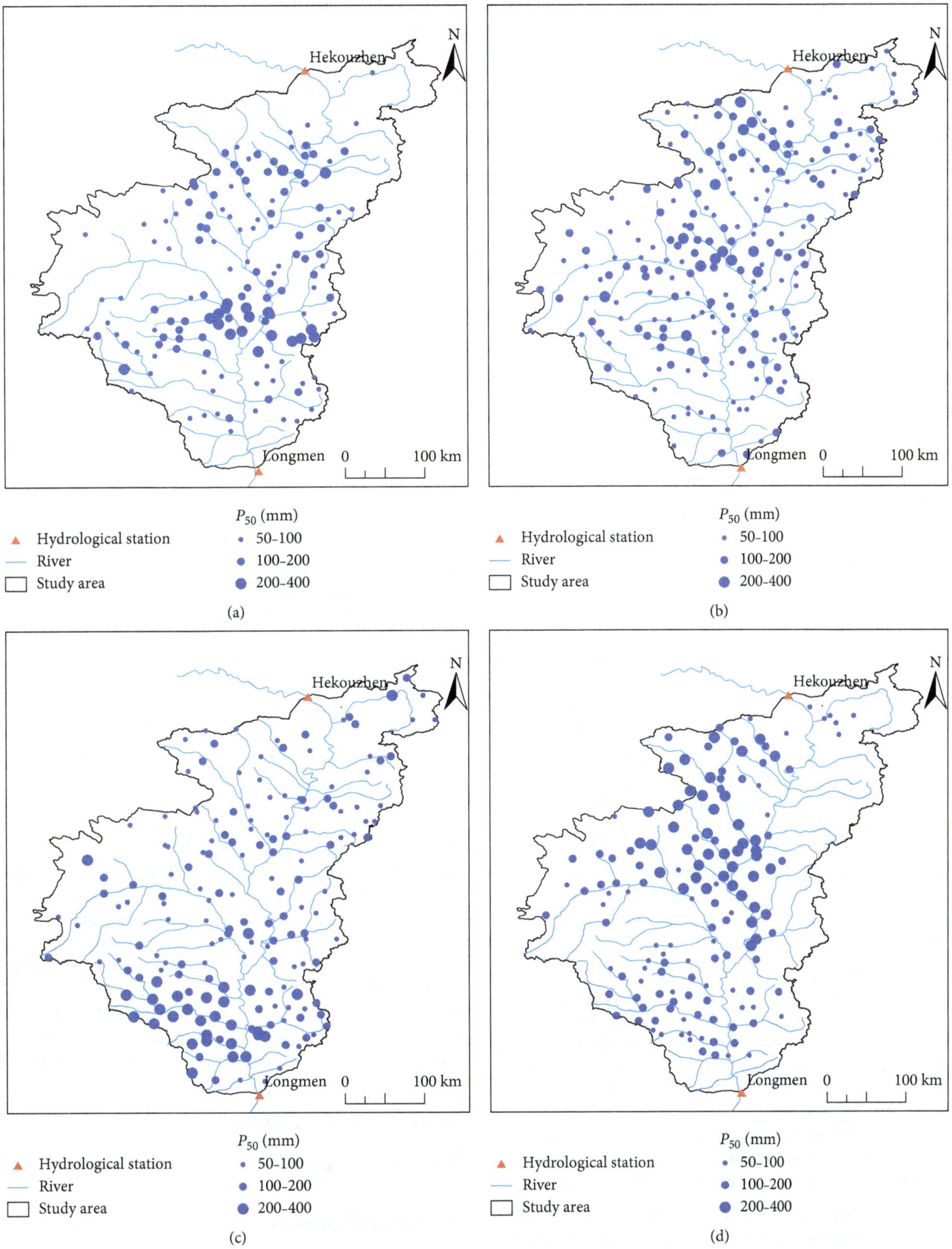

FIGURE 11: Spatial distribution of rainstorms in typical years: (a) 1977, (b) 2012, (c) 2013, and (d) 2016, in the HLR.

Table 7: Comparison of precipitation over the heavy rainfall events and the sediment yield of the HLR in typical years.

Year	P_{25} (mm)	P_{50} (mm)	P_{100} (mm)	S^* (million tons)
1977	160	71.0	22.4	1592
2012	222.4	85.6	22.3	142
2013	268.6	86.8	12.2	178
2016	241.7	125.1	42.8	108

*S represents the sediment yield of the HLR.

Table 8: Quantity of soil conservation practices by 2011 in the HLR.

Soil conservation practices		Quantity	Total storage capacity (10^8 m^3)	Area (km^2)
Reservoir	Large reservoir	4	14.0	—
	Medium-sized reservoir	44	19.24	—
	Small-sized reservoir	102	4.01	—
Check dam	Key check dam	3726	40.097	—
Level terrace		—	—	4716.5

Table 9: Statistics of the quantity of key check dams built in each decade in the HLR.

Decades	1950–1959	1960–1969	1970–1979	1980–1989	1990–1999	2000–2011
Quantity	73	261	1052	247	488	1605

6. Conclusions

In this study, based on the daily precipitation data collected at 301 rainfall stations in the HLR from 1966 to 2016, we investigated the spatial and temporal variations of annual precipitation and different classes of precipitation, and the impacts of precipitation on sediment yield were investigated. The main conclusions can be summarized as follows.

Spatially, the annual precipitation and P_{10} increased gradually from the northwest to the southeast of the HLR, and P_{25} and P_{50} were mainly concentrated in the north-western and southwestern parts of the HLR, suggesting that it was drier in the northwest region than the southeast region of the HLR, and heavy rain was more likely to occur in the northwest and southwest regions of the HLR. There was no significant trend in annual precipitation, P_{10}, P_{25}, P_{50}, and P_{100} in the HLR from 1966 to 2016.

Compared with the multiyear average precipitation from 1966 to 2016, the annual precipitation, P_{10}, P_{25}, P_{50}, and P_{100} in the HLR in the period of 2007–2016 were 14.4%, 20%, 26.4%, 33.5%, and 90.7% higher, respectively. The area where P_{25}, P_{50}, and P_{100} were more than 5% higher accounted for 84.40%, 80.11%, and 56.41% of the study area, respectively. The area with P_{25}, P_{50}, and P_{100} reduction of more than 5% accounted for 7.52%, 11.4%, and 41.67%, respectively. The occurrence frequency of rainstorms and heavy rainstorms was also higher in the period of 2007–2016 in the HLR.

The relationship between rainfall and sediment yield during 2007–2016 has changed compared with the period before the 1970s, and the analysis of the impacts of precipitation on sediment yield in typical years showed that precipitation is not the main cause of sediment yield reduction in recent years. The hydrological modeling method was used to quantitatively assess the attribution of precipitation and other factors to sediment yield changes in the HLR since 1980. The results showed that precipitation was the main driving factor for the sediment yield change during the period of 1990–1999, which accounted for 56.1% of the sediment yield reduction. Other factors were the main driving factor for the sediment yield change in the periods of 1980–1989, 2000–2009, and 2010–2016 and were responsible for 60.1%, 85.9%, and 119.8% of the sediment yield decrease, respectively, and other factors were playing a bigger role in the sediment yield change.

This study provided a comprehensive understanding of the variation in precipitation in the HLR and highlighted its effect on sediment yield. The precipitation in the HLR has been more abundant in recent years, although this increase does not explain the significant reduction in sediment yield. Further investigation is required to assess the impacts of underlying surface changes, especially vegetation restoration, on the sediment yield of the Yellow River. This study area is one of the main source areas for the sediment in the Yellow River Basin, and our results are helpful for understanding the cause of the significant reduction in sediment yield observed in recent years.

Conflicts of Interest

The authors declare that they have no conflicts of interest.

Acknowledgments

This research was supported by the National Key R&D Program of China (2016YFC0402400 and 2017YFC0403600) and the National Natural Science Foundation of China (41301030).

References

[1] X. Liu, *Causes of Sharp Reduction of Runoff and Sediment in the Yellow River in Recent Years*, Science China Press, Beijing, China, 2016, in Chinese.

[2] X. Jiongxin, *Influence of Soil and Water Conservancy Projects on Flood and Sediment in the Hekouzhen to Longmen Region of the Middle Yellow River*, Zhengzhou: The Yellow River Water Consercancy Press, Zhengzhou, China, 2009, in Chinese.

[3] G. Wang, *Yellow River Flood*, Zhengzhou: The Yellow River Water Conservancy Press, Zhengzhou, China, 1997.

[4] L. Kang, Y. Wang, G. Wang et al., "Analysis of rainfall distribution and the variation features in the region of Hekouzhen to Longmen in the middle Yellow River," *Yellow River*, vol. 21, no. 8, pp. 3–5, 1999, in Chinese.

[5] C. Hu, Y. Wang, Y. Zhang et al., "Variating tendency of runoff and sediment load in China major river and its causes," *Advances in Water Science*, vol. 21, no. 4, pp. 524–532, 2010, in Chinese.

[6] W. Yao, D. Ran, and J. Chen, "Recent changes in runoff and sediment regimes and future projections in the Yellow River basin," *Advances in Water Science*, vol. 24, no. 5, pp. 607–616, 2013, in Chinese.

[7] L. Luo, Z. J. Wang, X. Y. Liu et al., "Changes in characteristics of precipitation in flood season over five typical basins of middle reaches of the Yellow River in China," *Journal of Hydraulic Engineering*, vol. 44, no. 7, pp. 848–855, 2013, in Chinese.

[8] Q. Sun, C. Miao, Q. Duan et al., "Temperature and precipitation changes over the Loess Plateau between 1961 and 2011, based on high-density gauge observations," *Global and Planetary Change*, vol. 132, no. 1, pp. 1–10, 2015.

[9] G. Fu, S. Chen, C. Liu, and D. Shepard, "Hydro-climatic trends of the Yellow River basin for the last 50 years," *Climatic Change*, vol. 65, no. 1-2, pp. 149–178, 2004.

[10] Q. Zhang, J. Peng, V. P. Singh, J. Li, and Y. D. Chen, "Spatio-temporal variations of precipitation in arid and semiarid regions of China: the Yellow River basin as a case study," *Global and Planetary Change*, vol. 114, no. 2, pp. 38–49, 2014.

[11] F. Yuan, H. Yasuda, R. Berndtsson et al., "Regional sea-surface temperatures explain spatial and temporal variation of summer precipitation in the source region of the Yellow River," *Hydrological Sciences Journal*, vol. 61, no. 8, pp. 1383–1394, 2015.

[12] Y. He, X. Mu, P. Gao et al., "Spatial variability and periodicity of precipitation in the middle reaches of the Yellow River, China," *Advances in Meteorology*, vol. 2016, Article ID 9451614, 9 pages, 2016.

[13] F. F. Zhao, Z. X. Xu, J. X. Huang, and J. Y. Li, "Monotonic trend and abrupt changes for major climate variables in the headwater catchment of the Yellow River basin," *Hydrological Processes*, vol. 22, no. 23, pp. 4587–4599, 2010.

[14] E. Li, X. Mu, G. Zhao, P. Gao, and H. Shao, "Variation of runoff and precipitation in the Hekou-Longmen region of the Yellow River based on elasticity analysis," *Scientific World Journal*, no. 1, Article ID 929858, 11 pages, 2014.

[15] Y. He, P. Tian, X. Mu et al., "Changes in daily and monthly rainfall in the middle Yellow River, China," *Theoretical and Applied Climatology*, vol. 129, no. 1, pp. 1-2, 2016.

[16] X. Liu, S. Yang, S. Dang, Y. Luo, X. Y. Li, and X. Zhou, "Response of sediment yield to vegetation restoration at a large spatial scale in the Loess Plateau," *Science China Technological Sciences*, vol. 57, no. 8, pp. 1482–1489, 2014.

[17] J. Searcy and C. Hardison, *Double-Mass Curves*, U.S. Geological Survey Water Supply Paper.1541-B, Washington, DC, USA, 1960.

[18] J. M. Albert, "Hydraulic analysis and double mass curves of the Middle Rio Grande from Cochiti to San Marcial," M.S. thesis, Colorado State University, Fort Collins, CO, USA, 2004.

[19] X. Mu, X. Zhang, P. Gao et al., "Theory of double mass curve and its applications in hydrology and meteorology," *Journal of China Hydrology*, vol. 30, no. 4, pp. 47–51, 2010, in Chinese.

[20] D. Ran, B. Liu, L. Fu et al., "Methods of double mass curve calculation of effectiveness of water and sediment reduction of soil and water conservation measures," *Yellow River*, no. 6, pp. 24-25, 1996, in Chinese.

[21] X. Mu, C. Basang, and Z. Lu, "Impact of soil conservation measures on runoff and sediment in Hekou-Longmen region of the Yellow River," *Journal of Sediment Research*, no. 2, pp. 36–41, 2007, in Chinese.

[22] D. Zhang, H. Hong, Q. Zhang et al., "Attribution of the changes in annual streamflow in the Yangtze River Basin over the past 146 years," *Theoretical and Applied Climatology*, vol. 119, no. 1-2, pp. 323–332, 2015.

[23] H. B. Mann, "Nonparametric tests against trend," *Econometrica*, vol. 13, no. 3, pp. 245–259, 1945.

[24] M. G. Kendall, *Rank Correlation Methods*, Griffin, London, UK, 1955.

[25] J. M. Mitchell, B. Dzerdzeevskii, and H. Flohn, *Climate Change, World Meteorological Organization*, Geneva, Switzerland, 1966.

[26] C. Liu and J. Xia, "Water problems and hydrological research in the Yellow River and the huai and hai river basins of China," *Hydrological Processes*, vol. 18, no. 12, pp. 2197–2210, 2004.

[27] H. Zheng, Z. Lu, R. Zhu, C. Liu, Y. Sato, and Y. Fukushima, "Responses of streamflow to climate and land surface change in the headwaters of the Yellow River Basin," *Water Resources Research*, vol. 45, no. 7, pp. 641–648, 2009.

[28] M. Saifullah, Z. Li, Q. Li, M. Zaman, and S. Hashim, "Quantitative estimation of the impact of precipitation and land surface change on hydrological processes through statistical modeling," *Advances in Meteorology*, vol. 2016, Article ID 6130179, 15 pages, 2016.

[29] G. Zhao, G. Hörmann, N. Fohrer, Z. Zhang, and J. Zhai, "Streamflow trends and climate variability impacts in poyang lake basin, China," *Water Resources Management*, vol. 24, no. 4, pp. 689–706, 2010.

[30] Y. Chen, Y. Guan, G. Shao, and D. Zhang, "Investigating trends in streamflow and precipitation in huangfuchuan basin with wavelet analysis and the mann-kendall test," *Water*, vol. 8, no. 3, p. 77, 2016.

[31] X. Li, X. Liu, and Z. Li, "Effects of rainfall and underlying surface on sediment yield in the main sediment-yielding area of the Yellow River," *Journal of Hydraulic Engineering*, vol. 47, no. 10, pp. 1253–1259, 2016, in Chinese.

[32] C. Jiang, F. Wang, H. Zhang et al., "Quantifying changes in multiple ecosystem services during 2000–2012 on the Loess Plateau, China, as a result of climate variability and ecological restoration," *Ecological Engineering*, vol. 97, pp. 258–271, 2016.

[33] X. Yue, X. Mu, G. Zhao, H. Shao, and P. Gao, "Dynamic changes of sediment load in the middle reaches of the Yellow River basin, China and implications for eco-restoration," *Ecological Engineering*, vol. 73, pp. 64–72, 2014.

[34] Q. Yan, T. Lei, C. Yuan et al., "Effects of watershed management practices on the relationships among rainfall, runoff, and sediment delivery in the hilly-gully region of the Loess Plateau in China," *Geomorphology*, vol. 228, pp. 735–745, 2015.

[35] W. Wang, "Study on the relations between rainfall characteristics and loss of soil in loess region," *Bulletin of Soil and Water Conservation*, vol. 3, no. 4, pp. 7–13, 1983, in Chinese.

[36] P. Bai, X. Liu, K. Liang, and C. Liu, "Investigation of changes in the annual maximum flood in the Yellow River basin, China," *Quaternary International*, vol. 392, pp. 168–177, 2016.

[37] W. Sun, X. Mu, X. Song, D. Wu, A. Cheng, and B. Qiu, "Changes in extreme temperature and precipitation events in the Loess Plateau (China) during 1960–2013 under global warming," *Atmospheric Research*, vol. 168, no. 22, pp. 33–48, 2016.

[38] Y. Zhang, J. Xia, and D. She, "Spatiotemporal variation and statistical characteristic of extreme precipitation in the middle reaches of the Yellow River Basin during 1960–2013," *Theoretical and Applied Climatology*, no. 15, pp. 1–18, 2018.

[39] Q. X. Wang, X. H. Fan, Z. D. Qin et al., "Change trends of temperature and precipitation in the Loess Plateau region of China, 1961–2010," *Global and Planetary Change*, vol. 92-93, pp. 138–147, 2012.

[40] Z. Xin, X. Yu, Q. Li, and X. X. Lu, "Spatiotemporal variation in rainfall erosivity on the Chinese Loess Plateau during the period 1956–2008," *Regional Environmental Change*, vol. 11, no. 1, pp. 149–159, 2011.

[41] G. Zhao, J. Zhai, P. Tian et al., "Variations in extreme precipitation on the Loess Plateau using a high-resolution dataset and their linkages with atmospheric circulation indices," *Theoretical and Applied Climatology*, vol. 133, no. 3-4, pp. 1235–1247, 2017.

[42] X. Zhang, L. Zhang, J. Zhao, P. Rustomji, and P. Hairsine, "Responses of streamflow to changes in climate and land use/cover in the Loess Plateau, China," *Water Resources Research*, vol. 44, no. 7, pp. 2183–2188, 2008.

[43] G. Li and L. Sheng, "Model of water-sediment regulation in Yellow River and its effect," *Science China Technological Sciences*, vol. 54, no. 4, pp. 924–930, 2011.

[44] Z. Xu and J. Chen, "Analysis of soil and water conservation treatments on runoff in the Middle Yellow River," *Yellow River*, vol. 25, no. 7, pp. 125–129, 2010, in Chinese.

[45] D. Ran, Q. Luo, B. Liu et al., "Effect of soil-retaining dams on flood and sediment reduction in the middle reaches of the Yellow River," *Journal of Hydraulic Engineering*, vol. 35, no. 5, pp. 7–13, 2004, in Chinese.

[46] M. Huang and L. Zhang, "Hydrological responses to conservation practices in a catchment of the Loess Plateau, China," *Hydrological Processes*, vol. 18, no. 10, pp. 1885–1898, 2004.

[47] C. Miao, J. Ni, A. G. L. Borthwick, and L. Yang, "A preliminary estimate of human and natural contributions to the changes in water discharge and sediment load in the Yellow River," *Global and Planetary Change*, vol. 76, no. 3-4, pp. 196–205, 2011.

[48] Y. Luo, S. Yang, X. Liu et al., "Land use change in the reach from Hekouzhen to Tongguan of the Yellow River during 1998-2010," *Acta Geographica Sinica*, vol. 69, no. 1, pp. 42–53, 2014.

[49] K. Tang, *China Water and Soil Conservation*, Science China Press, Beijing, China, 2003, in Chinese.

[50] Q. Meng, *Water and Soil Conservation in Loess Plateau*, Yellow River Water Conservancy Press, Zhengzhou, China, 1996, in Chinese.

[51] K. Jing, W. Wang, and F. Zheng, *Chinese Soil Erosion and Environment*, Science China Press, Beijing, China, 2005, in Chinese.

PERMISSIONS

All chapters in this book were first published in AM, by Hindawi Publishing Corporation; hereby published with permission under the Creative Commons Attribution License or equivalent. Every chapter published in this book has been scrutinized by our experts. Their significance has been extensively debated. The topics covered herein carry significant findings which will fuel the growth of the discipline. They may even be implemented as practical applications or may be referred to as a beginning point for another development.

The contributors of this book come from diverse backgrounds, making this book a truly international effort. This book will bring forth new frontiers with its revolutionizing research information and detailed analysis of the nascent developments around the world.

We would like to thank all the contributing authors for lending their expertise to make the book truly unique. They have played a crucial role in the development of this book. Without their invaluable contributions this book wouldn't have been possible. They have made vital efforts to compile up to date information on the varied aspects of this subject to make this book a valuable addition to the collection of many professionals and students.

This book was conceptualized with the vision of imparting up-to-date information and advanced data in this field. To ensure the same, a matchless editorial board was set up. Every individual on the board went through rigorous rounds of assessment to prove their worth. After which they invested a large part of their time researching and compiling the most relevant data for our readers.

The editorial board has been involved in producing this book since its inception. They have spent rigorous hours researching and exploring the diverse topics which have resulted in the successful publishing of this book. They have passed on their knowledge of decades through this book. To expedite this challenging task, the publisher supported the team at every step. A small team of assistant editors was also appointed to further simplify the editing procedure and attain best results for the readers.

Apart from the editorial board, the designing team has also invested a significant amount of their time in understanding the subject and creating the most relevant covers. They scrutinized every image to scout for the most suitable representation of the subject and create an appropriate cover for the book.

The publishing team has been an ardent support to the editorial, designing and production team. Their endless efforts to recruit the best for this project, has resulted in the accomplishment of this book. They are a veteran in the field of academics and their pool of knowledge is as vast as their experience in printing. Their expertise and guidance has proved useful at every step. Their uncompromising quality standards have made this book an exceptional effort. Their encouragement from time to time has been an inspiration for everyone.

The publisher and the editorial board hope that this book will prove to be a valuable piece of knowledge for researchers, students, practitioners and scholars across the globe.

LIST OF CONTRIBUTORS

Fabio Nardecchia, Francesca Pagliaro and Luca Gugliermetti
Sapienza University of Rome, DAEEE, Via Eudossiana 18, Rome 00184, Italy

Annalisa Di Bernardino, Paolo Monti and Giovanni Leuzzi
Sapienza University of Rome, DICEA, Via Eudossiana 18, Rome 00184, Italy

Xiang Zhang, Xinming Tang and Xiaoming Gao
Satellite Surveying and Mapping Application Center, National Administration of Surveying, Mapping and Geo-information, Beijing 100048, China

Hui Zhao
National Geomatics Center of China, Beijing 100080, China

Todd W. Moore, Jennifer M. St. Clair and Tiffany A. DeBoer
Department of Geography and Environmental Planning, Towson University, Towson, MD 21252, USA

Janine A. Baijnath-Rodino and Claude R. Duguay
Department of Geography and Environmental Management and Interdisciplinary Centre on Climate Change, University of Waterloo, 200 University Avenue West, Waterloo, Ontario, Canada N2L 3G1

Adam Collingwood, Chen Shang and Paul Treitz
Department of Geography and Planning, Queen's University, Kingston, ON, Canada K7L 3N6

Adam Collingwood
Waterton Lakes National Park, Parks Canada, Box 200, Waterton Park, AB, Canada T0K 2M0

François Charbonneau
Canada Centre for Mapping and Earth Observation, Natural Resources Canada, 560 Rochester Street, Ottawa, ON, Canada K1A 0E4

Hong Zheng, Haibin Li, Xingjian Lu and Tong Ruan
Information Engineering and Computer Science College, East China University of Science and Technology, Shanghai 200237, China

Xingjian Lu
Smart City Collaborative Innovation Center, Shanghai Jiao Tong University, Shanghai 200240, China

Zhifang Wang, Fengjie Zheng and Wenhao Zhang
Institute of Remote Sensing and Digital Earth, Chinese Academy of Sciences, Beijing 100101, China

Zhifang Wang and Shutao Wang
Department of Instrument Science & Engineering, Yanshan University, Qinhuangdao, Hebei 066004, China

Haitao Xu and Zhengwei He
State Key Laboratory of Geohazard Prevention and Geoenvironment Protection (Chengdu University of Technology), Chengdu 610059, China
College of Earth Science, Chengdu University of Technology, Chengdu 610059, China

Peng Hou
State Environmental Protection Key Laboratory of Satellite Remote Sensing, Satellite Environment Center, Ministry of Environmental Protection of People's Republic of China, Beijing 100094, China

A. Duo
Satellization Application Centre for Disaster Reduction of the Ministry of Civil Affairs, National Disaster Reduction Center of China, Beijing 100124, China

Bing Zhang
College of Resource Environment and Tourism, Capital Normal University, Beijing 100048, China

Xingxing Ma, Hongnian Liu, Xueyuan Wang and Zhen Peng
School of Atmospheric Sciences, Nanjing University, Nanjing 210023, China

Agnieszka Wypych
Jagiellonian University, 7 Gronostajowa St., 30-387 Krakow, Poland

Bogdan Bochenek
Institute of Meteorology and Water Management—National Research Institute, 14 Piotra Borowego St., 30-215 Krakow, Poland

Yuying Zhu, Chengying Zhu, Fan Zu, Hongbin Wang, Chengsong Yuan, Shengming Jiao and Linyi Zhou
Key Laboratory of Transportation Meteorology, China Meteorological Administration, Nanjing 210009, China
Jiangsu Institute of Meteorological Sciences, Nanjing 210009, China

Qiuxia Xie, Qingyan Meng, Linlin Zhang and Chunmei Wang
Institute of Remote Sensing and Digital Earth, Chinese Academy of Sciences, Beijing 100101, China
Sanya Institute of Remote Sensing, Sanya 572029, China

Qiuxia Xie and Linlin Zhang
University of Chinese Academy of Sciences, Beijing 100101, China

Qiao Wang and Shaohua Zhao
Satellite Environment Center, Ministry of Environmental Protection, Beijing 100094, China

Daeun Kim
Center for Built Environment, Sungkyunkwan University, Suwon, Gyeonggi-do, Republic of Korea

Heewon Moon
Institute for Atmospheric and Climate Science, ETH Zürich, Zürich, Switzerland

Hyunglok Kim
School of Earth, Ocean and the Environment, University of South Carolina, Columbia, SC, USA

Jungho Im
School of Urban and Environmental Engineering, UNIST, Ulsan, Republic of Korea

Minha Choi
Graduate School of Water Resources, Sungkyunkwan University, Suwon, Gyeonggi-do, Republic of Korea

Suzhen Dang
Yellow River Institute of Hydraulic Research, Yellow River Conservancy Commission, Zhengzhou 450003, China
Key Laboratory of Soil and Water Loss Process and Control on the Loess Plateau, MWR, Yellow River Institute of Hydraulic Research, Zhengzhou 450003, China

Xiaoyan Liu
Yellow River Conservancy Commission, Zhengzhou 450003, China

Xiaoyu Li
Hydrology Bureau of Yellow River Conservancy Commission, Zhengzhou 450004, China

Manfei Yao
School of Water Conservancy, North China University of Water Resources and Electric Power, Zhengzhou 450045, China

Dan Zhang
Key Laboratory of Watershed Geographic Sciences, Nanjing Institute of Geography and Limnology, Chinese Academy of Sciences, Nanjing 210008, China

Index

www.ingramcontent.com/pod-product-compliance
Lightning Source LLC
Chambersburg PA
CBHW080530200326
41458CB00012B/4392
9781647401085